PEARSON EDEXCEL INTERNATIONAL AS/A LEVEL

BIOLOGY

Student Book 1

Ann Fullick
with Frank Sochacki

Published by Pearson Education Limited, 80 Strand, London, WC2R 0RL.

www.pearson.com/international-schools

Copies of official specifications for all Pearson Edexcel qualifications may be found on the website: https://qualifications.pearson.com

Text © Ann Fullick and Pearson Education Limited 2018
Edited by Deborah Webb and Penelope Lyons
Indexed by Judith Reading
Designed by © Pearson Education Limited 2018
Typeset by © Tech-Set Ltd, Gateshead, UK
Original illustrations © Pearson Education Limited 2018
Illustrated by © Tech-Set Ltd, Gateshead, UK
Cover design by © Pearson Education Limited 2018
Cover images: Front: **Getty Images**: Fernan Federici
Inside front cover: **Shutterstock.com**, Dmitry Lobanov

First published 2018

24
15

British Library Cataloguing in Publication Data
A catalogue record for this book is available from the British Library
ISBN 978 1 2922 4484 6

Printed in Slovakia by Neografia

Endorsement statement
In order to ensure that this resource offers high-quality support for the associated Pearson qualification, it has been through a review process by the awarding body. This process confirmed that this resource fully covers the teaching and learning content of the specification at which it is aimed. It also confirms that it demonstrates an appropriate balance between the development of subject skills, knowledge and understanding, in addition to preparation for assessment.

Endorsement does not cover any guidance on assessment activities or processes (e.g. practice questions or advice on how to answer assessment questions) included in the resource, nor does it prescribe any particular approach to the teaching or delivery of a related course.

While the publishers have made every attempt to ensure that advice on the qualification and its assessment is accurate, the official specification and associated assessment guidance materials are the only authoritative source of information and should always be referred to for definitive guidance.

Pearson examiners have not contributed to any sections in this resource relevant to examination papers for which they have responsibility.

Examiners will not use endorsed resources as a source of material for any assessment set by Pearson. Endorsement of a resource does not mean that the resource is required to achieve this Pearson qualification, nor does it mean that it is the only suitable material available to support the qualification, and any resource lists produced by the awarding body shall include this and other appropriate resources.

Acknowledgements

We are grateful to the following for permission to reproduce copyright material:

Figures
Figure on page 60 based on data from *American Journal of Epidemiology* similar to this http://healthhubs.net/heartdisease/waist-size-predicts-heart-disease-risk-better-than-bmi/ ;Figure on page 68 from pro.activ sponsored studies; Figure on page 63 data from http://annals.org/data/Journals/AIM/19985/10FF1.jpeg to show results of cohort study http://annals.org/aim/article/714567/effect-fruit-vegetable-intake-risk-coronary-heart-disease ;Figure on page 65 Dubai health survey 2014, http://gulfnews.com/polopoly_fs/1.1838305.1464708870!/image /754489788.jpg ;Figure on page 73 data from http://annals.org/data/Journals/AIM/19985/10FF1.jpeg to show results of cohort study http://annals.org/aim/article/714567/effect-fruit-vegetable-intake-risk-coronary-heart-disease; Figure on page 97 from 'Measuring the effect of an asthma attack' https://www.abpischools.org.uk/topic/breathingandasthma/7.

All other images © Pearson Education

Text
Extract on page 22 from Luyckx J., Baudouin C. Trehalose: an intriguing disaccharide with potential for medical application in ophthalmology. *Clinical Ophthalmology* (Auckland, NZ) 5 (2011): 577; Extract on page 70 from Mostafa Q Al-Shamiri. Heart failure in the Middle East. *Current Cardiology Reviews* 2013 May; 9(2):174–178 https://www.ncbi.nlm.nih.gov/pmc/articles/PMC3682400/#R10; Extract on page 96 from Qingling Zhang, Zhiming Qiu, Kian Fan Chung, Shau-Ku Huang. *Journal of Thoracic Disease*. 2015 Jan; 7(1): 14–22 https://www.ncbi.nlm.nih.gov/pmc/articles/PMC4311080/#r3; Extract on page 97 from Qingling Zhang, Zhiming Qiu, Kian Fan Chung, Shau-Ku Huang. *Journal of Thoracic Disease*. 2015 Jan; 7(1): 14–22 https://www.ncbi.nlm.nih.gov/pmc/articles/PMC4311080/#r3; Extract on page 122 based on a number of different websites promoting 'good health'; Extract on page 190 from *In vitro fertilisation* by Ann Fullick. 2nd ed. 2009. Heinemann series 'Science at the Edge'; Extract on page 218 from 'Me, myself, us. The human microbiome – looking at humans as ecosystems that contain many collaborating species could change the practice of medicine', *The Economist*, 18 August 2012; Extract on page 262 from Leonard, Jennifer A., Nadin Rohland, Scott Glaberman, Robert C. Fleischer, Adalgisa Caccone and Michael Hofreiter. 'A rapid loss of stripes: the evolutionary history of the extinct quagga.' *Biology Letters* 1, no. 3 (2005): 291–295; Extract on page 262 from 'Quagga rebreeding: a success story' by Keri Harvey, Copyright 2014 by *Farmer's Weekly Magazine*. Used by permission of the *Farmer's Weekly Magazine*; Extract on page 294 from Alexander Nater, et al. Morphometric, behavioral, and genomic evidence for a new orangutan species. *Current Biology* November 2017. http://www.cell.com/current-biology/fulltext/S0960-9822(17)31245-9.

TOPIC 1 MOLECULES, TRANSPORT AND HEALTH

TOPIC 2 MEMBRANES, PROTEINS, DNA AND GENE EXPRESSION

TOPIC 3
CELL STRUCTURE, REPRODUCTION AND DEVELOPMENT

TOPIC 4
PLANT STRUCTURE AND FUNCTION, BIODIVERSITY AND CONSERVATION

ABOUT THIS BOOK

This book is written for students following the Pearson Edexcel International Advanced Subsidiary (IAS) Biology specification. This book covers the full IAS course and the first year of the International A Level (IAL) course.

The book contains full coverage of IAS units (or exam papers) 1 and 2. Each unit in the specification has two topic areas. The topics in this book, and their contents, fully match the specification. You can refer to the Assessment Overview on page x for further information. Students can prepare for the written Practical Skills Paper (unit 3) by using the IAL Biology Lab Book (see page viii of this book).

Each topic is divided into chapters and sections to break the content down into manageable chunks. Each section features a mix of learning and activities.

Learning objectives
Each chapter starts with a list of key assessment objectives.

Specification reference
The exact specification references covered in the section are provided.

Exam hints
Tips on how to answer exam-style questions and guidance for exam preparation. Orange **Learning Tips** help you focus your learning and avoid common errors.

4B 1 PRINCIPLES OF CLASSIFICATION

SPECIFICATION REFERENCE 4.14(i) 4.15

LEARNING OBJECTIVES

- Understand that classification is a means of organising the variety of life based on relationships between organisms using differences and similarities in phenotypes and in genotypes, and is built around the species concept.

THE BACKGROUND TO BIODIVERSITY

Biodiversity is a measure of the variety of living organisms and their genetic differences. It is an important concept at the moment because the Earth's biodiversity is reducing rapidly. Many scientists think this may affect the future health of the planet. You will find out about biodiversity in more detail later in **Chapter 4C**. In this section you will be looking at some of the biology you need in order to understand biodiversity.

WHY CLASSIFY?

The result of millions of years of **evolution** is an enormous variety of living organisms. This great biodiversity (see **Chapter 4C**) means that there is a great variety of names. An organism may have different names not only in different countries, but even within different areas of the same country (see **fig A**). When biologists from different countries discuss an organism they need to be sure they are all referring to the same one. An internationally recognised way of referring to any living organism is essential. Biodiversity is a very important concept, and to quantify biodiversity we need a way of identifying the different groups of organisms. We classify the living world by putting organisms in groups based on their similarities and differences. Scientists can monitor changes in the populations of different types of organism if they know the numbers that there are in a particular habitat. It is also important for biologists to understand how different types of living organism are related to each other. A good classification system makes these ancestral relationships clear.

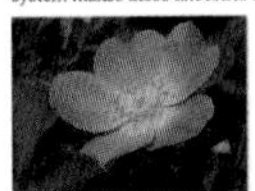

fig A This plant is a rose in English, English, ورد in Arabic, *ρόδο* in Greek, *rosa* in Spanish and *die Rosen* in German. The official classification Rosa is used and understood by biologists everywhere. The many different species of rose can be identified even more precisely, for example, Rosa canina **(the Wild** dog rose) and Rosa acicularis (the Arctic rose).

THE HISTORY OF TAXONOMY

Taxonomy is the science of describing, classifying and naming living organisms. This includes all of the plants, animals and microorganisms in the world and it is an enormous task. The aim of a classification system is to group organisms to accurately identify them and represent their ancestral relationships. From the time of the Greek philosopher Aristotle onwards, people put organisms into groups based mainly on their physical appearance or **morphology**. People often used **analogous features** to classify organisms. But such features may not have the same biological origin so this system can easily create misconceptions. For example, you might put wiggly, legless creatures including snakes, worms, slugs and eels in one classification group and flying animals such as bats, birds and flying insects in another group. A valid classification system must be based on careful observation and the use of **homologous structures** – that is, structures that genuinely show common ancestry.

In the 18th century, the Swedish botanist Carolus Linnaeus (1707–78) developed the first scientifically devised classification system. We still use many of his principles and his basic naming system today. However, we can now add many more modern techniques to the simple but detailed observation of organisms that he introduced.

THE MAIN TAXONOMIC GROUPS

The biggest taxonomic groupings are huge – the largest are the three **domains**, a grouping developed more recently which you will look at in more detail in **Section 4B.4**. The main taxonomic groups are, from the largest to the smallest: domain, **kingdom**, **phylum (division, for plants)**, **class**, **order**, **family**, **genus** and **species**.

The **Archaea** domain contains one kingdom:

- **Archaebacteria**: ancient bacteria thought to be early relatives of the eukaryotes. They were thought to be found only in extreme environments, but scientists are increasingly finding them everywhere – particularly in soil.

The Bacteria domain also contains one kingdom:

- **Eubacteria**: the true bacteria are what we normally think of when we are describing the bacteria that cause, for example, disease, and which are so useful in the digestive systems of many organisms and in recycling nutrients in the environment.

There are four kingdoms in the Eukaryota domain:

- **Protista**: a very diverse group of microscopic organisms. Some are heterotrophs – they need to eat other organisms – and some are autotrophs – they make their own food by photosynthesis. Some are animal-like, some are plant-like and some are more like fungi. Examples include *Amoeba*, *Chlamydomonas*, green and brown algae and slime moulds.

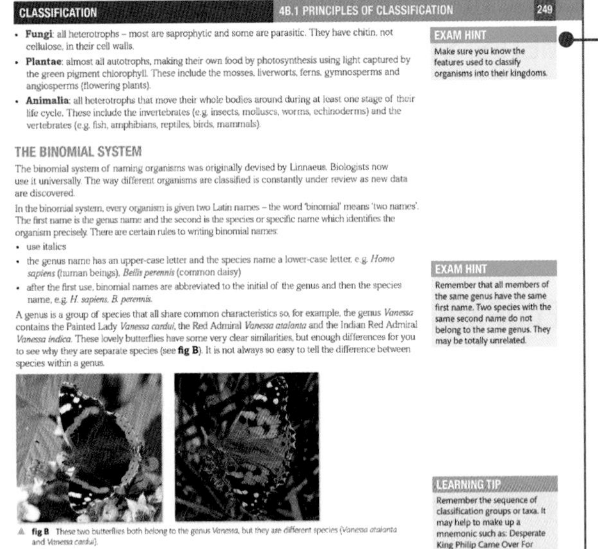

CLASSIFICATION 4B.1 PRINCIPLES OF CLASSIFICATION 249

- **Fungi**: all heterotrophs – most are saprophytic and some are parasitic. They have chitin, not cellulose, in their cell walls.
- **Plantae**: almost all autotrophs, making their own food by photosynthesis using light captured by the green pigment chlorophyll. These include the mosses, liverworts, ferns, gymnosperms and angiosperms (flowering plants).
- **Animalia**: all heterotrophs that move their whole bodies around during at least one stage of their life cycle. These include the invertebrates (e.g. insects, molluscs, worms, echinoderms) and the vertebrates (e.g. fish, amphibians, reptiles, birds, mammals).

EXAM HINT
Make sure you know the features used to classify organisms into their kingdoms.

THE BINOMIAL SYSTEM

The binomial system of naming organisms was originally devised by Linnaeus. Biologists now use it universally. The way different organisms are classified is constantly under review as new data are discovered.

In the binomial system, every organism is given two Latin names – the word 'binomial' means 'two names'. The first name is the genus name and the second is the species or specific name which identifies the organism precisely. There are certain rules to writing binomial names:

- use italics
- the genus name has an upper-case letter and the species name a lower-case letter, e.g. *Homo sapiens* (human beings), *Bellis perennis* (common daisy)
- after the first use, binomial names are abbreviated to the initial of the genus and then the species name, e.g. *H. sapiens*, *B. perennis*.

EXAM HINT
Remember that all members of the same genus have the same first name. Two species with the same second name do not belong to the same genus. They may be totally unrelated.

A genus is a group of species that all share common characteristics so, for example, the genus *Vanessa* contains the Painted Lady *Vanessa cardui*, the Red Admiral *Vanessa atalanta* and the Indian Red Admiral *Vanessa indica*. These lovely butterflies have some very clear similarities, but enough differences for you to see why they are separate species (see **fig B**). It is not always so easy to tell the difference between species within a genus.

fig B These two butterflies both belong to the genus *Vanessa*, but they are different species (*Vanessa atalanta* and *Vanessa cardui*).

LEARNING TIP
Remember the sequence of classification groups or taxa. It may help to make up a mnemonic such as: Desperate King Philip Came Over For Great Spaghetti.

Table A shows a number of different species with all of their levels of classification.

DOMAIN	Bacteria	Eukaryota	Eukaryota	Eukaryota
KINGDOM	Eubacteria	Animalia	Fungi	Plantae
PHYLUM/DIVISION	Proteobacteria	Chordata	Basidomycota	Magnoliophyta
CLASS	Gammaproteobacteria	Mammalia	Agaricomycetes	Liliopsida
ORDER	Enterobacteriales	Perissodactyla	Agaricales	Poales
FAMILY	Enterobacteriaceae	Equidae	Amanitaceae	Poaceae
GENUS	*Escherichia*	*Equus*	*Amanita*	*Oryza*
SPECIES	*Escherichia coli* *E. coli* common bacterium in the intestines	*Equus caballus* *E. caballus* domestic horse	*Amanita muscaria* *A. muscaria* fly agaric	*Oryza sativa* *O. sativa* rice

table A Full classification of four different organisms

Did you know?
Interesting facts help you remember the key concepts.

Worked examples show you how to work through questions and set out calculations.

Subject vocabulary
Key terms are highlighted in blue in the text. Clear definitions are provided at the end of each section for easy reference, and are also collated in a **glossary** at the back of the book.

Checkpoint
Questions at the end of each section check understanding of the key learning points in each chapter.

Your learning, chapter by chapter, is always put in context.

- Links to other areas of Biology include previous knowledge that is built on in the topic, and future learning that you will cover later in your course.
- A checklist details maths knowledge required. If you need to practise these skills, you can use the **Maths Skills** reference at the back of the book as a starting point.

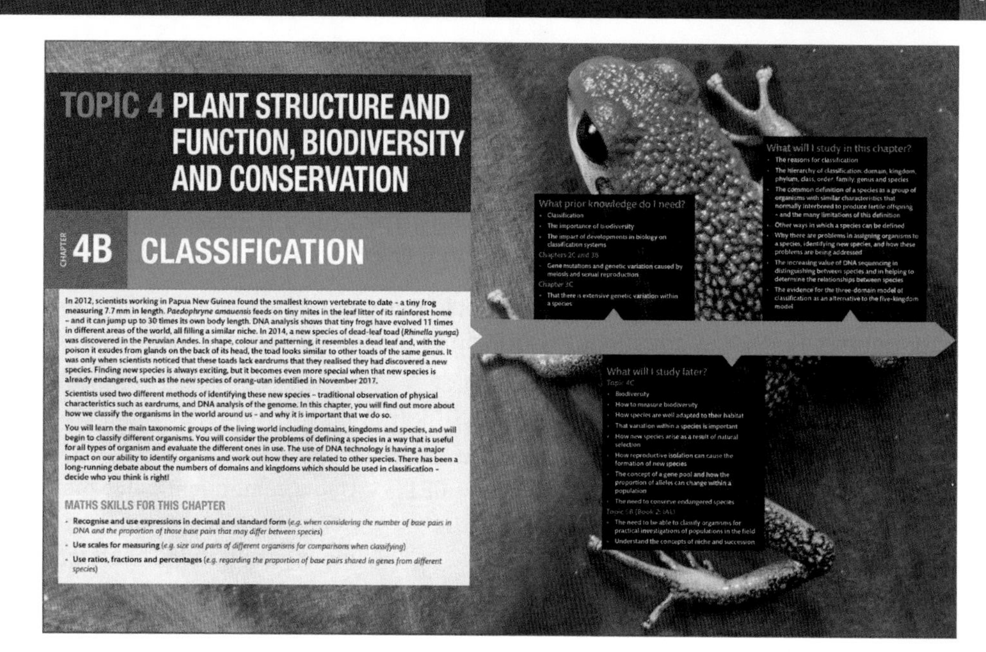

TOPIC 4 PLANT STRUCTURE AND FUNCTION, BIODIVERSITY AND CONSERVATION

CHAPTER 4B CLASSIFICATION

MATHS SKILLS FOR THIS CHAPTER

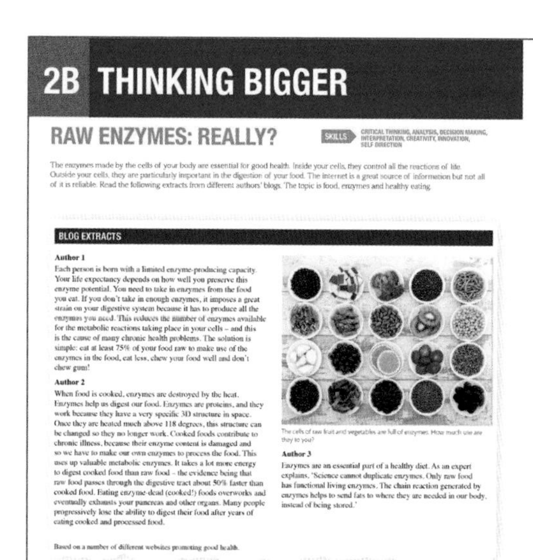

2B THINKING BIGGER

RAW ENZYMES: REALLY?

BLOG EXTRACTS

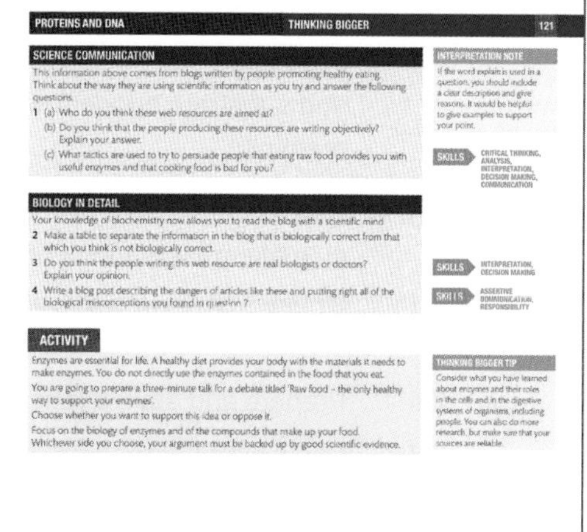

PROTEINS AND DNA — THINKING BIGGER — 121

SCIENCE COMMUNICATION

BIOLOGY IN DETAIL

ACTIVITY

Thinking Bigger
At the end of most chapters there is an opportunity to read and work with real-life research and writing about science. The activities help you to read real-life material that's relevant to your course, analyse how scientists write, think critically and consider how different aspects of your learning piece together.

Skills
These sections will help you develop transferable skills, which are highly valued in further study and the workplace.

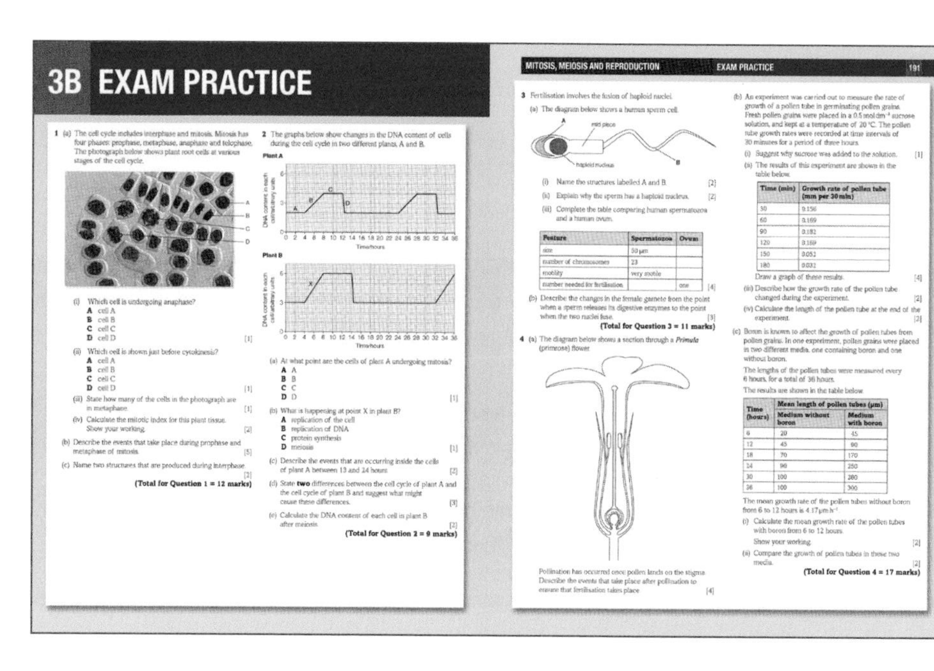

3B EXAM PRACTICE

MITOSIS, MEIOSIS AND REPRODUCTION — EXAM PRACTICE — 191

Exam Practice
Exam-style questions at the end of each chapter are tailored to the Pearson Edexcel specification to allow for practice and development of exam-writing technique. They also allow for practice responding to the command words used in the exams (see the **command words glossary** at the back of this book).

PRACTICAL SKILLS

Practical work is central to the study of biology. The Pearson Edexcel International Advanced Subsidiary (IAS) Biology specification includes nine Core Practicals that link theoretical knowledge and understanding to practical scenarios.

In order to develop practical skills, you should carry out a range of practical experiments related to the topics covered in your course. Further suggestions in addition to the Core Practicals are included below.

STUDENT BOOK TOPIC	IAS CORE PRACTICALS	
TOPIC 1 **MOLECULES, TRANSPORT AND HEALTH**	**CP1**	Use a semi-quantitative method with Benedict's reagent to estimate the concentrations of reducing sugars and with iodine solution to estimate the concentrations of starch, using colour standards.
	CP2	Investigate the vitamin C content of food and drink.
TOPIC 2 **MEMBRANES, PROTEINS, DNA AND GENE EXPRESSION**	**CP3**	Investigate membrane properties including the effect of alcohol and temperature on membrane permeability.
	CP4	Investigate the effect of temperature, pH, enzyme concentration and substrate concentration on the initial rate of enzyme-catalysed reactions.
TOPIC 3 **CELL STRUCTURE, REPRODUCTION AND DEVELOPMENT**	**CP5**	Use a light microscope to: (i) make observations and labelled drawings of suitable animal cells (ii) use a graticule with a microscope to make measurements and understand the concept of scale.
	CP6	Prepare and stain a root tip squash to observe the stages of mitosis.
TOPIC 4 **PLANT STRUCTURE AND FUNCTION, BIODIVERSITY AND CONSERVATION**	**CP7**	Use a light microscope to: (i) make observations, draw and label plan diagrams of transverse sections of roots, stems and leaves (ii) make observations, draw and label cells of plant tissues (iii) identify sclerenchyma fibres, phloem, sieve tubes and xylem vessels and their location.
	CP8	Determine the tensile strength of plant fibres.
	CP9	Investigate the antimicrobial properties of plants, including aseptic techniques for the safe handling of bacteria.

UNIT 1 (TOPICS 1 AND 2) MOLECULES, DIET, TRANSPORT AND HEALTH

Possible further practicals include:

- Investigate the structure of a mammalian heart by dissection.
- Investigate tissue water potentials using plant tissue and graded concentrations of a solute.
- Use a semi-quantitative method to estimate protein concentration using biuret reagent and colour standards.

UNIT 2 (TOPICS 3 AND 4) CELLS, DEVELOPMENT, BIODIVERSITY AND CONSERVATION

Possible further practicals include:

- Investigate factors affecting the growth of pollen tubes.
- Investigate plant mineral deficiencies.

Your knowledge and understanding of practical skills and activities will be assessed in all examination papers for the IAS Level Biology qualification.

- Papers 1 and 2 will include questions based on practical activities, including novel scenarios.
- Paper 3 will test your ability to plan practical work, including risk management and selection of apparatus.

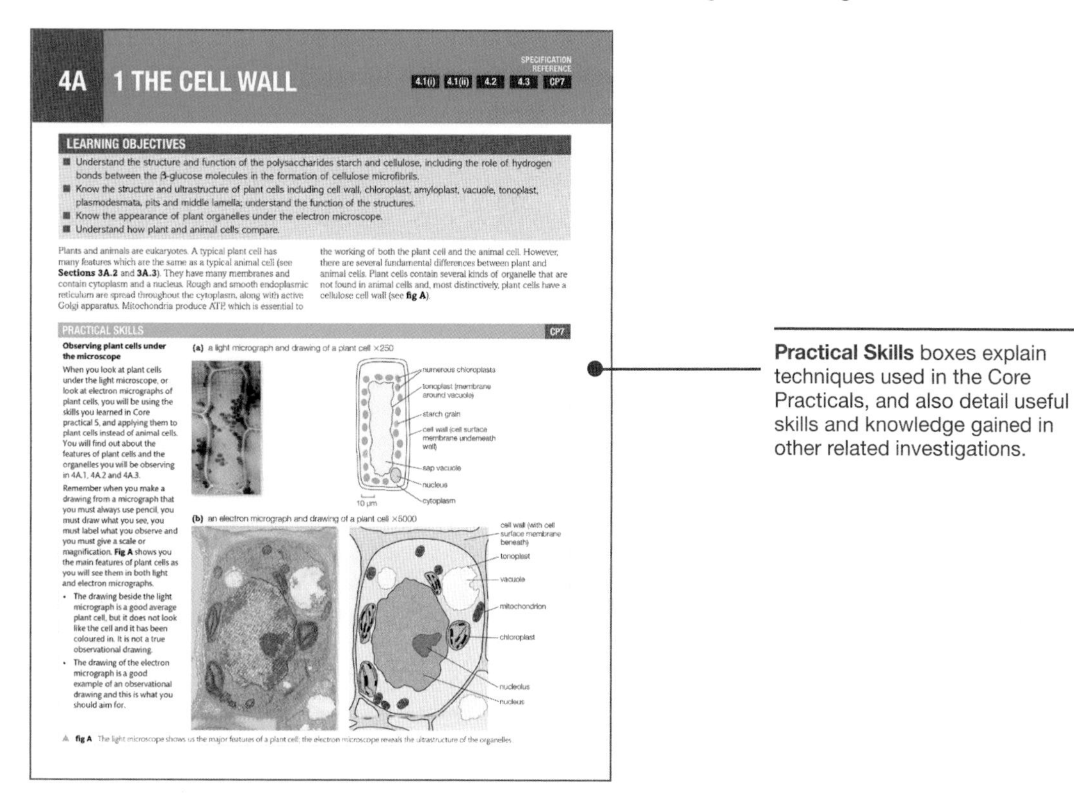

4A 1 THE CELL WALL

SPECIFICATION REFERENCE 4.1(i) 4.1(ii) 4.2 4.3 CP7

LEARNING OBJECTIVES

- Understand the structure and function of the polysaccharides starch and cellulose, including the role of hydrogen bonds between the β-glucose molecules in the formation of cellulose microfibrils.
- Know the structure and ultrastructure of plant cells including cell wall, chloroplast, amyloplast, vacuole, tonoplast, plasmodesmata, pits and middle lamella; understand the function of the structures.
- Know the appearance of plant organelles under the electron microscope.
- Understand how plant and animal cells compare.

Plants and animals are eukaryotes. A typical plant cell has many features which are the same as a typical animal cell (see **Sections 3A.2** and **3A.3**). They have many membranes and contain cytoplasm and a nucleus. Rough and smooth endoplasmic reticulum are spread throughout the cytoplasm, along with active Golgi apparatus. Mitochondria produce ATP, which is essential to the working of both the plant cell and the animal cell. However, there are several fundamental differences between plant and animal cells. Plant cells contain several kinds of organelle that are not found in animal cells and, most distinctively, plant cells have a cellulose cell wall (see **fig A**).

PRACTICAL SKILLS CP7

Observing plant cells under the microscope

When you look at plant cells under the light microscope, or look at electron micrographs of plant cells, you will be using the skills you learned in Core practical 5, and applying them to plant cells instead of animal cells. You will find out about the features of plant cells and the organelles you will be observing in 4A.1, 4A.2 and 4A.3.

Remember when you make a drawing from a micrograph that you must always use pencil, you must draw what you see, you must label what you observe and you must give a scale or magnification. **Fig A** shows you the main features of plant cells as you will see them in both light and electron micrographs.

- The drawing beside the light micrograph is a good average plant cell, but it does not look like the cell and it has been coloured in. It is not a true observational drawing.
- The drawing of the electron micrograph is a good example of an observational drawing and this is what you should aim for.

(a) a light micrograph and drawing of a plant cell ×250

(b) an electron micrograph and drawing of a plant cell ×5000

fig A The light microscope shows us the major features of a plant cell; the electron microscope reveals the ultrastructure of the organelles.

Practical Skills boxes explain techniques used in the Core Practicals, and also detail useful skills and knowledge gained in other related investigations.

This Student Book is accompanied by a **Lab Book**, which includes instructions and writing frames for the Core Practicals for students to record their results and reflect on their work. Practical skills practice questions and answers are also provided. The Lab Book records can be used as preparation and revision for the Practical Skills Paper.

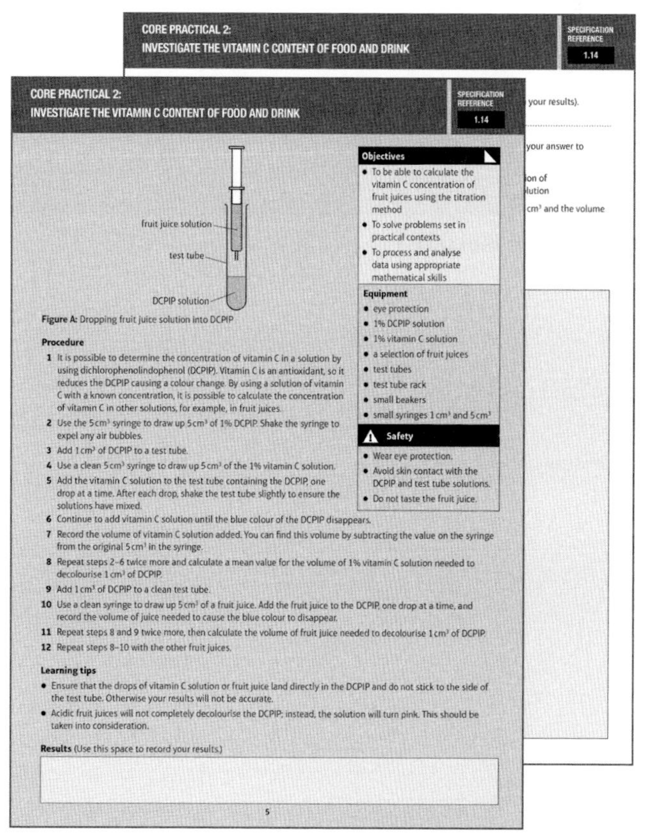

CORE PRACTICAL 2: INVESTIGATE THE VITAMIN C CONTENT OF FOOD AND DRINK SPECIFICATION REFERENCE 1.14

CORE PRACTICAL 2: INVESTIGATE THE VITAMIN C CONTENT OF FOOD AND DRINK

SPECIFICATION REFERENCE 1.14

Figure A: Dropping fruit juice solution into DCPIP

Objectives

- To be able to calculate the vitamin C concentration of fruit juices using the titration method
- To solve problems set in practical contexts
- To process and analyse data using appropriate mathematical skills

Equipment

- eye protection
- 1% DCPIP solution
- 1% vitamin C solution
- a selection of fruit juices
- test tubes
- test tube rack
- small beakers
- small syringes $1 cm^3$ and $5 cm^3$

Safety

- Wear eye protection.
- Avoid skin contact with the DCPIP and test tube solutions.
- Do not taste the fruit juice.

Procedure

1. It is possible to determine the concentration of vitamin C in a solution by using dichlorophenolindophenol (DCPIP). Vitamin C is an antioxidant, so it reduces the DCPIP causing a colour change. By using a solution of vitamin C with a known concentration, it is possible to calculate the concentration of vitamin C in other solutions, for example, in fruit juices.
2. Use the $5 cm^3$ syringe to draw up $5 cm^3$ of 1% DCPIP. Shake the syringe to expel any air bubbles.
3. Add $1 cm^3$ of DCPIP to a test tube.
4. Use a clean $5 cm^3$ syringe to draw up $5 cm^3$ of the 1% vitamin C solution.
5. Add the vitamin C solution to the test tube containing the DCPIP, one drop at a time. After each drop, shake the test tube slightly to ensure the solutions have mixed.
6. Continue to add vitamin C solution until the blue colour of the DCPIP disappears.
7. Record the volume of vitamin C solution added. You can find this volume by subtracting the value on the syringe from the original $5 cm^3$ in the syringe.
8. Repeat steps 2–6 twice more and calculate a mean value for the volume of 1% vitamin C solution needed to decolourise $1 cm^3$ of DCPIP.
9. Add $1 cm^3$ of DCPIP to a clean test tube.
10. Use a clean syringe to draw up $5 cm^3$ of a fruit juice. Add the fruit juice to the DCPIP, one drop at a time, and record the volume of juice needed to cause the blue colour to disappear.
11. Repeat steps 8 and 9 twice more, then calculate the volume of fruit juice needed to decolourise $1 cm^3$ of DCPIP.
12. Repeat steps 8–10 with the other fruit juices.

Learning tips

- Ensure that the drops of vitamin C solution or fruit juice land directly in the DCPIP and do not stick to the side of the test tube. Otherwise your results will not be accurate.
- Acidic fruit juices will not completely decolourise the DCPIP; instead, the solution will turn pink. This should be taken into consideration.

Results (Use this space to record your results.)

5

ASSESSMENT OVERVIEW

The following tables give an overview of the assessment for Pearson Edexcel International Advanced Subsidiary course in Biology. You should study this information closely to help ensure that you are fully prepared for this course and know exactly what to expect in each part of the examination. More information about this qualification, and about the question types in the different papers, can be found on page 302 of this book.

PAPER / UNIT 1	PERCENTAGE OF IAS	PERCENTAGE OF IAL	MARK	TIME	AVAILABILITY
MOLECULES, DIET, TRANSPORT AND HEALTH Written examination Paper code WBI11/01 Externally set and marked by Pearson Edexcel Single tier of entry	40%	20%	80	1 hour 30 minutes	January, June and October First assessment : January 2019

PAPER / UNIT 2	PERCENTAGE OF IAS	PERCENTAGE OF IAL	MARK	TIME	AVAILABILITY
CELLS, DEVELOPMENT, BIODIVERSITY AND CONSERVATION Written examination Paper code WBI12/01 Externally set and marked by Pearson Edexcel Single tier of entry	40%	20%	80	1 hour 30 minutes	January, June and October First assessment : June 2019

PAPER / UNIT 3	PERCENTAGE OF IAS	PERCENTAGE OF IAL	MARK	TIME	AVAILABILITY
PRACTICAL SKILLS IN BIOLOGY 1 Written examination Paper code WBI13/01 Externally set and marked by Pearson Edexcel Single tier of entry	20%	10%	50	1 hour 20 minutes	January, June and October First assessment : June 2019

ASSESSMENT OBJECTIVES AND WEIGHTINGS

ASSESSMENT OBJECTIVE	DESCRIPTION	% IN IAS	% IN IA2	% IN IAL
A01	Demonstrate knowledge and understanding of science	36–39	31–34	34–37
A02	(a) Application of knowledge and understanding of science in familiar and unfamiliar contexts.	34–36	33–36	33–36
	(b) Analysis and evaluation of scientific information to make judgments and reach conclusions.	9–11	14–16	11–14
A03	Experimental skills in science, including analysis and evaluation of data and methods	17–18	17–18	17–18

RELATIONSHIP OF ASSESSMENT OBJECTIVES TO UNITS

UNIT NUMBER	ASSESSMENT OBJECTIVE			
	A01	A02 (a)	A02 (b)	A03
UNIT 1	17–18	17–18	4.5–5.5	0
UNIT 2	17–18	17–18	4.5–5.5	0
UNIT 3	2–3	0	0	17–18
TOTAL FOR INTERNATIONAL ADVANCED SUBSIDIARY	**36–39**	**34–36**	**9–11**	**17–18**

TOPIC 1 MOLECULES, TRANSPORT AND HEALTH

CHAPTER 1A CHEMISTRY FOR BIOLOGISTS

Water is essential to life. Everyone knows this. Yet the jerboa, a small rodent found throughout Asia and Northern Africa, may never drink water in its life. The jerboa (family Dipodidae) is found in both hot and cold deserts from the Sahara Desert to the Gobi Desert. It is extremely well adapted for dry desert environments and gets the water it needs from the food it eats. This includes plant leaves, roots and seeds, and in some cases insects. Jerboas also produce tiny amounts of very concentrated urine to get rid of their waste products, another adaptation for saving the water needed for life.

Biology is the study of living things. The basic unit of life is the cell, but underpinning all life is chemistry. The way atoms are bonded together affects the way chemicals work in the cells – and that affects everything, from the way plants make food by photosynthesis to the way your eyes respond to light.

In this chapter, you will be looking at some of the important ways in which atoms and molecules interact to make up the chemistry of life. You will be using these basic principles throughout your biology course because they are fundamental to the structures and functions of all the organisms you will study.

You will see how the chemistry of water enables life to survive and chemical reactions to continue. You will look at carbohydrates, from the simplest sugars to the most complex polysaccharides. These molecules have a wide variety of uses in organisms, from the fuel for cellular respiration to the main structural material in plants. As you discover how the molecules are joined together, you will recognise the relationships between the structure of the molecules and their functions in the body.

The same links between structure and function are clear when you look at the structure of lipid molecules. For example, lipids are used as energy stores in both animals and plants. Lipids are non-polar molecules but you will discover how they can become polar in combination with other inorganic groups such as phosphates. This polarity has great importance for the characteristics of the cell membrane.

At the end of this chapter, you will study the structure of proteins. They are long chains of amino acids that are held together by chemical bonds to make complex structures. The bonds include the covalent bonds, ionic bonds and hydrogen bonds.

MATHS SKILLS FOR THIS CHAPTER

- **Recognise and make appropriate use of units in calculations** (*e.g. millimetres*)
- **Use ratios, fractions and percentages** (*e.g. representing the relationships between atoms in an ion or molecule*)

What prior knowledge do I need?

- Life processes depend on molecules whose structure is related to their function
- All living things are made up of cells
- Many processes in living cells, including diffusion and osmosis, depend on water
- Reactions in cells take place in solution in water
- Complex carbohydrates are made up of sugars joined together
- Complex carbohydrates can be broken down to give simple sugars that can be used by cells
- Plants make carbohydrates in photosynthesis
- Lipids are made up of fatty acids and glycerol
- Lipids are molecules used to store energy in the bodies of animals and plants
- Proteins are long chains of amino acids
- Enzymes are made of proteins

What will I study in this chapter?

- How ionic and covalent bonding affect the nature of the compound formed
- The formation of anions and cations in ionic bonding
- The formation of dipoles in some covalent molecules leading to intra- and intermolecular bonds (e.g. hydrogen bonds)
- The chemistry of water and how this affects its properties
- The importance of water to living things
- The structure of different types of monosaccharide
- The formation of disaccharides by the joining of two monosaccharides in a condensation reaction
- The structure of complex polysaccharides and how their structure is related to their functions as storage molecules
- The structure of lipids including the formation of ester bonds
- The primary, secondary, tertiary and quaternary structure of proteins and how the structure is related to the function of the protein
- The structure of amino acids, peptides and polypeptides and how they relate to each other
- The formation of peptide bonds between amino acids

What will I study later?

Chapter 2A

- The importance of polarity in the structure and function of phospholipids
- How the structure of phospholipids determines many of the characteristics of the cell membrane
- How proteins act as carrier systems in cell membranes
- How water is taken into and moved around plants
- The movement of water into and out of cells, tissues and vessels in animals, plants and fungi

Chapter 2B

- The importance of hydrogen bonding in the tertiary and quaternary structure of proteins and in the structure and function of enzymes

Chapter 4A

- The structure of cellulose in plant cell walls

Chapter 5A (Book 2: IAL)

- The role of water in the reactions of photosynthesis

Chapter 7A (Book 2: IAL)

- The importance of carbohydrates in cellular respiration
- The role of water in the reactions of cellular respiration

Chapter 8B (Book 2: IAL)

- The importance of water in plant movements

LEARNING OBJECTIVES

- Understand the importance of water as a solvent in transport, including its dipole nature.

IONIC AND COVALENT BONDING

Biology is the study of living things – but living things consist of chemical substances. The dragonfly and the plant it is resting on in **fig A** are all made of chemicals. So is the cow in **fig C** – and it needs the chemical known as salt which it is licking to stay alive. If you understand some of the basic principles of chemistry, you will develop a much better understanding of biological systems. The chemical bonds within and between molecules affect the properties of the compounds they form. This affects their functions within the cell and the organism. For example, if you want to understand the chemistry of water, you need to understand chemical bonds and how dipoles are created within molecules.

▲ **fig A** All life depends on some very fundamental chemistry.

The basic unit of all elements is the atom. When the atoms of two or more different elements react, they form a compound. An atom is made up of a nucleus containing positive protons and neutral neutrons. The nucleus is surrounded by negative electrons. We can show this in a model as electrons orbiting around the nucleus in shells. When an atom has a full outer shell of electrons, it is stable and does not react. However, most atoms do not have a full outer shell of electrons. In chemical reactions, these electrons are involved in changes that give the atom a stable outer shell. There are two ways they can achieve this.

- **Ionic bonding:** the atoms involved in the reaction give or receive electrons. One atom, or part of the molecule, gains one or more electrons and becomes an **anion** (a negative ion). The other atom, or part of the molecule, loses one or more electrons and becomes a **cation** (a positive ion). Strong forces of attraction called **ionic bonds** hold the oppositely charged ions together (see **fig B**).

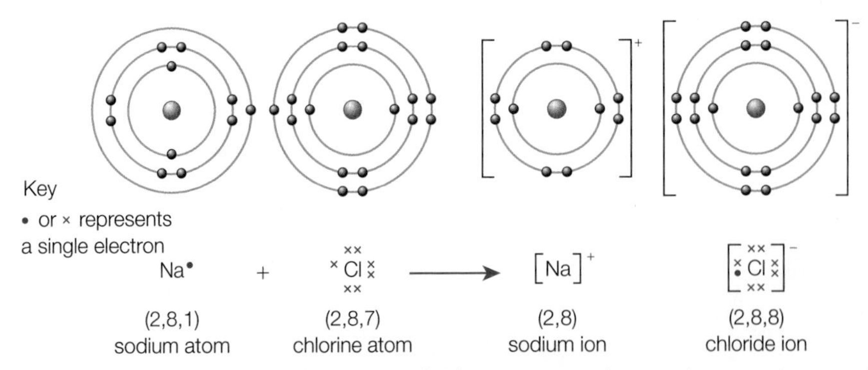

▲ **fig B** The formation of sodium chloride (salt), an inorganic substance that is very important in living organisms, is an example of ionic bonding.

- **Covalent bonding:** the atoms involved in the reaction share electrons (see **fig D**). **Covalent bonds** are very strong and the molecules formed are usually neutral. However, in some covalent compounds, the molecules are slightly polarised: this means that the electrons in the covalent bonds are not quite evenly shared. Consequently, the molecule has a part that is slightly negative and a part that is slightly positive. This separation of charge is called a **dipole**, and the tiny charges are represented as δ^+ and δ^- (see **fig F**). The molecule is described as a **polar molecule**. This polarity is particularly common if the bond involves one or more hydrogen atoms.

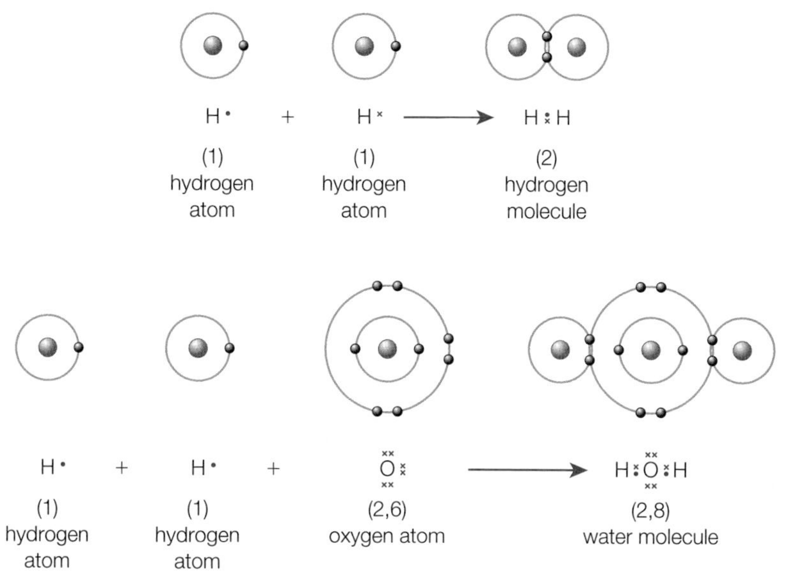

▲ **fig D** The formation of hydrogen molecules and water molecules are examples of covalent bonding.

▲ **fig C** Animals such as cows can use a mineral lick to get the salt they need to function.

THE IMPORTANCE OF INORGANIC IONS

When ionic substances dissolve in water, the ions separate in a process called **dissociation**. Cells are 60–70% water, so in living organisms most ionic substances exist as positive and negative ions. Many of these ions play specialised roles in individual cells and in the functioning of entire organisms. Here are some of the inorganic ions (and their roles) you will meet as you study biology.

EXAM HINT

Make sure you understand and can explain the difference between ionic substances, charged particles and polar molecules.

IMPORTANT ANIONS

- Nitrate ions (NO_3^-) – these are needed in plants to make DNA and also amino acids and, therefore, proteins from the products of photosynthesis (see **Sections 1A.5**, **2B.3** and **Book 2 Chapter 5A**).
- Phosphate ions (PO_4^{3-}) – these are needed in all living organisms to make ATP and ADP as well as DNA and RNA (see **Section 2B.3** and **Book 2 Chapter 5A**).
- Chloride ions (Cl^-) – these are needed in nerve impulses, sweating and many secretory systems in animals (see **Book 2 Chapters 7C** and **8A)**.
- Hydrogencarbonate ions (HCO_3^-) – these are needed to buffer blood pH to prevent it becoming too acidic (see **Section 1B.2**).

IMPORTANT CATIONS

- Sodium ions (Na^+) – these are needed in nerve impulses, sweating and many secretory systems in animals (see **Book 2 Chapter 8A**).
- Calcium ions (Ca^{2+}) – these are needed for the formation of calcium pectate for the middle lamella between two cell walls in plants, and for bone formation and muscle contraction in animals (see **Section 4A.1** and **Book 2 Chapters 7B** and **7C**).
- Hydrogen ions (H^+) – these are needed in cellular respiration and photosynthesis, and in numerous pumps and systems as well as pH balance (see **Section 2A.4** and **Book 2 Chapters 5A** and **7A**).
- Magnesium ions (Mg^{2+}) – these are needed for production of chlorophyll in plants (see **Book 2 Chapter 5A**).

THE CHEMISTRY OF WATER

All reactions in living cells take place in water. Without water, substances could not move around the body. Water is one of the reactants in the process of photosynthesis, on which almost all life depends (see **fig E**). Understanding the properties of water will help you understand many key systems in living organisms.

Water is also a major habitat – it supports more life than any other part of the planet.

▲ **fig E** Water is vital for life on Earth in many different ways – in a desert, the smallest amount of water allows plants to grow.

The simple chemical formula of water is H_2O. This tells us that two atoms of hydrogen are joined to one atom of oxygen to make up each water molecule. However, because the electrons are held closer to the oxygen atom than to the hydrogen atoms, water is a polar molecule (see **fig F**).

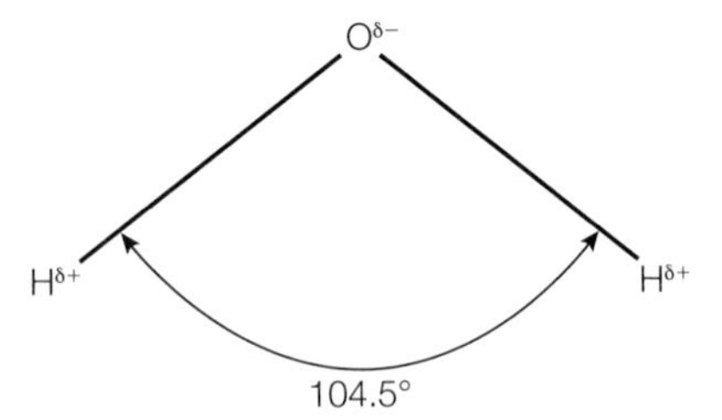

▲ **fig F** A model of a water molecule showing dipoles.

One major effect of this polarity is that water molecules form **hydrogen bonds**. The slightly negative oxygen atom of one water molecule will attract the slightly positive hydrogen atoms of other water molecules in a weak electrostatic attraction called a hydrogen bond. Each individual hydrogen bond is weak but there are many of them so the molecules of water 'stick together' more than you might expect (see **fig G**). Water has relatively high melting and boiling points compared with other substances that have molecules of a similar size because it takes a lot of energy to break all the hydrogen bonds that hold the molecules together. Hydrogen bonds are important in protein structure (see **Sections 1A.5** and **2B.1**) and in the structure and functioning of DNA (see **Section 2B.3**).

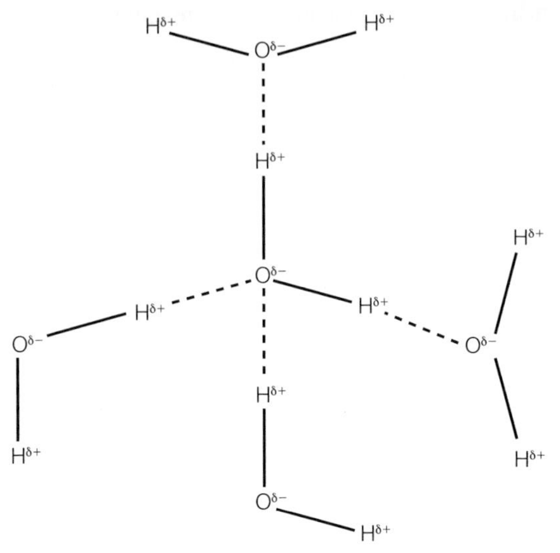

▲ **fig G** Hydrogen bonding in water molecules, based on attraction between positive and negative dipoles.

THE IMPORTANCE OF WATER

The properties of water make it very important in biological systems for many reasons.

- Water is a polar solvent. Because it is a polar molecule, many ionic substances like sodium chloride will dissolve in it (see **fig H**). Many covalently bonded substances are also polar and will dissolve in water, but often do not dissolve in other covalently bonded solvents such as ethanol. Water also carries other substances, such as starch. As a result, most of the chemical reactions within cells occur in water (in aqueous solution).

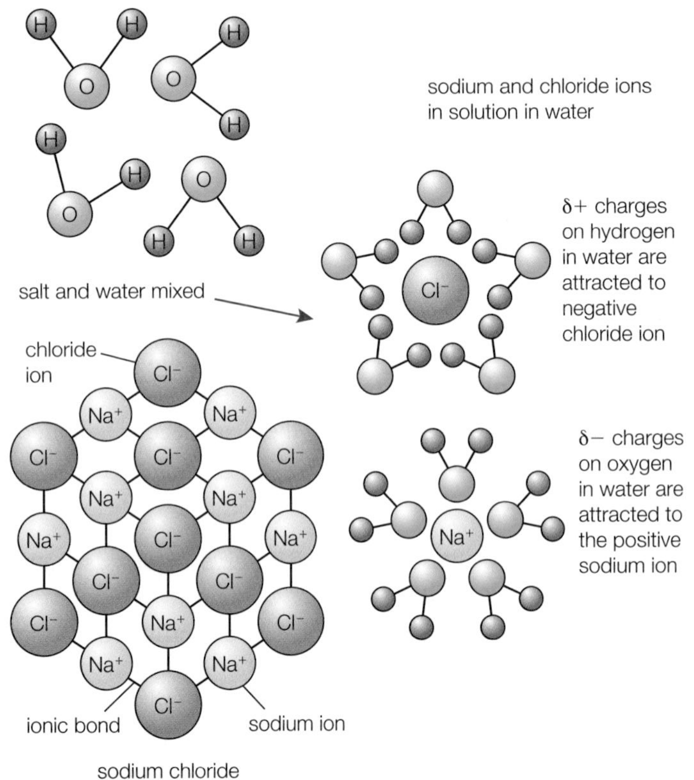

▲ **fig H** A model of sodium chloride dissolving in water as a result of the interactions between the charges on sodium and chloride ions and the dipoles of the water molecules.

- Water is an excellent transport medium because the dipole nature of water enables many different substances to dissolve in it (see **Sections 1B.2**, **4A.4** and **4A.5**).
- As water cools to 4 °C, it reaches its maximum density. As it cools further, the molecules become more widely spaced. As a result, ice is less dense than water and floats, forming an insulating layer and helping to prevent the water underneath it from freezing. It also melts quickly because, being at the top, it is exposed to the sun. It is very unusual for the solid form of a substance to be less dense than the liquid form. This unusual property enables organisms to live in water, even in countries where it gets cold enough to freeze in winter.
- Water is slow to absorb and release heat – it has a high specific heat capacity. The hydrogen bonds between the molecules need a lot of energy to separate them. This means the temperature of large bodies of water such as lakes and seas does not change much throughout the year. This makes them good habitats for living organisms.
- Water is a liquid – it cannot be compressed. This is an important factor in many hydraulic mechanisms in living organisms.
- Water molecules are cohesive – the forces between the molecules mean they stick together. This is very important for the movement of water from the roots to the leaves of plants (see **Sections 4A.3** and **4A.4**).
- Water molecules are adhesive – they are attracted to other different molecules. This is also important in plant transport systems and in surface tension.
- Water has a very high surface tension because the attraction between the water molecules, including hydrogen bonds, is greater than the attraction between the water molecules and the air. As a result, the water molecules hold together forming a thin 'skin' of surface tension. Surface tension is very important in plant transport systems, and also affects life at the surface of ponds, lakes and other water masses (see **fig I**).

LEARNING TIP

Remember that *co* means two similar things together, as in cohabit, and *ad* means two different things together.

▲ **fig I** Without surface tension, a raft spider like this could not move across the water and hunt.

EXAM HINT

All these properties are a result of dipoles and hydrogen bonding between water molecules. Make sure that you can explain the link between the property and the hydrogen bonding.

CHECKPOINT

SKILLS PROBLEM SOLVING

1. What is a dipole?
2. What are the differences between ionic substances and polar substances?
3. How are hydrogen bonds formed between water molecules and what effect do they have on the properties of water?
4. Discuss how the properties of water affect living organisms.

EXAM HINT

In exam questions, the command word *discuss* suggests that you may need to consider possible negative effects as well as the more obvious benefits to living organisms. You should identify the issue that is being assessed within the question. Explore all aspects of the issue. Investigate the issue by reasoning or argument.

SUBJECT VOCABULARY

anion a negative ion
cation a positive ion
ionic bonds bonds formed when atoms give or receive electrons; they result in charged particles called ions
covalent bonds bonds formed when atoms share electrons; covalent molecules may be polar if the electrons are not shared equally
dipole the separation of charge in a molecule when the electrons in covalent bonds are not evenly shared
polar molecule a molecule containing a dipole
dissociation splitting of a molecule into smaller molecules, atoms, or ions, especially by a reversible process
hydrogen bonds weak electrostatic intermolecular bonds formed between polar molecules containing at least one hydrogen atom

2 CARBOHYDRATES 1: MONOSACCHARIDES AND DISACCHARIDES

SPECIFICATION REFERENCE

1.2(i) 1.2(ii) 1.3 1.4 CP1

LEARNING OBJECTIVES

- Know the difference between monosaccharides and disaccharides.
- Know how to use Benedict's reagent.
- Know how monosaccharides (glucose, fructose and galactose) join to form disaccharides (maltose, sucrose and lactose) through condensation reactions forming glycosidic bonds, and how they can be split through hydrolysis reactions.

WHAT ARE ORGANIC COMPOUNDS?

Biological molecules are the key to the structure and function of living things. Biological molecules are often organic compounds. Organic compounds all contain carbon atoms. They also contain atoms of hydrogen, oxygen and, less frequently, nitrogen, sulfur and phosphorus. Most of the material in your body that is not water consists of these organic molecules. An understanding of why organic molecules are special will help you to understand the chemistry of biological molecules including carbohydrates, lipids and proteins.

Each carbon atom can make four bonds and so it can connect to four other atoms. Carbon atoms bond particularly strongly to other carbon atoms to make long chains. The four bonds of a carbon atom usually form a tetrahedral shape. This means carbon compounds can be rings, branched chains or any number of three-dimensional (3D) shapes (see **fig A**). In some carbon compounds small molecules (**monomers**) bond with many other similar units to make a very large molecule called a **polymer**. The ability of carbon to combine and make **macromolecules** (large molecules) is the basis of all biological molecules and provides the great variety and complexity found in living things.

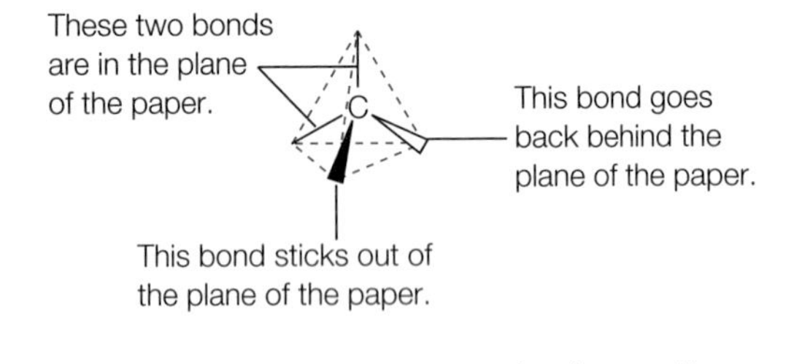

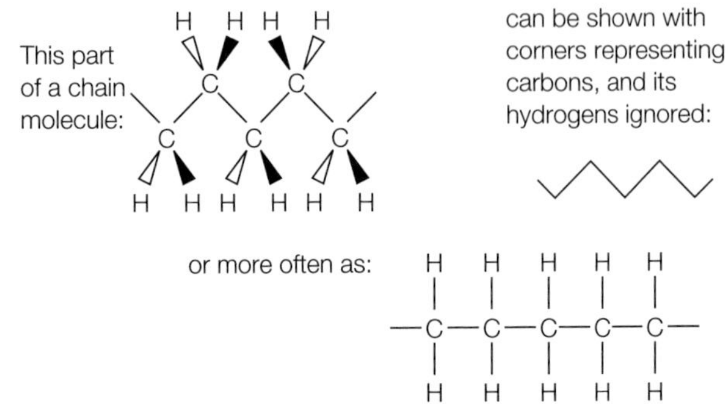

▲ **fig A** The bonds in a carbon atom have a complicated 3D shape. This is difficult to represent, so in most molecular diagrams we use one of several different ways to draw them.

CARBOHYDRATES

Carbohydrates are important in cells as a usable energy source and important in human foods around the world (see **fig B**). They are also important for storing energy and they form an important part of the cell wall in plants, fungi and bacteria. Sugars and **starch** are the best known carbohydrates. **Sucrose** is the familiar white crystalline table sugar; **glucose** is used as a fuel by the cells of our bodies. Starch is in rice, flour and potatoes. But the group of substances called carbohydrates contains many more compounds, as you will discover.

▲ **fig B** Carbohydrates are important molecules in both plants and animals – and carbohydrate foods like this bread play a major role in the human diet.

The basic structure of all carbohydrates is the same. They consist of carbon, hydrogen and oxygen. There are three main groups of carbohydrates: **monosaccharides**, **disaccharides** and **polysaccharides**. Some have more complex molecules than others (see **Section 1A.3**).

MONOSACCHARIDES: THE SIMPLE SUGARS

Monosaccharides are simple sugars in which there is one oxygen atom and two hydrogen atoms for each carbon atom in the molecule. A general formula for this can be written $(CH_2O)_n$. Here n can be any number, but it is usually low.

Triose sugars ($n = 3$) have three carbon atoms and the general formula $C_3H_6O_3$. They are important in mitochondria, where the respiration process breaks down glucose into triose sugars (see **Book 2 Chapter 7A**).

- **Pentose sugars** ($n = 5$) have five carbon atoms and the general formula $C_5H_{10}O_5$. **Ribose** and **deoxyribose** are important in the nucleic acids **deoxyribonucleic acid (DNA)** and **ribonucleic acid (RNA)**, which make up the genetic material (see **Section 2B.3**).
- **Hexose sugars** ($n = 6$) have six carbon atoms and the general formula $C_6H_{12}O_6$. They are the best known monosaccharides, often taste sweet and include glucose, galactose and fructose.

General formulae show you how many atoms there are in the molecule, and what type they are, but they do not tell you what the molecule looks like and why it behaves as it does. To show this, you can use displayed formulae. Although these do not show all the detailed shape in the carbon chain, they can give you a good idea of how the molecules are arranged in three dimensions. This can help to explain why biological systems behave as they do (see **figs C** and **D**).

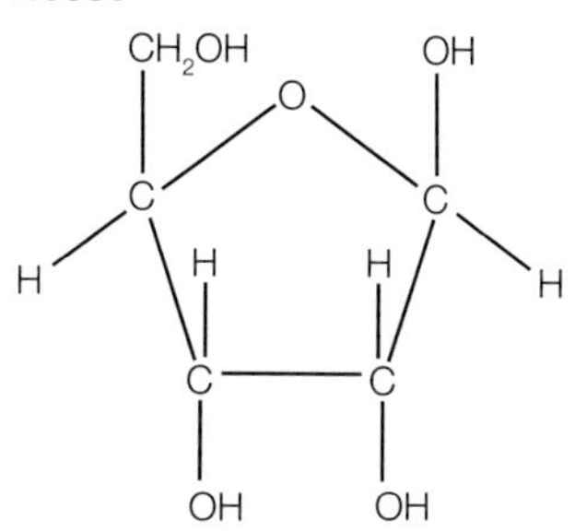

▲ **fig C** Pentose sugars such as ribose have 5 carbon atoms.

α-glucose β-glucose

or, even more simply:

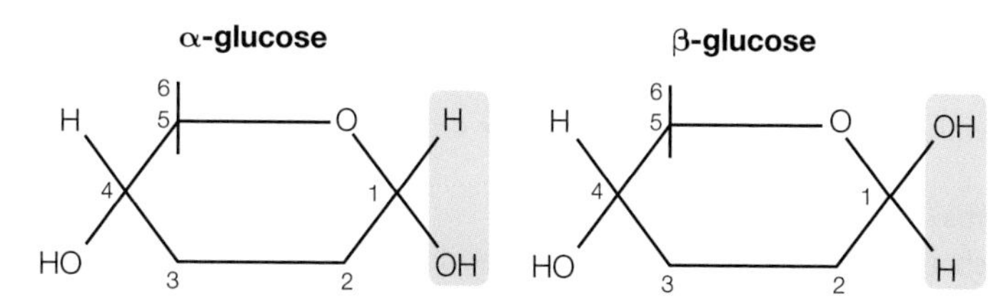

In these diagrams, the positions of carbon atoms are represented by their numbers only.
Note carefully the different arrangement of atoms around the carbon 1 atom in α-glucose and β-glucose.

▲ **fig D** Hexose sugars have a ring structure. The arrangement of the atoms on the side chains can make a significant difference to the way in which the molecule can be used by the body. The carbon atoms are numbered in order to identify the different arrangements.

Glucose has two **isomers** (different forms): α-glucose and β-glucose. The two isomers have different arrangements of the atoms on the side chains of the molecule. The different isomers form different bonds between neighbouring glucose molecules, and this affects the polymers that are made. You will learn more about α- and β-glucose in **Section 4A.5**.

> **DID YOU KNOW?**
> **Hydrogenating some sugars reduces the energy they provide. When glucose is hydrogenated, it forms sorbitol ($C_6H_{14}O_6$). Sorbitol tastes up to 60% sweeter than glucose but it provides less energy when it is used in the body (11 kJg^{-1} compared to 17 kJg^{-1}). The combination of the very sweet taste and the lower energy count makes it useful as a sweetener for people who want to lose weight. A small change in the chemical structure has a big effect on function.**

DISACCHARIDES: THE DOUBLE SUGARS

Disaccharides consist of two monosaccharides joined together – for example sucrose (table sugar) is formed by a molecule of α-glucose joining with a molecule of fructose. Two monosaccharides join in a **condensation reaction** to form a disaccharide, and a molecule of water (H_2O) is released. The link between the two monosaccharides results in a covalent bond known as a **glycosidic bond** (see **fig E**). We use numbers to show which carbon atoms are involved in the bond. If carbon 1 on one monosaccharide joins to carbon 4 on another monosaccharide, we call it a 1,4-glycosidic bond. If the bond is between carbon 1 and carbon 6, it is a 1,6-glycosidic bond.

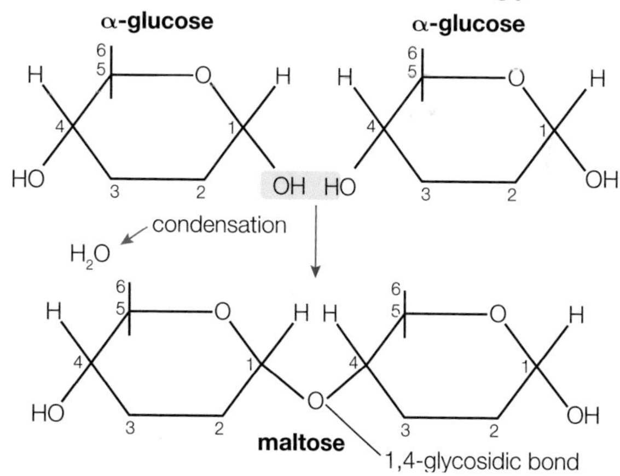

▲ **fig E** The formation of a glycosidic bond. The condensation reaction between two monosaccharides results in a disaccharide and a molecule of water.

When different monosaccharides join together, different disaccharides are made, and these have different properties. Many disaccharides taste sweet. **Table A** shows some of the more common ones.

DISACCHARIDE	SOURCE	MONOSACCHARIDE
sucrose	stored in plants such as sugar cane	glucose + fructose
lactose	milk sugar – this is the main carbohydrate found in milk	glucose + galactose
maltose	malt sugar – found in germinating seed such as barley	glucose + glucose

table A Three common disaccharides

EXAM HINT

A clearly labelled and annotated diagram will help your description of the formation of a glycosidic bond.

LEARNING TIP

All the common monosaccharides and disaccharides are reducing sugars, except sucrose. When testing for the presence of a non-reducing sugar you must test for reducing sugar first to ensure there is no reaction with Benedict's reagent. Why?

SKILLS PROBLEM SOLVING

LEARNING TIP

Remember that *iso* means same, the same atoms.

PRACTICAL SKILLS CP1

Testing for sugars

Benedict's solution is a chemical reagent for testing **reducing sugars**. It is a bright blue solution that contains copper(II) ions. Some sugars react readily with this solution when heated gently and reduce the copper(II) ions to copper(I) ions; a precipitate is formed and a colour change from blue to orange occurs (see **fig F**). Sugars that react in this way are known as reducing sugars. All of the monosaccharides and some disaccharides (but not sucrose) are reducing sugars.

▲ **fig F** Benedict's test for reducing sugars

Some sugars do not react with Benedict's solution. They are known as **non-reducing sugars**. You can heat a non-reducing sugar such as sucrose with a few drops of hydrochloric acid, allow it to cool and then neutralise the solution with sodium hydrogen carbonate to hydrolyse the glycosidic bonds. This produces the monosaccharide units of the sugar, which will give a positive Benedict's test.

CHECKPOINT

1. What are carbohydrates?
2. Describe how a glycosidic bond is formed between two monosaccharides to form a disaccharide.

SUBJECT VOCABULARY

monomer a small molecule that is a single unit of a larger molecule called a polymer
polymer a long-chain molecule made up of many smaller, repeating monomer units joined together by chemical bonds
macromolecule a very large molecule often formed by polymerisation
starch a long-chain polymer formed of glucose monomers
sucrose a sweet-tasting disaccharide formed by the joining of glucose and fructose by a 1,4-glycosidic bond
glucose a hexose sugar
monosaccharide a single sugar monomer
disaccharide a sugar made up of two monosaccharide units joined by a glycosidic bond, formed in a condensation reaction
polysaccharide a polymer consisting of long chains of monosaccharide units joined by glycosidic bonds
triose sugar a sugar with three carbon atoms
pentose sugar a sugar with five carbon atoms
ribose a pentose sugar that is part of the structure of RNA
deoxyribose a pentose sugar that is part of the structure of DNA
deoxyribonucleic acid (DNA) a nucleic acid that is the genetic material in many organisms
ribonucleic acid (RNA) a nucleic acid which is the genetic material in some organisms and is involved in protein synthesis
hexose sugar sugar with six carbon atoms
isomers molecules that have the same chemical formula, but different molecular structures
condensation reaction a reaction in which a molecule of water is removed from the reacting molecules as a bond is formed between them
glycosidic bond a covalent bond formed between two monosaccharides in a condensation reaction, which can be broken down by a hydrolysis reaction to release the monosaccharide units
reducing sugars sugars that react with blue Benedict's solution and reduce the copper(II) ions to copper(I) ions giving an orangey-red precipitate
non-reducing sugars sugars that do not react with Benedict's solution

1A 3 CARBOHYDRATES 2: POLYSACCHARIDES

SPECIFICATION REFERENCE 1.2(i) 1.2(ii) 1.3 1.4 CP1

LEARNING OBJECTIVES

- Know the difference between monosaccharides, disaccharides and polysaccharides, including glycogen and starch.
- Explain how monosaccharides join to form polysaccharides through condensation reactions forming glycosidic bonds, and how these can be split through hydrolysis reactions.
- Relate the structures of monosaccharides, disaccharides and polysaccharides to their roles in providing and storing energy.

The most complex carbohydrates are the polysaccharides. They are made of many monosaccharide units joined by condensation reactions that create glycosidic bonds (see **Section 1A.2 fig E**). Polysaccharides do not have the sweet taste of many mono- and disaccharides, but these complex polymers include some very important biological molecules.

Molecules with between 3 and 10 sugar units are known as **oligosaccharides**, while molecules containing 11 or more monosaccharides are known as true polysaccharides. The glycosidic bonds in the polysaccharide can be broken to release monosaccharide units for cellular respiration.

The glycosidic bond between two glucose units is split by a process known as **hydrolysis** (see **fig A**). The hydrolysis reaction is the opposite of the condensation reaction that created the molecule, so water is added to the bond. Starch and glycogen are gradually broken down into shorter and shorter chains and eventually single sugars are left. Disaccharides break down to form two monosaccharides. Hydrolysis takes place during digestion in the gut, and also in the muscle and liver cells when the carbohydrate stores are broken down to release sugars for use in cellular respiration (see **Book 2 Chapter 7A**).

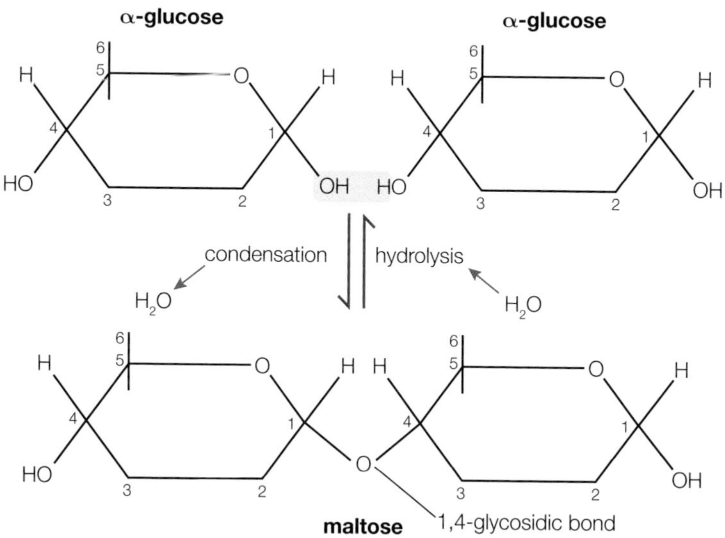

▲ **fig A** Glycosidic bonds are made by condensation reactions and broken down by hydrolysis.

LEARNING TIP

Glycosidic bonds are *formed* with the *removal* of a molecule of water in *condensation reactions.*

Glycosidic bonds are *broken* with the *addition* of a molecule of water in *hydrolysis reactions.*

CARBOHYDRATES AND ENERGY

MONOSACCHARIDES AND DISACCHARIDES

Every chemical reaction taking place in a cell needs energy. This energy is supplied by a substance called adenosine triphosphate, **ATP**. This ATP comes from the breakdown of the monosaccharide glucose, using oxygen, in the process of cellular respiration. You will learn much more about cellular respiration later in your course in **Book 2 Chapter 7A**.

The arrangement of atoms in a molecule of α-glucose means that it can be broken down completely in a series of reactions, if oxygen is available. The compounds that are produced, called the **end products**, are waste carbon dioxide and water, and lots of ATP. This supplies the energy needed for all the reactions in the cell.

EXAM HINT

Be careful not to say that this produces or creates energy for cell processes. Energy cannot be created – it is converted from one form to another. Here, chemical energy is transferred from the glucose molecule to the ATP molecules.

Any glucose in the food you eat can be absorbed and used directly in your cells. Other monosaccharides and disaccharides – for example, fructose, maltose and sucrose – are also easily absorbed in the body and rapidly converted to glucose. So foods containing monosaccharides and disaccharides are a good source of relatively instant energy (see **fig B**). However, these cannot be used to store energy because they are chemically active, and they are very soluble in water, so they affect the water balance of the cells. You will find out why that is so important in **Chapter 2A**.

fig B These medjool dates contain a lot of fructose. That is why they taste so sweet and give you instant energy.

POLYSACCHARIDES

The structure of polysaccharides makes them ideal as energy storage molecules within a cell.

- They can form very compact molecules, which take up little space.
- They are physically and chemically inactive, so they do not interfere with the other functions of the cell.
- They are not very soluble in water, so have almost no effect on water potential within a cell and cause no osmotic water movements.

STARCH

Starch is particularly important as an energy store in plants. The sugars produced by photosynthesis are rapidly converted into starch, which is insoluble and compact but can be broken down rapidly to release glucose when it is needed. Storage organs such as sweet potatoes (yams) are particularly rich in starch. You will find out a lot more about the importance of starch in plants in **Chapter 4A** of this book.

Starch consists of long chains of α-glucose. But if you look at it more closely you will see that it is a mixture of two compounds:

- **Amylose:** an unbranched polymer of between 200 and 5000 glucose units. As the chain lengthens the molecule spirals, which makes it more compact for storage.
- **Amylopectin:** a branched polymer of glucose units. The branching chains have many terminal glucose units that can be broken off rapidly when energy is needed.

Amylose and amylopectin are both long chains of α-glucose units – so why are the molecules so different? It all depends on the carbon atoms involved in the glycosidic bonds.

Amylose has only 1,4-glycosidic bonds, which is why the molecules are long unbranched chains.

In amylopectin, many of the glucose molecules are joined by 1,4-glycosidic bonds, but there are also a few 1,6-glycosidic bonds. This results in the branching chains that change the properties of the molecule.

EXAM HINT

Be sure you can relate the structure of the molecule to its function as a storage molecule.

Coiling makes it compact so it takes up less space and doesn't get in the way of organelles or substances moving around the cell.

Large molecules are insoluble so they do not interfere with the water potential of the cell.

PRACTICAL SKILLS CP1

Testing for starch

If you add a few drops of reddish-brown iodine solution to a sample containing starch (whether it is a solid sample or a sample in solution), the iodine solution will turn blue-black (see **fig C**).

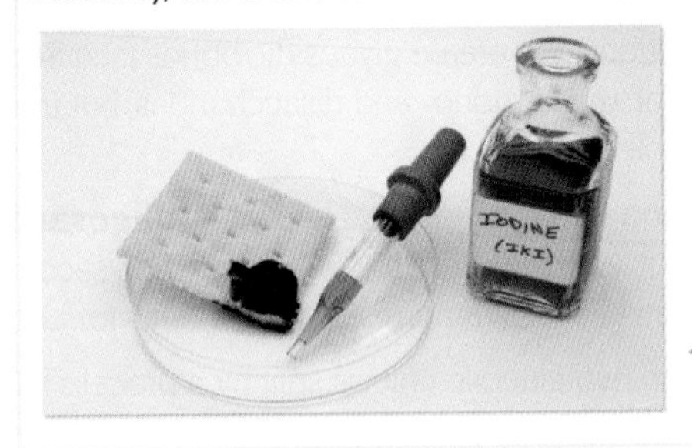

fig C The iodine test for starch

Starch has a combination of straight-chain amylose and branched-chain amylopectin molecules (see **fig D**). This combination explains why carbohydrate foods like rice and pasta are so good for you when you are doing sport or hard physical work. The amylopectin releases glucose for cellular respiration rapidly when needed. Amylose releases glucose more slowly over time, keeping you going longer.

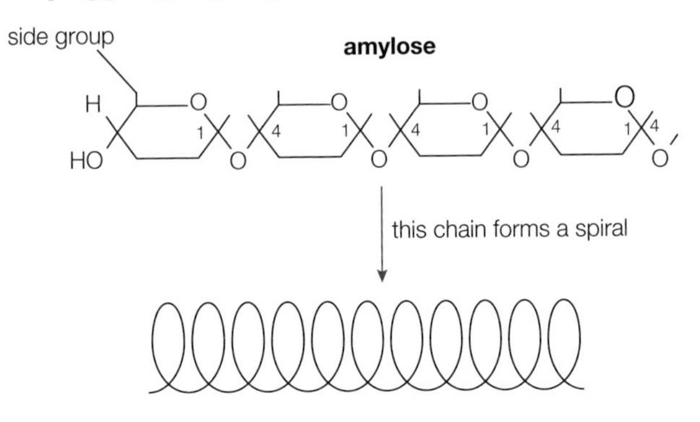

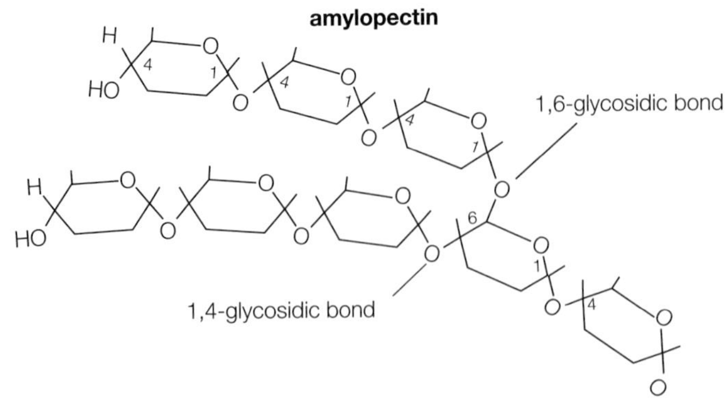

fig D Amylose and amylopectin – a small difference in the position of the glycosidic bonds in the molecule makes a big difference to the properties of the compounds.

GLYCOGEN

Glycogen is sometimes referred to as 'animal starch' because it is the only carbohydrate energy store found in animals. It is also an important storage carbohydrate in fungi. Chemically, glycogen is very similar to the amylopectin molecules in starch, and it also has many α-glucose units. Like starch, it is very compact, but the glycogen molecule has more 1,6-glycosidic bonds than the starch molecule, giving it many side branches. This means that glycogen can be broken down very rapidly. This makes it an ideal source of glucose for animals which may require rapid release of energy at certain times of high activity levels (see **fig E**).

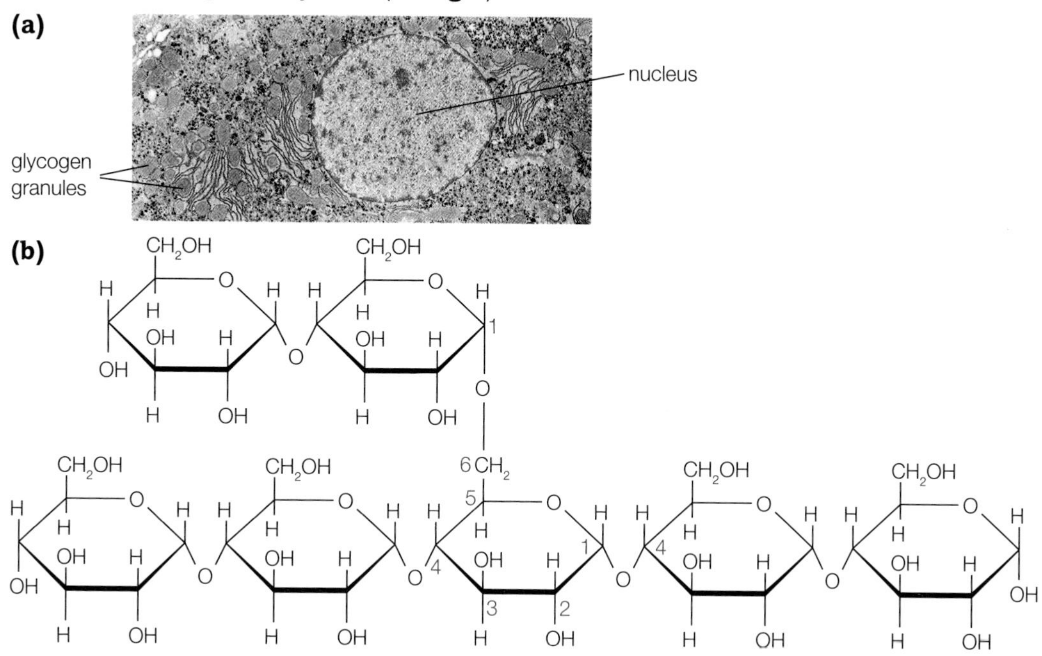

fig E In **(a)** you can see liver cells full of small glycogen granules, stained pink in this micrograph. If your blood glucose levels are low, this glycogen store in your liver can be broken down to provide the glucose you need for cellular respiration. In **(b)** you can see the structure of glycogen with 1,4 and 1,6-glycosidic bonds.

The chemical structure of glycogen shown in **fig E (b)** looks very similar to that of amylopectin. However, when you look at bigger sections of the molecules in **fig F** you can see that glycogen has many more branches than amylopectin.

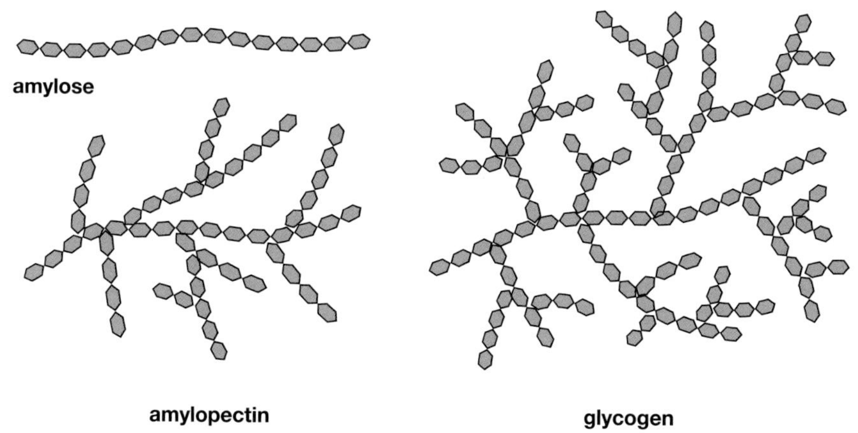

fig F You can clearly see the many side branches which allow glycogen to be broken down so quickly when you compare amylose, amylopectin and glycogen.

CHECKPOINT

SKILLS CREATIVITY

1. Explain why sugars such as glucose and sucrose are useful for immediate energy, but are not suitable as long-term energy stores.
2. Explain how the structure of carbohydrates is related to their function as storage molecules providing the fuel for cellular respiration in animals and plants.

DID YOU KNOW?

When starch is cooked all the coiled molecules unwind and get tangled together. This is why flour can thicken a gravy or sauce, and why you can make glue from flour. It also explains why the molecules must be coiled. If they were not coiled, the cytoplasm would be a solid tangled mass of starch molecules.

EXAM HINT

Be clear about the differences between 1,4-glycosidic bonds and 1,6-glycosidic links. It is easy to get them wrong and lose marks as a result.

SUBJECT VOCABULARY

oligosaccharides molecules with between 3 and 10 monosaccharide units
hydrolysis a reaction in which bonds are broken by the addition of a molecule of water
ATP adenosine triphosphate, the molecule that acts as a universal energy supply molecule in all cells
end products the final products of a chemical reaction
amylose a complex carbohydrate containing only α-glucose monomers joined together by 1,4-glycosidic bonds so the molecules form long unbranched chains
amylopectin a complex carbohydrate made up of α-glucose monomers joined by 1,4-glycosidic bonds with some 1,6-glycosidic bonds so the molecules branch repeatedly
glycogen a complex carbohydrate with many α-glucose units joined by 1,4-glycosidic bonds with many 1,6-glycosidic bonds, giving it many side branches

SPECIFICATION REFERENCE 1.5(i) 1.5(ii)

LEARNING OBJECTIVES

- Know how a triglyceride is synthesised by the formation of ester bonds during condensation reactions between glycerol and three fatty acids.
- Describe the differences between unsaturated and saturated fatty acids.

▲ **fig A** Olive oil comes from pressed olives and is widely used for food and cooking around the world.

The **lipids** are another group of organic substances that play a vital role in organisms. They are an integral part of all cell membranes and are also used as an energy store. Lipids contain many carbon–hydrogen bonds and almost no oxygen. When lipids are oxidised in the respiration process, the bonds are broken and carbon dioxide and water are the final products. This reaction can be used to drive the production of much ATP (see **Sections 1A.3**, **2A.2** and **2A.4** and **Book 2 Chapter 5A**). Lipids, especially triglycerides, store about three times as much energy as the same mass of carbohydrates. Many plants and animals convert spare food into oils or fats as an energy store to use when needed. For example, the seeds of plants contain lipids to provide energy for the seedling when it starts to grow, which is why seeds are such an important food source for many animals.

FATS AND OILS

Fats and oils are important groups of lipids. Chemically they are very similar, but fats such as butter are solids at room temperature whereas oils such as olive oil are liquids (see **fig A**). Fats come mainly from animal sources while oils are mainly from plant sources.

Like carbohydrates, the chemical elements that all lipid molecules contain are carbon, hydrogen and oxygen. However, lipids contain a much lower proportion of oxygen than carbohydrates. Fats and oils contain two types of organic chemical substance, **fatty acids** and **glycerol** (propane-1,2,3-triol). These are combined using **ester bonds**. Glycerol has the chemical formula $C_3H_8O_3$ (see **fig B**).

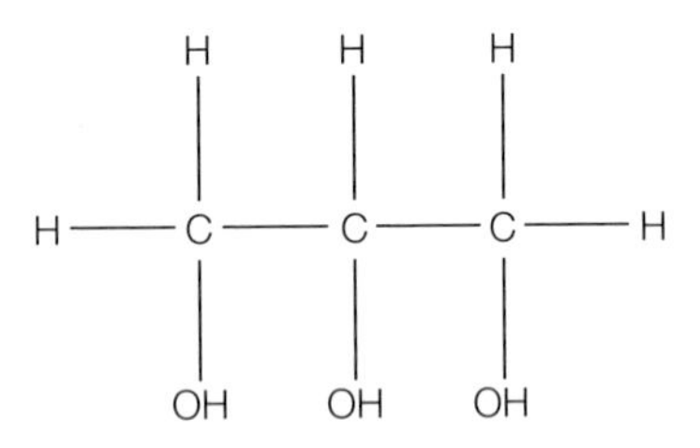

▲ **fig B** Displayed formula of glycerol (propane-1,2,3-triol)

All fatty acids have a long hydrocarbon chain – a folded backbone of carbon atoms with hydrogen atoms attached, and a carboxyl group (–COOH) at one end. Living tissues contain more than 70 different types of fatty acid. Fatty acids vary in two ways:

- the length of the carbon chain can differ (but is often 15–17 carbon atoms long)
- the fatty acid may be a **saturated fatty acid** or an **unsaturated fatty acid**.

In a saturated fatty acid, each carbon atom is joined to the one next to it by a single covalent bond. A common example is stearic acid (see **fig C**). In an unsaturated fatty acid, the carbon chains have one or more double covalent bonds between carbon atoms in them. A **monounsaturated fatty acid** has one carbon–carbon double bond and a **polyunsaturated fatty acid** has more than one carbon–carbon double bond (see **fig D**). Linoleic acid is an example of a polyunsaturated fatty acid. It is an essential fatty acid in our diet because we cannot make it from other substances.

CH_3 — $(CH_2)_{16}$ — COOH

▲ **fig C** Displayed formula of stearic acid, a saturated fatty acid found in both plant and animal fats

carbon–carbon double bond

▲ **fig D** Displayed formula of linoleic acid, a polyunsaturated fatty acid

FORMING ESTER BONDS

A triglyceride is made when glycerol combines with three fatty acids. A bond is formed in a condensation reaction between the carboxyl group (–COOH) of a fatty acid and one of the hydroxyl groups (–OH) of the glycerol. A molecule of water is removed and the bond created is called an ester bond. This type of condensation reaction is called **esterification** (see **fig E**). The nature of the lipid formed depends on which fatty acids are joined together. For example, lipids containing saturated fatty acids are more likely to be solid at room temperature than those containing unsaturated fatty acids. Longer chain fatty acids are also more likely to produce solid fats.

For simplicity, fatty acids are represented by this general formula where 'R' represents the hydrocarbon chain. The fatty acids below are drawn in reversed form.

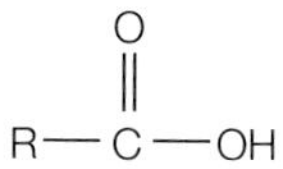

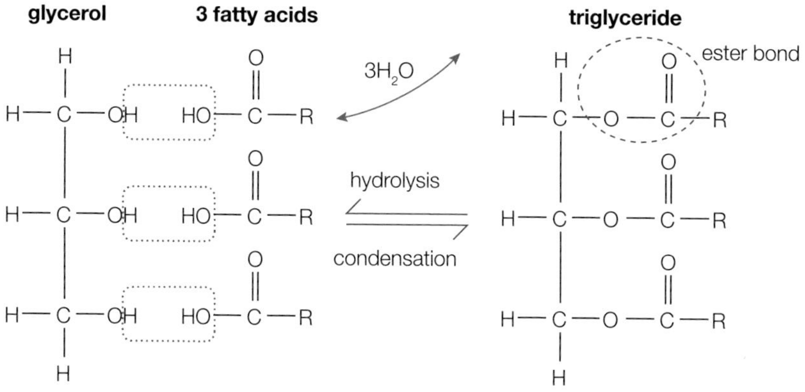

R—C(=O)—OH

glycerol
3 fatty acids
$3H_2O$
hydrolysis
condensation
triglyceride
ester bond

Note: there are only 6 atoms of oxygen in a triglyceride molecule.

▲ **fig E** The formation of ester bonds

EXAM HINT

When you discuss unsaturated fatty acids, make it clear that the double bonds are between carbon atoms. Refer to them as carbon–carbon double bonds, not just double bonds.

LEARNING TIP

Remember that animal fats are usually saturated fatty acids and are more likely to be solid at room temperature. This is why a spread made from plant oils is quite spreadable when you take it out of the fridge, but butter is not.

CHECKPOINT

1. Explain how triglycerides are formed.
2. Describe the main difference between a saturated and an unsaturated fatty acid, and the effect of this difference on the properties of the lipids formed from unsaturated fatty acids compared to lipids formed from saturated fatty acids.

SKILLS ADAPTIVE LEARNING

SUBJECT VOCABULARY

lipids a large family of organic molecules that are important in cell membranes and as an energy store in many organisms; they include triglycerides, phospholipids and steroids
fatty acids organic acids with a long hydrocarbon chain
glycerol propane-1,2,3-triol, an important component of triglycerides
ester bonds bonds formed in a condensation reaction between the carboxyl group (-COOH) of a fatty acid and one of the hydroxyl groups (-OH) of glycerol
saturated fatty acid a fatty acid in which each carbon atom is joined to the one next to it in the hydrocarbon chain by a single covalent bond
unsaturated fatty acid a fatty acid in which the carbon atoms in the hydrocarbon chain have one or more double covalent bonds in them
monounsaturated fatty acid a fatty acid with only one double covalent bond between carbon atoms in the hydrocarbon chain
polyunsaturated fatty acid a fatty acid with two or more double covalent bonds between carbon atoms in the hydrocarbon chain
esterification the process by which ester bonds are made

LEARNING OBJECTIVES

- Know the basic structure of an amino acid.
- Understand the formation of polypeptides and proteins, as amino acid monomers linked together by condensation reactions to form peptide bonds.
- Understand the significance of a protein's primary structure in determining its secondary structure, three-dimensional structure and properties, and the types of bond involved in its three-dimensional structure.
- Know the molecular structure of a globular protein and a fibrous protein and understand how their properties relate to their functions (including haemoglobin and collagen).

About 18% of your body is made up of protein. Proteins make hair, skin and nails, the enzymes needed for metabolism and digestion, and many of the hormones that control the different body systems. They enable muscle fibres to contract, make antibodies that protect you from disease, help clot your blood and transport oxygen in the form of **haemoglobin**. Understanding the structure of proteins helps you understand the detailed biology of cells and organisms. Like carbohydrates and lipids, proteins contain carbon, hydrogen and oxygen. In addition, they all contain nitrogen and many proteins also contain sulfur.

Proteins are a group of macromolecules made up of many small monomer units called **amino acids** joined together by condensation reactions. Amino acids combine in long chains to produce proteins. There are about 20 different naturally occurring amino acids that can combine in different ways to produce a wide range of different proteins.

AMINO ACIDS

All amino acids have the same basic structure, which is represented as a general formula. There is always an amino group ($-NH_2$) and a carboxyl group ($-COOH$) attached to a carbon atom (see **fig A**). The group known as the R group varies between amino acids. Some amino acids contain sulfur and selenium in their R group. The R groups are not involved in the reactions which join the amino acids together, but the structure of the R group does affect the way the amino acid interacts with others within the protein molecule. This will mainly depend on whether the R group is polar or not, and these interactions affect the tertiary structure of the protein formed (see **page 18**).

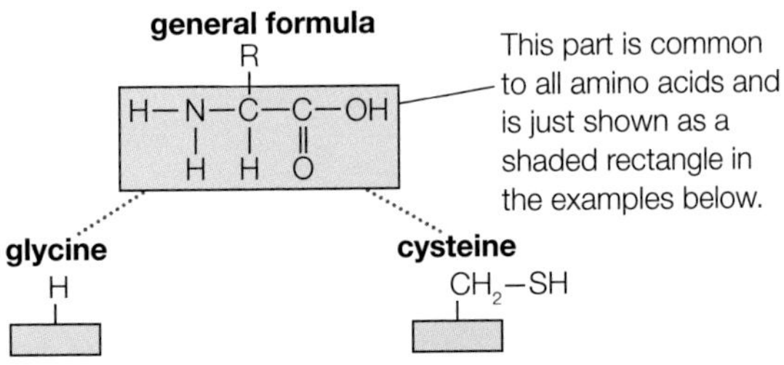

fig A Some different amino acids. In the simplest amino acid, glycine, R is a single hydrogen atom. In a larger amino acid such as cysteine, R is much more complex.

FORMING PROTEINS FROM AMINO ACIDS

Amino acids join by a reaction between the amino group of one amino acid, and the carboxyl group of another. They join in a condensation reaction and a molecule of water is released. A **peptide bond** is formed when two amino acids join, and a **dipeptide** is the result (see **fig B**). The R group is not involved in this reaction. More and more amino acids join to form **polypeptide** chains, which contain from about 100 to many thousands of amino acids. A polypeptide forms a protein when the structure of the chain changes by folding or coiling or associates with other polypeptide chains.

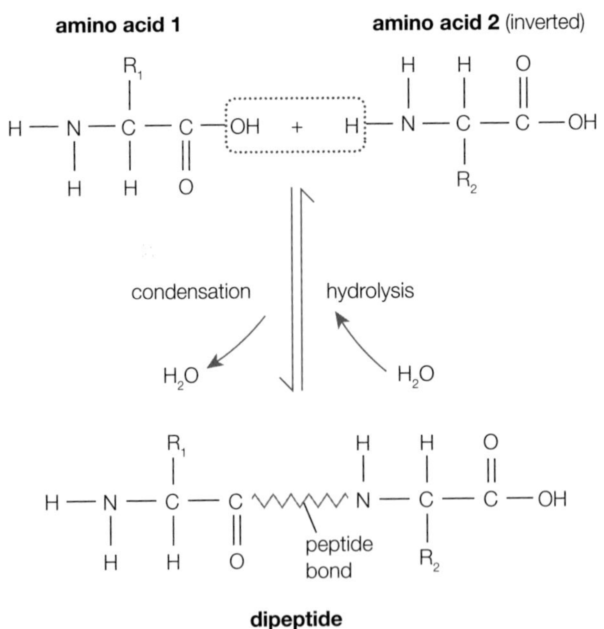

fig B Amino acids are the building blocks of proteins, joined together by peptide bonds.

BONDS IN PROTEINS

The peptide bond between amino acids is a strong bond. Other bonds are also made between the amino acids in a chain, to create the 3D structures of the protein. They depend on the atoms in the R group and include hydrogen bonds, **disulfide bonds** and ionic bonds.

HYDROGEN BONDS

You were introduced to hydrogen bonds in **Section 1A.1**. These same bonds are essential in protein structures. In amino acids, tiny negative charges are present on the oxygen of the carboxyl groups and tiny positive charges are present on the hydrogen atoms of the amino groups. When these charged groups are close to each other, the opposite charges attract, forming a hydrogen bond. Hydrogen bonds are weak but, potentially, they can be made between any two amino acids in the correct position, so there are many of them holding the protein together very firmly. They are very important in the folding and coiling of polypeptide chains (see **fig C**). Hydrogen bonds break easily and reform if pH or temperature conditions change.

DISULFIDE BONDS

Disulfide bonds form when two cysteine molecules are close together in the structure of a polypeptide (see **fig C**). An oxidation reaction occurs between the two sulfur-containing groups, resulting in a strong covalent bond known as a disulfide bond. These disulfide bonds are much stronger than hydrogen bonds but they happen much less often. They are important for holding the folded polypeptide chains in place.

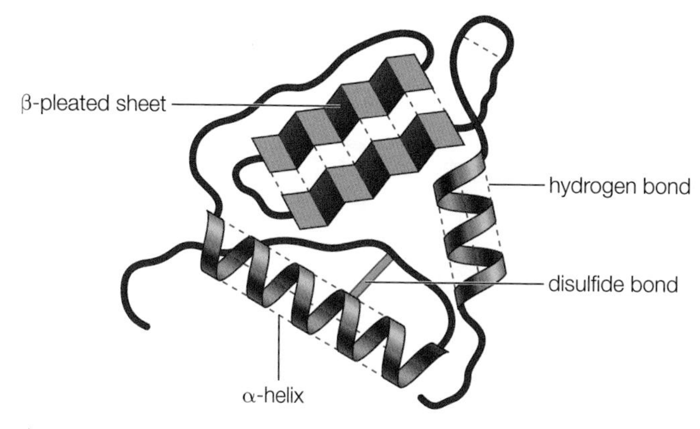

fig C Hydrogen bonds and disulfide bonds maintain the shape of protein molecules and this determines their function.

IONIC BONDS

Ionic bonds can form between some of the strongly positive and negative amino acid side chains which are sometimes found deep inside the protein molecules. They are strong bonds, but they are not as common as the other structural bonds.

Your hair is made of the protein keratin. Some methods of styling hair change the bonds within the protein molecules. Blow drying or straightening hair breaks the hydrogen bonds and temporarily reforms them with the hair curling in a different way until the hydrogen bonds reform in their original places.

Perming is a chemical treatment which is used in some hair salons to completely change the way hair looks for weeks or months. The chemicals break the disulfide bonds between the polypeptide chains and reform them in a different place. This effect is permanent – hair will stay styled in that particular way until it is cut off.

PROTEIN STRUCTURE

Proteins can be described by their primary, secondary, tertiary and quaternary structure (see **fig D**).

- The primary structure of a protein is the sequence of amino acids that make up the polypeptide chain, held together by peptide bonds.
- The secondary structure of a protein is the arrangement of the polypeptide chain into a regular, repeating three-dimensional (3D) structure, held together by hydrogen bonds. One example is the right-handed helix (α-helix), a spiral coil with the peptide bonds forming the backbone and the R groups protruding in all directions. Another is the ß-pleated sheet, in which the polypeptide chain folds into regular pleats held together by hydrogen bonds between the amino and carboxyl ends of the amino acids. Most **fibrous proteins** have this type of structure. Sometimes there is no regular secondary structure and the polypeptide forms a random coil.

LEARNING TIP

Remember that fibrous proteins have a simpler structure and so tend to be more stable to changes in temperature and pH.

LEARNING TIP

Remember that the primary structure of proteins is the result of peptide bonds between amino acids.

The secondary structure is the result of hydrogen bonding between nearby amino acids but R groups do not affect it.

Ionic bonds, hydrogen bonds and disulfide bridges are a result of interactions between the R groups and create the tertiary structure.

- The tertiary structure is another level of 3D organisation in addition to the secondary structure in many proteins. The amino acid chain, including any α-helices and β-pleated sheets, is folded further into complicated shapes. Hydrogen bonds, disulfide bonds and ionic bonds between the R groups of nearby amino acids hold these 3D shapes in place (see **page 17**). Globular proteins are an example of tertiary structures.
- The quaternary structure of a protein is only found in proteins consisting of two or more polypeptide chains. The quaternary structure describes the way these separate polypeptide chains fit together in three dimensions. Examples include some very important enzymes and the blood pigment haemoglobin.

Changes in conditions such as temperature or pH affect the bonds that keep the 3D shapes of proteins in place. Even small changes can cause the bonds to break, resulting in the loss of the 3D shape of the protein. This is called **denaturation**. Because the 3D structure of these proteins is important to the way they work, changing conditions inside the body can cause proteins such as enzymes to stop working properly.

Primary structure the linear sequence of amino acids in a peptide.

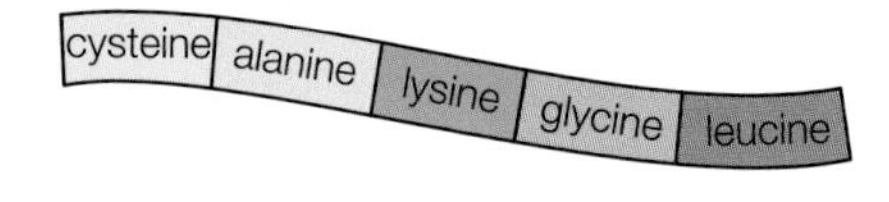

Secondary structure the repeating pattern in the structure of the peptide chains, such as an α-helix or β-pleated sheets.

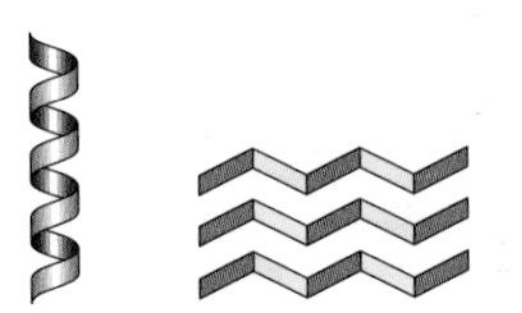

Tertiary structure the three-dimensional folding of the secondary structure.

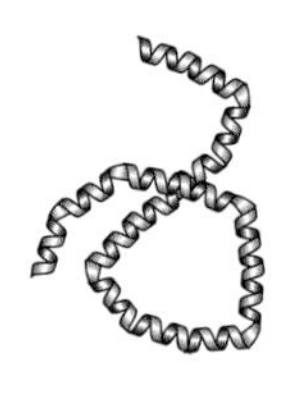

Quaternary structure the three-dimensional arrangement of more than one tertiary polypeptide.

▲ **fig D** The 3D structure of proteins

FIBROUS AND GLOBULAR PROTEINS

FIBROUS PROTEINS

The complex structures of large protein molecules relate closely to their functions in the body. Fibrous proteins have little or no tertiary structure. They are long, parallel polypeptide chains with occasional cross-linkages that form them into fibres. They are insoluble in water and are very tough, which makes them ideally suited to their structural functions within organisms. Fibrous proteins appear in the structure of connective tissue in tendons and the matrix of bones, as the silk of spiders' webs and silkworm cocoons, and as the keratin that makes up hair, nails, horns and feathers.

Collagen is a fibrous protein that gives strength to tendons, ligaments, bones and skin. It is the most common structural protein found in animals – up to 35% of the protein in your body is collagen. Collagen is extremely strong – the fibres have a tensile strength similar to that of steel. This is due to the unusual structure of the collagen molecule. Its quaternary structure has three polypeptide chains, each up to 1000 amino acids long. The primary structure of these chains is repeating sequences of glycine with two other amino acids – often proline and hydroxyproline. The three polypeptide α-chains are arranged in a unique triple helix, held together by a very large number of hydrogen bonds. Collagen molecules can be up to several millimetres long and are often found together in fibrils that are held together to form collagen fibres. You can see how collagen fibres are built up in **fig E**.

Collagen fibres combine with the bone tissue, giving it tensile strength, in the same way as the steel rods in reinforced concrete. In the genetic disease osteogenesis imperfecta, the collagen triple helix does not develop properly. Consequently, the bone does not have as much tensile strength; it is brittle and breaks very easily.

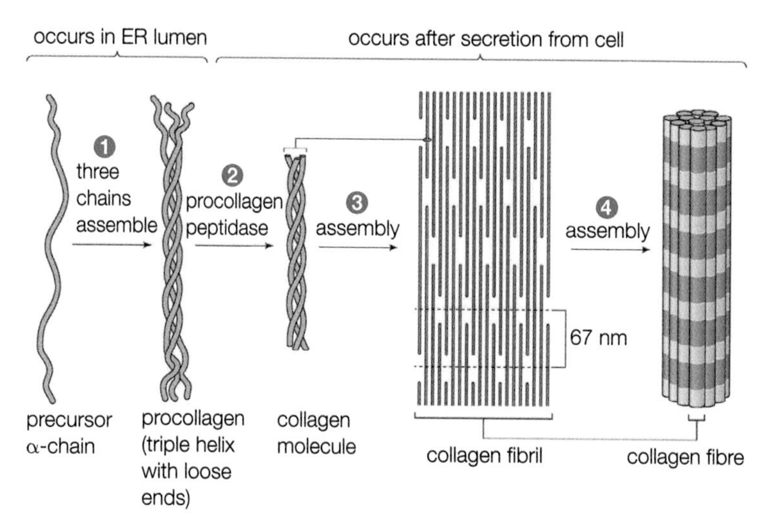

▲ **fig E** Collagen is a fibrous protein with an unusual triple helix structure and immense strength.

GLOBULAR PROTEINS

Globular proteins have complex tertiary and sometimes quaternary structures. They fold into spherical (globular) shapes. The character of the R groups on the amino acids plays an important role in the formation of globular proteins. Some R groups are **hydrophobic**. They repel water and will not mix or dissolve in it. They are usually found on the inside of globular proteins. Some R groups are **hydrophilic** – they have an affinity for water. These groups tend to be found on the outside of globular proteins. The large size of these globular protein molecules affects their behaviour in water.

The carboxyl and amino ends give them ionic properties, so you might expect them to dissolve in water and form a solution. Instead, the molecules are so big that they form a **colloid**. In a colloid, microscopic particles of one substance (in this case protein) are suspended throughout another substance (in this case water). They do not settle, and they cannot easily be separated. Globular proteins are important as they hold molecules in position in the cytoplasm. Globular proteins are also important in your immune system – for example, antibodies are globular proteins. Enzymes and some hormones are globular proteins and help maintain the structure of the cytoplasm (you will learn more about globular proteins as enzymes in **Sections 2B.1** and **2B.2**).

Haemoglobin is one of the best known globular proteins. It is a very large molecule with 574 amino acids arranged in four polypeptide chains which are connected by disulfide bonds. Each polypeptide chain surrounds an iron-containing haem group (see **fig F**). The iron enables the haemoglobin to bind and release oxygen molecules, and the arrangement of the polypeptide chains determines how easily the oxygen binds or is released (see **Section 1B.2** to find out how haemoglobin takes up and gives up oxygen in the tissues of your body).

▲ **fig F** The complex quaternary structure of haemoglobin produces a globular protein containing four haem groups which can carry oxygen to the tissues of the body.

LEARNING TIP

Remember that fibrous proteins are more stable than globular proteins and tend to create structures rather than being metabolically active.

EXAM HINT

Globular proteins have a specific 3D shape which means they can be metabolically active as enzymes or hormones. If the shape is altered slightly by changing conditions, they lose their ability to function.

CONJUGATED PROTEINS

The shape of a protein molecule is usually very important in its function. Some protein molecules are joined with (conjugated to) another molecule called a **prosthetic group** (see **fig F**). This structural feature usually affects the performance and functions of the molecules. These molecules are called **conjugated proteins**. Haemoglobin is a large protein with iron as the prosthetic group. It is a conjugated protein as well as a globular protein. **Lipoproteins** are formed when proteins are conjugated with lipids – you will find out more about these important biological molecules when you look at factors affecting the health of your heart in **Chapter 1B**.

Glycoproteins are proteins with a carbohydrate prosthetic group. The carbohydrate part of the molecule helps them to hold a lot of water and also makes it harder for protein-digesting enzymes (**proteases**) to break them down. Lots of lubricants used by the human body – such as mucus and the synovial fluid in the joints – are glycoproteins. Their water-holding properties make them slippery and viscous, which reduces friction. This also helps to explain why the mucus produced in the stomach protects the protein walls from digestion.

Lipoproteins are very important in the transport of cholesterol in the blood. The lipid part of the molecule enables it to combine with the lipid cholesterol. There are two main forms of lipoproteins in your blood – low-density lipoproteins (LDLs) (around 22 nm in diameter) and high-density lipoproteins (HDLs) (approximately 8–11 nm in diameter). The HDLs contain more protein than LDLs, which is partly why they are denser – proteins are more compact molecules than lipids. You will discover the impact of different lipoproteins on the risk of developing cardiovascular diseases in **Section 1C.4**.

PRACTICAL SKILLS CP2

Testing for protein (RAP)

To test for the presence of protein, add Biuret reagent (ready-mixed 5% (w/v) sodium hydroxide solution and 1% (w/v) copper sulfate solution). A purple colour indicates the presence of protein (see **fig G**).

▲ **fig G** Biuret test for protein

EXAM HINT

Remember that amino acids are joined together by peptide bonds to make dipeptides and then polypeptides. However, the 3D structures of proteins are the result of hydrogen bonds, disulfide bonds, hydrophobic links and ionic bonds between amino acids within the polypeptide chains.

CHECKPOINT

SKILLS REASONING

1. Explain how the order of amino acids in a protein affects the structure of the whole protein.
2. Hydrogen bonds are weaker than disulfide bonds and ionic bonds, but they are more important in maintaining protein structure. Why is this?
3. The body uses many resources to maintain a relatively constant internal environment. With reference to proteins, explain why constant internal conditions are so important.

SUBJECT VOCABULARY

haemoglobin a red pigment that carries oxygen and gives the erythrocytes their colour
amino acids the building blocks of proteins consisting of an amino group ($-NH_2$) and a carboxyl group ($-COOH$) attached to a carbon atom and an R group that varies between amino acids
peptide bond the bond formed by condensation reactions between amino acids
dipeptide two amino acids joined by a peptide bond
polypeptide a long chain of amino acids joined by peptide bonds
disulfide bond a strong covalent bond produced by an oxidation reaction between sulfur groups in cysteine or methionine molecules, which are close together in the structure of a polypeptide
fibrous proteins proteins that have long, parallel polypeptide chains with occasional cross-linkages that produce fibres; they have little tertiary structure
denaturation the loss of the 3D shape of a protein (e.g. caused by changes in temperature or pH)
collagen a strong fibrous protein with a triple helix structure
globular proteins large proteins with complex tertiary and sometimes quaternary structures, folded into spherical (globular) shapes
hydrophobic a substance that tends to repel water and that will not mix with or dissolve in water
hydrophilic a substance with an affinity for water that will readily dissolve in or mix with water
colloid a suspension of molecules that are not fully dissolved
prosthetic group the molecule incorporated in a conjugated protein
conjugated proteins protein molecules joined with or conjugated to another molecule called a prosthetic group
lipoproteins conjugated proteins with a lipid prosthetic group
glycoproteins conjugated proteins with a carbohydrate prosthetic group
proteases protein-digesting enzymes

1A THINKING BIGGER

TREHALOSE: A SUGAR FOR DRY EYES?

SKILLS CRITICAL THINKING, ANALYSIS, CONTINUOUS LEARNING, INTELLECTUAL INTEREST AND CURIOSITY, COMMUNICATION, CREATIVITY

Dry eyes is a condition that is caused by dry air and over-use of air conditioning. Both of these environmental factors are common in Middle Eastern countries. Biological molecules have an amazing number of different roles in living organisms, including some you would not expect. In this activity, you will discover how current research shows that the disaccharide trehalose can protect proteins from damage in stressful conditions. This property is being used to make dry eyes more comfortable – and possibly protect the brain from the damage that can result from ageing.

MEDICAL JOURNAL ARTICLE

TREHALOSE: AN INTRIGUING DISACCHARIDE WITH POTENTIAL FOR MEDICAL APPLICATION IN OPHTHALMOLOGY

Abstract

Trehalose is a naturally occurring disaccharide comprising two molecules of glucose. The sugar is widespread in many species of plants and animals, where its function appears to be to protect cells against desiccation, but it is not found in mammals. Trehalose has the ability to protect cellular membranes and labile proteins against damage and denaturation as a result of desiccation and oxidative stress. Trehalose appears to be the most effective sugar for protection against desiccation. Although the exact mechanism by which trehalose protects labile macromolecules and lipid membranes is unknown, credible hypotheses do exist. As well as being used in large quantities in the food industry, trehalose is used in the biopharmaceutical preservation of labile protein drugs and in the cryopreservation of human cells. Trehalose is under investigation for a number of medical applications, including the treatment of Huntington's chorea [disease] and Alzheimer's disease. Recent studies have shown that trehalose can also prevent damage to mammalian eyes caused by desiccation and oxidative insult. These unique properties of trehalose have thus prompted its investigation as a component in treatment for dry eye syndrome. This interesting and unique disaccharide appears to have properties which may be exploited in ophthalmology and other disease states.

Trehalose, a naturally occurring alpha-linked disaccharide formed of two molecules of glucose (**fig A**) … is synthesized by many living organisms, including insects, plants, fungi, and micro-organisms as a response to prolonged periods of desiccation. This very useful property, known as anhydrobiosis, confers on an organism the ability to survive almost complete dehydration for prolonged periods and subsequently reanimate.

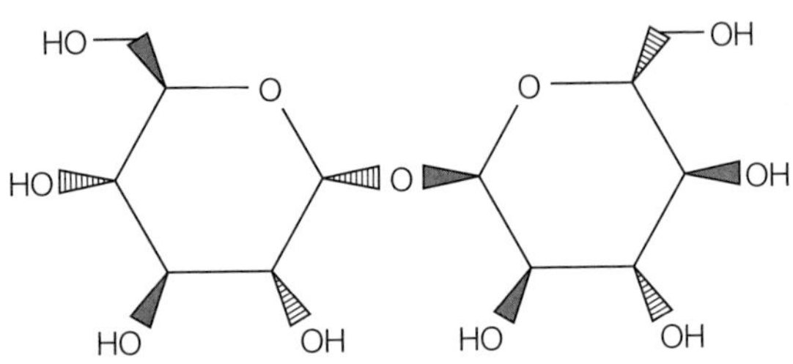

fig A Structure of trehalose; Registry number: 99-20-7; Molar mass: 342.296 g/mol (anhydrous); 378.33 g/mol (dihydrate); molecular structure: α-D-glucopyranosyl α-D-glucopyranoside (α,α-trehalose)

References

1 Iturriaga G., Suárez R., Nova-Franco B. Trehalose metabolism: From osmoprotection to signaling. Int J Mol Sci 2009; 10:3793–3810 […]

8 Elbein A.D., et al. New insights on trehalose: A multifunctional molecule. Glycobiology 2003; 13:17R–27R

11 Jain N.K., Roy I. Effect of trehalose on protein structure. Protein Sci 2009; 18:24–36 […]

20 Matsuo T. Trehalose protects corneal epithelial cells from death by drying. Br J Ophthalmol 2001; 85:610–612 […]

30 Matsuo T., Tsuchida Y., Morimoto N. Trehalose eye drops in the treatment of dry eye syndrome. Ophthalmology 2002; 109:2024–2029

31 Matsuo T. Trehalose versus hyaluronan or cellulose in eyedrops for the treatment of dry eye. Jpn J Ophthalmol 2004; 48:321–327

From: Luyckx J., Baudouin C. Trehalose: an intriguing disaccharide with potential for medical application in ophthalmology. *Clinical ophthalmology* (Auckland, NZ) 5 (2011): 577

SCIENCE COMMUNICATION

This extract comes from a paper published in *Clinical Ophthalmology*, an online journal. Think about the type of writing being used and the audience it is intended for as you try and answer the following questions.

1 (a) What aspects of this writing tell you it is more like a scientific paper than a general interest article in a magazine?

(b) Many words in this article may be unfamiliar. But words are often made up of familiar components. Break up the word anhydrobiosis and it becomes an-hydro-biosis (an = non, hydro = water and biosis = life). So anhydrobiosis means life without water. Choose two other unfamiliar words used in the article. Find out what they mean and suggest why they have been used by the authors.

(c) How do you think these ideas about trehalose and the way it may be used to help human health might be presented in a newspaper or on a news or science website? Have a go at writing an article for a public interest website yourself.

(d) If trehalose can really help protect people's sight and prevent brain diseases such as Huntington's and Alzheimer's this would make a big difference to people's lives. Notice how careful the author is. Why are scientific papers so cautious in the way they report things?

SKILLS COMMUNICATION, CREATIVITY

INTERPRETATION NOTE

Think about the level of scientific detail that is suitable for your expected audience. How will you ensure your article is eye-catching and interesting?

BIOLOGY IN DETAIL

Now you are going to think about the science in the article. You will be surprised how much you know already, but if you choose to do so, you can return to these questions later in your course.

2 What do you know about the chemical nature of trehalose from the article? Can you work out what type of bond joins the subunits together?

3 Desiccation (drying out) is a major problem for living organisms. Suggest reasons why drying out is so hard to survive.

4 Scientists think that trehalose protects both lipid membranes and certain proteins from damage, both from drying out and oxidation. Explain why it is so important biologically to protect cell membranes and protein structures.

THINKING BIGGER TIP

Think about the chemistry of biological molecules you have learned already and use it to help you understand how trehalose works.

ACTIVITY

Which aspect of trehalose would you like to know more about? The way it prevents desiccation in many groups of organisms? The way it can protect human eyes from damage? The evidence that it could help reduce brain diseases in people?

Choose the area that interests you most and use as many resources as you can to produce a 3-minute presentation about that aspect of trehalose biology. Find interesting images and list all the references to help your colleagues decide if they can rely on the information you present.

fig B The desert plant *Selaginella lepidophylla* is a 'resurrection' plant – it can withstand almost complete dehydration and recover within about 24 hours, thanks to high levels of trehalose in the plant cells.

THINKING BIGGER TIP

You can refer to the full version of this paper, to the references listed at the end, to online encyclopaedias, to other scientific papers and to books. In each case, judge the reliability of your source before you use it.

1A EXAM PRACTICE

1 Water is one of the most essential molecules for life.

(a) Which diagram most accurately represents a water molecule?

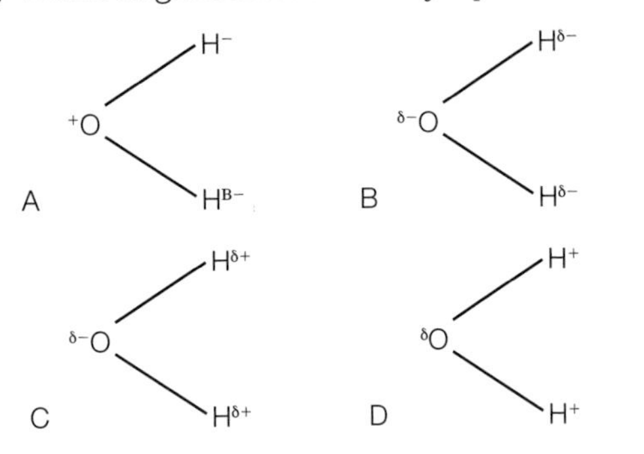

[1]

(b) (i) Name the bond that occurs between two water molecules. [1]

(ii) State the property of water that enables these bonds to form. [1]

(iii) These intermolecular bonds give water a property called cohesion. Describe two ways in which cohesion is important to living things. [2]

(c) Water has a high specific heat capacity. Assess how important this property is to living things. [2]

(d) Explain how a molecule of sodium chloride can dissolve in water. [3]

(Total for Question 1 = 10 marks)

2 Water is a good solvent.

(a) Which of the following particles will not dissolve in water?

A Na^+ ion

B oxygen molecule

C maltose molecule

D starch molecule [1]

(b) Mosquito larvae live in water. They appear to be attached to the surface of the water. Which property of water enables the larvae to do this?

A Water is most dense at 4 °C

B Hydrogen bonds hold the water molecules together

C Water is a polar molecule

D Water cannot be compressed [1]

(c) Words written in water-based ink will smudge when water is spilled on the paper. Words written in ballpoint pen are not affected.

Explain this using your knowledge of the properties of water. [2]

(d) It is better to use a pencil for writing if the paper may get wet. Use your knowledge of solvents to justify this statement. [2]

(e) Draw two water molecules showing the bonding between the water molecules. [3]

(Total for Question 2 = 9 marks)

3 Carbohydrates include monosaccharides, disaccharides and polysaccharides.

(a) A disaccharide can be split by:

A hydrolysis of glycosidic bonds

B condensation of glycosidic bonds

C hydrolysis of ester bonds

D condensation of ester bonds. [1]

(b) Amylose is an example of a:

A monosaccharide

B disaccharide

C polysaccharide

D trisaccharide. [1]

(c) Complete this table to show the components and bonding within each carbohydrate.

	Lactose	**Amylose**	**Glycogen**
Component monosaccharides			
Bonds between monosaccharides			

[6]

(Total for Question 3 = 8 marks)

4 Disaccharides and polysaccharides consist of monosaccharides joined together.

(a) Name the bond holding the monosaccharides together.

A ionic

B ester

C glycosidic

D hydrogen [1]

(b) What is the function of starch molecules?

A provide a source of energy for plants

B store energy in all living organisms

C store energy in plants

D store energy in animals [1]

(c) A disaccharide can be hydrolysed to its two monosaccharides. Explain the term hydrolysis. [2]

(d) State the role of glycogen molecules and explain why they are well suited to the role. [5]

(Total for Question 4 = 9 marks)

5 (a) Which is the best description of an amino acid?

A contains the elements carbon, hydrogen and oxygen

B an amino group at one end and a carboxyl group at the other end

C a small molecule containing peptide bonds

D an amino group with an R group attached [1]

(b) Proteins are polymers of amino acids joined by peptide bonds formed between the:

A R groups

B R group and the amino group

C R group and the carboxyl group

D carboxyl group and the amino group. [1]

(c) Collagen and haemoglobin are both proteins that have a primary structure consisting of amino acids joined together by peptide bonds.

(i) Explain what is meant by the term *primary structure* of a protein. [1]

(ii) Name the type of reaction that occurs when a peptide bond is broken causing a dipeptide to split into two amino acids. [1]

(d) Collagen contains the amino acids glycine and serine. The diagram below shows a dipeptide formed from these two amino acids.

H
N—C—C—N—C—C
H
O
OH
CH$_2$
OH

H
H
N—C—C
H
H
O
OH
+

Glycine Serine

Complete the diagram to show the structure of serine when the peptide bond breaks. [1]

(e) In the table below give three structural differences between the molecules of collagen and haemoglobin.

Difference	Collagen	Haemoglobin
1		
2		
3		

[3]

(Total for Question 5 = 8 marks)

6 (a) Complete the table below describing the different levels of protein structure (primary, secondary, tertiary, or quaternary structure). [4]

Description	Level of protein structure (primary, secondary, tertiary, or quaternary)
Hydrophobic amino acids such as proline are not found on the surface of protein molecule	
The molecule contains three polypeptide chains	
The protein molecule contains short helical sections separated by pleated sheets	
40% of the amino acids in the molecule are glutamine	

(b) Amino acids contain a residual or R group. Describe how the R group can affect the structure of a protein. [4]

(c) Explain, using examples, why globular proteins are more metabolically active than fibrous proteins. [4]

(Total for Question 6 = 12 marks)

TOPIC 1 MOLECULES, TRANSPORT AND HEALTH

CHAPTER 1B MAMMALIAN TRANSPORT SYSTEMS

If a car breaks down, mechanics can replace worn-out parts, put in new oil and transmission fluid, or change perished or worn-out pipes. We do not expect doctors to be able to do the same for our bodies. But they can do a lot to replace or repair the various parts of the circulatory system. The heart can have new valves, new blood vessels to supply the muscle and can even be replaced in a transplant. The blood vessels can be opened up, unblocked or replaced with grafts from other healthy areas of the body. Blood can be replaced by transfusions, and the bone marrow that makes the blood cells can be replaced by transplants. Doctors have even developed techniques by which they can operate on the circulatory system of a fetus in the uterus, to give blood transfusions or repair some heart conditions long before birth.

In this chapter, you will be looking at mammalian transport systems. This involves studying the general principles of circulatory systems and why larger organisms need a complex circulatory system. You will learn about the details of the human blood, blood vessels and heart.

You will consider how blood fluid and blood cells help to transport gases and other substances in the blood and how haemoglobin - the pigment which carries oxygen - attracts oxygen and then releases it when and where it is needed. You will discover the way that the blood vessels are well adapted to their roles in different parts of the circulatory system and what can go wrong if they are not healthy. Finally, you will learn about the heart as a complex organ with a well-coordinated cycle of contraction.

MATHS SKILLS FOR THIS CHAPTER

- **Recognise and make use of appropriate units in calculations** (*e.g. work out the unit for the heart rate*)
- **Find arithmetic means** (*e.g. measuring mean heart rate*)
- **Construct and interpret frequency tables and diagrams, bar charts and histograms** (*e.g. explain volume and pressure changes in the heart chambers*)
- **Substitute numerical values into algebraic equations using appropriate units for physical quantities** (*e.g. calculating the volumes of cubes and spheres*)
- **Solve algebraic equations in a biological context** (*e.g. surface area to volume ratios*)
- **Calculate the circumferences, surface areas and volumes of regular shapes** (*e.g. work out the approximate surface area and volume of a single cell*)
- **Use ratios, fractions and percentages** (*e.g. calculate surface area to volume ratio of a single cell, compare the thickness of the wall of the heart chambers*)

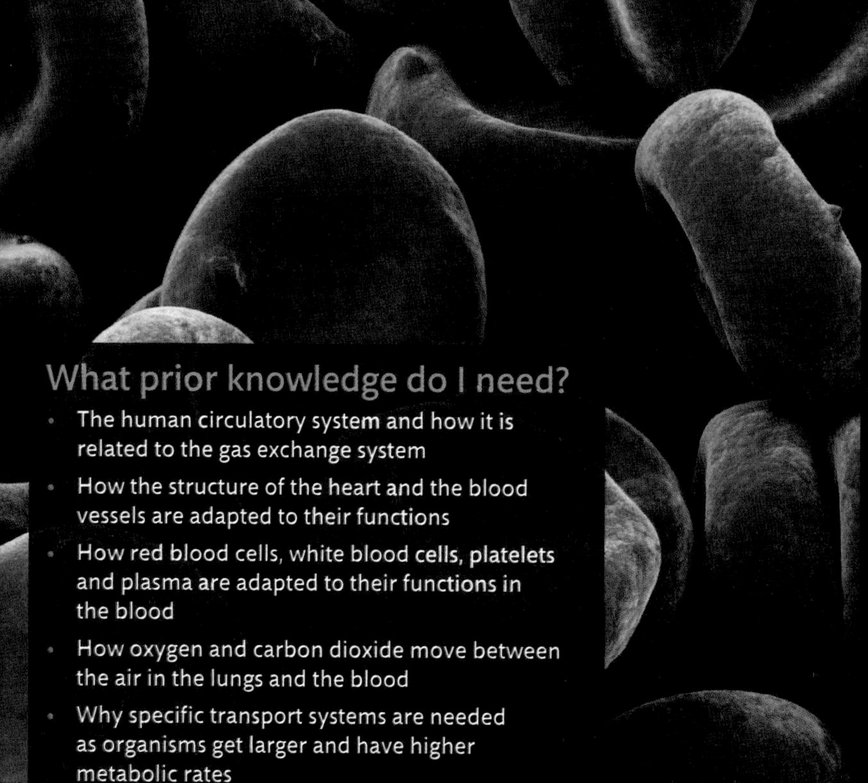

What prior knowledge do I need?

- The human circulatory system and how it is related to the gas exchange system
- How the structure of the heart and the blood vessels are adapted to their functions
- How red blood cells, white blood cells, platelets and plasma are adapted to their functions in the blood
- How oxygen and carbon dioxide move between the air in the lungs and the blood
- Why specific transport systems are needed as organisms get larger and have higher metabolic rates

What will I study in this topic?

- The need for a circulatory system in larger and more complex organisms
- The importance of the surface area to volume ratio of larger organisms
- The advantages of a double circulation in mammals over the single circulatory system in fish
- The structure of the blood including the different cells, plasma and platelets and their functions in transport
- The role of platelets and plasma proteins in blood clotting
- The transport of oxygen in the blood, the roles of haemoglobin, fetal haemoglobin and the effect of carbon dioxide concentrations on the transport of oxygen
- The structure of the heart, arteries, veins and capillaries related to their functions
- The sequence of events of the cardiac cycle
- How blood vessels can become damaged by atherosclerosis and its effect on health

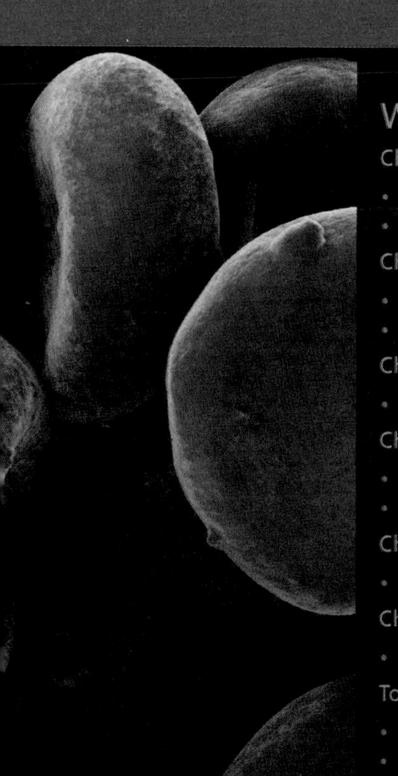

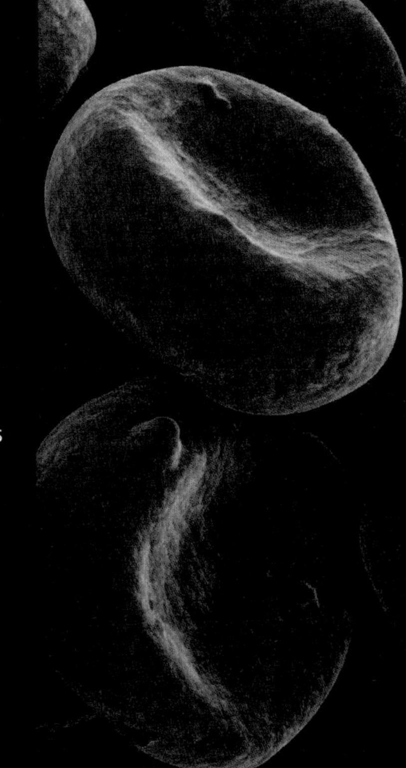

What will I study later?

Chapter 1C

- The risk factors for cardiovascular disease
- Treatments for cardiovascular disease

Chapter 2A

- Movement of gases into and out of the blood in the lungs
- Diffusion and osmosis

Chapter 2C

- Inheritance of blood groups

Chapter 3C

- Bone marrow as a source of blood cells and stem cells
- How stem cells from the bone marrow can have therapeutic uses

Chapter 4A

- Transport systems in plants

Chapter 6B (Book 2: IAL)

- Role of blood in immunity

Topic 7 (Book 2: IAL)

- The role of the blood in homeostasis
- The transport of heat in thermoregulation
- The movement of substances out of and into the blood in the nephrons of the kidney
- The role of the blood in transporting hormones
- The effect of hormones such as adrenaline on the circulatory system
- Coordination of the cardiac cycle
- The control of the heart rate by the nervous system
- Cardiac output

1B 1 THE PRINCIPLES OF CIRCULATION

SPECIFICATION REFERENCE 1.6

LEARNING OBJECTIVES

- Know why many animals have a heart and circulation which act as a mass transport system to overcome the limitations of diffusion.

THE NEED FOR TRANSPORT

- Within any organism, substances need to be moved from one place to another. One of the main ways substances move into and out of cells is by **diffusion**. Diffusion is the free movement of particles in a liquid or a gas down a **concentration gradient**. This movement is from an area where the particles are at a relatively high concentration to an area where they are at a relatively low concentration (see **Section 2A.2**).

In single-celled organisms and microscopic multicellular organisms, diffusion is sufficient to supply all their needs. However, when organisms reach a certain size, diffusion alone is not enough.

TRANSPORT IN SMALL ORGANISMS

For a single-celled organism like an amoeba and for very small multicellular organisms including many marine larvae, the nutrients and oxygen that they need can diffuse directly into the cells from the external environment and waste substances can diffuse directly out. This works well for the following reasons.

- The diffusion distances from the outside to the innermost areas of the cells are very small.
- The surface area in contact with the outside environment is very large when compared to the volume of the inside of the organism. Its **surface area to volume ratio (sa : vol)** is large, so there is a relatively big surface area over which substances can diffuse into or out of the organism (see **figs A** and **B**).
- The metabolic demands are low – the organisms do not regulate their own temperature and the cells do not use much oxygen and food or produce much carbon dioxide.

Single-celled organisms and very small multicellular organisms do not need specialised transport systems because diffusion is enough to supply their needs.

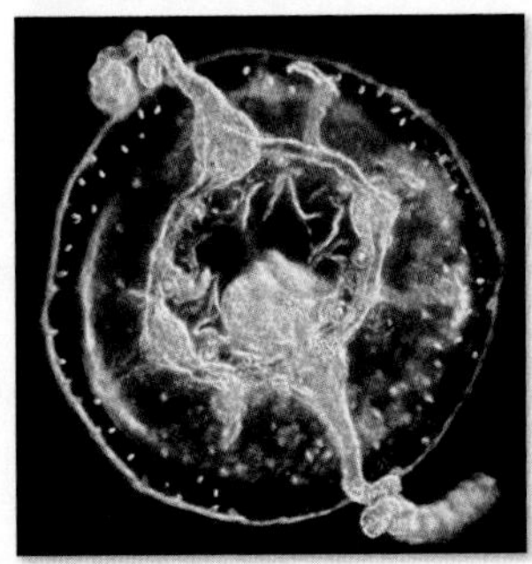

fig A The surface area : volume ratio of this tiny jellyfish larva is relatively large so simple diffusion can supply all its needs.

MODELLING SURFACE AREA : VOLUME RATIOS

The surface area to volume ratio of an organism is the key factor that determines whether diffusion alone will allow substances to move into and out of all the cells rapidly enough. However, it is not easy to calculate the surface area to volume ratio of organisms such as elephants, people and palm trees. It is difficult even for a single-celled *Amoeba* because of its irregular shape.

So scientists use models to help show what happens in the real situation (see **fig B**). A simple cube makes surface area to volume calculations easy. The bigger the organism gets, the smaller the surface area to volume ratio becomes. The distance from the outside of the organism to the inside gets longer, and there is proportionately less surface for substances to enter through. So it takes longer for substances to diffuse in, and they may not reach the individual cells quickly enough to supply all their needs.

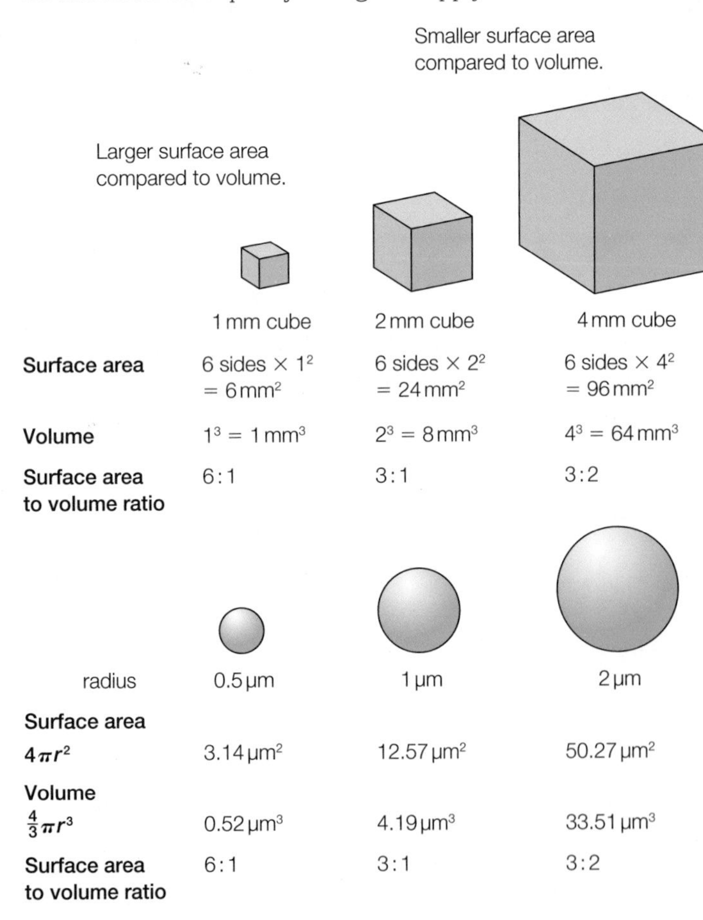

fig B In this diagram, the cubes and spheres represent models of organisms.

LEARNING TIP

Remember that small organisms have a small surface area. But this surface area is large compared to the volume inside the organism, so it has a large surface area to volume ratio.

THE NEED FOR TRANSPORT IN MULTICELLULAR ANIMALS

Within a large multicellular organism, many chemical reactions take place inside every microscopic cell. These cells require a supply of chemical substances such as glucose and oxygen for cellular respiration. These must be transported from outside a large organism into the cells. Respiration supplies energy for the other reactions of life, but it also produces the toxic waste product carbon dioxide. This and other waste products need to be removed from the cells before they cause damage to them.

Large multicellular organisms have internal transport systems that carry substances to every cell in the body. These systems deliver oxygen and nutrients and remove waste so that cells can carry out their functions efficiently. In large complex animals such as humans, chemicals made in a cell in one part of the body – such as a hormone like insulin or adrenaline – may influence a different type of cell elsewhere in the body. So substances made internally need to be moved around the body as well.

In many animals, including all the **vertebrates**, this transport system is the heart and circulatory system and the fluid that flows through it. This is an example of a **mass transport system** – substances are transported in the flow of a fluid with a mechanism for moving it around the body. All large complex organisms have some form of mass transport system which overcomes the limits of diffusion between the internal and external environments. Substances are delivered over short distances from the mass transport system to individual cells deep in the body by processes such as diffusion, osmosis and active transport.

FEATURES OF MASS TRANSPORT SYSTEMS

Mass transport systems are very effective for moving substances around the body. Most mass transport systems have certain features which are the same. They have:

- exchange surfaces to get materials into and out of the transport system
- a system of vessels that carry substances – these are usually tubes, sometimes following a very specific route, sometimes widespread and branching
- a way of making sure that substances are moved in the right direction (e.g. nutrients in and waste out)
- a way of moving materials fast enough to supply the needs of the organism – this may involve mechanical methods such as the pumping of the heart or ways of maintaining a concentration gradient so that substances move quickly from one place to another (e.g. using active transport)
- a suitable transport medium (e.g. fluid)
- in many cases, a way of adapting the rate of transport to the needs of the organism.

CIRCULATION SYSTEMS

Many animals have a circulatory system in which a heart pumps blood around the body. Insects have an open circulatory system with the blood circulating in large open spaces. However, most larger animals, including mammals, have a closed circulatory system with the blood contained within tubes. The blood makes a continuous journey out to the most distant parts of the body and back to the heart.

LEARNING TIP

The main advantages of a closed system are:

- the pressure can be increased to make the blood flow more quickly
- the flow can be directed more precisely to the organs that need most oxygen and nutrients.

Animals such as fish have a **single circulation system** (see **fig C**). The heart pumps deoxygenated blood to the gills, the organs of gas exchange where the blood takes in oxygen (becomes oxygenated) and gives up carbon dioxide at the same time. The blood then travels on around the rest of the body of the fish, giving up oxygen to the body cells before returning to the heart.

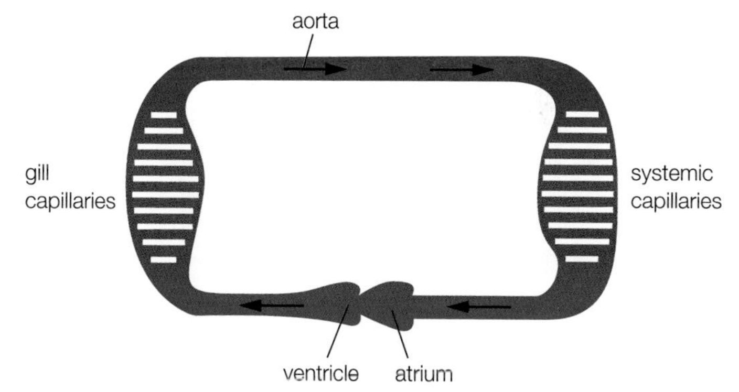

▲ **fig C** The single circulation of a fish

LEARNING TIP

Remember that mass flow is like a stream with all the components flowing together.

Birds and mammals need much more oxygen than fish. Not only do they have to move around without the support of water, but they also maintain a constant body temperature that may be higher or lower than their surroundings. This takes a lot of resources, so their cells need plenty of oxygen and glucose and make waste products that need to be removed quickly. Birds and mammals have evolved the most complex type of transport system, known as a **double circulation** because it involves two separate circulation systems. The **systemic circulation** carries **oxygenated blood** (oxygen-rich blood) from the heart to the cells of the body where the oxygen is used. It also carries the **deoxygenated blood** (blood that has given up its oxygen to the body cells) back to the heart. The **pulmonary circulation** carries deoxygenated blood from the heart to the lungs to be oxygenated and then carries the oxygenated blood back to the heart (see **fig D**).

The separate circuits of a double circulatory system ensure that the oxygenated and deoxygenated blood cannot mix, so the tissues receive as much oxygen as possible. Another big advantage is that the fully oxygenated blood can be delivered quickly to the body tissues at high pressure. The blood going through the tiny blood vessels in the lungs is at relatively low pressure, so it does not damage the vessels and allows gas exchange to take place. If this oxygenated blood at low pressure went straight into the big vessels that carry it around the body, it would move very slowly. However, the oxygenated blood returns to the heart, so it can be pumped hard and sent around the body at high pressure. This means it reaches all the tiny capillaries between the body cells quickly, supplying oxygen for an active way of life.

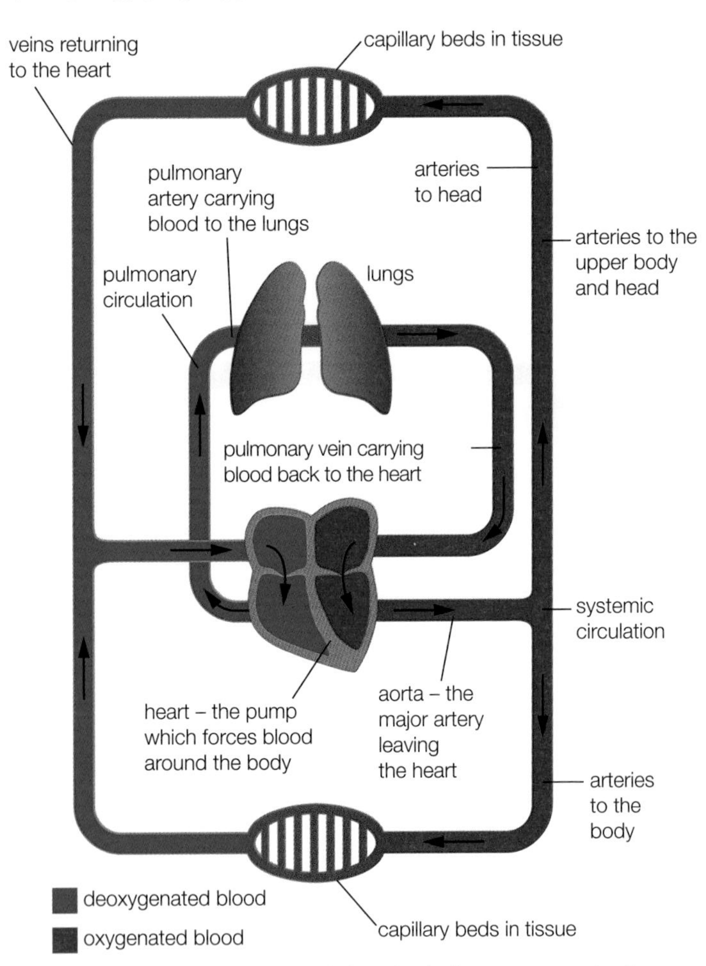

fig D A double circulation sends blood at high pressure, carrying lots of oxygen, to the active cells of the body. Take note: this is a schematic diagram. In a real double circulation, all of the blood vessels enter and leave from the top of the heart.

EXAM HINT

When you write about mass transport systems, make it clear that they are needed to overcome the limits of diffusion in organisms with a small surface area : volume ratio.

CHECKPOINT

SKILLS ADAPTIVE LEARNING

1. Explain why large animals cannot take in all the substances they need from outside the body through their skin.
2. What are the main characteristics of a mass transport system?
3. In fish, blood is supplied to the body tissues at low pressure. Why is low pressure sufficient in organisms such as fish?
4. Explain why a double circulation is ideal for an active animal that maintains its own body temperature independently of the environment.

SUBJECT VOCABULARY

diffusion the movement of the particles in a liquid or a gas down a concentration gradient from an area where they are at a relatively high concentration to an area where they are at a relatively low concentration
concentration gradient the change in the concentration of solutes present in a solution between two regions; in biology, this typically means across a cell membrane
surface area to volume ratio (sa : vol) the relationship between the surface area of an organism and its volume
vertebrates animals with a backbone or spinal column; they include mammals, birds, reptiles, amphibians and fish
mass transport system an arrangement of structures by which substances are transported in the flow of a fluid with a mechanism for moving it around the body
single circulation system a circulation in which the heart pumps the blood to the organs of gas exchange and the blood then travels on around the body before returning to the heart
double circulation system a circulation that involves two separate circuits, one of deoxygenated blood flowing from the heart to the gas exchange organs to be oxygenated before returning to the heart, and one of oxygenated blood leaving the heart and flowing around the body, returning as deoxygenated blood to the heart
systemic circulation carries oxygenated blood from the heart to the cells of the body where the oxygen is used, and carries the deoxygenated blood back to the heart
oxygenated blood blood that is carrying oxygen
deoxygenated blood blood that has given up its oxygen to the cells in the body
pulmonary circulation carries deoxygenated blood to the lungs and oxygenated blood back to the heart

2 THE ROLES OF THE BLOOD

SPECIFICATION REFERENCE 1.9(i) 1.9(ii) 1.11

LEARNING OBJECTIVES

- Understand the role of haemoglobin in the transport of oxygen and carbon dioxide.
- Understand the oxygen dissociation curve of haemoglobin, the Bohr effect and the significance of the oxygen affinity of fetal haemoglobin compared with adult haemoglobin.
- Understand the blood clotting process.

In mammals, the mass transport system is the **cardiovascular system**. This is made up of a series of vessels with the heart as a pump to move blood through the vessels. The blood is the transport medium and its passage through the vessels is called the **circulation**. The system delivers the materials needed by the cells of the body, and carries away the waste products of their metabolism. Substances move between the plasma or red blood cells and the body cells by diffusion or by **active transport** (see **Section 2A.4**). The tiniest blood vessels have walls only one cell thick, so diffusion distances are short and substances pass easily across these into other cells. Every cell in the body is near one of these small vessels. The blood also carries out other functions, such as carrying hormones (chemical messages) from one part of the body to another, forming part of the defence system of the body and distributing heat.

THE COMPONENTS OF THE BLOOD AND THEIR MAIN FUNCTIONS

You are going to study all three parts of the cardiovascular system, starting with the transport medium – the blood. Your blood is a complex mixture carrying a wide variety of cells and substances to all areas of your body (see **fig A**).

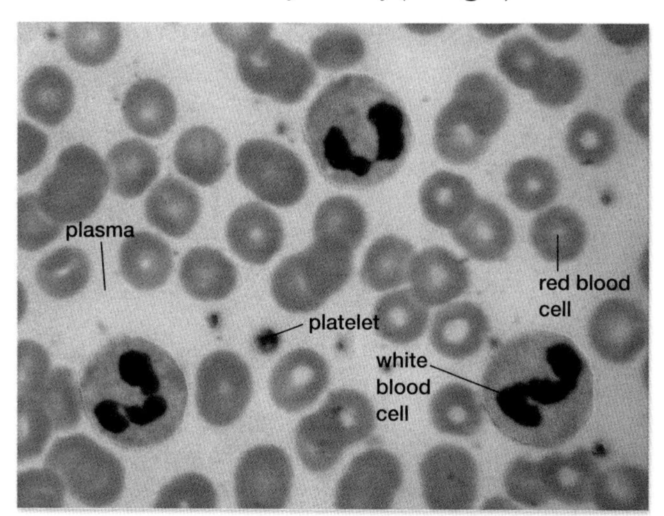

fig A This light micrograph shows red blood cells, white blood cells and platelets.

PLASMA

Your blood plasma is the fluid part of your mass transport system. Over 50% of your blood volume is plasma, and it carries all of your blood cells and everything else that needs transporting around your body. This includes:

- digested food products (e.g. glucose and amino acids) from the small intestine to the liver and then to all the parts of the body where they are needed either for immediate use or storage
- nutrient molecules from storage areas to the cells that need them
- excretory products (e.g. carbon dioxide and urea) from cells to the organs such as the lungs or kidneys that excrete them from the body
- chemical messages (hormones) from where they are made to where they cause changes in the body.

The plasma helps to maintain a steady body temperature by transferring heat around the system from internal organs (e.g. the gut) or very active tissues (e.g. leg muscles in someone running) to the skin, where it can be lost to the surroundings. It also acts as a **buffer** to regulate pH changes.

ERYTHROCYTES (RED BLOOD CELLS)

There are approximately 5 million erythrocytes per mm^3 of blood (4–5 million per mm^3 in women, 5–6 million per mm^3 in men). They contain haemoglobin, a red pigment that carries oxygen and gives them their colour (see **Section 1A.5**). They are made in the bone marrow. Mature erythrocytes do not contain a nucleus and have a limited life of about 120 days.

The erythrocytes transport oxygen from the lungs to all the cells. They are well adapted for their function. The biconcave disc shape of the cells means that they have a large surface area to volume ratio, so oxygen can diffuse into and out of them rapidly (see **fig B**). Having no nucleus leaves much more space inside the cells for the haemoglobin molecules that carry the oxygen. In fact, each red blood cell contains around 250–300 million molecules of haemoglobin and can carry approximately 1000 million molecules of oxygen. Haemoglobin also carries some of the carbon dioxide produced in respiration back to the lungs. The rest is transported in the plasma.

▲ **fig B** Healthy red blood cells have a biconcave disc shape.

LEUCOCYTES (WHITE BLOOD CELLS)

Leucocytes or white blood cells are much larger than erythrocytes, but can also squeeze through tiny blood vessels because they can change their shape. There are around 4000–11 000 per mm^3 of blood and there are several different types. They are made in the bone marrow, although some mature in the thymus gland. Their main function is to defend the body against infection. Leucocytes are also very important in the inflammatory response of the body when an area of tissue is damaged. They all contain a nucleus and have colourless cytoplasm, although some types contain granules which can be stained. There are several different types of leucocyte, which you will study further in **Book 2 Chapter 6B**.

PLATELETS

Platelets are tiny fragments of large cells called **megakaryocytes**, which are found in the bone marrow. There are about 150 000–400 000 platelets per mm^3 of blood. They are involved in blood clotting (see **page 35**).

TRANSPORT OF OXYGEN

The many haemoglobin molecules that are in the red blood cells transport oxygen. Each haemoglobin molecule is a large globular protein consisting of four peptide chains, each with an iron-containing prosthetic group. Each group can collect four molecules of oxygen in a reversible process to form **oxyhaemoglobin**:

$$Hb + 4O_2 \rightleftharpoons Hb.4O_2$$

$$\text{haemoglobin} + \text{oxygen} \rightleftharpoons \text{oxyhaemoglobin}$$

The first oxygen molecule that binds to the haemoglobin changes the arrangement of the molecule making it easier for the following oxygen molecules to bind. The final oxygen molecule binds several hundred times faster than the first. The same process happens in reverse when oxygen dissociates from haemoglobin – it gets progressively harder to remove the oxygen.

EXAM HINT

Always use suitable scientific terms. This is important because it can help to focus a response clearly.

When the blood enters the lungs, the concentration of oxygen in the red blood cells is relatively low. Oxygen moves into the red blood cells from the air in the lungs by diffusion. The oxygen is collected and bound to the haemoglobin, so the free oxygen concentration in the cytoplasm of the red blood cells stays low. This maintains a steep concentration gradient from the air in the lungs to the red blood cells, so more and more oxygen diffuses in and joins onto the haemoglobin.

The oxygen levels are relatively low in the body tissues. The concentration of oxygen in the cytoplasm of the red blood cells is higher than in the surrounding tissue. As a result, oxygen moves out into the body cells by diffusion down its concentration gradient. The haemoglobin molecules give up some of their oxygen. When you are at rest or exercising gently, only about 25% of the oxygen carried by the haemoglobin is released into your cells. There is another 75% in reserve in the transport system for when you are very active.

The strong affinity of haemoglobin for oxygen means that a small change in the proportion of oxygen in the surrounding environment can have a big effect on the saturation of the blood with oxygen. So in the lungs, the haemoglobin rapidly gains oxygen and in the tissues, as the oxygen saturation of the environment falls, oxygen is released rapidly (see **fig C**).

As deoxygenated blood approaches the lungs, the steep part of the curve means that a *small* increase in partial pressure causes a *large* increase in % saturation.

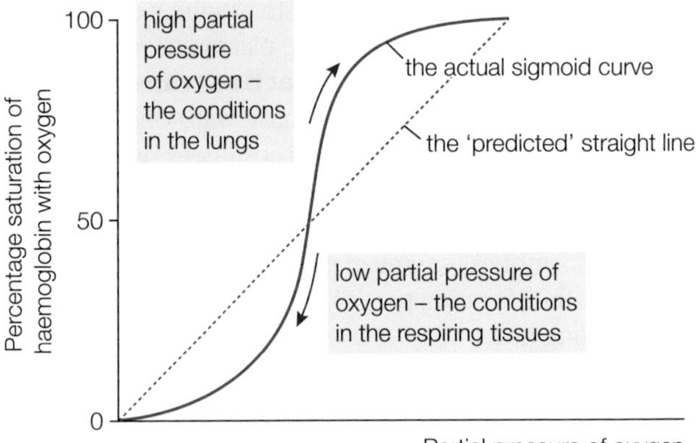

As oxygenated blood approaches the tissues, a *small* decrease in partial pressure causes a *large* decrease in % saturation (i.e. a large release of oxygen).

▲ **fig C** Oxygen dissociation curve for human haemoglobin

TRANSPORT OF CARBON DIOXIDE

Waste carbon dioxide diffuses from the respiring cells of the body tissues into the blood along a concentration gradient. The reaction of the carbon dioxide with water is crucial. When carbon dioxide is dissolved in the blood it reacts slowly with the water to form carbonic acid, H_2CO_3. The carbonic acid separates to form hydrogen ions H^+ and hydrogencarbonate ions HCO_3^-:

$$CO_2 + H_2O \rightleftharpoons H_2CO_3 \rightleftharpoons HCO_3^- + H^+$$

About 5% of the carbon dioxide is carried in solution in the plasma. A further 10–20% combines with haemoglobin molecules to make **carbaminohaemoglobin**. Most of the carbon dioxide is transported in the cytoplasm of the red blood cells as hydrogencarbonate ions. The enzyme **carbonic anhydrase** controls the rate of the reaction between carbon dioxide and water to produce carbonic acid.

In the body tissues, there is a high concentration of carbon dioxide in the blood, so carbonic anhydrase catalyses the formation of carbonic acid.

In the lungs, the carbon dioxide concentration is low, so carbonic anhydrase catalyses the reverse reaction and free carbon dioxide diffuses out of the blood and into the lungs (see **fig D**).

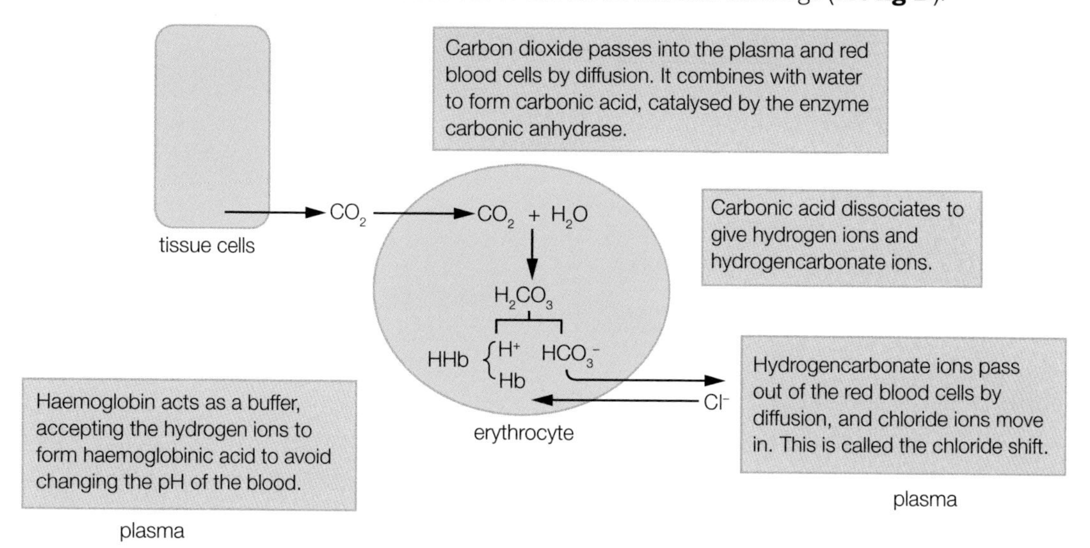

▲ **fig D** The transport of carbon dioxide from the tissues to the lungs depends on the reaction of carbon dioxide with water, controlled by an enzyme in the red blood cells.

THE BOHR EFFECT

The way in which haemoglobin collects and releases oxygen is also affected by the proportion of carbon dioxide in the tissues (see **fig E**). When the proportion of carbon dioxide in the tissues is high, the affinity of haemoglobin for oxygen is reduced. In other words, haemoglobin needs higher levels of oxygen to become saturated and releases oxygen much more easily. So in active tissues with high carbon dioxide levels, haemoglobin releases oxygen more readily. Carbon dioxide levels in the lung capillaries are relatively low, which makes it easier for oxygen to bind to the haemoglobin. The changes in the oxygen dissociation curve that result as the carbon dioxide level changes are known as the **Bohr effect**.

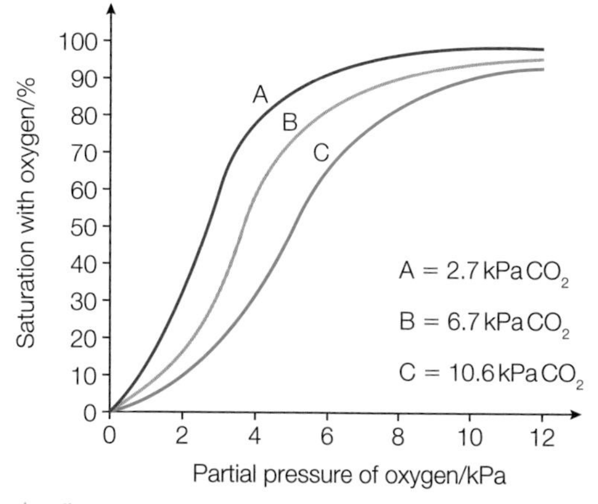

▲ **fig E** As the proportion of carbon dioxide in the environment rises, the haemoglobin curve moves down and to the right, so it gives up oxygen more easily. This is known as the Bohr effect.

LEARNING TIP

Remember that as CO_2 builds up it affects the pH and this has an effect on the protein structure, so haemoglobin does not work as well (i.e. it has a lower affinity for oxygen).

FETAL HAEMOGLOBIN

A fetus in the uterus depends on its mother to supply it with oxygen. Oxygenated blood from the mother flows through the placenta close to the deoxygenated fetal blood. If the blood of the fetus had the same affinity for oxygen as the blood of the mother, very little oxygen would be transferred. Fortunately, the blood of the fetus contains a special form of the oxygen-carrying pigment called **fetal haemoglobin**. Fetal haemoglobin has a higher affinity for oxygen than the adult haemoglobin of the mother. Consequently, the fetal haemoglobin can remove oxygen from the maternal blood even when the proportion of oxygen is relatively low (see **fig F**). The maternal and fetal blood also run in opposite directions. This makes the oxygen concentration gradient between the mother's blood and that of her fetus as steep as possible, maximising the oxygen transfer to the blood of the fetus.

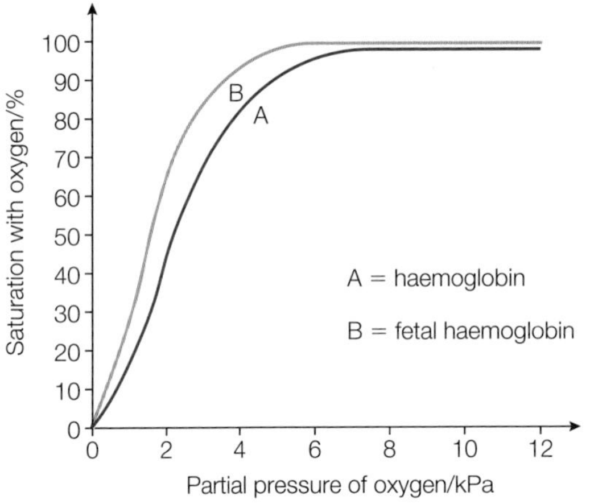

fig F Fetal haemoglobin has a higher affinity for oxygen than the adult haemoglobin of the mother, so it can take oxygen from the mother's blood and deliver it to the cells of the growing fetus.

EXAM HINT

Be precise with your descriptions e.g. fetal *haemoglobin* has a higher affinity for oxygen than maternal haemoglobin - it is not the fetus or the blood that has a higher affinity.

DID YOU KNOW?

Down into the depths!

Elephant seals can dive to depths of almost 2 km and stay underwater for up to 2 hours, swimming and hunting, although most do not dive so deep or for so long (see **fig G**). While underwater, the seals cannot breathe but they have three adaptations of the blood that allow them to stay underwater for a long time.

- They have up to twice the blood volume of a land mammal of the same size, with extra spaces in their circulatory system to store oxygenated blood.
- They have more erythrocytes per unit of blood than land mammals, and scientists think the erythrocytes also contain more haemoglobin. As a result, they have up to three times as much haemoglobin as a land mammal of a similar size.
- They have over 10 times more myoglobin in their muscles than humans. Myoglobin is another pigment with an affinity for oxygen which is higher than either haemoglobin or fetal haemoglobin. In elephant seals, the myoglobin is so dense their muscles look almost black.

These adaptations mean that elephant seals have an enormous oxygen store in their bodies when they dive, which helps to explain why they are such masters of the underwater world.

fig G Elephant seal underwater

THE CLOTTING OF THE BLOOD

You have a limited volume of blood. In theory, a minor cut could endanger life as the torn blood vessels allow blood to escape. First, your blood volume will reduce and if you lose too much blood, you will die. Second, pathogens can get into your body through an open wound. In normal circumstances, your body protects you through the clotting mechanism of the blood. This mechanism seals damaged blood vessels to minimise blood loss and prevent pathogens getting in.

FORMING A CLOT

Plasma, blood cells and platelets flow from a cut vessel. Contact between the platelets and components of the tissue (e.g. collagen fibres in the skin) causes the platelets to break open in large numbers. They release several substances, two of which are particularly important.

- **Serotonin** causes the smooth muscle of the blood vessel to contract. This narrows the blood vessels, cutting off the blood flow to the damaged area.
- **Thromboplastin** is an enzyme that starts a sequence of chemical changes that clot the blood (see **fig H**).

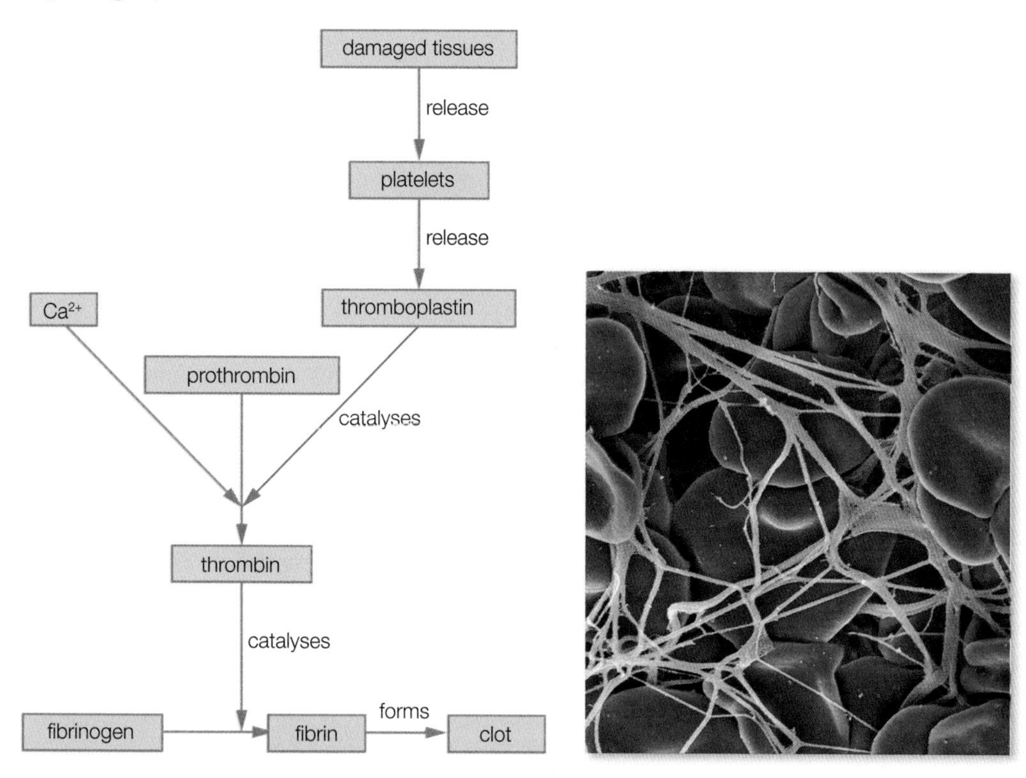

▲ **fig H** The cascade of events that results in a life-saving or life-threatening clot. When you cut yourself, this is the process which seals the blood vessels and protects the delicate new tissues that form underneath.

THE BLOOD CLOTTING PROCESS

The blood clotting process is a very complex sequence of events in which there are many different clotting factors. Vitamin K is important in the production of many of the compounds needed for the blood to clot, including prothrombin. Here is a simple version of the events in the blood clotting process.

1 Thromboplastin catalyses the conversion of a large soluble protein called **prothrombin** found in the plasma into another soluble protein, the enzyme called **thrombin**. Prothrombin is biologically inactive while thrombin is biologically active – prothrombin is a **precursor** of thrombin. This conversion happens on a large scale at the site of a wound. Calcium ions need to be present in the blood at the right concentration for this reaction to happen.

2 Thrombin acts on another soluble plasma protein called **fibrinogen**, converting it to an insoluble substance called **fibrin**. Again, fibrinogen is the biologically inactive precursor of biologically active fibrin. The fibrin forms a mesh of fibres to cover the wound.

LEARNING TIP

Remember that *pro* means before. Prothrombin is a precursor that will form an active molecule.

3 More platelets and red blood cells pouring from the wound get trapped in the fibrin mesh. This forms a clot.
4 Special proteins in the structure of the platelets contract, making the clot tighter and tougher to form a scab that protects the skin and vessels underneath as they heal.

In a sequence such as clot formation, a small event is amplified through a series of steps. However, sometimes the body's clotting mechanism is started in the wrong place, and this can lead to serious problems in the blood vessels. A clot in the vessels that supply your heart muscle with blood can cause a heart attack and a clot in the brain can cause a stroke (see **Section 1B.5**).

CHECKPOINT

SKILLS CRITICAL THINKING

1. Red blood cells are unusual because they do not have a nucleus. Explain how this is an adaptation for their role in carrying oxygen, and why they have a limited life.
2. Describe how oxygen is transported in the blood.
3. Explain why fetal haemoglobin needs to have a higher affinity for oxygen than adult haemoglobin.
4. Prothrombin and fibrinogen are both precursors. Discuss the similarities and differences between these two proteins.
5. There is a rare condition in babies that causes excessive internal bleeding, which can cause brain damage and even death. Newborn babies in most countries in the world are routinely given vitamin K either by injection or orally. Suggest how these two facts might be linked.

SUBJECT VOCABULARY

cardiovascular system the mass transport system of the body made up of a series of vessels with a pump (the heart) to move blood through the vessels
circulation the passage of blood through the blood vessels
active transport the movement of substances into or out of the cell using ATP produced during cellular respiration
buffer a solution which resists changes in pH
leucocytes white blood cells; there are several different types which play important roles in defending the body against the entry of pathogens and in the immune system
platelets cell fragments involved in the clotting mechanism of the blood
megakaryocytes large cells that are found in the bone marrow and produce platelets
oxyhaemoglobin the molecule formed when oxygen binds to haemoglobin
carbaminohaemoglobin the molecule formed when carbon dioxide combines with haemoglobin
carbonic anhydrase the enzyme that controls the rate of the reaction between carbon dioxide and water to produce carbonic acid
Bohr effect the name given to changes in the oxygen dissociation curve of haemoglobin that occur due to a rise in carbon dioxide levels and a reduction of the affinity of haemoglobin for oxygen
fetal haemoglobin a form of haemoglobin found only in the developing fetus with a higher affinity for oxygen than adult haemoglobin
serotonin a chemical that causes the smooth muscle of the blood vessels to contract, narrowing them and cutting off the blood flow to the damaged area
thromboplastin an enzyme that sets in progress a cascade of events that leads to the formation of a blood clot
prothrombin a large, soluble protein found in the plasma that is the precursor to an enzyme called thrombin
thrombin an enzyme that acts on fibrinogen, converting it to fibrin during clot formation
precursor a biologically inactive molecule which can be converted into a closely related biologically active molecule when needed
fibrinogen a soluble plasma protein which is the precursor of the insoluble protein fibrin
fibrin an insoluble protein formed from fibrinogen by the action of thrombin that forms a mesh of fibres that trap erythrocytes and platelets to form a blood clot

1B 3 CIRCULATION IN THE BLOOD VESSELS

SPECIFICATION REFERENCE 1.7

LEARNING OBJECTIVES

- Understand how the structures of blood vessels (arteries, veins and capillaries) relate to their functions.

THE BLOOD VESSELS

The blood vessels that make up the circulatory system can be thought of as the biological equivalent of a road transport system. The **arteries** and **veins** are like the large roads carrying heavy traffic while the narrow town streets and tracks are represented by the vast area of branching and spreading **capillaries** called the capillary network. In the capillary network, substances carried by the blood are exchanged with cells in the same way that products are transported from factories, oil refineries or farms and distributed into shops and homes. The structures of the different types of blood vessel closely reflect their functions in your body.

ARTERIES

Arteries carry blood away from your heart towards the cells of your body. The structure of an artery is shown in **fig A**. Almost all arteries carry oxygenated blood. The exceptions are:

- the pulmonary artery – carrying deoxygenated blood from the heart to the lungs
- the umbilical artery – during pregnancy, this carries deoxygenated blood from the fetus to the placenta.

The arteries leaving the heart branch off in every direction, and the diameter of the **lumen**, the central space inside the blood vessel, gets smaller the further away it is from the heart. The very smallest branches of the **arterial system**, furthest from the heart, are the **arterioles**.

LEARNING TIP

Remember that all arteries carry blood away from the heart, so they have thick walls and lots of collagen to withstand the high pressure.

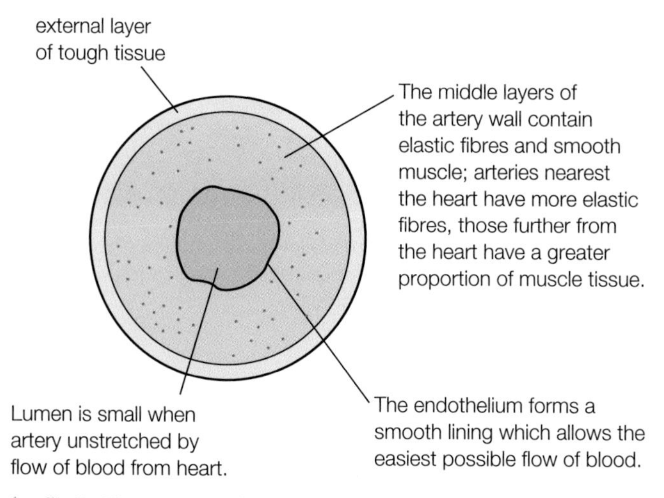

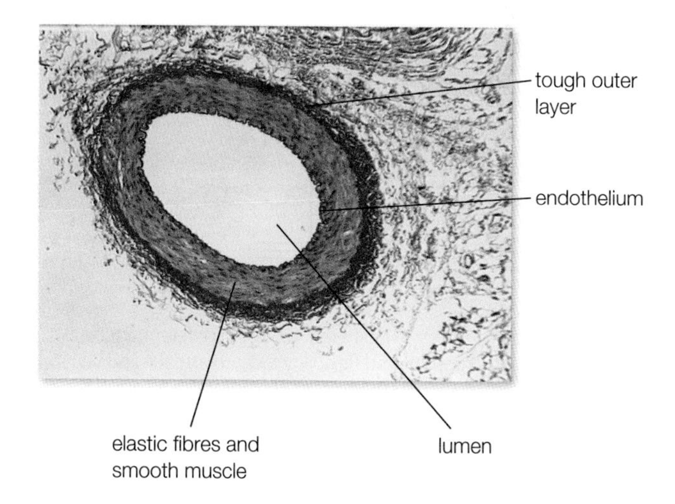

▲ **fig A** The structure of an artery means it is adapted to cope with the surging of the blood as the heart pumps.

Blood is pumped out from the heart in a regular rhythm, about 70 times a minute. Each heartbeat sends a high-pressure flow of blood into the arteries. The major arteries close to the heart must withstand these pressure surges. Their walls contain a lot of elastic fibres, so they can stretch to accommodate the greater volume of blood without being damaged (see **fig B**). Between surges, the elastic fibres return to their original length, squeezing the blood to move it along in a continuous flow. The pulse you can feel in an artery is the effect of the surge each time the heart beats. The blood pressure in all arteries is relatively high, but it falls in arteries further away from the heart. These are known as the **peripheral arteries**.

EXAM HINT

You will study the structure and the function of the types of blood vessel separately. However, you should remember that the vessels do not exist separately – they are all interlinked within the whole circulatory system.

EXAM HINT

The role of the elastic fibres in artery walls is to return to their original length to help maintain the pressure. This is called recoil. The elastic recoil does not help to increase pressure, it simply helps to maintain the pressure - so do not suggest that the recoil helps pump blood along.

In the peripheral arteries, the muscle fibres in the vessel walls contract or relax to change the size of the lumen, controlling the blood flow. The smaller the lumen, the harder it is for blood to flow through the vessel. This controls the amount of blood that flows into an organ, so regulating its activity. You will find out more about this important response in **Book 2 Topic 7**.

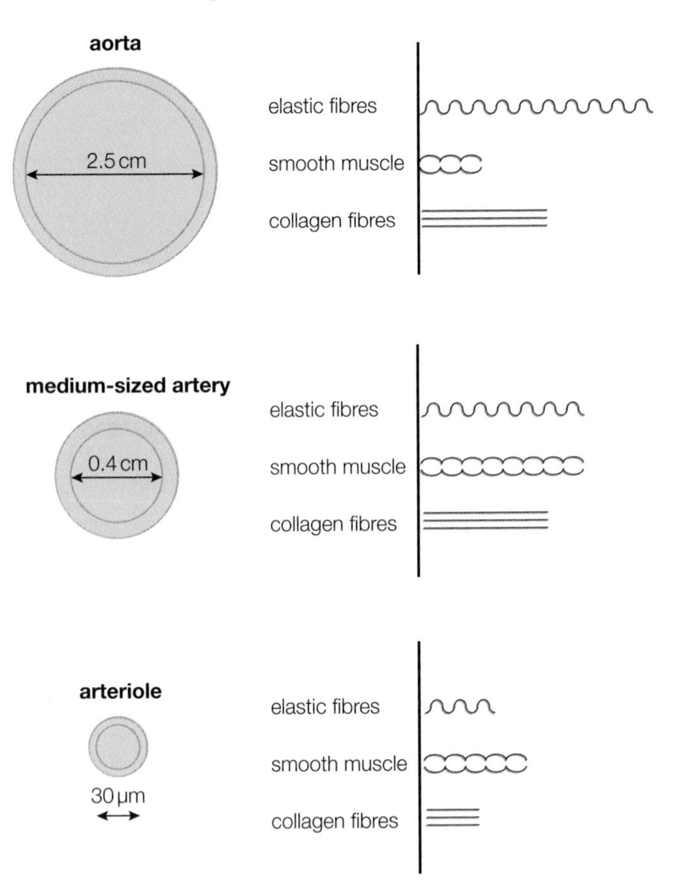

▲ **fig B** The relative proportions of different tissues in different arteries. Collagen gives general strength and flexibility to both arteries and veins.

LEARNING TIP

The role of the muscles in the wall of the arterioles is to reduce the size of the lumen to increase resistance - this can reduce blood flow to areas that do not need so much blood and will cause the oxygenated blood to flow to other tissues. Remember to link this to things you may learn later such as how blood flow to the skin changes when you are too hot or too cold.

CAPILLARIES

Arterioles lead into networks of capillaries. These are very small vessels that spread throughout the tissues of the body. The capillary network links the arterioles and the **venules**. Capillaries branch between cells – no cell is far from a capillary, so substances can diffuse between cells and the blood quickly. Also, because the diameter of each individual capillary is small, the blood travels relatively slowly through them, giving more opportunity for diffusion to occur (see **fig C**). The smallest capillary is no wider than a single red blood cell.

Capillaries have a very simple structure which is well adapted to their function. Their walls are very thin and contain no elastic fibres, smooth muscle or collagen. This helps them fit between individual cells and allows rapid diffusion of substances between the blood and the cells. The walls consist of just one very thin cell. Oxygen and other molecules, such as digested food molecules and hormones, quickly diffuse out of the blood in the capillaries into the nearby body cells, and carbon dioxide and other waste molecules diffuse into the capillaries. Blood entering the capillary network from the arteries is oxygenated. When it leaves, it carries less oxygen and more carbon dioxide.

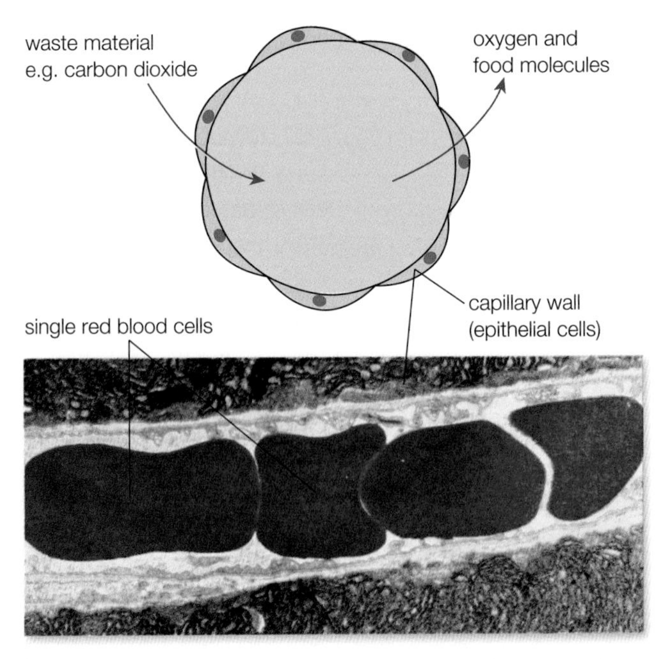

▲ **fig C** The very thin walls of capillaries allow rapid diffusion of oxygen, carbon dioxide and digested food molecules. The lumen is just wide enough for red blood cells to pass through.

VEINS

Veins carry blood back towards the heart. Most veins carry deoxygenated blood. The exceptions are:

- the pulmonary vein – carrying oxygen-rich blood from the lungs back to the heart for circulation around the body
- the umbilical vein – during pregnancy, it carries oxygenated blood from the placenta into the fetus.

Tiny venules lead from the capillary network, combining into larger and larger vessels going back to the heart (see **fig D**).

LEARNING TIP

Remember that all veins carry blood back to the heart so they have low pressure and do not need a thick wall.

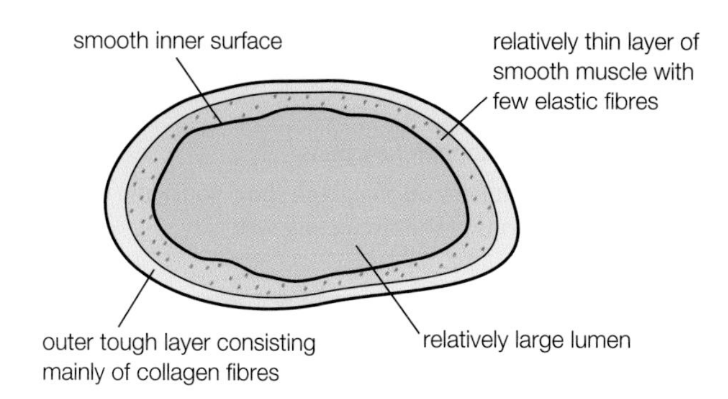

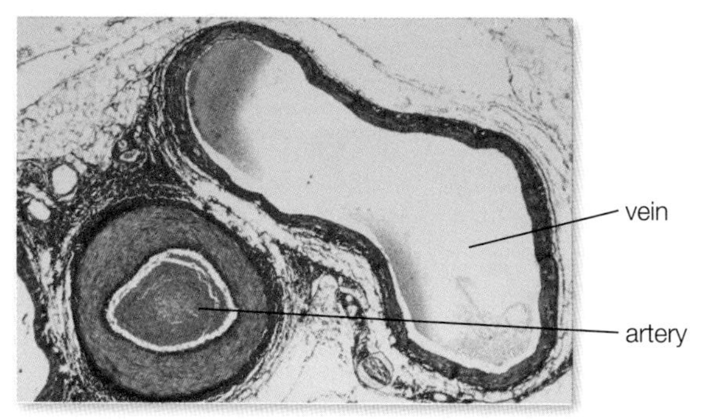

▲ **fig D** The arrangement of tissues in a vein reflects the pressure of blood in the vessel.

Eventually only two veins (sometimes called the great veins) carry the blood from the body tissues back to the heart – the **inferior vena cava** from the lower parts of the body and the **superior vena cava** from the upper parts of the body.

Veins can hold a large volume of blood – in fact more than half of the body's blood volume is in the veins at any one time. They act as a blood reservoir. The blood pressure in the veins is relatively low – the pressure surges from the heart are eliminated before the blood reaches the capillary system. This blood at low pressure must be returned to the heart and lungs to be oxygenated again and recirculated.

The blood is not pumped back to the heart, it returns to the heart by means of muscle pressure and one-way valves.

- Many of the larger veins are situated between the large muscle blocks of the body, particularly in the arms and legs. When the muscles contract during physical activity they squeeze these veins. The valves (see below) keep the blood travelling in one direction and this squeezing helps to return the blood to the heart.
- There are one-way valves at frequent intervals throughout the **venous system**. These are called **semilunar valves** because of their half-moon shape. They develop from infoldings of the inner wall of the vein. Blood can pass through towards the heart, but if it starts to flow backwards the valves close, preventing any backflow (see **fig E**).

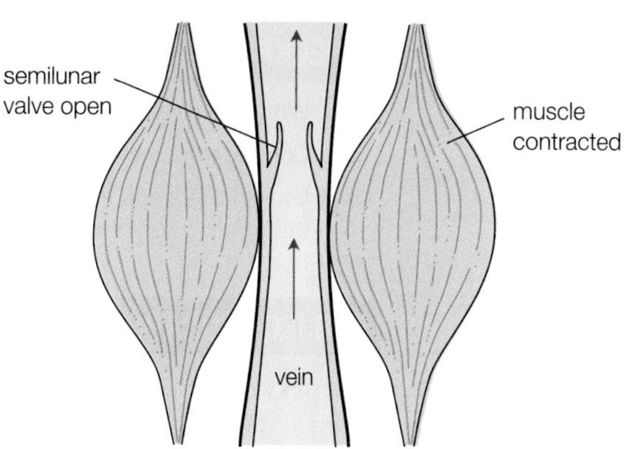

Blood moving in the direction of the heart forces the valve open, allowing the blood to flow through.

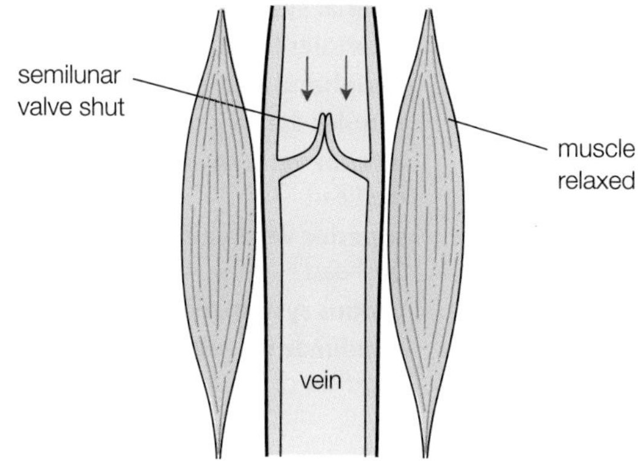

A backflow of blood will close the valve, ensuring that blood cannot flow away from the heart.

▲ **fig E** Valves in the veins make sure blood only flows in one direction – towards the heart. The contraction of large muscles encourages blood flow through the veins.

The main types of blood vessel – the arteries, veins and capillaries – have very different characteristics. These affect the way the blood flows through the body, and what the vessels do in the body. Some of these differences are summarised in **fig F**.

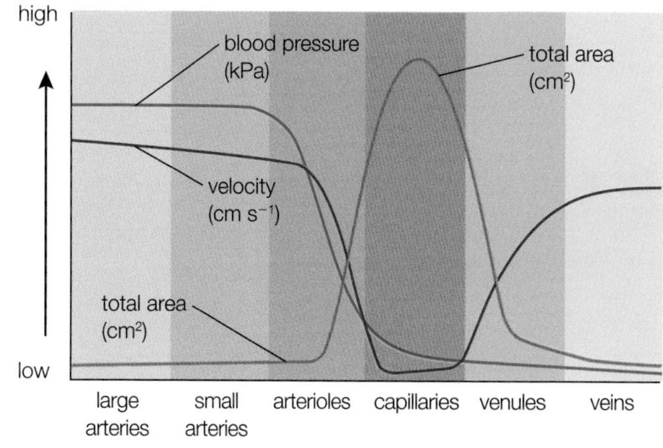

▲ **fig F** Graph to show the surface area of each major type of blood vessel in your body, along with the velocity and pressure of the blood travelling in them.

CHECKPOINT

1. Why are valves important in veins but unnecessary in arteries?
2. Compare the main structures of arteries, veins and capillaries to their functions.
3. Look at the graph in **fig F**. Explain carefully what the different lines on the graph show you. How is this information linked to the functions of the different regions of the circulatory system?

SKILLS ANALYSIS

SUBJECT VOCABULARY

arteries vessels that carry blood away from the heart
veins vessels that carry blood towards the heart
capillaries tiny vessels that spread throughout the tissues of the body
lumen the central space inside the blood vessel
arterial system the system of arteries in the body
arterioles the very smallest branches of the arterial system, furthest from the heart
peripheral arteries arteries further away from the heart but before the arterioles
venules the very smallest branches of the venous system, furthest from the heart
inferior vena cava the large vein that carries the returning blood from the lower parts of the body to the heart
superior vena cava the large vein that carries the returning blood from the upper parts of the body to the heart
venous system the system of veins in the body
semilunar valves half-moon shaped, one-way valves found at frequent intervals in veins to prevent the backflow of blood

4 THE MAMMALIAN HEART

LEARNING OBJECTIVES

- Relate the structure and operation of the mammalian heart, including the major blood vessels, to its function.
- Know the cardiac cycle.

In most animal transport systems, the heart is the organ that moves the blood around the body. In mammals, the heart is a complex, four-chambered muscular organ that sits in the chest protected by the ribs and sternum. In an average lifetime, the heart beats about 3 000 000 000 (3×10^9) times and will pump over 200 million litres of blood – quite a workload.

THE STRUCTURE OF THE HEART

The human heart, like other mammalian hearts (see **fig A**), is not a single muscular pump but two pumps, joined and working in time together. The right side of the heart receives blood from the body and pumps it to the lungs. The left side of the heart receives blood from the lungs and pumps it to the body. The blood in each side of the heart does not mix with the blood from the other side. The two sides are separated by a thick, muscular **septum**. The heart is made of a unique type of muscle, known as **cardiac muscle**, which has special properties – it can carry on contracting regularly without resting or getting fatigued. You will study this in more detail in **Book 2 Topic 7**. Cardiac muscle has a good blood supply – the coronary arteries bring oxygenated blood to the tissue (see **fig B**). It also contains lots of **myoglobin**, a respiratory pigment which has a stronger affinity for oxygen than haemoglobin. This myoglobin stores oxygen for the respiration needed to keep the heart contracting regularly.

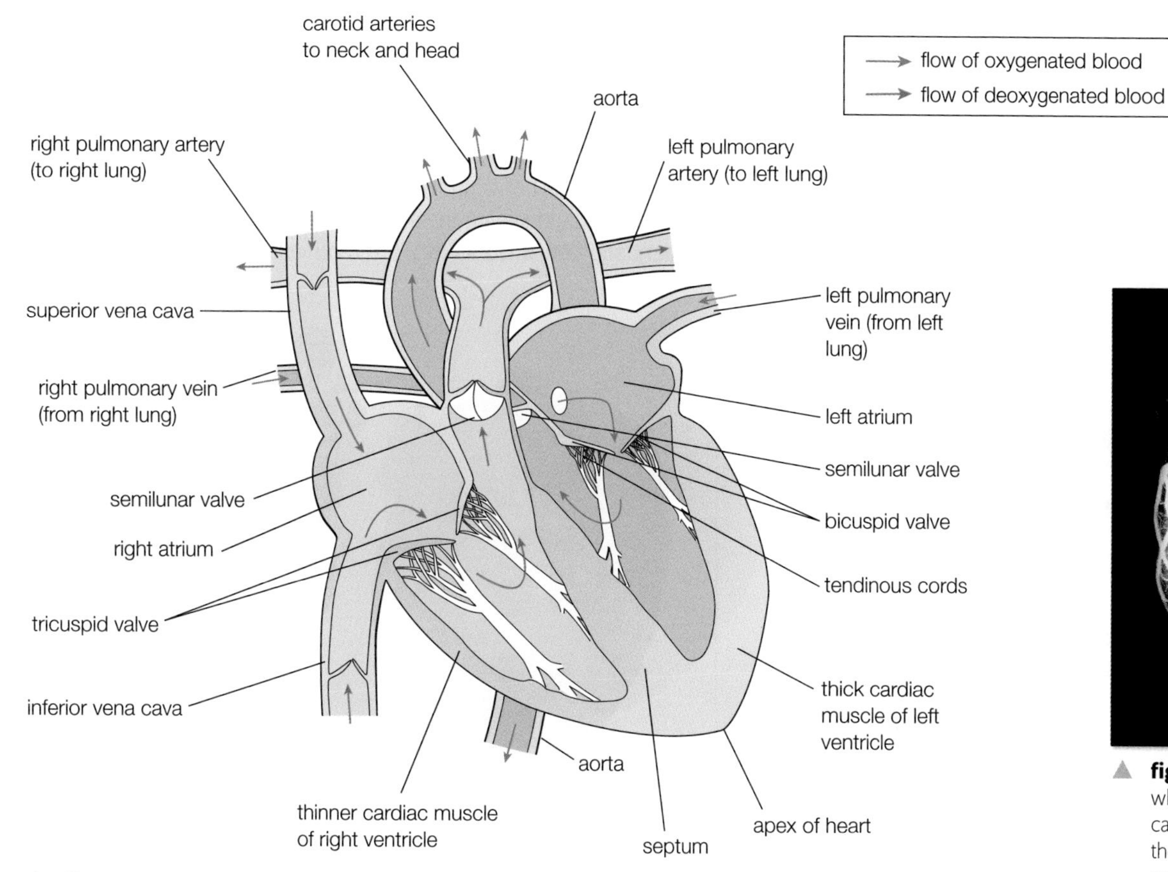

fig A The structure of the human heart

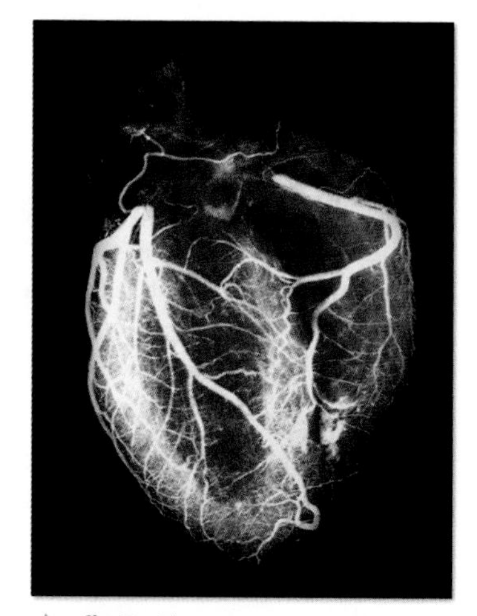

fig B The coronary arteries, which you can clearly see here, carry oxygenated blood from the aorta to the heart muscle, providing it with oxygen and digested food and removing carbon dioxide.

THE ACTION OF THE HEART

EXAM HINT

When describing the action of the heart, remember that blood flows through both sides at the same time. Make it clear that both atria pump at the same time and both ventricles pump at the same time.

1 The inferior vena cava collects deoxygenated blood from the lower parts of the body, while the superior vena cava receives deoxygenated blood from the head, neck, arms and chest. Deoxygenated blood is delivered to the **right atrium**.

2 The right atrium receives the blood from the great veins. As it fills with blood, the pressure builds up and opens the tricuspid valve, so the **right ventricle** starts to fill with blood too. When the atrium is full it contracts, forcing more blood into the ventricle. The atrium has thin muscular walls because it receives blood at low pressure from the inferior vena cava and the superior vena cava and it needs to exert relatively little pressure to move the blood into the ventricle. One-way semilunar valves (like the valves in veins described in **Section 1B.3**) at the entrance to the atrium stop a backflow of blood into the veins.

3 The **tricuspid valve** consists of three flaps and is also known as an **atrioventricular valve** because it separates an atrium from a ventricle. The valve allows blood to pass from the atrium to the ventricle, but not in the other direction. The tough **tendinous cords**, also known as valve tendons or heartstrings, make sure the valves are not turned inside out by the pressure exerted when the ventricles contract.

4 The right ventricle is filled with blood under some pressure when the right atrium contracts, then the ventricle contracts. Its muscular walls produce the pressure needed to force blood out of the heart into the **pulmonary arteries**. These carry the deoxygenated blood to the capillaries in the lungs. As the ventricle starts to contract, the tricuspid valve closes to prevent blood flowing into the atrium. Semilunar valves, like those in veins, prevent the blood flowing back from the arteries into the ventricle.

5 The blood returns from the lungs to the left side of the heart in the **pulmonary veins**. The blood is at relatively low pressure after passing through the extensive capillaries of the lungs. The blood returns to the **left atrium**, another thin-walled chamber that performs the same function as the right atrium. It contracts to force blood into the **left ventricle**. Backflow is prevented by another atrioventricular valve known as the **bicuspid valve**, which has only two flaps.

6 As the left atrium contracts, the bicuspid valve opens and the left ventricle is filled with blood under pressure. As the left ventricle starts to contract the bicuspid valve closes to prevent backflow of blood to the left atrium. The left ventricle pumps the blood out of the heart and into the **aorta**, the major artery of the body. This carries blood away from the heart at even higher pressure than the major arteries that branch off from it.

The muscular wall of the left side of the heart is much thicker than that of the right. The right side pumps blood to the lungs, which are relatively close to the heart. The delicate capillaries of the lungs need blood delivered at relatively low pressure. The left side must produce sufficient force to move the blood under pressure to all the extremities of the body and overcome the elastic recoil of the arteries. Semilunar valves prevent the blood flowing back from the aorta into the ventricle.

LEARNING TIP

Remember that the valves do not operate on their own. They open and close as blood pressure changes in the chambers on either side of the valve. When the pressure is higher on one side it will push the valve open. When it is higher on the other side, it will close the valve.

The septum is a thick wall of muscle and connective tissue between the two sides of the heart. It prevents the oxygenated blood mixing with the deoxygenated blood.

DID YOU KNOW?

In an embryo, there is a gap in the septum called the *foramen ovale* and the blood from the two sides of the heart can mix. This does not matter because the lungs of the fetus do not function and little blood flows to them. At birth, this hole closes over as the lungs begin to function. If it does not, the baby has a condition called patent *foramen ovale* or 'hole in the heart'. This may be so small it does not really matter. If it is large, it must be closed by surgery.

DID YOU KNOW?

Inside the heart

The diagrams we use of the inside of a mammalian heart make it look very clean and simple (see fig A). But when you dissect the heart of a mammal such as a sheep or a cow, you can see that it is much more complicated. Cutting open a heart and exploring the connections between the blood vessels and the chambers of the heart, and seeing the structure of the valves, helps you understand how it works as a three-dimensional pump (see fig C).

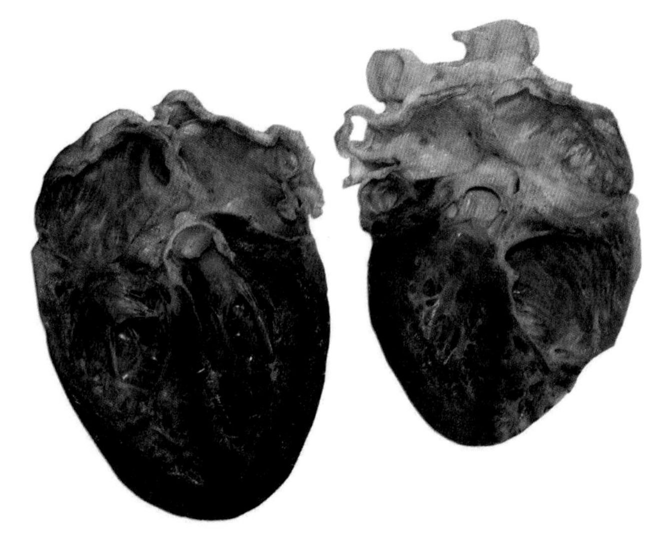

▲ **fig C** A dissected mammalian heart

HOW YOUR HEART WORKS

The beating of your heart produces the sounds that are your heartbeat. The sounds are not made by the contracting of the heart muscle but by the heart valves closing. The two sounds of a heartbeat are often described as 'lub-dub'. The first sound ('lub') comes when the ventricles contract and the blood is forced against the atrioventricular valves. The second sound ('dub') comes when the ventricles relax and a backflow of blood hits the semilunar valves in the pulmonary artery and aorta. The rate of your heartbeat shows how frequently your heart is contracting.

THE CARDIAC CYCLE

Your heart is continuously contracting then relaxing. The contraction of the heart is called **systole**. Systole can be divided into **atrial systole**, when the atria contract together forcing blood into the ventricles, and **ventricular systole**, when the ventricles contract. Ventricular systole happens about 0.13 seconds after atrial systole, and forces blood out of the ventricles into the pulmonary artery and the aorta. Between contractions the heart relaxes and fills with blood. This relaxation stage is called **diastole**. One cycle of systole and diastole makes up a single heartbeat, which lasts about 0.8 seconds in humans. This is known as the **cardiac cycle** (see **fig D**). You will learn more about how the rate of the heartbeat is controlled in **Book 2 Topic 7**.

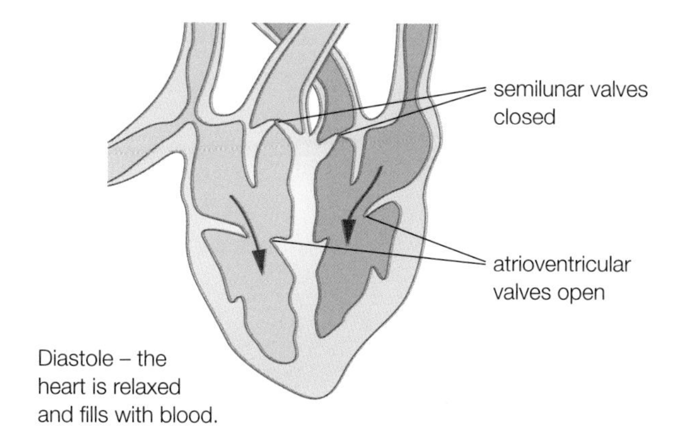

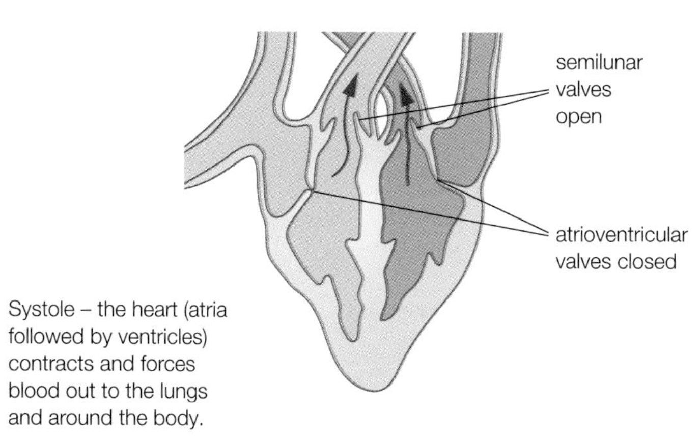

▲ **fig D** The cardiac cycle

CHECKPOINT

1. Describe the path of blood around the human body, identifying at which points the blood is oxygenated and where it is deoxygenated. Explain how this system efficiently supplies cells with the oxygen they need.
2. Discuss the relationship between structure and function for these parts of the heart:
   (a) semilunar valves
   (b) atria
   (c) ventricles including the thickness of the muscle walls
   (d) tendinous cords.

SUBJECT VOCABULARY

septum the thick muscular dividing wall through the centre of the heart that prevents oxygenated and deoxygenated blood from mixing
cardiac muscle the special muscle tissue of the heart, which has an intrinsic rhythm and does not fatigue
myoglobin a respiratory pigment with a stronger affinity for oxygen than haemoglobin.
right atrium the upper right-hand chamber of the heart that receives deoxygenated blood from the body
right ventricle the lower chamber that receives deoxygenated blood from the right atrium and pumps it to the lungs
tricuspid valve (atrioventricular valve) the valve between the right atrium and the right ventricle that prevents backflow of blood from the ventricle to the atrium when the ventricle contracts
tendinous cords (valve tendons, heartstrings) cord-like tendons that make sure the valves are not turned inside out by the large pressure exerted when the ventricles contract
pulmonary arteries the blood vessels that carry deoxygenated blood from the heart to the lungs
pulmonary veins the blood vessels that carry oxygenated blood back from the lungs to the heart
left atrium the upper left-hand chamber of the heart that receives oxygenated blood from the lungs
left ventricle the chamber that receives oxygenated blood from the left atrium and pumps it around the body
bicuspid valve (atrioventricular valve) the valve between the left atrium and the left ventricle that prevents backflow of blood into the atrium when the ventricle contracts
aorta the main artery of the body; it leaves the left ventricle of the heart carrying oxygenated blood under high pressure
systole the contraction of the heart
atrial systole when the atria of the heart contract
ventricular systole when the ventricles of the heart contract
diastole when the heart relaxes and fills with blood
cardiac cycle the cycle of contraction (systole) and relaxation (diastole) in the heart

LEARNING OBJECTIVES

- Understand the course of events that lead to atherosclerosis.

CARDIOVASCULAR DISEASES

Problems with the cardiovascular system have serious consequences. Globally, almost 18 million people die from cardiovascular diseases each year. World Health Organization (WHO) data from 2017 show that **cardiovascular diseases** were responsible for 31% of all global deaths – it is the single biggest cause of death and disability (see **fig A**). What is more, around a third of these deaths were in people younger than 70.

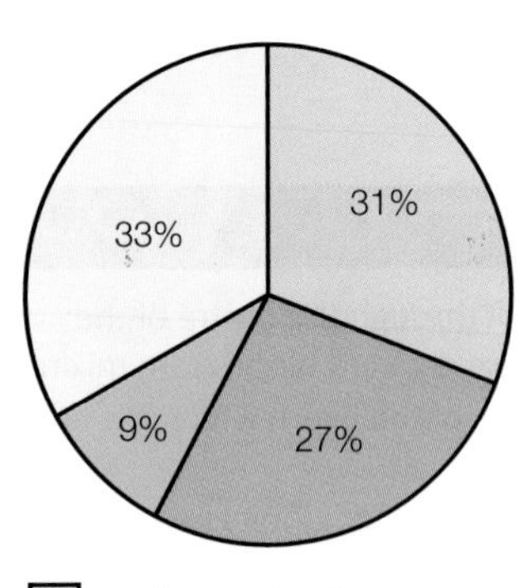

fig A This WHO data from 2017 shows cardiovascular disease is the biggest single cause of death around the world.

Many cardiovascular diseases are linked to a condition called **atherosclerosis**.

ATHEROSCLEROSIS

Atherosclerosis, a hardening of the arteries, is a disease in which **plaques** (made of a yellowish fatty substance) build up on the inside of arteries. It can begin in late childhood and continues throughout life. A plaque can continue to develop until it restricts the flow of blood through the artery or even blocks it completely. Plaques are most likely to form in the arteries of the heart (coronary arteries) and neck (carotid arteries). The typical development of a plaque is summarised in **fig B**.

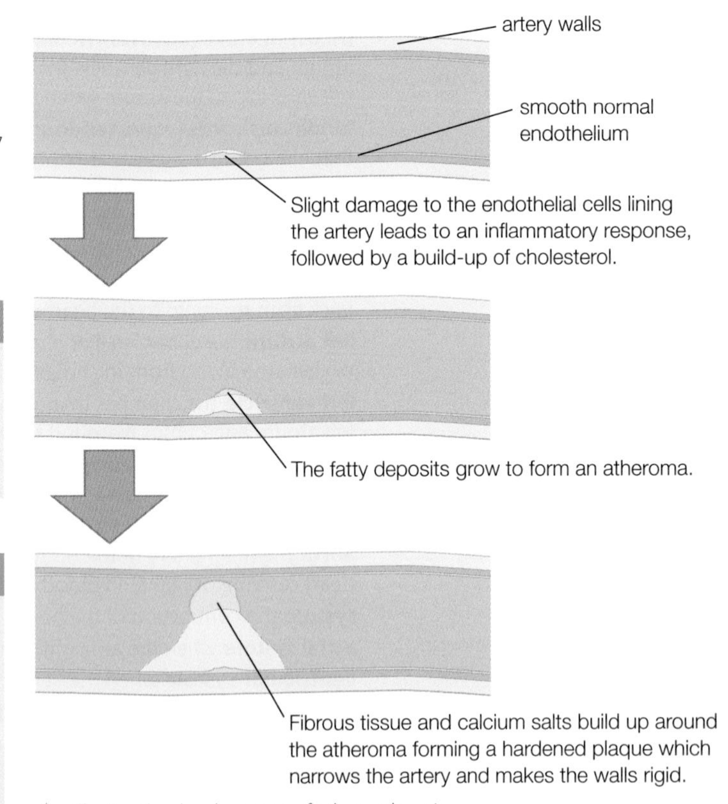

fig B The development of atherosclerosis

We can now look at this development in more detail. Atherosclerosis begins with damage to the endothelial lining of blood vessels. This damage can be caused by several factors, including high blood pressure and substances in tobacco smoke. Atherosclerosis usually occurs in arteries rather than in veins. This is because the blood in the arteries flows fast under relatively high pressure, which puts more strain on the endothelial lining of the vessels and can cause small areas of damage. In the veins, the pressure is lower so damage to the endothelium is much less likely.

LEARNING TIP

Remember that damage to the endothelium occurs first and this is often caused by high blood pressure, often as a result of smoking.

EXAM HINT

Learn the stages of the development of atherosclerosis: damage to the endothelium of the arteries → inflammatory response → accumulation of cholesterol → atheroma → fibrous tissue/calcium salts → plaque → narrowing/loss of elasticity of the artery.

Once damage to the endothelium has occurred, the body's inflammatory response begins and white blood cells arrive at the site of the damage. These cells accumulate chemicals from the blood, especially cholesterol. This leads to a plaque (also known as an **atheroma**) forming on the endothelial lining of the artery (see **fig C**). Fibrous tissue and calcium salts also build up (increase in amount) around the atheroma, turning it into a hardened plaque. This hardened area means that part of the artery wall is less elastic and narrower than it should be. This is atherosclerosis and is summarised in **fig B**.

The plaque causes the lumen of the artery to become much smaller. This increases the blood pressure, making it harder for the heart to pump blood around the body. The raised blood pressure makes damage more likely in other areas of the endothelial lining and more plaques will form. This will make the blood pressure even higher, and so the problem gets worse. There are many factors that are linked to the development of atherosclerosis. You will look at these in more detail in **Chapter 1C**.

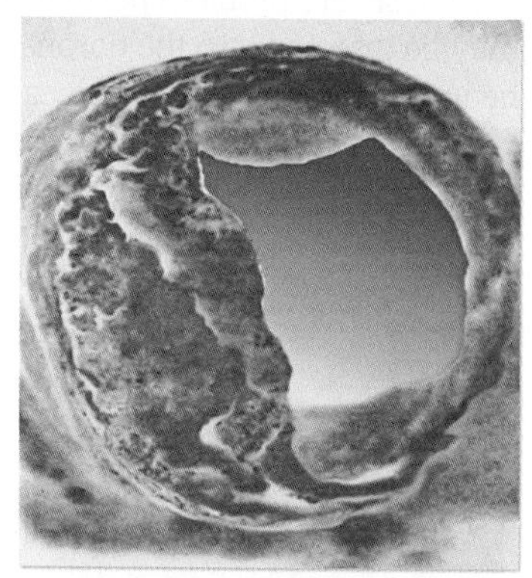

▲ **fig C** Fatty deposits like these in an artery cause disease and death in millions of people every year.

EFFECT OF ATHEROSCLEROSIS ON HEALTH

Atherosclerosis can have many serious effects on the health of an individual. The development of atherosclerosis can be summarised as: damage to the endothelium of the arteries → inflammatory response → accumulation of cholesterol → atheroma → fibrous tissue/calcium salts → plaque → narrowing/loss of elasticity of the artery.

ANEURYSMS

If an area of an artery is narrowed by plaque, blood tends to collect behind the blockage. The artery bulges and the wall is put under more pressure than usual, so it becomes weakened. This is known as an **aneurysm**. The weakened artery wall may split open, leading to massive internal bleeding. Aneurysms frequently happen in the blood vessels supplying the brain or in the aorta, especially when it passes through the abdomen. The massive blood loss and drop in blood pressure are often fatal, but if aneurysms are diagnosed they can be treated by surgery before they burst.

EXAM HINT

Make sure you are clear about all of the different terms. Atherosclerosis can cause aneurysms, angina and myocardial infarctions.

RAISED BLOOD PRESSURE

The arteries narrowed due to plaques on the walls cause raised blood pressure. This can lead to severe damage in a number of organs, including the kidneys, the eyes and the brain. The high pressure damages the tiny blood vessels where your kidney filters out urea and other substances from the blood. If the vessels feeding the kidney tubules become narrowed, the pressure inside them gets even higher and proteins may be forced out through their walls. If you have high blood pressure, your doctors can test for protein in your urine as a sign of kidney damage.

Similarly, the tiny blood vessels supplying the retina of your eye are easily damaged. If they become blocked or leak, the retinal cells are starved of oxygen and die and this can cause blindness.

Bleeding from the capillaries into the brain results in one type of stroke (see below).

HEART DISEASE

There are many kinds of heart disease, but the two most common ones are **angina** and **myocardial infarction (heart attack)**; both are closely linked to atherosclerosis (see **figs B** and **D**).

In angina, plaques build up slowly in the coronary arteries, reducing blood flow to the parts of the heart muscle beyond the plaques. Often symptoms are first noticed during exercise, when the cardiac muscle is working harder and needs more oxygen. The narrowed coronary arteries cannot supply enough oxygenated blood and the heart muscle resorts to **anaerobic respiration**. This causes a gripping pain in the chest that can extend into the arms, particularly the left one, and the jaw, and often also causes breathlessness. The symptoms of angina subside once exercise stops, but the experience is painful and frightening.

Fortunately, most angina is relatively mild. It can be helped by taking regular exercise, losing weight and not smoking. The symptoms can be treated by drugs that cause rapid dilation of the coronary blood vessels so that they supply the cardiac muscle with the oxygen it needs. However, if the blockage of the coronary arteries continues to get worse, so will the symptoms of the angina. Other drugs are then used to dilate the blood vessels and reduce the heart rate. Unfortunately, drugs cannot solve a severe problem permanently. A small tube called a **stent** may be inserted into the coronary arteries to hold them open, or heart bypass surgery may be carried out.

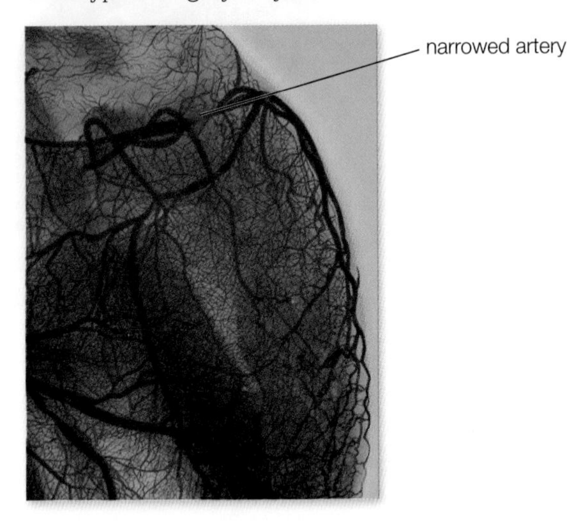

fig D Injecting the blood vessels with special dye allows doctors to see where the coronary arteries are narrowing due to atherosclerosis so they can treat the problem.

EXAM HINT

Be clear that angina and myocardial infarction are caused by reduced blood flow to the cardiac muscle - if you say 'to the heart' this suggests reduced flow in the veins carrying blood into the atria rather than arteries carrying oxygenated blood to the muscle itself.

In a myocardial infarction, often called a heart attack, one of the branches of the coronary artery becomes completely blocked and part of the heart muscle is permanently starved of oxygen.

Many heart attacks are caused by a blood clot resulting from atherosclerosis. As you have seen, the wall of an artery affected by a plaque is stiffened, making it much more likely to suffer cracks or damage. Platelets touch the damaged surface of the plaque and the clotting process is triggered (see **Section 1B.2**). The plaque itself may rupture and break open, and the cholesterol that is released will also cause the platelets to trigger the blood clotting process. A clot may also develop because the endothelial lining is damaged, for example by high blood pressure or smoking.

A clot that forms in a blood vessel is known as a **thrombosis**. The clot can rapidly block the whole blood vessel, particularly if it is already narrowed by a plaque. A clot that gets stuck in a coronary artery is known as a coronary thrombosis. The clot can block the artery, starving the heart muscle beyond that point of oxygen and nutrients, and this often leads to a heart attack (see **fig E**).

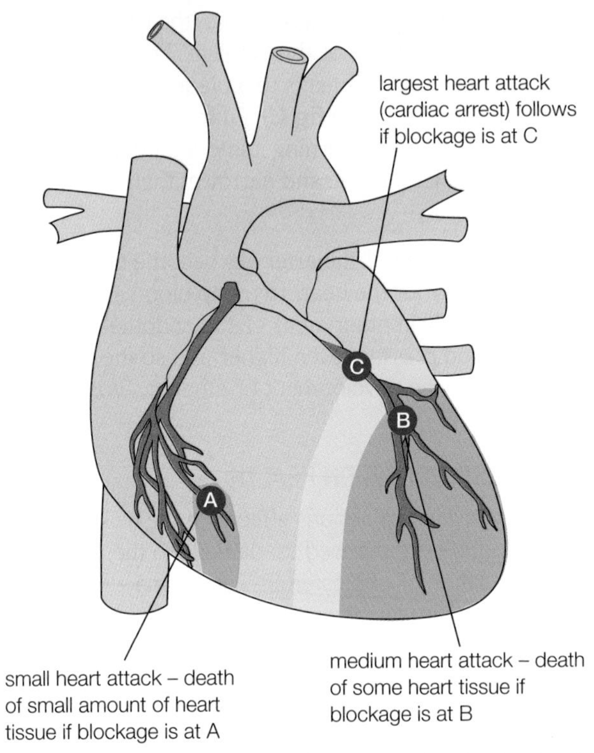

fig E The size and severity of a heart attack is closely related to the position of the blockage in the coronary artery.

During a heart attack, there is chest pain in the same areas as in an angina attack, but it is much more severe. The pain may occur at any time, although exercise may start it, and it often lasts for several hours. Death may occur very rapidly with no previous symptoms, or it may happen after several days of feeling tired and suffering symptoms mistaken for indigestion.

It is very important to react quickly if you suspect someone is having a heart attack. Give them two full-strength aspirin tablets to help stop the blood clotting, and get them to hospital as fast as you can.

STROKES

A **stroke** is caused by an interruption to the normal blood supply to an area of the brain. This may be due to bleeding from damaged capillaries or a blockage cutting off the blood supply to the brain. A blockage is usually caused by a blood clot, an atheroma or a combination of the two. Sometimes, the blood clot forms somewhere else in the body and is carried in the bloodstream until it gets stuck in an artery in the brain. The damage happens very quickly. A blockage in one of the main arteries leading to the brain causes a very serious stroke that may lead to death. In one of the smaller arterioles leading into the brain, the effects may be less disastrous.

The symptoms of strokes vary, depending on how much of the brain is affected. Very often, the blood is cut off from one part or one side of the brain only (see **fig F**). Symptoms include dizziness, confusion, slurred speech, blurred vision or partial loss of vision (usually one eye) and numbness. In more severe strokes, there can be paralysis, usually on one side of the body.

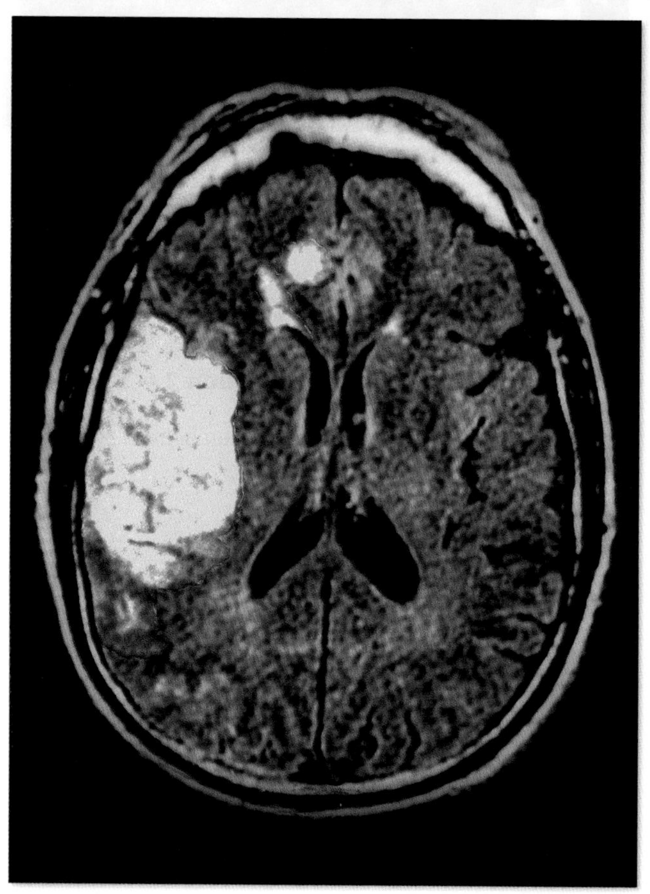

▲ **fig F** The damage caused in the brain by a major stroke resulting from a blood clot in the wrong place can be seen on the left of this MRI scan. The healthy part of the brain is shown in blue.

The outcome of either a heart attack or a stroke usually depends on how soon the person is treated. The sooner the patient is given treatment, including clot-busting drugs that break down or dissolve the blood clot, the more likely they are to survive. For example, if treatment is given rapidly, 75% of patients who survive the first week after a heart attack can expect to be alive five years later.

CHECKPOINT

1. When a plaque starts to form on the endothelium (lining) of an artery, it usually gets worse and worse. Unless the person changes their lifestyle or gets medical treatment, the lining of the artery does not return to normal. Explain why this happens, and why it is so dangerous.
2. (a) Describe in detail the role of atherosclerosis in cardiovascular disease.

   (b) Summarise the similarities and differences between a heart attack and a stroke.

SKILLS INNOVATION

3. The build-up of a fatty plaque in the artery leads to changes in the blood flow and an increase in the blood pressure. Plan a way of modelling this that could be used on a television programme to explain high blood pressure to young people.

SUBJECT VOCABULARY

cardiovascular diseases diseases of the heart and circulatory system, many of which are linked to atherosclerosis

atherosclerosis a condition in which yellow fatty deposits build up (increase in amount) on the lining of the arteries, causing them to be narrowed and resulting in many different health problems

plaques yellowish fatty deposits that form on the inside of arteries in atherosclerosis

atheroma another term for a plaque formed on the arterial lining

aneurysm a weakened, bulging area of artery wall that results from blood collecting behind a blockage caused by plaques

angina a condition in which plaques are deposited on the endothelium of the arteries and reduce the blood flow to the cardiac muscle through the coronary artery; it results in pain during exercise

myocardial infarction (heart attack) the events which take place when atherosclerosis leads to the formation of a clot that blocks the coronary artery entirely and deprives the heart muscle of oxygen, so it dies; it can stop the heart functioning

anaerobic respiration cellular respiration that takes place in the absence of oxygen

stent a metal or plastic mesh tube that is inserted into an artery affected by atherosclerosis to hold it open and allow blood to pass through freely

thrombosis a clot that forms in a blood vessel

stroke an event caused by an interruption to the normal blood supply to an area of the brain which may be due to bleeding from damaged capillaries or a blockage cutting off the blood supply to the brain, usually caused by a blood clot

1B EXAM PRACTICE

1 (a) Which statement is incorrect?

A All arteries carry oxygenated blood.

B All veins carry blood at low pressure.

C All veins carry blood towards the heart.

D All arteries carry blood away from the heart. [1]

(b) Draw a labelled diagram to show the structure of an artery wall. [3]

(c) Explain how the structure of an artery wall relates to its function. [2]

(d) Give **two** differences between the structure of a vein and the structure of a capillary. [2]

(Total for Question 1 = 8 marks)

2 Many animals have hearts that pump blood through a network of blood vessels.

(a) What is the correct term for the circulation of a mammal?

A closed single circulation

B open single circulation

C closed double circulation

D open double circulation [1]

(b) The table below refers to blood flow in the four major blood vessels of the human heart. If the statement is correct, place a tick (✓) in the appropriate box and if the statement is incorrect, place a cross (✗) in the appropriate box.

Name of blood vessel	Carries blood away from the heart	Carries oxygenated blood
aorta		
vena cava		
pulmonary artery		
pulmonary vein		

[4]

(c) The diagram below shows a section through the heart of a mammal.

Name the parts labelled A and B. [2]

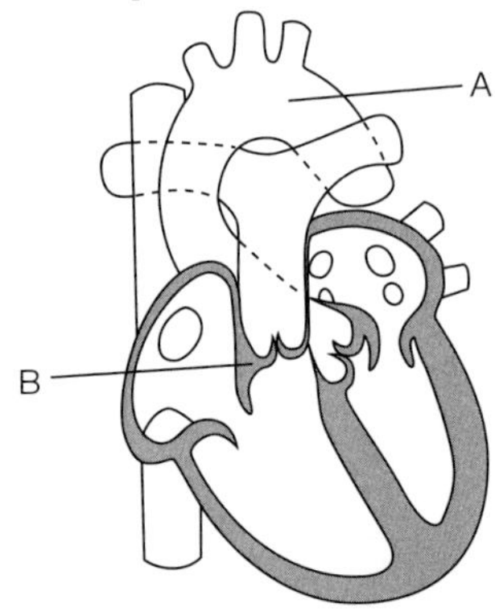

(d) Heart muscle has a high demand for oxygen. Describe how heart muscle is supplied with oxygen. [3]

(e) The volume of blood pumped by the ventricles is $0.07\,dm^3$. Calculate the cardiac output when the heart rate is 72 bpm. [2]

(Total for Question 2 = 12 marks)

3 (a) The diagram below shows a mammalian heart during atrial systole.

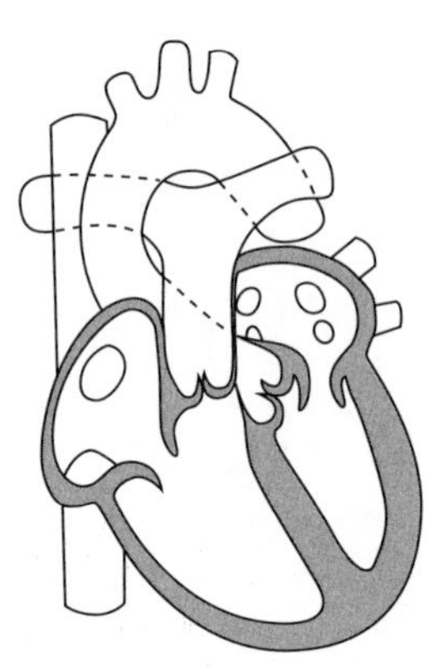

What evidence can be seen in the diagram to confirm it is in atrial systole?

A Both the semilunar valves and the atrioventricular valves are open.

B Both the semilunar valves and the atrioventricular valves are closed.

C The semilunar valves are open and the atrioventricular valves are closed.

D The semilunar valves are closed and the atrioventricular valves are open. [1]

(b) Humans and fish are both animals that have a heart and a network of blood vessels. However, there are some differences in their circulatory systems. The diagrams below illustrate a human circulatory system and a fish circulatory system.

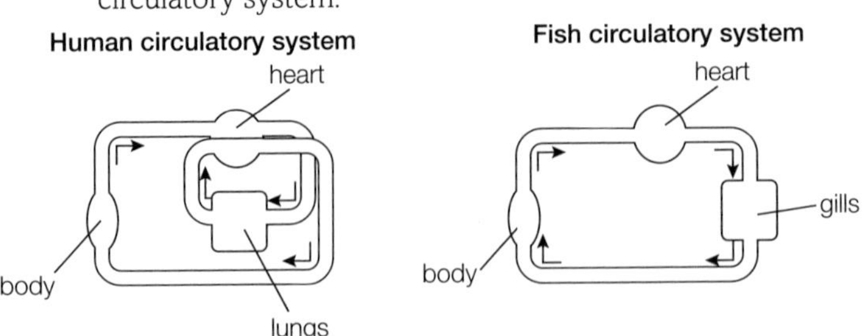

The arrows show the direction of blood flow.

(i) Describe the circulation of blood in a fish using the information in the diagram. [3]

(ii) Using the information in both diagrams, evaluate the advantages that the human circulatory system has compared with that of the fish. [2]

(c) Describe the cardiac cycle. [5]

(Total for Question 3 = 11 marks)

4 (a) The graph below shows oxygen dissociation curves for human myoglobin and human haemoglobin.

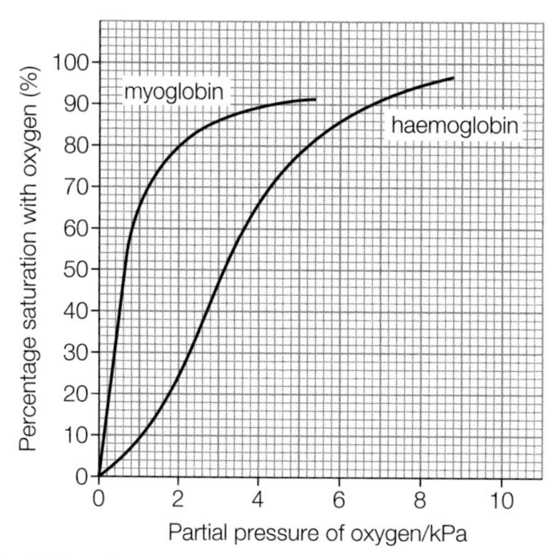

(i) Using the graph, state the partial pressures of oxygen at which myoglobin and haemoglobin are 50% saturated with oxygen. [2]

(ii) Calculate the increase in percentage saturation for haemoglobin between 2 kPa of oxygen and 8 kPa of oxygen. Show your working. [2]

(b) The muscle of diving mammals such as elephant seals contains a lot of myoglobin. Use the information in the graph to explain how the myoglobin can help an elephant seal to dive for longer. [3]

(c) At increased partial pressures of carbon dioxide, the oxygen dissociation curve for haemoglobin moves to the right. This is known as the Bohr effect.

Explain the importance of the Bohr effect. [4]

(Total for Question 4 = 11 marks)

5 (a) During a dissection of a mammalian heart a student measured the thickness of the ventricular walls. The left ventricle wall was 32 mm thick and the right wall was 11 mm thick.

(i) Calculate the thickness of the right ventricle wall as a percentage of the left ventricle wall thickness. Show your working and state your answer to three significant figures. [2]

(ii) Explain the difference in the thickness of the ventricle walls. [2]

(b) Describe and explain the function of the semilunar valves. [3]

(c) Explain the function of the tendinous cords in the atrioventricular valves. [2]

(Total for Question 5 = 9 marks)

6 (a) Platelets are fragments of cells that are involved in blood clotting. Describe how blood clots. [4]

(b) Describe two conditions that are caused by blood clots that form inside the blood vessels. [4]

(c) State the meaning of the term *aneurysm* and explain how an aneurysm is caused. [3]

(Total for Question 6 = 11 marks)

7 (a) The diagram below shows how most carbon dioxide is transported.

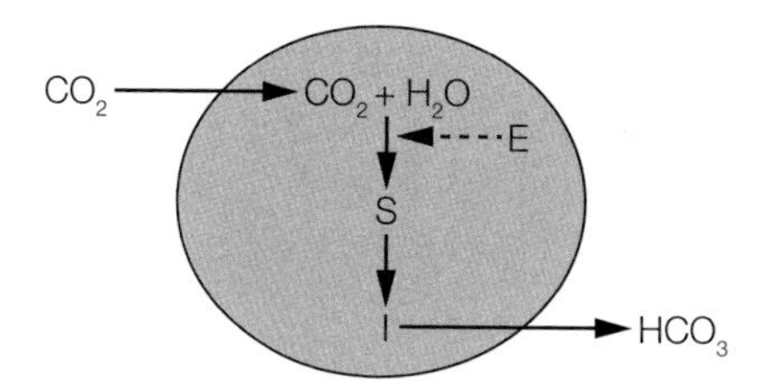

(i) What is the name of the cell represented by the circle. [1]

A leucocyte
B neutrophil
C erythrocyte
D stem cell

(ii) What is the name of enzyme E. [1]

A catalase
B carbonase
C anhydrous carbonase
D carbonic anhydrase

(iii) Name substance S. [1]

(iv) Identify ion I. [1]

(b) If ion I is allowed to accumulate inside the cell it could affect the functioning of the cell.

(i) Describe how the accumulation of ion I is prevented. [2]

(ii) Explain the effect that accumulation of ion I has on the transport of oxygen. [2]

(Total for Question 7 = 8 marks)

TOPIC 1 MOLECULES, TRANSPORT AND HEALTH

CHAPTER 1C CARDIOVASCULAR HEALTH AND RISK

What do you do with your spare time? Do you like to watch TV, read, use the internet or spend time on social media? According to a recent poll of professionals in the Middle East and North Africa, a third of people spend more than five hours a day using the internet for leisure. One fifth of people spend at least one hour a day relaxing online. Compare that to only 4.6% who prefer to play sport and only 2.2% who prefer outdoor pursuits. Over 35% of people eat out nearly every day. People enjoy an affluent lifestyle – eating well, using a car instead of walking, relaxing at home or spending time with family and friends. Unfortunately, many people do not realise that these are also risk factors that increase the probability that they will develop a serious disease of the heart or blood vessels.

In this topic, you will learn about the meaning of risk and how we perceive risk. Cardiovascular diseases are those of the heart and blood vessels. You will learn how we can determine what causes cardiovascular diseases and what increases the chances of us developing such diseases. Diet is an important factor and the things we choose to eat can increase the risk of developing these life-threatening diseases. Fortunately, making the right choices can also significantly decrease the risks. You will learn how scientists and health professionals can use evidence to analyse the risks and how these diseases can be treated.

MATHS SKILLS FOR THIS CHAPTER

- **Recognise and make use of appropriate units in calculations** (*e.g. calculating a BMI in* $kg\,m^{-2}$)
- **Recognise and use expressions in decimal and standard form** (*e.g. calculating a person's BMI*)
- **Use ratios, fractions and percentages** (*e.g. calculating proportions of the population with certain risk factors*)
- **Construct and interpret frequency tables and diagrams, bar charts and histograms** (*e.g. interpreting data about incidence of disease or the effect of reducing risk factors*)
- **Understand simple probability** (*e.g. considering the chances of developing a particular cardiovascular disease*)
- **Use a scatter diagram to identify a correlation between two variables** (*e.g. comparing risk of developing a disease with risk factors such as diet or blood pressure*)
- **Translate information between graphical, numerical and algebraic forms** (*e.g. draw graphs from data tables or select data from graphs about various risk factors*)
- **Calculate rate of change from a graph showing a linear relationship** (*e.g. calculate how quickly the proportion of people who are obese has increased over the last 50 years*)

What will I study in this chapter?

- The meaning of risk, correlation and cause
- The causes of cardiovascular disease
- The risk factors that contribute to developing cardiovascular disease
- The effect of diet on cardiovascular health
- How scientists and health professionals use data as evidence about risk factors
- How we can reduce the risk of developing cardiovascular disease
- How cardiovascular diseases can be treated

What prior knowledge do I need?

Chapter 1B

- Components of a healthy diet
- Structure of the circulatory system
- Structure of blood vessels

What will I study later?

Chapter 2C

- Genetic screening for risk factors

Chapter 3C

- Potential for using stem cells in treatment

Topic 6 (Book 2: IAL)

- The immune response

Topic 7 (Book 2: IAL)

- Control of the cardiovascular system

1 RISK, CORRELATION AND CAUSE

SPECIFICATION REFERENCE 1.15 1.17

LEARNING OBJECTIVES

- Understand why people's perception of risks is often different from the actual risks, including underestimating and overestimating the risks due to diet and other lifestyle factors in the development of heart disease.
- Be able to distinguish between correlation and causation.

Every country has diseases which affect its people and may even kill them. Some of these diseases affect you randomly – there is nothing you can do to change whether you are affected or not. However, for many diseases, especially **non-communicable** (non-infectious) **conditions** such as heart disease and cancer, you can increase or lower your **risk** of becoming ill, based on factors in your lifestyle. If you understand the risk factors, you can help to make yourself and your family healthier.

WHAT IS RISK?

The word *risk* is used regularly in everyday conversation, but in science it has a very specific meaning. In science, risk describes the **probability** that an event will happen. Probability means the chance or likelihood of the event, calculated mathematically. For example, imagine you have six coloured balls – red, blue, green, yellow, orange and purple – in a black cloth bag (see **fig A**). If you reach in and pull out a single ball, the probability (risk) of getting, say, a green ball can be expressed in one of three ways:

- 1 in 6
- 0.166 66 recurring (0.17)
- 17%.

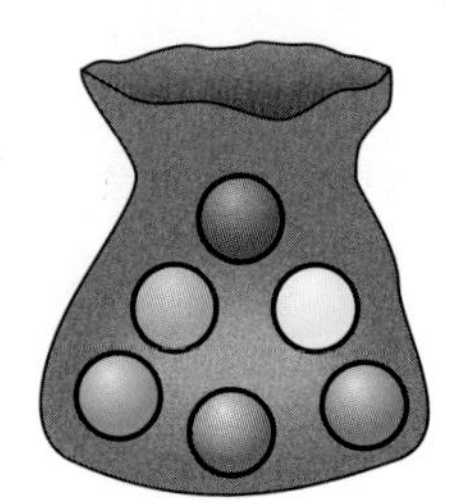

fig A The risk, chance or probability that you will pick a blue ball out of the black bag is 1 in 6. If you return the ball to the bag each time, you will have exactly the same probability of picking a blue ball again the next time. The probability will always be 1 in 6.

This is the case for any one of the six colours in the bag. In the same way, it is possible to work out your risk of developing certain specified diseases or of dying from a specified cause.

HOW DO WE PERCEIVE RISK?

The actual risk of doing something is not always the same as the sense of risk one feels. Most people don't think twice before getting into their car – but globally you have an annual risk of 1 in 5747 of being killed in a road traffic accident. On the other hand, many people get very worried before flying, but commercial flights have a 1 in 4.5–5.5 million risk of crashing. Personal perception of risk is based on a variety of factors which include:

- how familiar you are with the activity
- how much you enjoy the activity
- whether or not you approve of the activity.

The actual mathematical risk may play very little part in developing your personal perception of risk. People often overestimate the benefits, or minimise the risk, of behaviour that they want to continue. For example, there is now strong evidence from around the world that obesity is linked to a range of diseases such as diabetes, cardiovascular disease and some cancers. However, people like eating and so they still become overweight. On the other hand, they will over-emphasise the risks of activities if they want to avoid them or prevent others from doing them. For example, parents over-emphasise the risk of wandering away to small children, to help make the child behave and stay close.

In another example, there is good scientific evidence that smoking affects our risk of developing diseases such as atherosclerosis, as well as lung cancer. However, knowledge of the mathematical risk of an early death if you smoke cigarettes doesn't always stop people from smoking.

EPIDEMIOLOGY

If you know the number of people in a population who are affected by a disease, it is possible to calculate the average risk of developing that disease for a person within that population. But the risk is higher for some people than others, depending on their lifestyle and which genes they inherit.

It is possible to identify the **risk factors** that may contribute to the cause of a disease. You can look at people who have the same factors (e.g. smoking) and compare their risk of the disease with the average risk for the whole population. Using these techniques, it appears that there are a number of factors that increase the likelihood that a person will develop atherosclerosis. If many factors influence your chance of having a disease, it is called a **multifactorial disease**. The study of the patterns of diseases and their causes is called **epidemiology**.

When two different sets of data change together, there may be a link, which is called a **correlation**. For example, mortality data from a disease such as atherosclerosis may change in a similar pattern to a lifestyle factor such as smoking or lack of exercise. However, this does not prove that one is the cause of the other. They could both be caused by something else which would explain why they change in the same way. Correlation is not the same as **causation** – further research is always needed to demonstrate a causal link.

For example, **fig B** shows the percentage of deaths in the UAE from cardiovascular disease each year. It also shows the statistics for obesity and diabetes. This data suggests a possible correlation between obesity and heart disease.

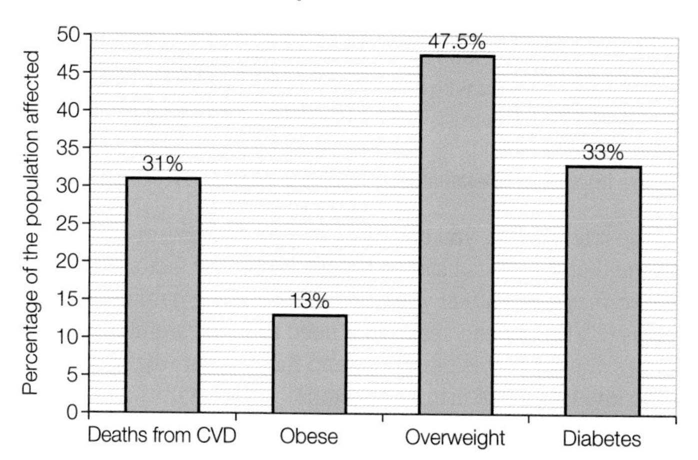

▲ **fig B** This data shows a possible correlation between obesity, diabetes and death from cardiovascular disease (CVD). It needs more data from other sources to show that obesity can cause diabetes and CVD.

EXAM HINT

When provided with data in the form of a graph, make sure you read the axis titles carefully so that you know what the data refers to.

You cannot base conclusions on a single set of data. However, these UAE data agree with findings from countries around the world that suggest obesity is closely linked to the development of diabetes and cardiovascular diseases. Dr Raghib Ali, principle investigator in the UAE Healthy Future Study, has set up a long-term study in Abu Dhabi. The team is aiming to collect the data needed to confirm the pattern of increased disease risk in the UAE – and so to find ways of reducing the risk of diabetes and cardiovascular diseases.

CHECKPOINT

1. Explain the difference between risk, correlation and causation.
2. In most areas of the world, the risk of dying from heart disease is about three times greater for smokers than for non-smokers. Explain why this does not mean that an individual person who smokes will die from heart disease.

SKILLS PERSONAL AND SOCIAL RESPONSIBILITY

3. ▶ The risks of developing diabetes and heart disease linked to obesity are well known. Suggest reasons why people stay obese, and why more people are becoming obese.
4. In the Maldives, 39% of deaths are the result of cardiovascular disease. 25% of the population smoke and around 12.9% are obese. Draw a bar chart to show this information. How does it compare with the information in **fig B** on the UAE? Suggest any other information you might need to make accurate comparisons between the causes of disease in the two countries.

SUBJECT VOCABULARY

non-communicable conditions diseases which are not caused by pathogens and cannot be spread from one person to another
risk the probability that an event will take place
probability a measure of the chance or likelihood that an event will take place
risk factors factors which affect the risk of an event happening
multifactorial disease a disease which results from the interactions of many different factors - not from one simple cause
epidemiology the study of patterns of health and disease, to identify causes of different conditions and patterns of infection
correlation a strong tendency for two sets of data to change together
causation when a factor directly causes a specific effect

2 INVESTIGATING THE CAUSES OF CVDs

LEARNING OBJECTIVES

- Be able to evaluate the design of studies used to determine health risk factors, including sample selection and sample size used to collect data that are both valid and reliable.
- Know how factors such as genetics, age and gender increase the risk of cardiovascular diseases (CVDs).

Everywhere you look, on television, in newspapers and on the internet, there are reports of factors which affect your health. Eat fruit and vegetables, drink orange juice, take lots of exercise, enjoy these foods – how do we know which advice is based on good science, and which is given to us because someone wants to sell us something? There are many ways in which you can evaluate the design of studies to decide if the data are meaningful.

DESIGNING STUDIES

Most epidemiological studies are based on a very big sample size – usually, the bigger the study, the more meaningful the results.

The ideal is to investigate one factor or variable, keeping all other variables the same (controlled). However, controlling variables is almost impossible when you are working with human beings. The way people live is complex and varies a lot, so it is hard to detect how any one factor affects people. When a larger number of people are studied, it is more likely that patterns may emerge, even with all the other differences between the people involved. Evidence based on large amounts of data is more likely to be statistically significant than evidence based on small studies.

Some epidemiological studies are carried out over a long time. These **longitudinal studies** are very valuable because they follow the same group of individuals over many years (see **fig A**). This means the impact of their known lifestyle on their health can be tracked over time. For example, the Münster Heart Study looked at cardiovascular disease in 10 856 men aged 36–65 in Europe, following them from the start of the study well into the 21st century. The results from this study are still seen as important because so many people were involved over a long period of time. The Framingham Study in the US also provided much data – but was limited because they were all from similar American citizens. The study started in 1948 and it is still going on – the scientists have widened the population they gather data from, so it is more relevant now.

An ambitious new study called the National Children's Study has been set up in the US to follow 100 000 children from birth until they are 21 years old. From 2008–2012, children were selected to be representative of the whole of the US population. One major objective of the study is to examine how environmental inputs and genetic factors interact to affect the health and development of children. This is believed to be the biggest longitudinal study ever set up. Similarly, the UAE Healthy Future Study will be longitudinal, looking at the same group of volunteers over a number of years.

Sometimes, scientists look at all the available studies in a subject area and analyse the available data in a massive literature study. This combines small and large studies and can give more reliable evidence than any one of the studies alone. This is called a **metadata analysis (meta-analysis)**.

EVALUATING SCIENTIFIC STUDIES

When considering a study, you need to examine the methodology to see if it is **valid**. That means that it is properly designed to answer the question or questions being asked. You also need to see if the measurements have been carried out with **precision**. It is important to find out if other scientists have been able to repeat the methodology and have had similar results – if so, the results are considered more **reliable**.

It is also important to know who carried out the research, who funded it and where it was published. Then to decide whether or not any of these factors might have affected or **biased** the study. You need to **evaluate** the data and conclusions from the study in the light of all these factors.

EXAM HINT

Remember the meaning of the following terms.
Valid: answers the question the scientists are asking
Precise: measurements with little difference between them
Reliable: the investigation is repeatable by other scientists who get similar results
They are important in evaluating all practical work and research.

In the next few pages, you are going to look at some of the evidence that scientists have collected suggesting factors that may – or may not – affect your risk of developing cardiovascular diseases (CVDs). In each case, you need to look carefully at the type of evidence that is presented and think about what else you need to know to make firm conclusions.

RISK FACTORS FOR CVDs

The results from many epidemiological studies have identified a range of risk factors linked to CVDs. These factors divide into two main groups – those you can't change and those you can do something about (see **Section 1C.3**).

NON-MODIFIABLE RISK FACTORS FOR ATHEROSCLEROSIS

There are three main risk factors for CVDs which cannot (at the present time) be changed.

- **Genes**: studies show that there is a genetic tendency (trend) in some families, and also in some ethnic groups, to develop CVDs. These trends can include
  - arteries which are easily damaged

- a tendency to develop hypertension which can cause arterial damage and make CVDs more likely
- problems with the cholesterol balance of the body.

- **Age**: as you get older, your blood vessels begin to lose their elasticity and to narrow slightly. This can make you more likely to suffer from CVDs, particularly heart disease.
- **Gender**: statistically, under the age of 50, men are more likely to suffer from heart disease (and other CVDs) than women. The female hormone oestrogen, which is an important factor in the woman's menstrual cycle, appears to reduce the build-up of plaque. This gives women some protection against CVDs until they go through the menopause when oestrogen levels fall.

LEARNING TIP

Remember that risk factors do not cause the disease, they contribute to the chances of developing the disease.

LOOKING AT THE DATA

Identical twin studies are an excellent resource when investigating whether there is a genetic factor at work, because identical twins have exactly the same genes. Any differences should therefore be due to the environment in which they live. A major twin study was conducted in Sweden and was based on over 21 000 pairs of twins, both identical and non-identical. This study showed that (for male twins) if one twin died of heart disease between the ages of 36 and 55, then the risk of the other twin also dying of heart disease was eight times higher than if neither was affected (see **fig A**). However, as the twins got older, one dying of heart disease had less of a correlation with the other twin also dying of heart disease. In other words, there appears to be a clear genetic link to heart disease in younger men, but it gets less in much older men.

Epidemiological studies have also identified several lifestyle factors linked to CVDs, some of which you will look at in the following pages. These lifestyle factors are important for health because they are the factors that we can change.

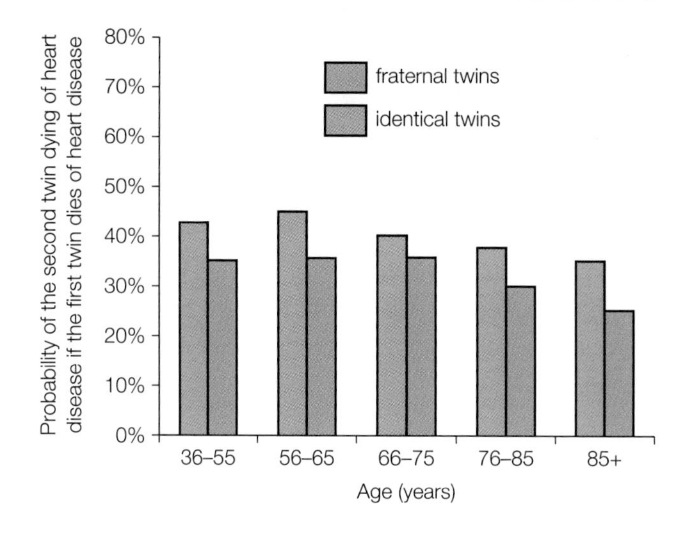

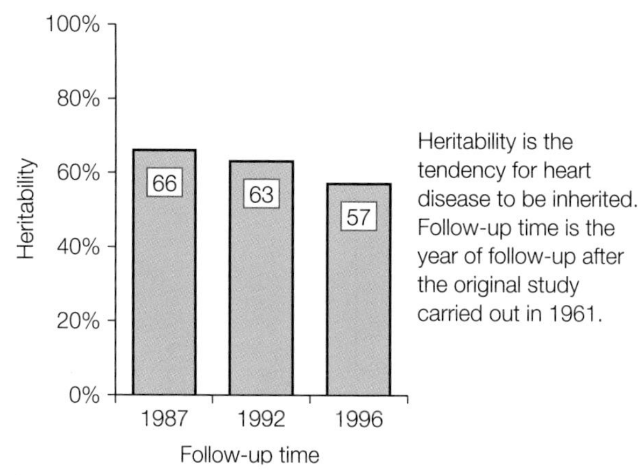

▲ **fig A** These are results from an epidemiological study of male twins in Sweden published in 1994. Although this study was carried out a long time ago, the findings are still important, because of the large number of twin pairs who took part and the length of the study.

CHECKPOINT

SKILLS EXECUTIVE FUNCTION

1. When scientists design a major study, what can they do to try and make sure their results will be both valid and reliable?
2. Using the data in **fig A**, answer the following questions.
   (a) How do the figures for identical and fraternal twins differ in the top graph? What does this suggest about a genetic link to heart disease?
   (b) What does the bottom graph show you about the apparent heritability of heart disease in men? What might affect the fall in apparent heritability as the men got older?

SUBJECT VOCABULARY

longitudinal studies scientific studies which follow the same group of individuals for many years
metadata analysis (meta-analysis) when data from all the available studies in a particular area are analysed
valid an investigation which is well designed to answer the question being asked
precision measurements with only slight variation between them
reliable evidence which can be repeated by several different scientists
biased when someone is unfairly for or against an idea (e.g. when a scientist is paid by someone with a vested interest in a specific result - they may receive benefit from the outcome)
evaluate to assess or judge the quality of a study and the significance of the results

1C 3 RISK FACTORS FOR CARDIOVASCULAR DISEASE

SPECIFICATION REFERENCE 1.12 1.16

LEARNING OBJECTIVES

- Be able to evaluate the design of studies used to determine health risk factors, including sample selection and sample size used to collect data that are both valid and reliable.
- Know how factors such as diet, high blood pressure, smoking and inactivity increase the risk of cardiovascular disease (CVD).

The non-modifiable factors affecting your risk of developing CVDs – age, genetics and gender – are the same all over the world. However, the numbers of people who die of CVDs varies enormously, depending on where you live, as you can see in **fig A**. This tells us that other factors are involved – factors which vary with your lifestyle. In the rest of this topic, you will find out more about the lifestyle factors which affect us and influence our risk of developing – or dying from – heart disease.

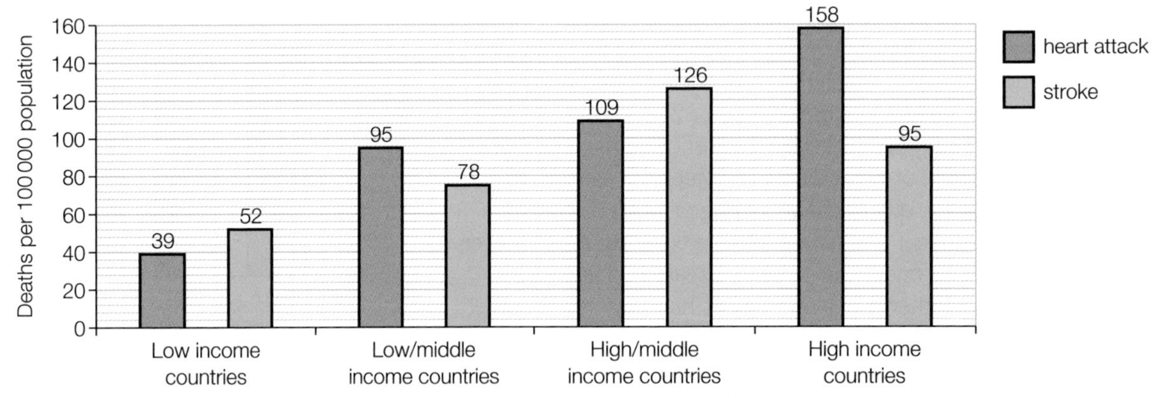

▲ **fig A** Deaths from CVDs in different countries (based on 2012 data)

EXAM HINT

When you are given a graph, try to analyse it briefly before looking at the question. Find trends and obvious comparison points. This will help you to understand what the question is asking for.

MODIFIABLE (LIFESTYLE) RISK FACTORS FOR CVDs

The development of atherosclerosis is linked to many types of CVD. Your lifestyle can affect your risk of developing atherosclerosis in the future. Epidemiological studies have shown links with smoking, diet and weight, lack of activity and high blood pressure. These are the factors we can change, so we can change our risk of developing CVDs by the lifestyle choices we make.

SMOKING AND ATHEROSCLEROSIS

Studies have shown that smokers are far more likely to develop atherosclerosis than non-smokers with a similar lifestyle. Nine out of ten people who need heart bypass surgery or stents as a result of atherosclerosis are smokers. In 2007, a Spanish study showed a clear correlation between smoking and the incidence of death from atherosclerotic heart disease. Causation was established by further research. For example, studies found that the substances in tobacco smoke:

- can damage the artery linings, which makes the build-up of plaques more likely
- can cause the arteries to narrow, raise the blood pressure and increase the risk of atherosclerosis.

Similar findings were made in a study on adults with heart disease in Jordan in 2017 (see **fig B**).

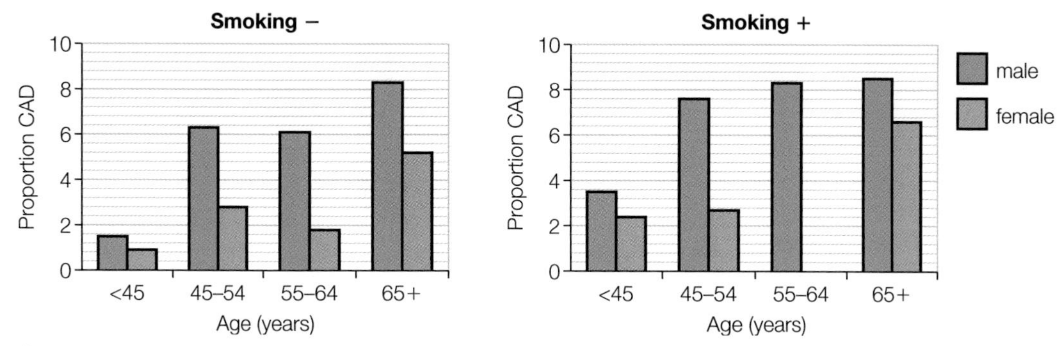

fig B Proportion of adults with coronary artery disease (CAD) depending on whether they smoke or not based on a Jordanian study in 2017.

INACTIVITY AND CVDs

Regular exercise helps lower blood pressure, prevent obesity and diabetes, lower blood cholesterol levels, balance lipoproteins and reduce stress. These also lower your risk of developing atherosclerosis and CVDs. A study of 10 269 male Harvard University (USA) graduates aged between 45 and 84 showed that the men who changed from being inactive to taking regular exercise had a 23% lower mortality over the life of the study than those who did not exercise. Moreover, the main cause of the deaths was atherosclerosis and the linked CVDs. A study of 72 488 female nurses showed the same benefits for women – the more active women had a significantly lower risk of developing atherosclerosis and other CVDs.

A 2013 study from Sri Lanka showed a similar pattern – high levels of inactivity were linked to an increased risk of obesity, diabetes, **high blood pressure** and CVDs. High levels of activity were linked to a reduced risk of all these conditions. Several studies have shown that exercise both reduces the formation of plaques in the arteries and also keeps plaques that are present more stable and less likely to break.

HIGH BLOOD PRESSURE AND ATHEROSCLEROSIS

As you saw in **Section 1B.4**, the heart pumps blood out into your arteries in a regular rhythm. The blood travels through your arterial system at pressures which change as your heart beats and which are easily measured (see **fig C**). At systole, when the blood is forced out of the heart, a healthy blood pressure is around 120 mmHg. When the heart is relaxed and filling during diastole, a healthy blood pressure is around 80 mmHg. Measuring blood pressure is used as an indicator of the health of both your heart and your blood vessels. Your blood pressure goes up and down naturally during the day – but it shouldn't be constantly raised. If your blood pressure is regularly above 140/90 mmHg, you have high blood pressure or **hypertension**. Raised blood pressure can be a sign of atherosclerosis. The blood pressure goes up when the walls of the arteries become less flexible due to the build-up of plaque, and when the lumen of the arteries get narrower as they are blocked by the plaques. This means that raised blood pressure can be the result of atherosclerosis and can be used to help diagnose the disease.

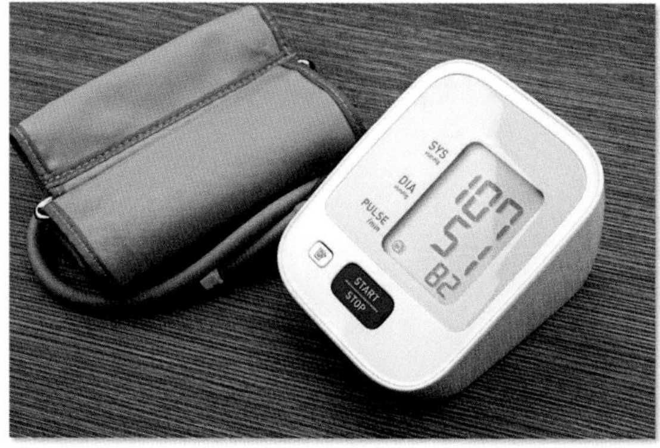

fig C Doctors can use a blood pressure monitor like this to check that your blood pressure is in the healthy range. Some people buy and use their own monitor, so they can check their blood pressure regularly to help prevent them getting CVDs.

However, other factors can also raise the blood pressure. For example, smoking narrows the blood vessels and raises the blood pressure. Obesity, inactivity, a high level of salt in the diet and stress can also narrow the arteries or affect the way the heart is pumping and raise the blood pressure. When the blood pressure is constantly high, the lining of the arteries is more likely to be damaged, leading to atherosclerosis and ultimately CVDs. So high blood pressure can also contribute to CVDs. If a doctor discovers you have high blood pressure, they will try to help you reduce the level by making lifestyle changes or with medication, to try to reduce your risk of developing CVDs.

DIET, OBESITY AND ATHEROSCLEROSIS

An increasing number of studies suggest that being overweight does not directly affect your risk of developing CVDs, but it is a very important indicator of risk. Most scientists think that the best predictors of future CVDs are:

- where fat is stored on your body
- how much exercise you do
- the levels of different fats in your blood.

Two other factors which are often a direct result of being overweight do increase the risk of atherosclerosis and CVDs. These are:

- high blood pressure – increases the risk of damage to blood vessel linings, and so of plaque formation
- type 2 diabetes – this can result in damage to the lining of the blood vessels which increases the risk of plaque formation.

There have been many studies on how diet is linked to atherosclerosis and CVDs, some looking at general diet, some looking at the role of diet in becoming overweight or obese, and some looking at specific foods. The evidence is very mixed and very difficult to interpret. You can find out more about this in **Section 1C.4**.

EXAM HINT

Remember that atherosclerosis is a multifactorial disease - there is no single cause but there are many factors that contribute to the chances of it occurring.

Smoking is one important factor that increases the chances of atherosclerosis. Inactivity and diet are other factors.

LINKS BETWEEN FACTORS

Many epidemiological studies are starting to find that an increased risk of developing a disease is often due to a combination of factors. For example, it is known that smoking increases your risk of atherosclerosis because of its effect on your blood vessels and blood pressure. Evidence now suggests that smoking also changes the balance of lipoproteins in your blood in a way which raises your risk of dying from atherosclerosis-related CVDs. You will find out more about lipoproteins and their effect on cardiovascular risk in **Section 1C.4**.

PREVENTING ATHEROSCLEROSIS AND CVDs

The advice about what is 'good' for us and what is 'bad' changes. This happens because epidemiological studies of links between risk factors and CVDs become more sophisticated and scientific research discovers more reasons why some factors can contribute to atherosclerosis. Current evidence suggests that eating a balanced diet with a variety of fats and plenty of fruit and vegetables helps prevent atherosclerosis. It helps not to smoke, to maintain a healthy weight to avoid high blood pressure and type 2 diabetes, to reduce constant stress and get plenty of exercise. It is important to take action as early as possible because there is clear evidence of the early signs of atherosclerosis in teenagers and even young children, if known risk factors are already in place.

CHECKPOINT

1. (a) What is the difference between modifiable and non-modifiable risk factors for CVDs?

(b) Give **two** non-modifiable and **two** modifiable risk factors for atherosclerosis.

(c) For each factor you have chosen, explain how it increases the risk of developing atherosclerosis and CVDs.

2. (a) Take the data from **fig A** and combine the information on deaths from heart attacks and strokes for each group of countries. Record your answers in a table. This will give you the total number of deaths from CVDs in each.

(b) Using your answer to part (a), draw a bar chart of the total numbers of deaths from CVDs in the different groups of countries in 2012.

SKILLS PERSONAL AND SOCIAL RESPONSIBILITY

3. The numbers of deaths from CVDs in poorer countries are much lower than in wealthier countries. Using what you know about the risk factors for CVDs, suggest at least **three** reasons for these differences.

SUBJECT VOCABULARY

high blood pressure blood pressure that is regularly more than 140/90 mmHg; this increases your risk of developing CVDs

hypertension high blood pressure, regularly measuring over 140/90 mmHg, which increases your risk of developing CVDs

4 DIET AND CARDIOVASCULAR HEALTH

SPECIFICATION REFERENCE 1.12 1.18(i) 1.18(ii)

LEARNING OBJECTIVES

- Know how factors such as diet increase the risk of cardiovascular disease.
- Be able to analyse data on the possible significance for health of blood cholesterol levels and levels of high-density lipoproteins (HDLs) and low-density lipoproteins (LDLs).
- Know the evidence for a causal relationship between blood cholesterol levels (total cholesterol and LDL cholesterol) and cardiovascular disease.
- Understand how people use obesity indicators such as BMI and waist-to-hip ratios.

There is strong evidence from around the world that the food we eat has a big effect on our health in many different ways. It certainly has a big impact on the health of our cardiovascular system. However, our understanding of what the effect is and how our food affects our risk of developing cardiovascular diseases keeps changing as scientists learn more.

WEIGHT ISSUES

There is plenty of food in the developed world and people can easily eat more than they need to supply the metabolic needs of the body. This means that many people have a positive energy balance. The excess food energy is converted into a store of fat so these people become overweight and then obese. All the evidence suggests that being obese increases your risk of developing many different diseases, including CVDs.

MEASURING A HEALTHY WEIGHT: THE BODY MASS INDEX

What do we mean by 'overweight'? It isn't just how much you weigh. Doctors and scientists look at your **body mass index (BMI)** to decide if you are unhealthily heavy (see **fig A**). This compares your weight to your height in a simple formula:

$$\text{BMI} = \frac{\text{weight in kilograms}}{(\text{height in metres})^2}$$

For an adult, the following definitions apply:

- a BMI of less than 18.5 $kg\,m^{-2}$ means you are underweight
- a BMI of 18.5–25 $kg\,m^{-2}$ is the ideal range
- a BMI over 25 and up to 30 $kg\,m^{-2}$ means you are overweight
- a BMI of 30–40 $kg\,m^{-2}$ is considered obese
- a BMI over 40 $kg\,m^{-2}$ defines you as morbidly obese.

EXAM HINT

Using numbers in examinations has always been a weak point with biology students.

If you can apply your maths skills accurately, you will gain marks easily. When calculating BMI, candidates often forget to square the height, or they use height in centimetres rather than metres.

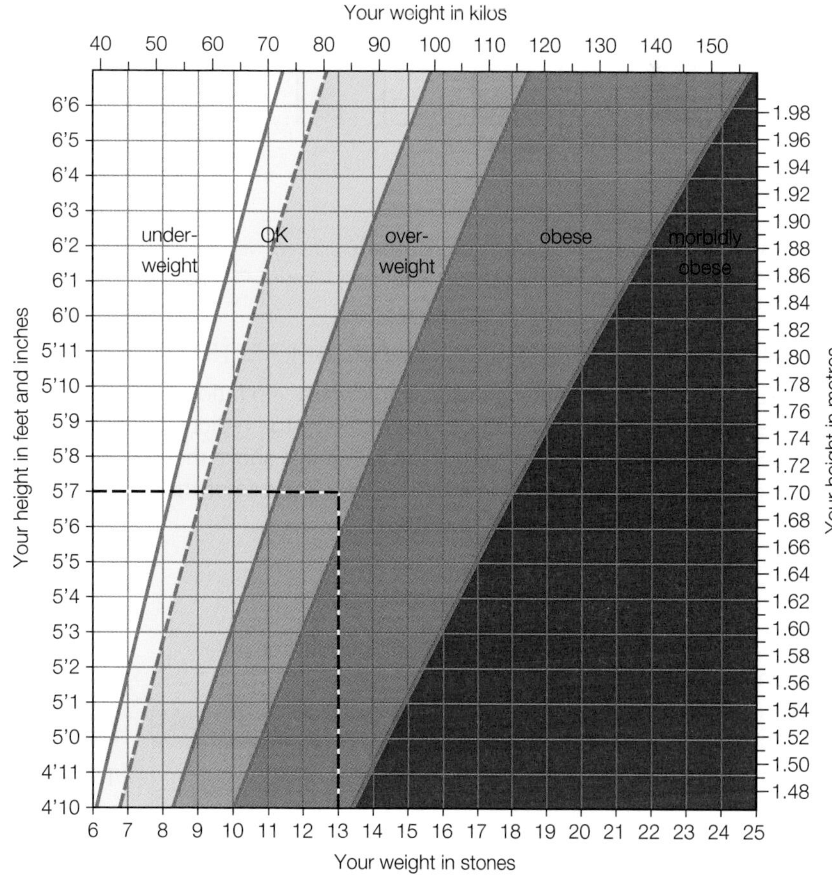

fig A Using a graph like this gives an adult a good idea of whether or not their BMI is in the healthy range.

The BMI measure was developed in the mid-1800s and it was originally used to classify normal, relatively inactive people of average body composition. The normal charts apply to adults only – there are special charts for children and teenagers. Young people grow and their body composition changes as they mature, so both age and gender are important in calculating what is normal until they become adults. The BMI became widely used for deciding whether people are a healthy weight for their height and even for predicting the likelihood of CVDs – but doctors increasingly feel it is a very limited tool. Most top athletes would have BMIs in the obese range, because BMI makes no allowance for the difference in composition of people's bodies. The reason athletes often have BMIs that suggest they are obese is because BMI does not recognise the difference between fat and muscle. BMI values also underestimate body fat in older people who have lost a lot of their muscle mass. There are also international differences, with some groups having a greater or lower than average risk of obesity-related diseases. More and more, the evidence suggests that BMI is not a good predictor of CVDs on its own – but combined with other factors it is part of the picture (see **fig B**).

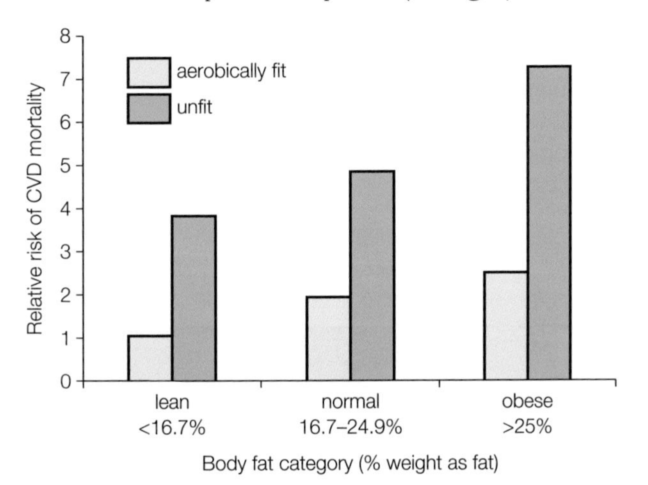

fig B The risk of dying of CVDs differs for people in different weight and fitness categories. Being overweight or obese does increase the risk – but it is not the whole story.

MEASURING A HEALTHY WEIGHT: THE WAIST-TO-HIP RATIO

Increasingly, scientists are finding that a simple waist:hip ratio is the best measure of obesity, and also the best way to predict an increased risk of CVDs. The waist is measured just above the navel, and the hips at the widest point of the hips. The size of the waist is then divided by the size of the hips:

$$\frac{\text{waist size (cm)}}{\text{hip size (cm)}}$$

Waist size gives a good indication of the amount of fat a person is carrying.

GENDER	WAIST:HIP RATIO INDICATING OBESITY
Male	>0.9
Female	>0.85

A simple measurement like this can be used easily by individuals to monitor their own health and well-being. Simple measures can lower the waist:hip ratio. These include eating less and taking more exercise to reduce the fat stores and reduce the size of the waist. Keeping the waist:hip ratio at a healthy level reduces your overall risk of CVDs, as well as the many other conditions linked to obesity.

Being seriously underweight is not good for you either and can lead to muscle wasting, heart damage and other health problems. However, it is the other end of the scale that is causing the most concern. The available data show that around 61% of all adults in England (that's almost 24 million people) are either overweight or obese, and that the proportion of the population affected is continuing to rise. The trend towards obesity is being seen across the developed world. For example, Gulf Cooperation Council countries face challenges with health problems related to obesity. Saudi Arabia (35.2%), Qatar (33.1%) and UAE (33%) face similar concerns.

TACKLING OBESITY

Evidence from around the world suggests that the change in energy balance is linked to modern lifestyles rather than simply to poor individual choices. Energy-rich food is widely available and cheap. The 21st century way of life in many countries involves almost no exercise – relatively few jobs now require manual labour, household tasks are often automated and people drive instead of walking or cycling. So, as the average energy input has increased (or stayed the same), the energy output has decreased and people are gaining weight. Solutions include taxes on fatty foods, town planning to make walking and cycling easier and educating children to prevent childhood obesity – but there is no clear scientific evidence that any of these solutions work. One thing we do know – since 2006, the number of overweight people in the world is greater than the number of people who do not get enough to eat. The importance of reducing obesity is shown in the graph in **fig C**. This also shows that waist:hip ratio is a better predictor of heart disease than BMI.

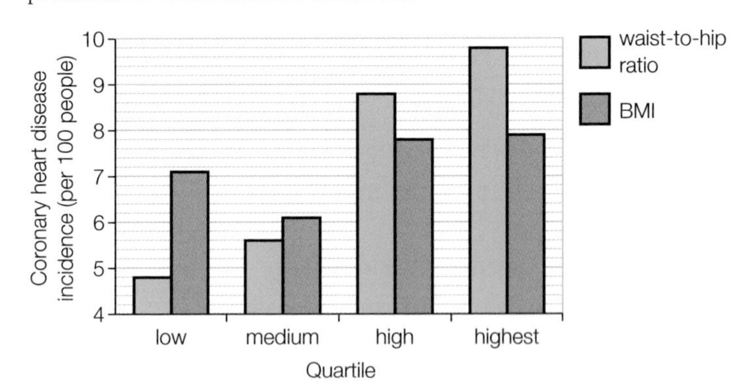

fig C This graph shows that as men become more obese, their risk of developing coronary heart disease also increases. It also shows that waist:hip ratio is a better predictor than BMI.

DIET AND CVDs

The effect of diet on the risk of developing CVDs isn't simply about becoming obese. What you eat, as well as how much you eat, seems to be very important. Many studies have looked at the general diet people eat and at the incidence of heart disease. For example, one study produced the graph in **fig D**. This shows that in countries where people eat a lot of fatty meat and dairy foods (mostly saturated fats), many people die of heart disease. This suggests that high levels of saturated fats in the diet may be a risk factor.

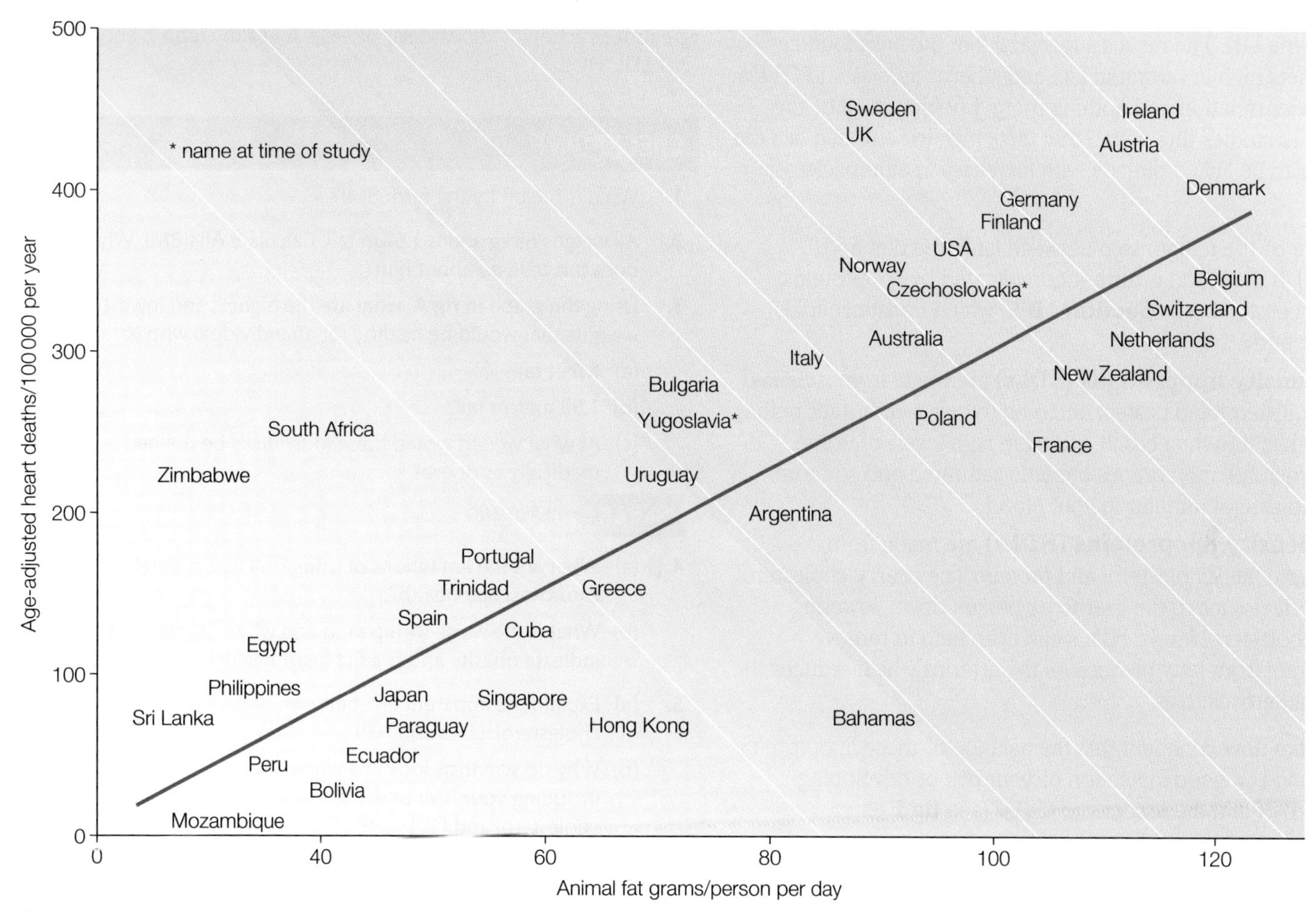

▲ **fig D** Data from the early 21st century showing the death rates per 1000 men and women from heart attacks in different countries compared with the average intake of animal (saturated) fats.

The link between a diet high in saturated fats and a raised incidence of CVDs shows a correlation, but not a cause. Over the last 50 years or so, many scientific studies showed that a high intake of saturated fats was often associated with high blood cholesterol levels. Cholesterol is involved in plaque formation in atherosclerosis, so this suggested a cause for the link between a high-fat diet and CVDs (see **fig E**).

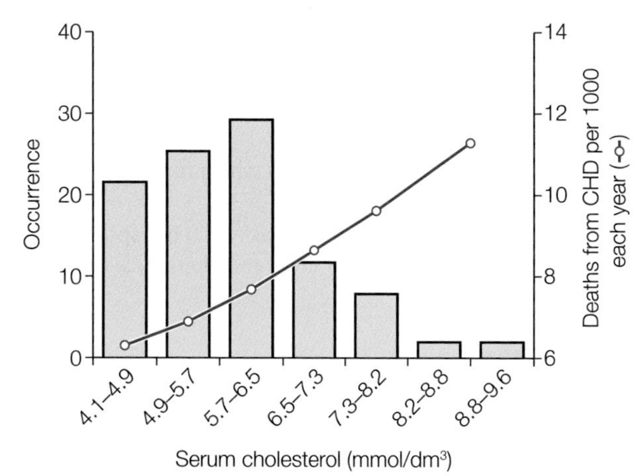

▲ **fig E** The relationship between blood cholesterol levels and death from coronary heart disease (CHD) in men in the UK. The bars show the frequency with which the different cholesterol concentrations are found, while the line graph shows the number of heart attacks per 1000 men each year.

LEARNING TIP

Remember that it is saturated fats that are harmful. These are the animal fats that are solid at room temperature.

Then, in 2014, a major study was published by scientists from prestigious institutions including the universities of Cambridge, Oxford and Bristol, Imperial College and the Medical Research Council in the UK. The results suggested that the links found between diets high in saturated fats and atherosclerosis and CVDs had been a correlation and nothing more. Looking at all of the data from 72 studies they found that diets high in saturated fats did not appear to be linked directly with increases in atherosclerosis and CVDs.

Our picture of the relationship between fat in the diet and cholesterol in the blood is further complicated by lipoproteins, conjugated proteins (see **Section 1B.5**) which transport lipids around the body.

- **Low-density lipoproteins (LDLs)** are made from *saturated* fats, cholesterol and protein and bind to cell membranes before being taken into the cells. If there are high levels of some LDLs, your cell membranes become saturated and so more LDL cholesterol remains in your blood.
- **High-density lipoproteins (HDLs)** are made from *unsaturated* fats, cholesterol and protein. They carry cholesterol from body tissues to the liver to be broken down, lowering blood cholesterol levels. HDLs can even help to remove cholesterol from fatty plaques on the arteries which reduces the risk of atherosclerosis.

Scientists are now confident that the balance of these lipoproteins in your blood is a good indication of your risk of developing atherosclerosis and the associated CVDs (see **fig F**).

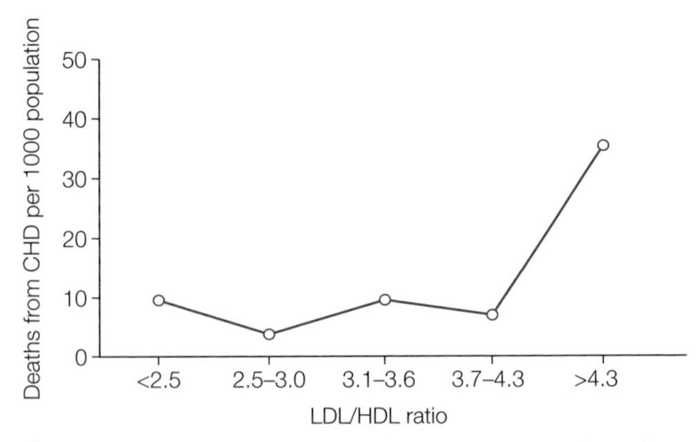

▲ **fig F** This evidence from a well-known European study into heart disease appears to show a clear link between the LDL/HDL ratio and deaths from coronary heart disease (CHD).

Blood cholesterol and LDL/HDL levels are not simply related to diet. The way your body metabolises the fats you eat and manages the levels of cholesterol and balance of lipoproteins in your blood are all linked to your genetic make-up. Some people can metabolise almost any amount of fat and maintain a good balance of LDLs and HDLs. Other people cannot cope so well and even small amounts of fat in the diet are reflected in raised blood cholesterol levels.

LEARNING TIP

Remember that it is the ratio of LDL:HDL that seems to have the greatest effect on cardiovascular disease. A healthy ratio is about 3:1 LDL:HDL.

CHECKPOINT

1. What is meant by the term BMI?
2. Ali weighs 65 kg and is 1.68 m tall. Calculate Ali's BMI. What does this tell you about him?
3. Using the graph in **fig A**, what are the highest and lowest weights that would be healthy for an individual who is:
   (a) 6 feet tall
   (b) 1.58 metres tall?
   (c) At what weight would these individuals be defined medically as obese?

SKILLS REASONING

4. (a) What are the limitations of using BMI as a predictor of cardiovascular health?
   (b) What is the waist-to-hip ratio and why is it often used to indicate obesity and predict heart health?
5. (a) Explain the apparent link between dietary fats, blood cholesterol, LDLs and HDLs.
   (b) Why do scientists look at a whole range of indicators including your BMI or waist-to-hip ratio, your blood cholesterol and HDLs and LDLs, and your history of smoking and exercise when they try to decide your risk of developing heart disease?
6. In 2014, a report was published which suggested that the link between fat in the diet and the risk of atherosclerosis and CVDs was a correlation but that it was **not** causative. Many ordinary people felt confused and upset when this report was discussed on news programmes. Others were very pleased. Suggest reasons for both of these responses.

SUBJECT VOCABULARY

body mass index (BMI) a calculation to determine if you are a healthy weight by comparing your weight to your height in a simple formula
low-density lipoproteins (LDLs) lipoproteins which transport lipids around the body
high-density lipoproteins (HDLs) lipoproteins which transport cholesterol from body tissues to the liver and can help reduce risks of CVDs

5 DIETARY ANTIOXIDANTS AND CARDIOVASCULAR DISEASE

SPECIFICATION REFERENCE 1.13 1.14 1.15 CP2

LEARNING OBJECTIVES

- Understand the link between dietary antioxidants and the risk of cardiovascular disease.
- Be able to distinguish between correlation and causation and recognise conflicting evidence.

Your diet is not all about the fats you eat. Lots of studies show that eating lots of fruit and vegetables benefits your health in many ways – including reducing your risk of developing CVDs. The graph in **fig A** is one piece of evidence which shows how eating five or more portions of fruit or vegetables a day can lower your risk of having a heart attack. It was based on data from a longitudinal study of more than 84 000 women and 42 000 men over eight years, looking at their fruit and vegetable intake and cardiac health.

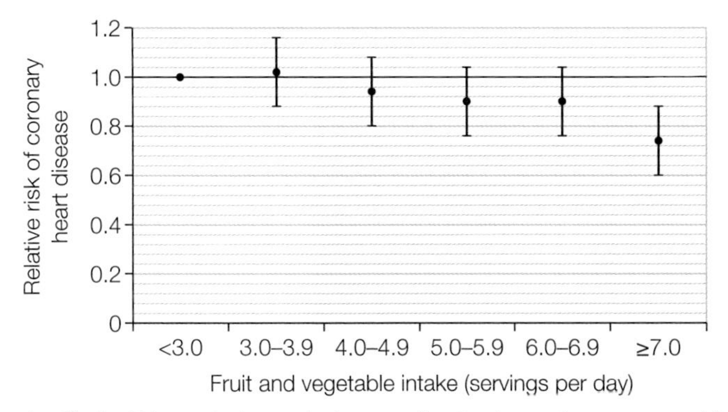

fig A This graph shows the impact of eating increasing amounts of fruit and vegetables on the risk for coronary heart disease.

EXAM HINT

Remember to look closely at the data you are given. This graph shows the relative risk of developing coronary heart disease compared to a person eating fewer than three servings of fruit and vegetables per day.

ANTIOXIDANTS AND HEART HEALTH

However, as you know, it isn't enough to show a correlation between two things. We need to show how one thing causes another and we still don't really know how fruit and vegetables have their effect. They are very varied in their chemistry. For some time, it was thought that the **antioxidants** found in fruit and vegetables might be the answer. Vitamins, such as vitamin A found in carrots, vitamin C from citrus fruits and vitamin E from leafy green vegetables, almonds and sunflower oil, are antioxidants and they are found in fruits and vegetables. Several studies appeared to show that antioxidants were the answer and many people started taking antioxidant supplements to protect their hearts. However, recent studies, including some very large metadata analyses (studies where scientists have looked at the results of many different investigations), have shown that the evidence for antioxidants being good for your heart is inconclusive. There is some evidence that some antioxidants may cause harm.

VITAMIN C: A CASE STUDY

Vitamin C is important in the formation of connective tissue in the body, such as in the bones, teeth, skin and many internal body surfaces including the endothelial lining of blood vessels. A severe lack of vitamin C in the diet causes scurvy, which can result in bleeding gums, bruising easily and painful joints. As you have seen, if the lining of an artery is damaged, atherosclerosis is more likely to develop. So, it makes sense, in theory, that if your diet is low in vitamin C, your arteries are more likely to be damaged and you are more likely to be affected by CVD.

A study published in the *British Medical Journal* in 1997 looked at the association between concentration of vitamin C in the blood and risk of heart attack in 1605 men from eastern Finland. The men had no sign of coronary artery disease when they were tested between 1984 and 1989. Their vitamin C levels were also tested. Between 1984 and 1992 a total 70 of the men had a heart attack (some fatal, some not). Of the men who showed low vitamin C levels, 13.2% had heart attacks, compared with 3.8% of the men who showed no sign of vitamin C deficiency. Many people began eating lots of vitamin C rich foods, and taking vitamin C supplements, in the belief that they were reducing their risk of having heart disease.

Then, in 2016, a major metadata study was published in the *International Journal of Molecular Sciences*. It looked at all the evidence for the antioxidant properties of vitamin C as an explanation of the known benefits of fruit and vegetables on heart health. The conclusions were that there was no relationship between them. The study even showed that taking vitamin C supplements could damage heart health. This is a good example of where there is contradictory evidence – and where scientists must look at all of that evidence to avoid coming to the wrong conclusions.

PRACTICAL SKILLS CP2

Testing for vitamin C

There is a simple laboratory test for the presence of vitamin C in foods. It involves a reagent called DCPIP, which stands for 2,6-dichlorophenol-indophenol. DCPIP solution is blue. When it reacts with vitamin C, it turns colourless (although in a very acidic solution such as lemon juice, it may turn pink).

You can estimate the concentration of vitamin C in different foods and drinks by recording the volume of DCPIP which is added before it remains blue - at this point all the vitamin C has been used up.

▲ **fig B** Citrus fruit contains vitamin C, which can be tested for with DCPIP.

EXAM HINT

Ensure that you know about measuring vitamin C levels - there could be questions directly about the experimental procedure and how to make it valid, precise and reliable.

CHECKPOINT

1. Look at **fig A**. Calculate the percentage reduction in deaths from heart disease if everyone ate:
   (a) five portions of fruit and vegetables every day
   (b) seven or more portions of fruit and vegetables every day.
2. What is the difference between the data that show a link between the amount of fruit and vegetables eaten and heart health, and the evidence looking at the effect of antioxidants on heart health?
3. Plan an investigation to compare the vitamin C content of two fruits.

SKILLS EXECUTIVE FUNCTION

SUBJECT VOCABULARY

antioxidants molecules that inhibit the oxidation of other molecules which can lead to chain reactions that may damage cells

LEARNING OBJECTIVES

- Understand how people use scientific knowledge about the effects of diet, including obesity indicators, exercise and smoking to reduce their risk of coronary heart disease.

There is much scientific evidence about the main factors that increase the risk of heart disease. A lot of that evidence is used by governments and health organisations to produce advice on how to improve our health. Why do they do this?

PREVENTION IS BETTER THAN CURE

Cardiovascular disease has a negative effect on individuals, on families and on society. It costs a lot of money to treat people in hospital. When people are too ill to work, they are losing money for their families, and also for the companies where they work. Treating people with drugs to prevent them from needing surgery is cheaper for health service providers. It is even cheaper (and better for the individual) if we can stop ourselves needing the drugs. So, prevention is better than treatment for CVDs for many reasons (see **fig A**). However, persuading people to change their lifestyle habits is often difficult.

For example, there is a lot of reliable evidence to show that smoking is one of the highest risk factors for CVDs. However, if you stop smoking, your risk of developing heart disease is almost halved after just one year. Moreover, research carried out by a team led by Azra Mahmud from Trinity College Dublin and published in 2007 suggests that after 10 years, the arteries of smokers who stop smoking are the same as if they had never smoked. There is a lot of support available for people who want to stop smoking. Yet, despite all this, almost 1 billion people around the world smoke cigarettes and millions of them die each year of CVDs and cancers linked to their smoking.

Health education programmes in schools and communities can help to make sure that everyone is aware of the risks associated with different lifestyle choices. However, each individual has to make their own choices and take their own risks.

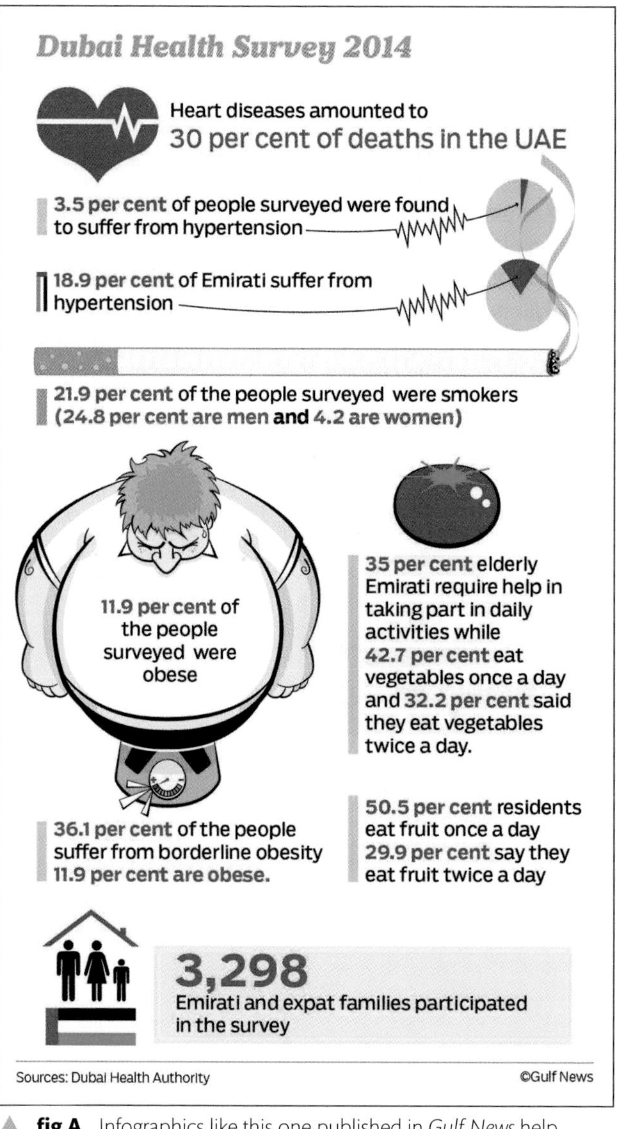

fig A Infographics like this one published in *Gulf News* help people understand the factors which affect their heart health.

OVERWEIGHT OR UNDERFIT?

Most people are aware that obesity is linked to CVDs, and many go on slimming diets to try to lose weight. Most people also know that taking regular exercise helps protect against CVDs – but more people choose to go on a diet than choose to take regular exercise. The results of a study carried out over an average of eight years on 20 000 men aged from 30 to 83 years are given in **Section 1C.4 fig B**. Fitness was defined by how much oxygen they used during exercise. The results show that if you are obese and fit you have a lower risk of dying from CVDs than someone who is not obese but unfit. Obviously, being the correct weight and fit is best of all!

The problem is that many people enjoy their food, and don't take sufficient exercise. It takes a lot of effort to cut down the amount of food you eat and change to eating healthier foods including lots of fruit and vegetables.

LEARNING TIP

Remember to read all the information in graphics.

DID YOU KNOW?

SALT AND CVDs

The Yanomami Indians in Brazil eat far less salt than people in the developed world and have much lower blood pressure levels. This evidence was used to draw a link between high salt levels, high blood pressure and CVDs. After much more research, there is wide agreement among scientists that high levels of dietary salt increase blood pressure in many people in the developed world. Processed foods typically have high levels of salt, and often people add salt to their food. If people eat less processed food, add less salt to their own cooking, and food manufacturers add less salt to their products, we might be able to lower our salt intake from the current daily average of 9–12 g to the recommended level of less than 5–6 g of salt per day.

▲ **fig B** If everyone in richer countries ate less salt, it would result in a global drop in high blood pressure and cardiovascular disease.

SO WHY DON'T PEOPLE CHANGE THEIR LIFESTYLE?

Part of the problem is that people find it difficult to distinguish between perceived risk and actual risk. They can see a risk applies to an average of a group, but not to themselves as individuals. If you see people smoking, eating a high-fat, high-salt diet, never exercising and yet appearing well, the evidence of your own experience contradicts and overrides the evidence from research reported in the media. People then underestimate the risk of CVDs associated with smoking, obesity, lack of exercise or a high-salt diet.

There are other reasons that lead to mistakes when people assess risk. Sometimes people will continue smoking because they don't want to gain weight. Smoking speeds up the metabolism and reduces the appetite, which both help to control body mass. Here, the health risks of obesity are overestimated in comparison with those of smoking. In other words, the risks of smoking are ignored because people do not want to get fat. Also, the nicotine in tobacco smoke is addictive to many people, and this makes it very difficult to give up smoking.

When people calculate their personal risk/benefit situation, it is easy to think that the immediate benefit (pleasure in eating high-fat food, smoking, not wanting to make the effort to exercise) is more important than the apparently low risk of heart disease.

CHECKPOINT

1. Why do you think people are more likely to try to lose weight than to take more exercise?

SKILLS DECISION MAKING

2. What are the limitations of drawing conclusions about the effect of salt on blood pressure from a study comparison of the Yanomami in Brazil and the population of a country such as the UK or Qatar? Do you think people in developed countries under- or overestimate the risk of eating too much salt, and what influences these perceptions?
3. Many governments spend millions of pounds on health advertising each year. Discuss whether or not this is a waste of money.

1C 7 THE BENEFITS AND RISKS OF TREATMENT

SPECIFICATION REFERENCE 1.20

LEARNING OBJECTIVES

- Know the benefits and risks of treatments for CVDs including antihypertensives, statins, anticoagulants and platelet inhibitors.

Once a patient has signs of cardiovascular disease, there are a number of different treatment options available. Changing lifestyle, such as improving diet, giving up smoking and taking more exercise can help but there are also various drugs that can be given. The drugs aim to reduce the risks associated with CVDs by helping to prevent problems developing. However, all medicines carry some risk.

CONTROLLING BLOOD PRESSURE

As you have seen, hypertension or high blood pressure is a major risk factor for cardiovascular diseases.

ANTIHYPERTENSIVES

Drugs that reduce blood pressure are known as **antihypertensives**. Some commonly prescribed antihypertensive drugs are described below.

- Treatment often begins with **diuretics**, which increase the volume of urine produced. This eliminates excess fluids and salts, so that the blood volume decreases. With less blood, a smaller volume is pumped from the heart and the blood pressure falls.
- **Beta blockers** interfere with the normal system for controlling the heart. They block the response of the heart to hormones such as adrenaline, which normally act to speed up the heart and increase the blood pressure (you will find out more in **Book 2 Topic 7**). So, beta blockers make the heart rate slower and the contractions less strong, so the blood pressure is lower.

Sympathetic nerve inhibitors affect the sympathetic nerves which go from your central nervous system to all parts of your body (you will find out more in **Book 2 Topic 7**). Sympathetic nerves stimulate your arteries to constrict, which raises your blood pressure. The inhibitors prevent these nerves signalling to the arteries, which helps to keep the arteries dilated and your blood pressure lower.

- Angiotensin is a hormone which stimulates the constriction of your blood vessels and so causes the blood pressure to rise. **ACE inhibitors** block the production of angiotensin, which reduces the constriction of your blood vessels and so keeps your blood pressure lower.

The benefits of these drugs in reducing blood pressure are clear. They reduce the risk of CVDs, and also reduce the risk of damage to the kidneys and eyes from the high blood pressure.

But there are risks. The risks of these treatments are twofold. If the treatment is not monitored carefully, your blood pressure may become too low. That can lead to falls and injuries which, particularly in elderly patients, can be serious and even life-threatening. The second major risk is the **side-effects** that may result from the way your body reacts to the drugs. Each type of drug has its own possible side-effects (see **fig A**). For a drug to be given a licence for use, the benefits of the treatment must be judged to outweigh any side-effects.

The side-effects from commonly used antihypertensives include coughs, swelling of the ankles, impotence, tiredness and fatigue, and constipation. These are not serious compared with the health risks from high blood pressure – but to the patient they may feel very significant. High blood pressure often doesn't make you feel ill, but the medication needed to control it can affect your quality of life. Doctors find many patients stop taking their medication – the side-effects make them ignore the much larger but invisible risk of CVDs.

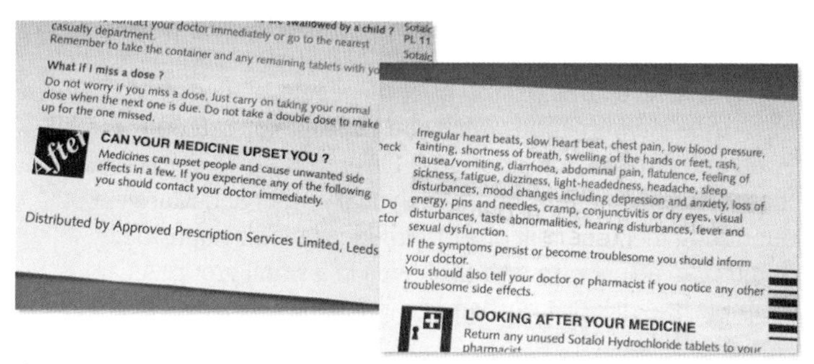

fig A All medically licensed drugs come with instructions and information which includes possible side-effects known to be caused by the drug.

EXAM HINT

If you are asked to discuss or evaluate the use of medication or some other treatment, you must remember to include both the benefits and the potential risks associated with the treatment.

STATINS

Statins are a group of drugs that lower the level of cholesterol in your blood. They block the enzyme in the liver that is responsible for making cholesterol, and are very effective at blocking the production of LDLs. Statins also improve the balance of LDLs to HDLs and reduce inflammation in the lining of the arteries. Both functions reduce the risk of atherosclerosis developing.

Fig B shows the results from a trial using statins with a group of 6605 Asian Indians in the US. This shows the results for men and women, and other groups who are high-risk categories for cardiovascular disease. Statins reduce the incidence of serious cardiovascular disease in all categories, but they seem to have a greater effect for some groups than for others.

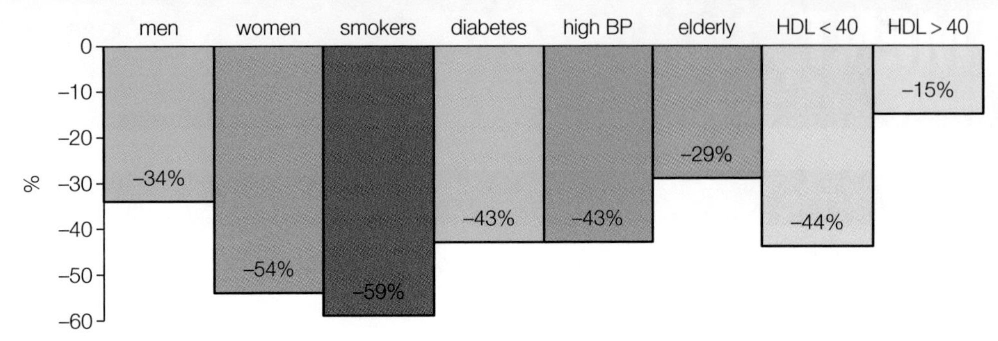

fig B These data show the benefits of statins in reducing the risk of CVDs in Asian Indians, who are a particularly high-risk group. Statins had a strong positive benefit to a range of patients.

A UK study showed that men who took a particular statin for five years had a lower risk of death or heart attack, even 10 years after they stopped taking the drug. The study involved 6595 middle-aged men. It showed that for the first five years, the overall risk of heart attack or death from any type of heart disease was 11.8% for the men who took the statin, compared with 15.5% for men who took a **placebo** (an inactive substance that resembles the drug but has no action in the body). The risk was reduced the most while the men were taking the drug, but some level of protection lasted for up to 10 years afterwards.

Most people use statins with little or no ill effect. Side-effects of muscle and joint aches and nausea, constipation and diarrhoea are sometimes reported. However, there are two serious but very rare side-effects. In a tiny number of people, statins trigger a form of muscle inflammation which can be fatal. For example, the US Food and Drug Administration (FDA) reported 3339 cases of these muscle reactions between January 1990 and March 2002, but during this time millions of Americans took statins daily. Statins can also cause liver problems in a small group. As an example, the risk of liver damage in people taking lovastatin is two in a million. Out of 51 741 liver transplant patients in the US between 1990 and 2002, liver failure appeared to be caused by statins in only three cases.

Another risk is more subtle: there is a risk that, if people take statins to lower their blood cholesterol, they will no longer try to eat a healthy diet, and statins give no protection against the other ill effects of a bad diet.

Plant stanols and sterols are now widely sold in spreads and yoghurts. These compounds are very similar in structure to cholesterol. They reduce the amount of cholesterol absorbed from your gut into your blood, which can make it easier for your body to metabolise cholesterol and reduce the levels of LDLs in the blood. Products like these are sold as a food, not as a drug. While there is scientific evidence that they are effective in many people, they have not undergone the levels of testing that drugs such as statins have. Metadata analysis has shown that these products do work if they are eaten regularly in the recommended amounts (2 g per day of plant sterols and stanols). It has been estimated that these products can lower your risk of heart disease by about 25% if they are used correctly (see **fig C**).

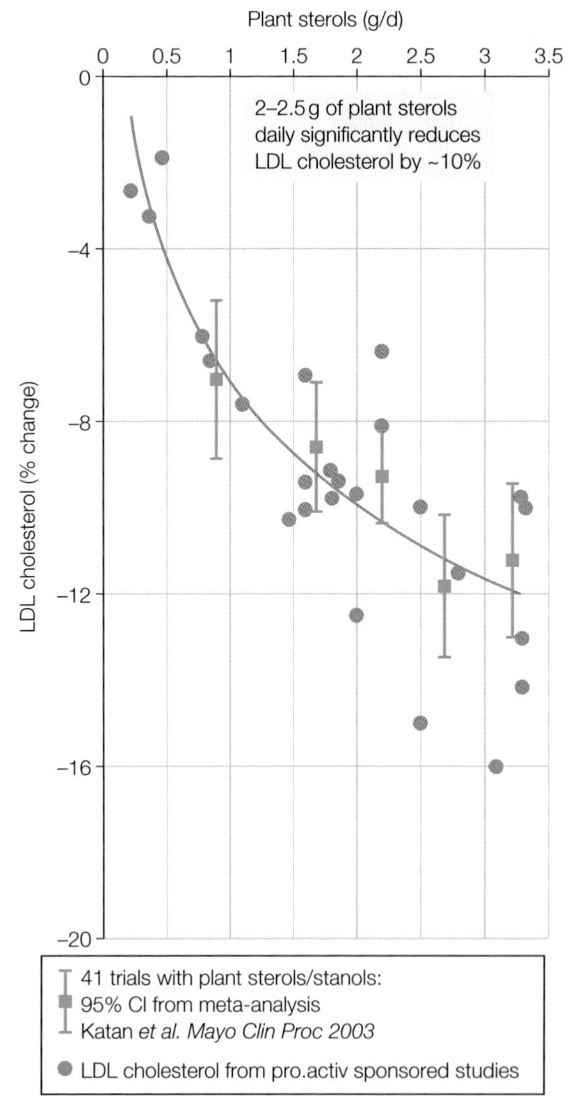

fig C Scientific evidence from a number of studies shows that plant sterols can reduce harmful LDL cholesterol levels in the blood if used correctly.

ANTICOAGULANTS AND PLATELET INHIBITORY DRUGS

Following heart surgery, or after suffering from a blood clot (thrombosis), drug treatments are used to help prevent the blood clotting too easily. Here are two examples.

Warfarin is an **anticoagulant** that interferes with the manufacture of prothrombin in the body. Low prothrombin levels make the blood clot less easily (see **Section 1B.2**). Warfarin has been used in rat poison – in high doses the blood will not clot at all and the rats bleed to death after the slightest injury. In humans, the dose is carefully monitored to make sure that the clotting of the blood is reduced but not prevented completely.

Platelet inhibitory drugs make the platelets less sticky, and so reduce the clotting ability of the blood. The cheapest and most common of these is aspirin (**fig D**) but clopidogrel is also commonly used.

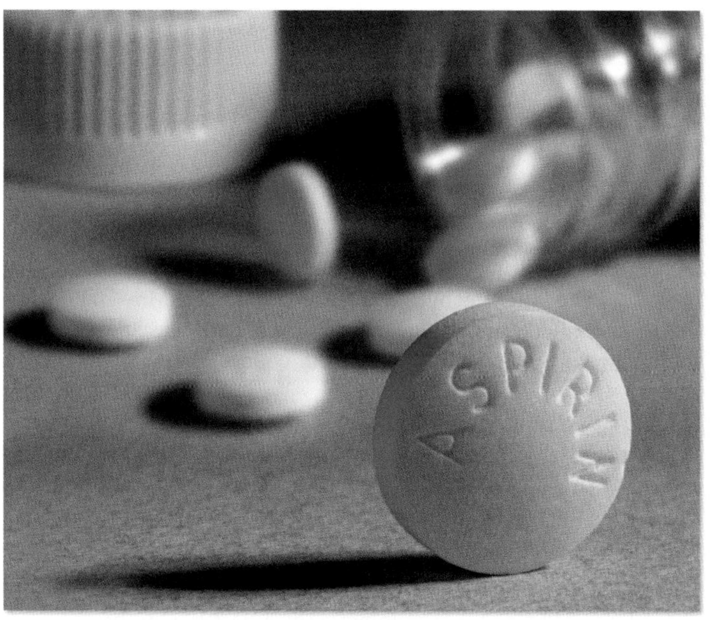

▲ **fig D** Aspirin is a relatively cheap drug. It has been used traditionally as a painkiller. It is also a very effective way of preventing many cardiovascular problems.

The risks of taking aspirin are well known – it irritates the stomach lining and causes bleeding in the stomach which can become serious. A combination of aspirin and clopidogrel can reduce the risk of developing a range of cardiovascular diseases by 20–25% in some low-risk patients. However, based on data from several studies it appears that, for some patients, the risk of side-effects is much higher when the two drugs are combined. For example, for every 1000 patients at high risk of CVDs treated for 28 months, five cardiovascular events would be avoided – but three major stomach bleeds would be caused. In lower-risk patients, 23 cardiovascular events would be avoided while 10 major bleeds would be caused.

It is difficult to achieve the correct balance between preventing the blood from clotting too easily while allowing it to clot when necessary. For example, when people are treated with anticoagulant drugs such as warfarin, they must be monitored very carefully to make sure that they do not bleed internally, particularly in the brain. The decision whether to give warfarin will depend on many factors, including the patient's age and condition as well as other medication they may be taking.

CHECKPOINT

1. Explain why the side-effects of medication may result in a patient giving up on the treatment. Use the terms *perceived risk* and *actual risk* in your answer.

SKILLS ▶ DECISION MAKING

2. ▶ The graph in **fig C** comes from the website of a company that makes products containing plant stanols. However, the data appear scientifically acceptable - why? What do they show you about the effect of plant stanols on blood cholesterol levels?
3. (a) Explain why placebos may be used in drug trials.
   (b) The study shown in **fig B** was stopped 2 years early because it was deemed unfair to the patients taking the placebo. Why do you think it was unfair? Is it ever unethical to use a placebo in a trial?
4. Look at **fig B** and answer these questions.
   (a) Explain why statins have a greater effect on reducing the risk of CVDs in people with a lower HDL level.
   (b) Considering that all medical drugs have associated side-effects, what does this graph suggest about which groups should be targeted with statins to reduce CVDs overall in the population?

SUBJECT VOCABULARY

antihypertensive drug which reduces high blood pressure
diuretics drugs which increase the volume of urine produced
beta blockers drugs which block the response of the heart to hormones such as adrenaline
sympathetic nerve inhibitors drugs which inhibit sympathetic nerves, keeping arteries dilated
ACE inhibitors drugs which block the production of angiotensin
side-effect a secondary, usually undesirable effect of a drug or medical treatment
statins drugs that lower the level of cholesterol in the blood
placebo an inactive substance resembling a drug being trialled which is used as an experimental control
plant stanols and sterols similar in structure to cholesterol, these compounds can help reduce blood cholesterol in those consuming them
anticoagulant a substance that interferes with the manufacture of prothrombin in the body
platelet inhibitory drugs drugs used to prevent blood clots forming by preventing platelets clumping together

HEART FAILURE IN THE MIDDLE EAST

SKILLS CRITICAL THINKING, PROBLEM SOLVING, ANALYSIS, DECISION MAKING, CREATIVITY, INNOVATION, PERSONAL AND SOCIAL RESPONSIBILITY, CONTINUOUS LEARNING, INTELLECTUAL INTEREST AND CURIOSITY, COMMUNICATION, EMPATHY/PERSPECTIVE TAKING

One in four adults in Saudi Arabia are expected to suffer a heart attack within the next 10 years. In the Middle East, the average age for onset of heart failure is 10 years lower than in Western Countries. Heart disease is a multifactorial disease – many factors contribute to the risks of developing heart disease. However, the final failure of the heart can often be attributed to one contributing factor. While 70% of heart failure in the West can be attributed to coronary artery disease, this is not the case in the Middle East.

MEDICAL REVIEW ARTICLE

ABSTRACT

The clinical syndrome of heart failure is the final pathway for a myriad of diseases that affect the heart, and is a leading and growing cause of morbidity and mortality worldwide. Evidence-based guidelines have provided clinicians with valuable data for better applying diagnostic and therapeutic tools, particularly the overwhelming new imaging technology and other, often expensive, therapies and devices, in heart failure patients. In the Middle East, progress has recently been made with the development of regional and multi-centre registries to evaluate the quality of care for patients with heart failure. A new heart function clinic recently began operation and has clearly resulted in a reduced readmission rate for heart failure patients. Many Middle Eastern countries have observed increases in the prevalence of the risk factors for the development of heart failure, including diabetes mellitus, obesity, and hypertension, with heart failure in the Middle Eastern population developing earlier than it is in their Western counterparts by at least 10 years. The earlier onset of disease is the result of the earlier onset of coronary artery disease, highlighting the need for Middle Eastern countries to establish prevention programs across all age groups. The health systems across the Middle East need to be modified in order to provide improved evidence-based medical care. Existing registries also need to be expanded to include long-term survey data, and additional funding for heart failure research is warranted.

References

Hazebroek M., Dennert R., Heymans S. Idiopathic dilated cardiomyopathy: possible triggers and treatment strategies. Neth Heart J 2012 Aug; 20(7–8)332–335 https://www.ncbi.nlm.nih.gov/pmc/articles/PMC3402574/

Arjen Radder. MENA has a heart disease problem. The solution is both basic and high-tech. World Economic Forum May 2017 https://www.weforum.org/agenda/2017/05/cardiovascular-disease-is-threatening-the-middle-easts-health-connected-technology-is-the-answer/

Hala Khalef. Heart disease causes 45% of early deaths in Middle East. The National 2010 January https://www.thenational.ae/uae/heart-disease-causes-45-of-early-deaths-in-middle-east-1.499807

Why the number of heart attack cases are growing. *The Kahjeel Times* 2015 October https://www.khaleejtimes.com/nation/uae-health/why-the-number-of-heart-attack-cases-are-growing

In Middle East and North Africa, health challenges are becoming similar to those in western countries. The World Bank 2013 September http://www.worldbank.org/en/news/press-release/2013/09/04/middle-east-north-Africa-health-challenges-similar-western-countries

From: Mostafa Q Al-Shamiri. Heart failure in the Middle East. *Current Cardiology Reviews* 2013 May; 9(2):174–178 https://www.ncbi.nlm.nih.gov/pmc/articles/PMC3682400/#R10

SCIENCE COMMUNICATION

SKILLS ANALYSIS, INTERPRETATION

1 (a) Who do you think the intended audience is?

(b) What is the intended message the author is trying to give?

(c) There are a number of terms used in the article. Select **three** unfamiliar terms from the article and research their meaning. Suggest why those terms have been used.

(d) An abstract is meant to indicate to the reader what is contained in the rest of the document. From the abstract, a researcher might decide whether or not to read the rest of the article. What sort of information would you expect to find in the rest of this article?

INTERPRETATION NOTE

This article is an abstract from a longer article in the journal *Current Cardiology Reviews*. Think about the type of writing used.

BIOLOGY IN DETAIL

SKILLS REASONING

Now you are going to think about the science in the article. You will be surprised how much you know already, but if you choose to do so, you can return to these questions later in your course.

2 (a) What are the main contributing factors to development of heart failure mentioned by the author?

(b) Name **two** more contributing factors.

3 Changes in lifestyle and diet have been blamed for huge increases in the risk factors. For example, diabetes has increased by 87% between 1990 and 2010.

(a) What lifestyle changes are likely to cause such an increase in diabetes?

(b) What changes in diet may contribute to such an increase in diabetes?

4 The author is writing about epidemiology. That is the branch of medicine which deals with the incidence, distribution and possible control of diseases and other factors relating to health. How does understanding the incidence of heart failure and the factors that contribute to it help the medical profession to combat heart failure?

5 Common causes of heart failure are ischaemic heart disease, uncontrolled hypertension and valvular disease. However, in up to 50% of the cases its exact cause remains unknown; this condition is called idiopathic cardiomyopathy. The table shows the results of an epidemiological study.

CAUSE OF HEART FAILURE	% OF CASES IN EACH COUNTRY			
	OMAN	EGYPT	SAUDI ARABIA	YEMEN
ischaemic heart disease	52	66	52	52
valvular disease	8.5	22.5	10.5	7
hypertension	25			25
idiopathic cardiomyopathy	8.3			11

(a) Which cause of heart failure should be the focus of most research?

(b) Suggest why the exact cause of heart failure may be unknown in up to 50% of cases.

ACTIVITY

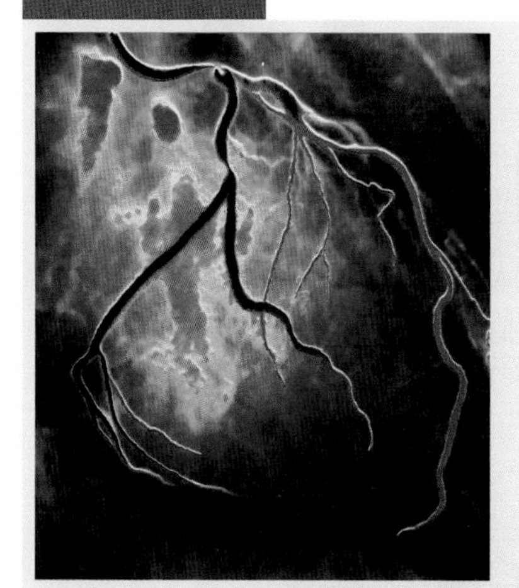

▲ **fig A** Heart disease, in particular coronary artery disease, is an increasing problem in the Middle East.

Research the risk factors for heart disease. Find out what can be done to reduce these factors.

Prepare a leaflet designed to help people understand how their lifestyle and diet affect their chances of developing heart disease.

Provide advice to someone who has been told that they are at risk of developing heart disease.

THINKING BIGGER TIP

You can refer to the full version of this paper, to the references listed at the end, to online encyclopaedias, to other scientific papers and to books. In each case, judge the reliability of your source before you use it.

1C EXAM PRACTICE

1 (a) What is an atheroma?

A a hardened part of the artery wall
B a swelling in the artery wall
C a fatty deposit in the artery wall
D a narrowing of the artery wall [1]

(b) Justify the use of the term *multifactorial* when used to describe atherosclerosis. [2]

(c) (i) Which of the following is **not** a risk factor for atherosclerosis?

A obesity
B high-salt diet
C high-fibre diet
D lack of exercise [1]

(ii) One weight loss plan suggests eating a diet with no carbohydrate but allows eating as much protein and fat as you like. Discuss the merits of such a diet. [6]

(Total for Question 1 = 10 marks)

2 There are many factors that increase the risk of developing atherosclerosis.

(a) Discuss whether correlation means cause. [4]

(b) Many studies have been conducted to investigate the causes of atherosclerosis and cardiovascular disease. State **three** factors in the design of a study which make the findings more valid. [3]

(c) The photograph below shows a small artery with a large plaque or atheroma. If the internal diameter of the artery is 1 mm, determine the area blocked by the plaque. [2]

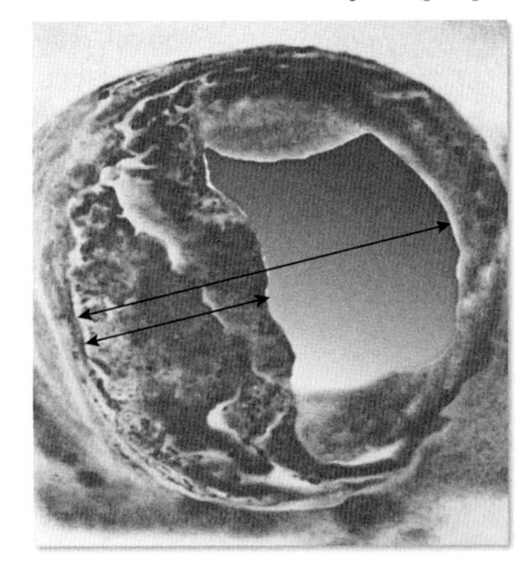

(Total for Question 2 = 9 marks)

3 (a) What is the role of the coronary arteries?

A to deliver oxygenated blood to the atria
B to deliver oxygenated blood to the ventricle walls
C to carry deoxygenated blood into the atria
D to carry oxygen and nutrients to the rest of the body [1]

(b) The illustration below shows the coronary arteries in the heart with the position of three possible blockages labelled A, B and C.

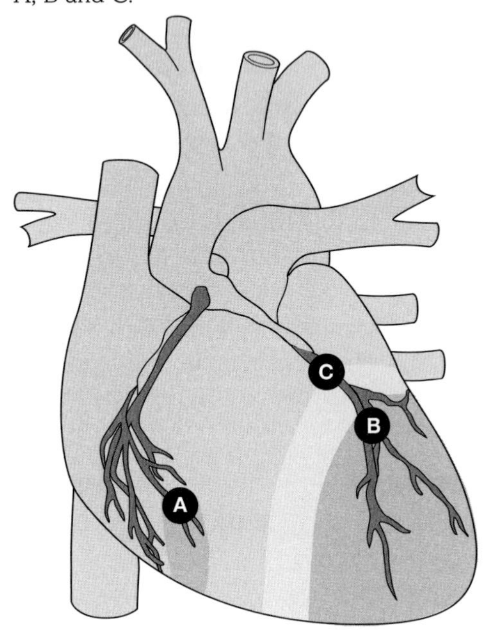

Explain why a blockage at C would have far greater consequences than a blockage at either A or B. [3]

(c) The bar charts below show the proportion of people with coronary artery disease based on a Jordanian study in 2017.

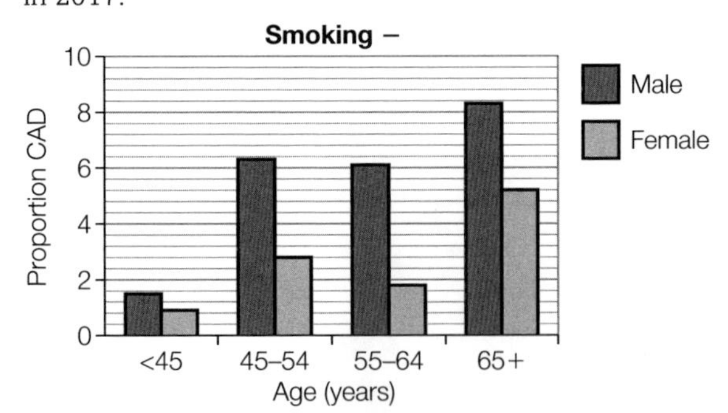

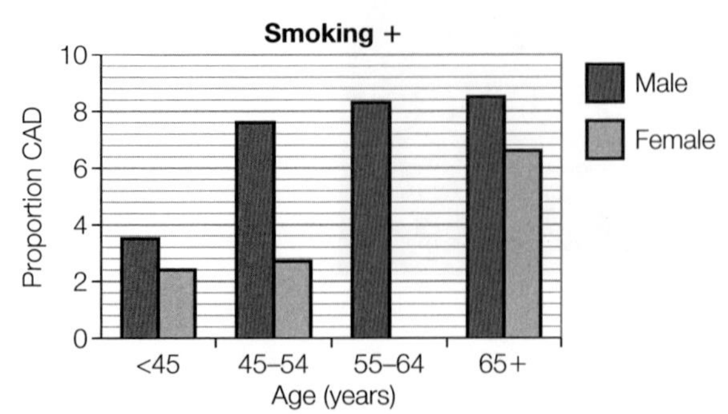

(i) Describe the effect that smoking has on the risk of developing coronary artery disease in men of different ages. [4]

(ii) Suggest why there is no data for women in the 55–64 age bracket. [1]

(iii) What does the lack of data for women in the 55–64 age bracket suggest about the study? That it is:

A not valid

B not reliable

C not precise

D not repeatable. [1]

(iv) Assuming that smoking affects women to the same extent as men, what proportion of women would you expect to have CAD in the 55–64 age bracket of smokers? [3]

(Total for Question 3 = 13 marks)

4 (a) Describe the main stages in the formation of an atheroma. [3]

(b) Explain how atherosclerosis can lead to a heart attack. [3]

(c) The average age for a first heart attack in the UAE is 20 years younger than the global average. A recent survey of 850 heart attack patients in the UAE gave the following results.

Risk factor	Diabetes	High blood pressure	High blood cholesterol
Number of patients with risk factor		380	212
% of patients with risk factor	38	44.7	

Calculate the number of patients in the survey who had diabetes. [2]

Calculate the % of heart attack patients in the survey who had high blood cholesterol. [2]

(d) State **two** other risk factors that increase the chance of a heart attack. [2]

(Total for Question 4 = 12 marks)

5 (a) Explain why perceived risk is often not the same as actual risk. [2]

(b) One risk factor for cardiovascular disease is obesity. The table below shows the BMI scale.

BMI	Category
18.5–24.9	normal
25–29.9	overweight
30–34.9	obese
35–39.9	dangerously obese

(i) What is the BMI for a man who is 1.8 metres tall and weighs 115 kg?

A 0.03

B 30.6

C 35.5

D 63.9 [1]

(ii) Suggest three changes in lifestyle that a health professional might advise for this man. [3]

(iii) Name another quick and easy measure that people could use to indicate if they are obese. [1]

(c) Despite increasing evidence of the risks caused by obesity, approximately one-third of the population in the Middle East are obese. Discuss why so many people are obese. [3]

(Total for Question 5 = 10 marks)

6 A person at risk of cardiovascular disease may be given medication to reduce their risk.

(a) How might anticoagulants or platelet inhibitors reduce the risk of a heart attack or stroke? [2]

(b) Factors such as high blood pressure and type 2 diabetes are a direct result of obesity. Evaluate the use of medication such as antihypertensives and statins to treat people who are at risk of cardiovascular disease. [4]

(c) The graph below shows the relative risk of coronary heart disease plotted against the number of servings of fruit and vegetables eaten per day.

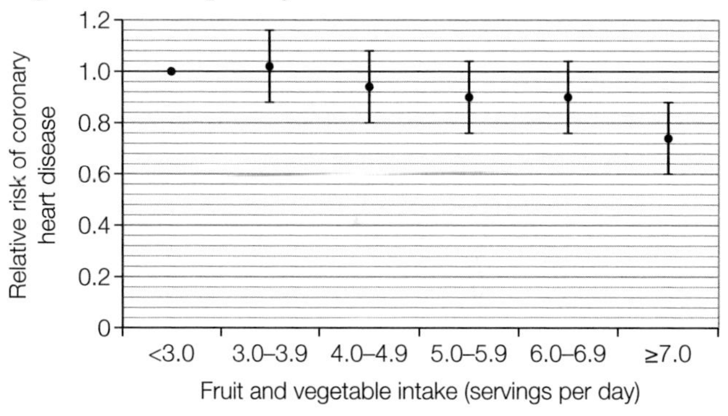

(i) People are advised to eat at least five servings of fruit and vegetables per day. Justify this advice using the information in the graph. [3]

(ii) Suggest what evidence there is in the graph to modify the advice. [2]

(Total for Question 6 = 11 marks)

TOPIC 2 MEMBRANES, PROTEINS, DNA AND GENE EXPRESSION

CHAPTER 2A MEMBRANES AND TRANSPORT

The single-celled *Amoeba proteus* has a problem. It lives in fresh water. The cytoplasm of the cell contains lots of dissolved minerals and sugars, and the outer membrane is partially permeable. As a result, water moves in by osmosis, a passive process common to all living things. *Amoeba proteus* also has a contractile vacuole, a specialised organelle for expelling the excess water. Water moves into the cytoplasm passively, but it is then moved into the contractile vacuole. This process involves the active transport of ions into the vacuole. Once the vacuole is full, it contracts emptying the water out of the cell. Scientists have been studying *Amoeba* for many years, but they are still not sure exactly how this happens. However, they do know that osmosis and active transport all take place within this single-celled organism.

In this chapter, you will learn how substances are transported into and out of cells. Diffusion is an important concept – the net movement of particles down a concentration gradient from an area of high concentration to an area of lower concentration. This happens due to the random movement of particles. It is an important form of transport in all organisms.

You will also consider osmosis. This is a specialised form of diffusion that involves the movement of water through a partially permeable membrane. However, sometimes things need to be moved against a concentration gradient – and this is where active transport comes in. Active transport, endocytosis and exocytosis use energy from ATP to move substances against their concentration gradients. This occurs in many different places in the cell.

You will also consider why organisms need specialised gas exchange surfaces as they get bigger and their metabolic rate increases. All successful gas exchange surfaces have certain features in common. You will be discovering these features and looking for them in the gas exchange systems of mammals using humans as an example.

MATHS SKILLS FOR THIS CHAPTER

- **Recognise and make use of appropriate units in calculations** (*e.g. nm^2, nm^3, molar quantities for serial dilutions*)
- **Substitute numerical values into algebraic equations using appropriate units for physical quantities** (*e.g. water potential calculations*)
- **Solve algebraic equations in a biological context** (*e.g. water potential equations*)
- **Recognise and use expressions in standard and decimal form** (*e.g. use of magnification*)
- **Use ratios, fractions and percentages** (*e.g. calculate surface area to volume ratio*)
- **Find arithmetic means** (*e.g. the mean lung volume of a group of people*)
- **Understand the terms mean, median and mode** (*e.g. the lung volumes of a group of people*)
- **Calculate the circumferences, surface areas and volumes of regular shapes** (*e.g. work out the approximate surface area and volume of model organisms*)

What prior knowledge do I need?

- How substances are moved into and out of cells by diffusion, osmosis and active transport
- The nature of ATP as a source of energy for cell processes

Chapter 1A

- The structure of lipids

Chapter 1B

- The concept of surface area : volume ratio in determining if a transport system is needed
- That larger multicellular organisms need a transport system because their surface area : volume ratio makes direct diffusion of substances in and out of all cells impossible on a realistic timescale
- The need for exchange surfaces in multicellular organisms in terms of surface area : volume ratio
- Why oxygen and carbon dioxide need to be moved into and out of cells and organisms
- How oxygen is carried in the blood from the lungs to the cells in mammals
- How carbon dioxide is carried from the cells to the lungs in the blood of mammals

What will I study in this chapter?

- The fluid mosaic model of the cell surface membrane, including protein pores and active pumps
- The role of passive transport in the cell, including diffusion, facilitated diffusion and osmosis
- The importance of osmosis and water potentials in animal cells and plant cells
- The process of active transport into and out of the cell, including the hydrolysis of ATP
- The transport of large molecules into and out of the cell using vesicles in the processes of endocytosis and exocytosis
- Why organisms need specialised gas exchange surfaces as they increase in size and metabolic rate
- Calculating the rate at which substances of a given size will diffuse at a known temperature (Fick's Law)
- The main features of a successful gas exchange surface
- The structure and function of the gas exchange system in humans, including the adaptations that make it effective

What will I study later?

Chapter 4A

- The role of osmosis in moving water into a plant through the root hair cells and through the plant to the xylem and phloem
- The importance of osmosis in support (turgor), growth, movement, tropisms and photosynthesis in plants
- The role of active transport in the movement of minerals into plant cells and in the transport of sugars in the phloem
- How gas exchange in plants is linked to water movements in transpiration

Chapter 5A (Book 2: IAL)

- Why it is important for plants to take in carbon dioxide for the biochemistry of photosynthesis and how oxygen is a waste product of the process

Chapter 6A (Book 2: IAL)

- How *Mycobacterium tuberculosis* infects human cells and affects the lungs

Chapter 7A (Book 2: IAL)

- Why oxygen is needed in cells, how it is used in the biochemistry of cellular respiration and how carbon dioxide is produced in the same process

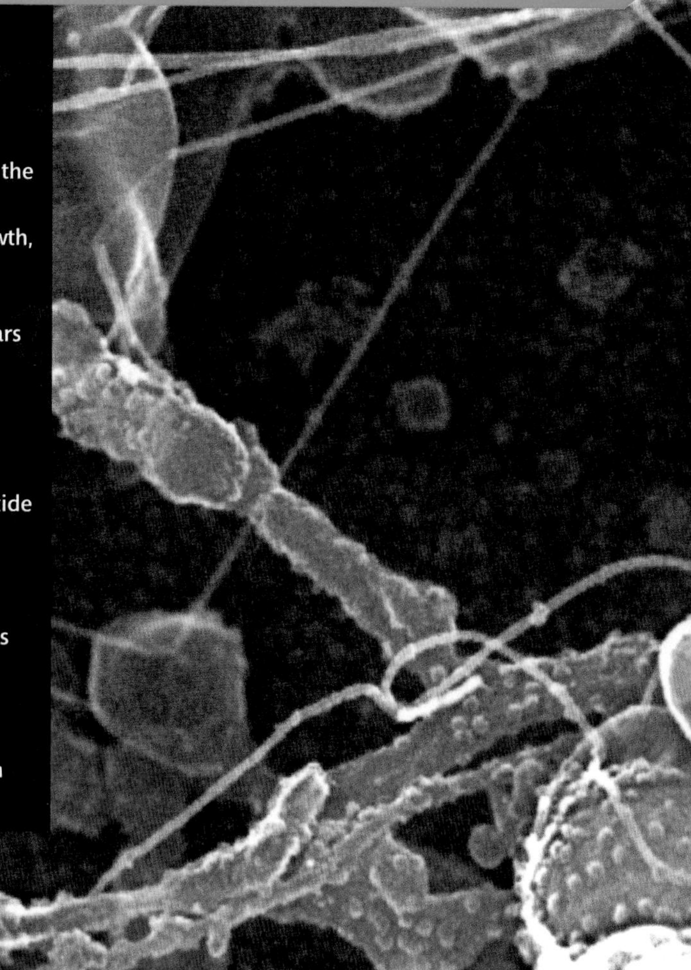

1 CELL MEMBRANES

SPECIFICATION REFERENCE 2.2(i) 2.2(ii) CP3

LEARNING OBJECTIVES

- Know the structure and properties of cell membranes.
- Understand how models such as the fluid mosaic model of membrane structure are interpretations of data used to develop scientific explanations of the structure and properties of cell membranes.

In this chapter, you are going to look at some of the ways in which cells and organisms work, focussing on the role of membranes in the movement of substances into and out of cells. Every process of life in a whole organism depends on reactions happening in its cells. Many of these basic cellular processes are affected by the **cell membrane**, so it is important to understand the best available model of this important structure.

MEMBRANES IN CELLS

There are many membranes within cells, such as those that surround **organelles** like the nucleus and mitochondria. But the most obvious membrane is the cell surface membrane (outer cell membrane) which forms the boundary of the cell. Anything that leaves or enters the cell must pass through this membrane. All membranes act as barriers, controlling what passes through them and allowing the fluids either side of them to have different compositions. This makes it possible to have the right conditions for a particular reaction in one part of a cell and different conditions to suit other reactions elsewhere in the cell.

Membranes perform many other functions too. Many chemical processes take place on membrane surfaces; for example, some of the reactions of respiration happen on the inner mitochondrial membrane. **Enzymes** and all the factors that are needed to make the reactions happen, are held closely together so that the process can go easily from one reaction to the next. The cell surface membrane must also be flexible to allow the cell to change shape; this might be very slightly as its water content changes, or quite dramatically, for example, when a white blood cell engulfs a bacterium. Chemical secretions made by the cell are contained in membrane bags, called **vesicles**, which must be able to combine with the cell surface membrane to release their contents.

THE STRUCTURE OF MEMBRANES

Our current model of the structure of membranes has been developed over many years. The model developed as microscopy improved, from the light microscope to the electron microscope and then the scanning electron microscope. In the future, there will probably be further adjustments to the model presented here but overall the ideas are unlikely to change much. The membrane is composed of mainly two types of molecule – **phospholipids** and proteins – arranged in a very specific way.

THE PHOSPHOLIPID BILAYER

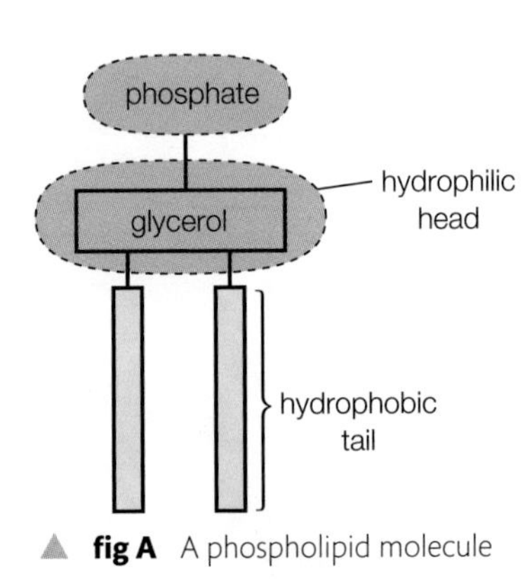

▲ **fig A** A phospholipid molecule

The lipids in the membrane are called **polar lipids**. These are lipid molecules with one end joined to a polar group. Many of the polar lipids in the membrane are phospholipids, with a phosphate group forming the polar part of the molecule (see **fig A**).

The fatty acid chains of a phospholipid are neutral and insoluble in water. In contrast, the phosphate head carries a negative charge and is soluble in water.

When the phospholipids contact water, the two parts of the molecule behave differently. The polar phosphate part is **hydrophilic** (water-loving) and dissolves easily in water. The lipid tails are **hydrophobic** (water-hating) and insoluble in water. If the molecules are tightly packed in water they form either a **monolayer**, with the hydrophilic heads in the water and the hydrophobic lipid tails in the air, or clusters which are called **micelles**. In a micelle, all the hydrophilic heads point outwards and all the hydrophobic tails are hidden inside (see **fig B**).

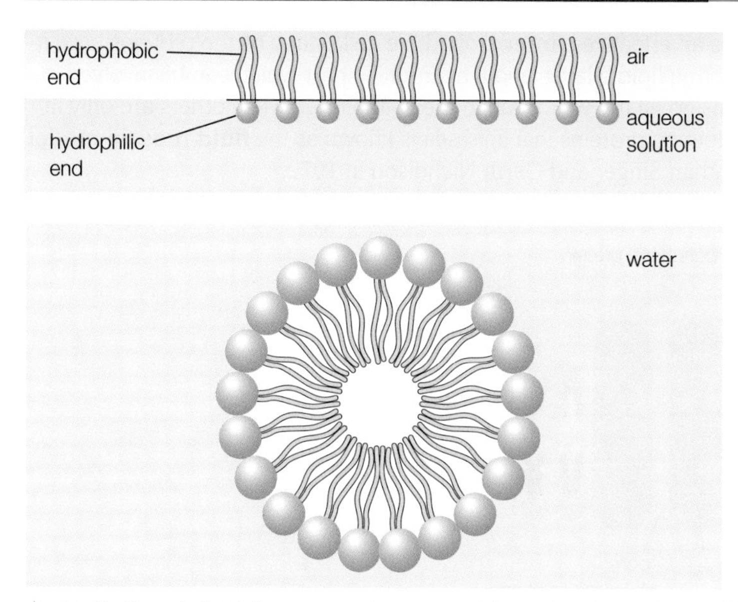

▲ **fig B** Phospholipids form a monolayer at an air/water junction and a micelle when water surrounds them.

A monolayer may develop at a surface between air and water, but this does not happen often in living cells where there are water-based solutions on either side of membranes. With water on each side the phospholipid molecules form a **bilayer**, with the hydrophilic heads pointing into the water while the hydrophobic tails are protected in the middle (see **fig C**). This structure, the **unit membrane**, is the basis of all membranes. However, a simple phospholipid bilayer alone does not explain either the microscopic appearance of membranes or the way in which they behave.

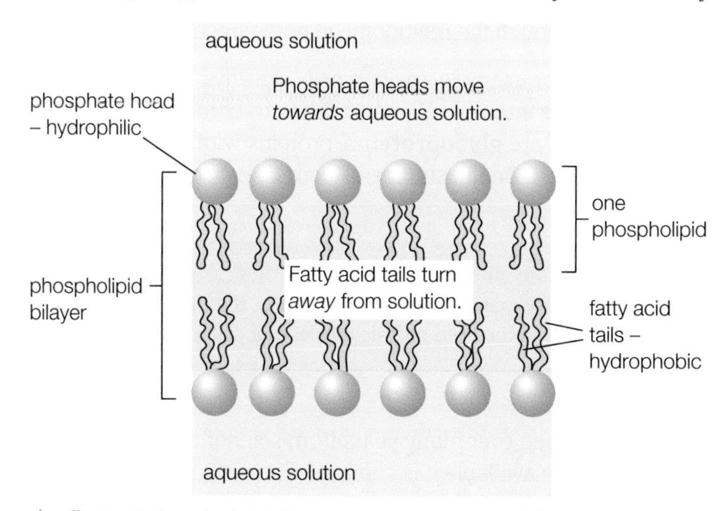

▲ **fig C** A phospholipid bilayer is the key structure of all membrane structures in a cell.

A simple phospholipid bilayer allows fat-soluble organic molecules to pass through it, but many vital molecules needed in cells are ionic. While these dissolve in water, they cannot dissolve in or pass through lipids, even polar lipids. They can enter cells because the membrane consists not only of lipids but also of proteins and other molecules. You will learn more about the structure and properties of proteins in **Chapter 2B**.

THE FLUID MOSAIC MODEL OF THE CELL MEMBRANE

The best model of a membrane we have today has the phospholipid bilayer as a fluid system. Many proteins and other molecules are floating within it like icebergs, together with others that are fixed in place (see **fig D**). The proportion of phospholipids containing unsaturated fatty acids (see **Section 1A.4**) in the bilayer seems to affect how freely the proteins move within the membrane. Another important lipid in the cell membrane is cholesterol (see **Chapters 1B** and **1C** in relation to heart disease). Cholesterol is a more rigid molecule than many of the phospholipids and so makes the membrane more stable and stronger. This makes it harder for small molecules and ions to pass

EXAM HINT

Remember the presence of cholesterol in the phospholipid bilayer makes it more stable, stronger and a more effective barrier to the movement of small molecules and ions.

through the membrane, so it acts as an effective barrier around the cell. Many of the proteins have a hydrophobic part, which is buried in the lipid bilayer, and a hydrophilic part which can be involved in a variety of activities. Some proteins go all the way through the lipid bilayer, while others are only in part of the bilayer. This model of floating proteins in a lipid sea is known as the **fluid mosaic model** and it was first proposed by S. Jonathan Singer and Garth Nicholson in 1972.

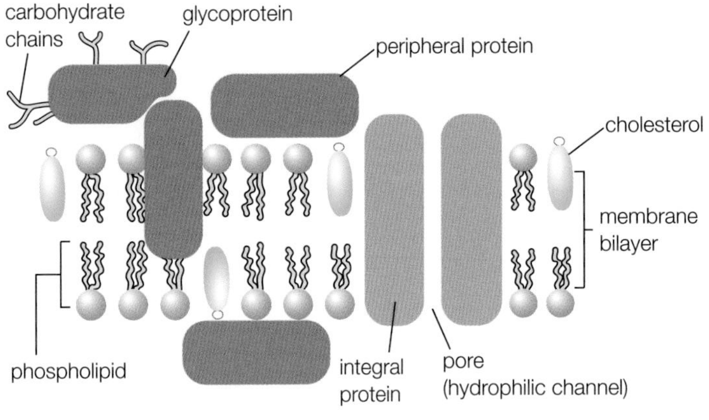

fig D The fluid mosaic model of the cell membrane – a phospholipid sea with associated proteins, which may be floating or anchored within the membrane.

LEARNING TIP

Remember the phospholipid bilayer is not permeable to polar substances. It is the proteins that make the membrane selectively permeable.

One of the main functions of the membrane proteins is to help substances move across the membrane. The proteins may make pores or channels – some permanent, some temporary – which allow specific molecules to move through the membrane. Some of these channels can be open or shut, depending on conditions in the cell. These are known as **gated channels**. Some of the protein pores are active carrier systems, using energy to move molecules, as you will see later. Others are simply gaps in the lipid bilayer which allow ionic substances to move through the membrane in both directions.

LEARNING TIP

An important aspect of biology is relating structure to function. Here we can easily describe the relationship between certain structures in the membrane and their function, such as the presence of channel proteins to enable the passage of charged particles.

Proteins may act as specific receptor molecules – for example, making cells sensitive to a specific hormone. They may be enzymes, particularly on the internal cell membranes, to control reactions linked to that membrane. Some membrane proteins are **glycoproteins**, proteins with a carbohydrate part added to the molecule. These are very important on the surface of cells as part of the way cells recognise each other.

EXAM HINT

The proteins and glycoproteins in a membrane have very many roles - be clear about the different roles of glycoproteins, peripheral proteins and integral proteins in the fluid mosaic model of a cell membrane.

BUILDING A MODEL OF THE MEMBRANE

Our ideas about membrane structures have developed over many years from scientific observations. These observations have changed as the technology available has changed. Here is a summary of the story of where our ideas of the cell membrane have come from.

- The first indications that lipids are important components of cell membranes came at the end of the 19th century when Charles Ernest Overton (1865–1933) made a series of observations on how easily substances passed through cell membranes. Lipid-soluble substances entered more easily than any others, so he concluded that a large part of the membrane structure must be lipid.
- The idea that cell membranes are not rigid but are fluid structures came from observations of the behaviour of cell surface membranes when cells join, together with the way in which most membranes seal themselves if they are punctured with a fine needle.
- In 1917, Irving Langmuir (1881–1957) demonstrated the lipid monolayer mentioned earlier. In collaboration with Katharine Blodgett, he also developed a piece of equipment for collecting lipid monolayers called the Langmuir–Blodgett trough.
- In 1925, two Dutch scientists, Evert Gorter and François Grendel, decided to measure the total size of the monolayer film formed by lipids extracted from human red blood cells (erythrocytes). They also estimated the total surface area of an erythrocyte, and found that their measured area of monolayer was about twice the estimated surface area of the cell. They concluded that the cell membrane was a lipid bilayer. We now know that their results were wrong in two ways – they

didn't extract all the lipid, and they miscalculated the surface area of the erythrocytes because they thought they were flat rather than biconcave discs. Nevertheless, their conclusions were correct – by lucky coincidence, the two errors cancelled each other out.

- By 1935, Hugh Davson and James Danielli had produced a further model of the membrane with a lipid centre coated on each side by protein. This is the basis of our current ideas.
- The Davson–Danielli hypothesis was backed up in the 1950s by work on the electron microscope by James D. Robertson. He found ways of staining the membrane which revealed it as a three-layered structure – two distinct lines with a gap in the middle. When the membrane was treated with propanone (acetone) to extract the lipid, the two lines remained intact, suggesting that they were the protein layers.
- More recently, techniques such as X-ray diffraction and new electron microscopy methods have added to our knowledge of the structure of cell membranes. This gives more detail of the layers, the pores and the carrier molecules – **fig E** shows the pores in the nuclear membrane, for example.

▲ **fig E** The membrane pores through which mRNA leaves the nucleus are clearly visible in this freeze fracture electron micrograph of the nuclear membrane of a cell.

EXAM HINT

This is a very good example of how technical developments over time enable better scientific understanding.

CHECKPOINT

1. Which kinds of molecule are in the structure of a membrane and how do their properties affect the properties of the membrane itself?
2. Explain why a membrane may be more fluid when it contains more unsaturated fatty acids. (You may need to refer back to **Section 1A.4**.)

SKILLS ADAPTIVE LEARNING

3. Choose **three** different pieces of evidence that have been important in developing our models of the cell membrane. For each one, explain how the evidence moved the model forward towards the fluid mosaic model.

PRACTICAL SKILLS CP3

Investigating membrane properties

Cell membranes act as barriers keeping some substances in and other substances out. The effectiveness of the barrier depends on the structure of the cell membrane. You can investigate the properties of the cell membrane by measuring its permeability to particular molecules. For example, the cell membranes of samples of red beet in water keep the red colour inside the cells. If the permeability of the membrane is changed, the red colour will appear in the water. You can measure the amount of red colour in the water by eye or by using a colorimeter. Alcohol dissolves lipids, and raised temperatures make lipids more fluid and can denature proteins. How do you think these factors might affect the permeability of the cell membranes? You can use red beet to find out.

SUBJECT VOCABULARY

cell membrane the selectively permeable membrane which surrounds the cytoplasm of a cell, acting as a barrier between the cell contents and their surroundings
organelles sub-cellular bodies found in the cytoplasm of cells
enzymes proteins that act as biological catalysts for a specific reaction or group of reactions
vesicles membrane 'bags' that hold secretions made in cells
phospholipids chemicals in which glycerol bonds with two fatty acids and an inorganic phosphate group
polar lipids lipids with one end attached to a polar group (e.g. a phosphate group) so that one end of the molecule is hydrophilic and one end is hydrophobic
hydrophilic a substance with an affinity for water that will readily dissolve in or mix with water
hydrophobic a substance that tends to repel water and that will not mix with or dissolve in water
monolayer a single closely packed layer of atoms or molecules
micelles a spherical aggregate of molecules in water with hydrophobic areas in the middle and hydrophilic areas outside
bilayer a double layer of closely packed atoms or molecules
unit membrane a lipoprotein membrane which is composed of two protein layers enclosing a less dense lipid
fluid mosaic model the current model of the structure of the cell membrane including floating proteins forming pores, channels and carrier systems in a lipid bilayer
gated channels protein channels through the lipid bilayer of a membrane that are opened or closed, depending on conditions in the cell
glycoproteins conjugated proteins with a carbohydrate prosthetic group

2 CELL TRANSPORT AND DIFFUSION

SPECIFICATION REFERENCE 2.5(i) 2.5(ii)

LEARNING OBJECTIVES

- Understand what is meant by passive transport (diffusion, facilitated diffusion), active transport (including the role of ATP as an immediate source of energy), endocytosis and exocytosis.
- Understand the involvement of carrier and channel proteins in membrane transport.

We know from scientific analysis that the concentration of substances on either side of a membrane can be very different. This suggests that a membrane exercises control over which substances move across it. The properties of the membrane affect the transport of substances into and out of the cell, but the properties of the molecules to be transported also have an effect. The size of a molecule is important in how it is transported through cell membranes and so is its solubility in lipids and water. The presence or absence of charge on a molecule also affects how it is transported. For example, some substances simply pass through the membrane in the process of diffusion. This especially applies to those that dissolve very easily in lipids. Other very small molecules, such as the gases oxygen and carbon dioxide, also pass freely in and out of cells through the membrane. Some large molecules, such as steroid hormones, are not transported through the membrane. Many charged particles, such as sodium ions, need specific carriers and pores to get from one side to the other. The movement of particles across membranes is vital in living organisms for everything from cellular respiration to rapid movements of all or part of an organism (see **fig A**).

THE MAIN TYPES OF TRANSPORT ACROSS MEMBRANES

Substances are transported into, out of and around cells by a variety of different mechanisms. **Passive transport** takes place when there are concentration, pressure or electrochemical gradients, and it involves no energy from the cell. **Active transport** involves moving substances into or out of the cell by using adenosine triphosphate (ATP) which is produced during cellular respiration.

PASSIVE TRANSPORT MECHANISMS

There are three main types of passive transport in cells.

- **Diffusion** – the movement of particles in a liquid or gas down a concentration gradient. They move from an area where they are at a relatively high concentration to an area where they are at a relatively low concentration by random movements. Cell membranes are no barrier to the diffusion of small particles such as the gases oxygen and carbon dioxide.
- **Facilitated diffusion** – diffusion that takes place through carrier proteins or protein channels. The protein-lined pores of the cell membrane make facilitated diffusion possible.
- **Osmosis** – a specialised form of diffusion that involves the movement of solvent molecules (in cells, this is free water molecules) down a water potential gradient through a **partially permeable membrane**. You will learn more about water potential gradients in **Section 2A.3**. The partially permeable nature of the cell membrane means **solutes** (dissolved substances) can be accumulated either side of the membrane and this results in the movement of water by osmosis across the membrane.

ACTIVE TRANSPORT MECHANISMS

Active transport is the movement of substances across the membrane of cells using ATP as an immediate source of energy. Active transport always involves a **carrier protein** which carries molecules or ions through the membrane using energy supplied by the breakdown of ATP.

There are two other mechanisms for moving substances into and out of cells. These also use energy from ATP. They are:

- **Endocytosis** – the movement of large molecules into cells through vesicle formation. The fluid nature of the cell membrane makes it possible to form vesicles.
- **Exocytosis** – the movement of large molecules out of cells through fusion of vesicles to the membrane.

You will learn more about active transport, endocytosis and exocytosis in **Section 2A.4**.

fig A The rapid movements of the sensitive plant *Mimosa pudica* are the result of different types of transport within the cells of the plant.

You will learn more about the ways in which the membrane is adapted to its transport functions as you look at the different transport mechanisms in more detail, beginning with diffusion.

DIFFUSION

In physical terms, diffusion is the movement of the molecules of a liquid or gas from an area where they are at a high concentration to an area where they are at a lower concentration. We say that they move down their concentration gradient (see **Section 1B.1**). This happens because of the random motion of molecules due to the energy they have, which depends on the temperature. If you have many molecules tightly packed together, random motion means that they spread out and eventually reach a uniform distribution. The molecules do not stop moving once they reach a uniform distribution, but the movement no longer causes a net change in concentration because equal numbers are moving in all directions (see **fig B**).

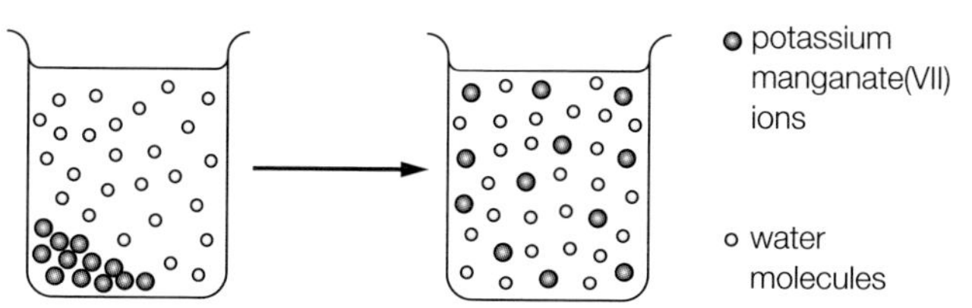

▲ **fig B** If the beaker is left to stand, diffusion takes place as the random movement of both the water and the potassium manganate(VII) ions ensures that they become evenly mixed.

For many small molecules, like oxygen and carbon dioxide, the membrane is not a barrier and they can diffuse freely across it. This movement, by diffusion alone, is a form of passive transport. However, hydrophilic molecules and ions which are larger than carbon dioxide molecules cannot move across the membrane by simple diffusion.

EXAM HINT

Do not say that no energy is required for diffusion - it requires no metabolic energy from ATP, but it depends on the kinetic energy of the molecules.

FACILITATED DIFFUSION

Substances with a strong positive or negative charge and large molecules cannot cross cell membranes by simple diffusion. Nevertheless, they may move into and out of the cell down a concentration gradient by a specialised form of diffusion. Facilitated diffusion involves proteins in the membrane that allow specific substances to move passively down their concentration gradient (see **fig C**). The proteins may be channel proteins that form pores through the membrane. Each type of channel protein allows one particular type of molecule to pass through; this depends on the molecule's shape and charge. For example, some are sodium ion channels and others are potassium ion channels. Some channels open only if a specific molecule is present, or if there is an electrical change across the membrane, such as during the passage of nerve impulses along neurones. These are called gated channels.

Another type of facilitated diffusion depends on carrier molecules floating on the surface of the membrane. The carriers will be found on the *outside* surface of the membrane structure when a substance is to be moved *into* the cell or organelle, and on the *inside* for transport *out* of the cell or organelle. The protein carriers are specific for particular molecules or groups of molecules, depending on the shape of the protein carrier and the substance to be carried. Once a carrier has picked up a molecule it rotates through the membrane to the other side, carrying the molecule with it, and then releases the molecule. The movement through the membrane takes place because the carrier changes shape once it is carrying something. The process can only take place down a concentration gradient – from a high concentration of a molecule to a low concentration. It does not use energy, so it is considered to be a form of diffusion. For example, red blood cells have a carrier to help glucose move into the cells rapidly.

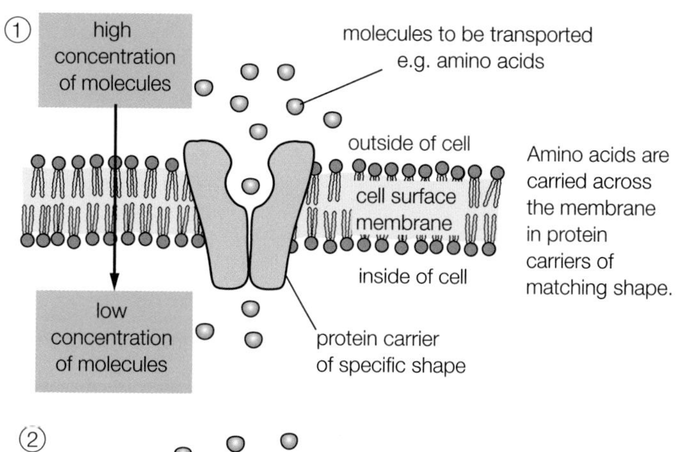

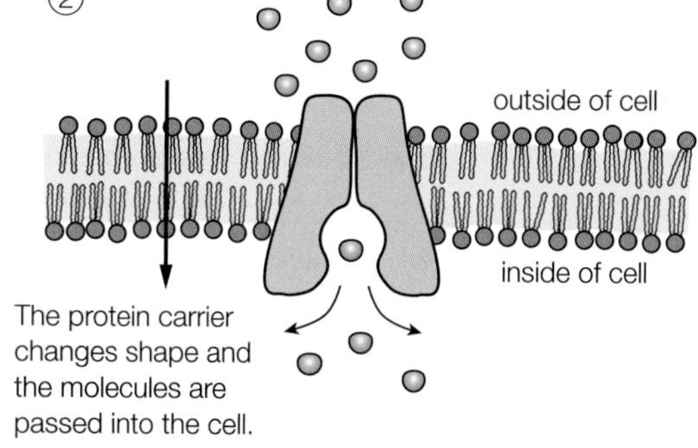

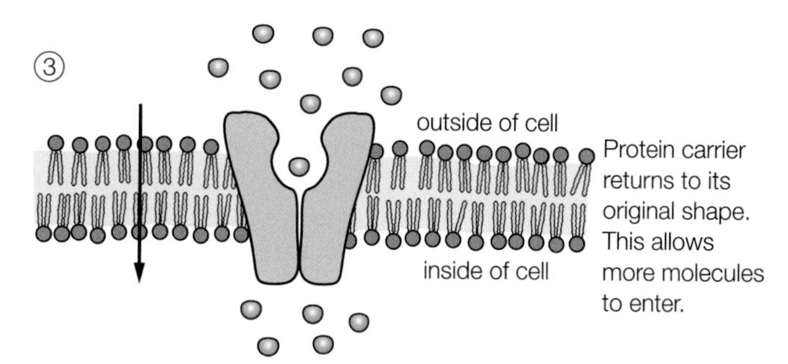

▲ **fig C** Facilitated diffusion acts as a ferry across the lipid membrane sea. It is not an active process, so it can only work when the concentration gradient is in the right direction.

IN SUMMARY

The three types of passive transport are summarised in **fig D**.

EXAM HINT

Remember that facilitated diffusion does not use energy from ATP – it needs a concentration gradient.

LEARNING TIP

Remember that facilitated means helped. The proteins help diffusion to occur by providing a pathway.

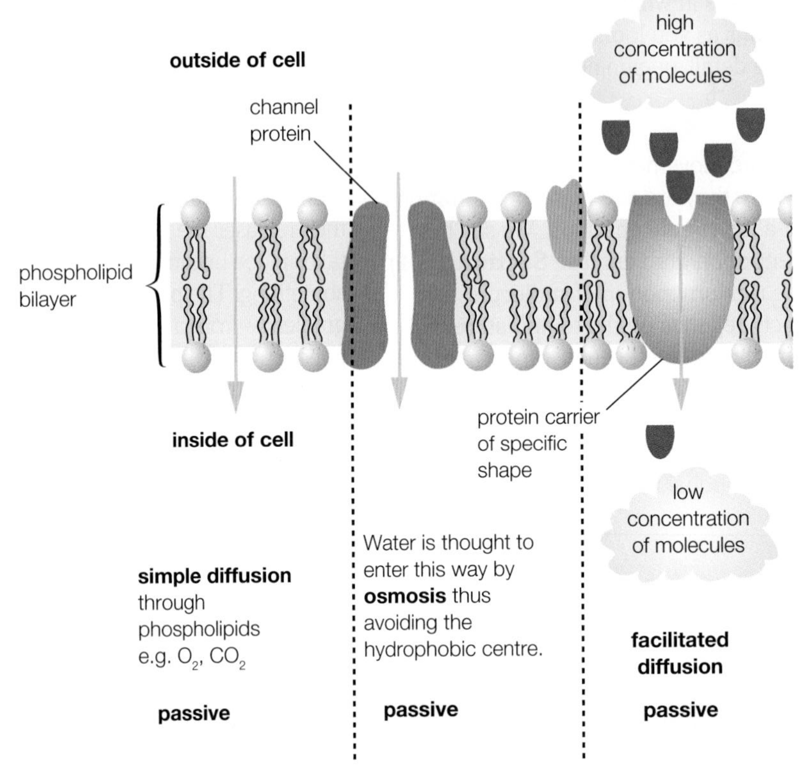

▲ **fig D** The main passive transport routes through a cell surface membrane

SKILLS CRITICAL THINKING

CHECKPOINT

1. Explain why transport systems are needed across membranes.
2. Describe the conditions needed for the passive transport of molecules into a cell.
3. Water and ions often enter the cell through protein pores but they cannot pass through the lipid layer in the same way that oxygen and carbon dioxide do. Why not?
4. Explain the differences between simple diffusion and facilitated diffusion.

SUBJECT VOCABULARY

passive transport transport that takes place as a result of concentration, pressure or electrochemical gradients and involves no energy from a cell
active transport the movement of substances into or out of the cell using ATP which is produced during cellular respiration
diffusion the movement of the particles in a liquid or gas down a concentration gradient from an area where they are at a relatively high concentration to an area where they are at a relatively low concentration
facilitated diffusion diffusion that takes place through carrier proteins or protein channels
osmosis a specialised form of diffusion that involves the movement of solvent molecules down their water potential gradient
partially permeable membrane a membrane which only allows specific substances to pass through it
solute a substance in a solution, dissolved in the solvent
carrier protein a protein that moves a substance through the membrane in active transport using energy from the breakdown of ATP or in passive transport such as facilitated diffusion down a concentration gradient
endocytosis the movement of large molecules into cells through vesicle formation
exocytosis the movement of large molecules out of cells by the fusing of a vesicle containing the molecules with the surface cell membrane; the process requires ATP

3 OSMOSIS: A SPECIAL CASE OF DIFFUSION

LEARNING OBJECTIVES

- Understand what is meant by *osmosis* in terms of the movement of free water molecules through a partially permeable membrane, down a water potential gradient.

As you have learned, diffusion takes place when molecules or ions can move freely. Free water molecules (those not associated with a solute) can move easily even through partially permeable membranes. So, as a result of random motion, water molecules will tend to move across a partially permeable membrane down their water potential gradient.

LEARNING TIP

Remember that water always flows downhill, so it goes down the potential gradient.

WHAT IS OSMOSIS?

Osmosis in cells can be defined as the net movement of free water molecules through a partially permeable membrane, down a water potential gradient. In living organisms, the solvent is always water, and membranes in cells are generally partially permeable. This means that they let some molecules through, but not others. But what is a water potential gradient? The **water potential** of a solution measures the concentration of free water molecules – in other words, the water molecules that are ***not*** associated with solute molecules. More free water molecules mean a higher water potential. Water molecules will move from an area of high water potential to an area with a low water potential, where there are fewer free water molecules. This is a water potential gradient. So, osmosis in cells involves the movement of water from a region of high water potential to a region of lower water potential down a water potential gradient across a partially permeable membrane such as the cell surface membrane or nuclear membrane. **Fig A** summarises the processes of diffusion and osmosis.

LEARNING TIP

Remember that a gradient must be between two places. Don't write that water moves from a high gradient to a lower gradient.

If the solution bathing the outside of a cell has a higher water potential than the solution inside the cell, there will be a water potential gradient that encourages water molecules to move into the cell. If the opposite is true and the solution bathing the cell has a higher concentration of solute than the cell contents, the water potential gradient will be from the inside to the outside and free water molecules will move out of the cell. The **osmotic concentration** of a solution concerns only those solutes that have an osmotic effect. Many large insoluble molecules found in the cytoplasm, for example starch and lipids, do not affect the movement of water and so are ignored when considering osmotic concentration. Only soluble particles are considered, including the big plasma proteins such as albumin and fibrinogen.

In the context of living cells, the movement of water by osmosis, and the control of this process, is very important. In animal cells, it is essential that water does not simply move continuously into the cells from a dilute external solution. If this were to happen, the cells would swell up and burst.

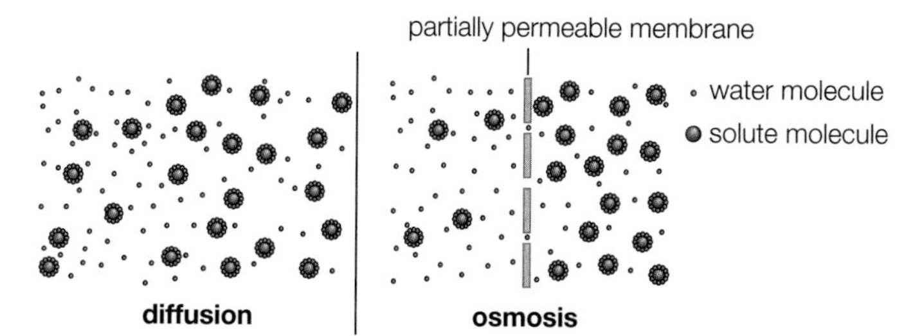

fig A In diffusion, the random movement of particles results in an even distribution of both solute and solvent particles. In osmosis, a partially permeable membrane means only solvent molecules and very small solute particles can move freely.

MODELLING OSMOSIS IN CELLS

You can make a model cell using an artificial membrane that is permeable to water molecules and impermeable to others such as sucrose. There are many experiments showing the movement of water in these circumstances, and one of the simplest is illustrated in **fig B**. The presence or absence of sucrose in the different regions of the model can be shown by carrying out Benedict's test for non-reducing sugars on the solutions (see **Section 1A.2**).

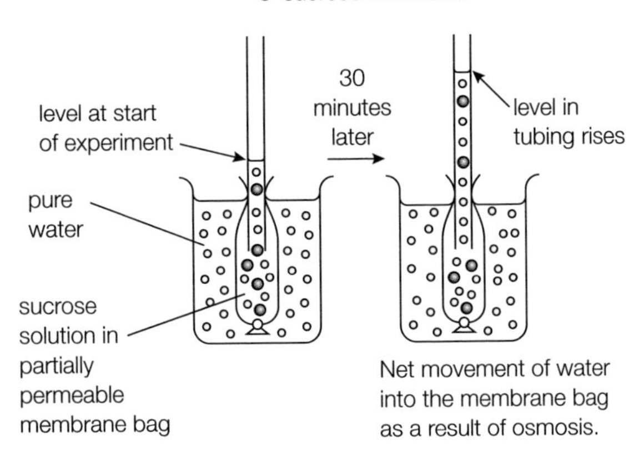

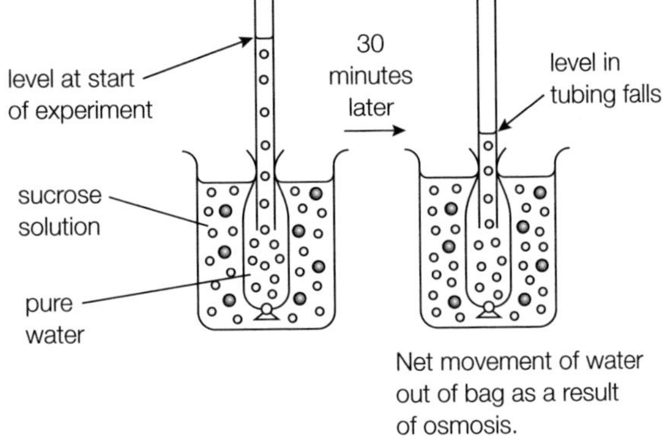

▲ **fig B** The artificial partially permeable membrane in this experiment provides a model for the cell surface membrane. It allows water molecules to pass through freely, but not the solute molecules.

LEARNING TIP

Imagine that the partially permeable membrane has tiny holes in it, which are just big enough for the water molecules to pass through.

OSMOTIC CONCENTRATIONS

During osmotic experiments, cells are often immersed in solutions of different osmotic concentration. The osmotic concentration of a solution is a measure of only those dissolved substances that have an osmotic effect. This is especially important in living things because many of the large molecules found in the cytoplasm of a cell do not affect the movement of water into or out of the cell and so we ignore them when calculating osmotic concentration.

- In an **isotonic** solution, the osmotic concentration of the solutes in the solution is the same as that in the cells.
- In a **hypotonic** solution, the osmotic concentration of solutes in the solution is lower than that in the cytoplasm of the cells.
- In a **hypertonic** solution, the osmotic concentration of solutes in the solution is higher than that in the cytoplasm.

OSMOSIS IN ANIMAL CELLS

Osmosis needs to be carefully controlled in animal cells. The net movement of water in or out needs to be kept to a minimum. Animal cells are effectively like fragile balloons filled with jelly. When too much water moves in, the cells burst; when too much moves out, the cells shrivel as the concentrated cytoplasm loses its internal structure (see **fig C**). This means that the chemical reactions that normally take place in the cell stop working.

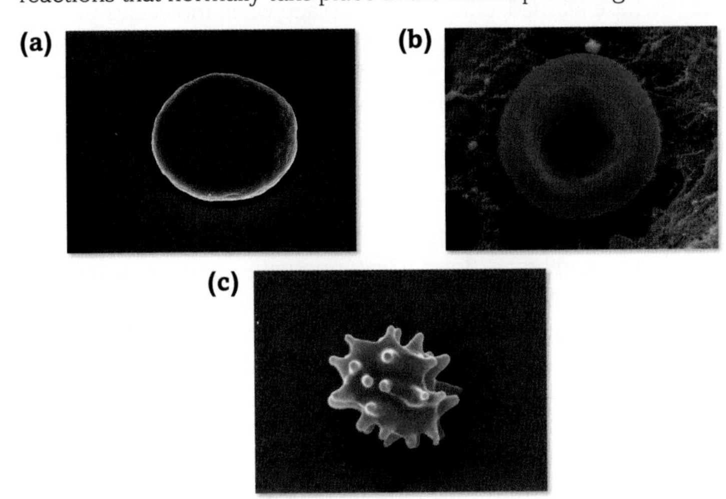

▲ **fig C** The effects of osmosis on red blood cells show why the systems of the body that maintain solute concentrations and water balance are so important. **(a)** In hypotonic solution, water moves in and the cell swells and bursts; **(b)** in isotonic solution the red blood cell maintains its normal shape; **(c)** in hypertonic solution, water moves out and the cell shrivels.

OSMOSIS IN PLANT CELLS

Plant cells are also like fragile balloons filled with jelly – but the balloon is inside the rigid 'box' of the cellulose cell wall. In plant cells, the cellulose cell wall prevents cells bursting. If the surrounding fluid is hypotonic to the cytoplasm of a plant cell, water will enter the cell by osmosis – but not indefinitely. As the cytoplasm swells and presses on the cell walls, it generates **hydrostatic pressure**. The inward pressure of the cell wall on the cytoplasm increases until it cancels out the tendency for water molecules to move in. This inward pressure is called the pressure potential. When the osmotic force moving water into the plant cell is balanced by the pressure potential forcing it out, the plant cell is rigid, in a state known as **turgor**. Most plant cells are in a state of turgor most of the time – the rigid structure supports the stems and leaves of the plant.

Most of the work on osmosis is done on plant cells. They are generally bigger and easier to see with a light microscope than animal cells, and the changes due to osmosis are easier to see and measure than those in animal cells. If plant cells are put in a solution which is slightly hypertonic, water moves out of the cell by osmosis and turgor is lost. The cell membrane begins to pull away from the cell wall as the protoplasm shrinks. This is called **incipient plasmolysis**. We measure incipient plasmolysis using serial dilutions, looking for the point at which 50% of the cells are plasmolysed and 50% are not. This is the concentration that is equivalent to the solute potential of the cell sap. If the cell is placed in a hypertonic solution, so much water will leave the cell that the vacuole will be reduced and the protoplasm will shrink away from the cell walls completely – the cells suffer **plasmolysis**. However, because of the cell wall, the size and shape of plant cells does not change much whether they are fully turgid or fully plasmolysed. They do not swell and burst, nor do they become very small. It is only the contents that change (see **fig D**).

(a) 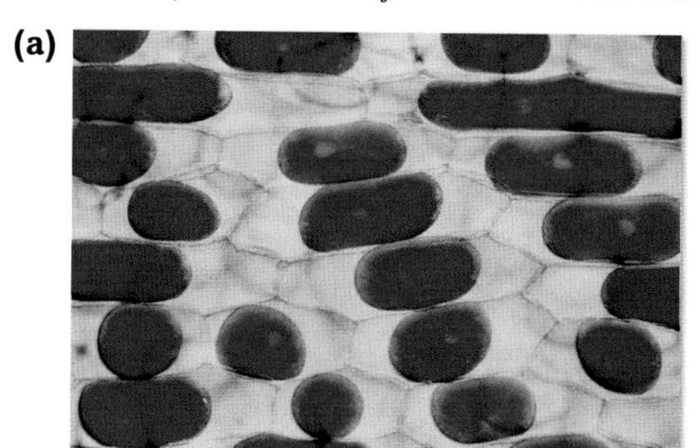 **(b)** 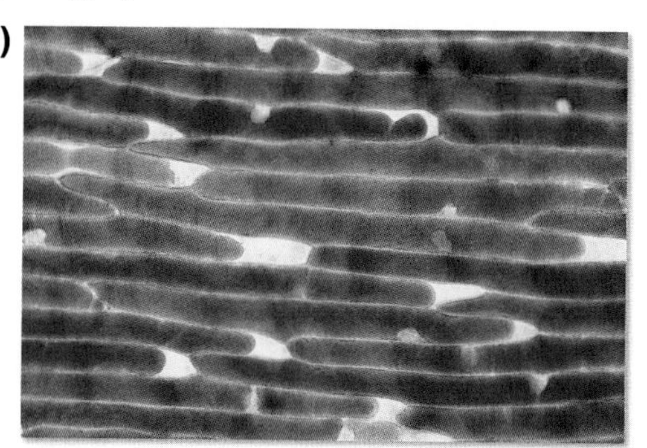

fig D Plant cells from red beet showing **(a)** plasmolysis; and **(b)** turgor.

EXAM HINT

Always use the terms *water potential* and *water potential gradient* to explain osmosis and the movement of water in and out of cells.

CHECKPOINT

1. Write a definition of osmosis using the term *water potential* and give **two** examples of where osmosis is important in living organisms.
2. In an experiment, human cheek cells were placed in three solutions: an isotonic solution, a hypertonic solution and a hypotonic solution. Describe what you would expect to happen in each case. Explain your answers in terms of water potential and osmosis.
3. Without osmosis, plants as we know them would not survive. True or false? Discuss.

SKILLS INTERPRETATION

SKILLS CREATIVITY

SUBJECT VOCABULARY

water potential a measure of the potential for water to move out of a solution by osmosis
osmotic concentration a measure of the concentration of the solutes in a solution that have an osmotic effect
isotonic solution a solution in which the osmotic concentration of the solutes is the same as that in the cells
hypotonic solution a solution in which the osmotic concentration of solutes is lower than that in the cell contents
hypertonic solution a solution in which the osmotic concentration of solutes is higher than that in the cell contents
hydrostatic pressure the pressure exerted by a fluid in an equilibrium
turgor the state of a plant cell when the solute potential causing water to be moved into the cell by osmosis is balanced by the force of the cell wall pressing on the protoplasm
incipient plasmolysis the point at which so much water has moved out of the cell by osmosis that turgor is lost and the cell membrane begins to pull away from the cell wall as the protoplasm shrinks
plasmolysis the situation when a plant cell is placed in hypertonic solution when so much water leaves the cell by osmosis that the vacuole is reduced and the protoplasm is concentrated and shrinks away from the cell walls

4 ACTIVE TRANSPORT

SPECIFICATION REFERENCE 2.5(i) 2.5(ii)

LEARNING OBJECTIVES

- Understand what is meant by active transport, including the role of ATP as an immediate source of energy, endocytosis and exocytosis.
- Understand the involvement of carrier and channel proteins in membrane transport.

As you have already learned, diffusion and facilitated diffusion are passive processes that allow small molecules to move across membranes. Cells can maintain steep concentration gradients by simply 'mopping up' the substance as soon as it arrives inside the cell. They can do this by immediately starting to metabolise the substance – by chemically changing it to something else – or by using a carrier molecule on the surface of an organelle to take it into the organelle.

Both diffusion and facilitated diffusion rely on a concentration gradient in the right direction to move a substance into the cell. However, cells have another system called active transport. This enables them to move substances across membranes against a concentration or electrochemical gradient using energy supplied by the cell.

HOW DOES ACTIVE TRANSPORT WORK?

Active transport involves a carrier protein, which often spans the whole membrane (see **fig A**). It may be very specific, picking up only one type of ion or molecule, or it may work for several relatively similar substances that have to compete with each other for a place on the carrier.

The energy needed for active transport is provided by molecules of adenosine triphosphate (ATP). Cells that carry out a lot of active transport generally have many mitochondria to supply the ATP they need. The active transport carrier system in the membrane involves the enzyme **ATPase**. This enzyme catalyses the hydrolysis of ATP by breaking one bond and forming two more. This provides the energy needed to move carrier systems in the membrane or to release the transported substances and return the system to normal.

Active transport is a one-way system for each specific substance – the carriers will not transport a substance back through the membrane. An active transport system moves substances only in the direction required by the cell – these transport systems are also known as gated channels because they act as a gate and control what comes into and out of the cell. In some cases, the substances will move out again through open channels, down the concentration or electrochemical gradient that has just been overcome, but active transport can move substances in faster than they can move out by diffusion.

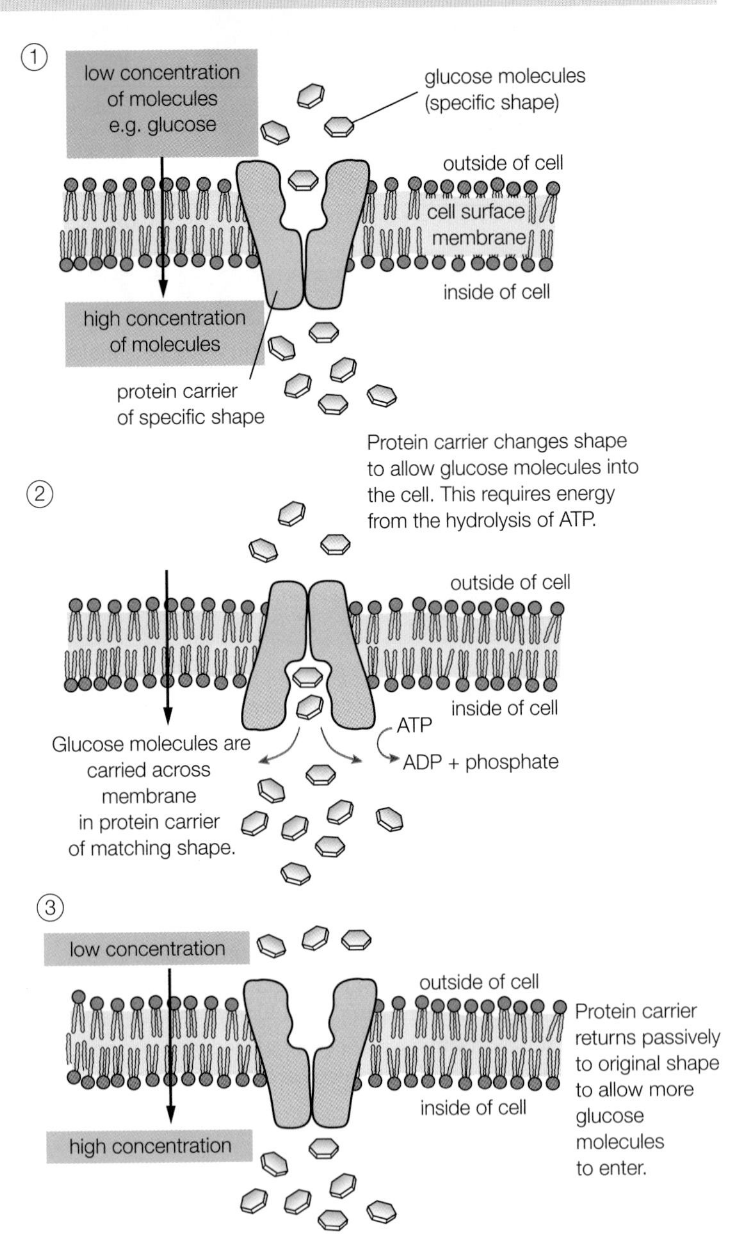

▲ **fig A** Using active transport, cells can move selected substances into or out of the cell, even when the concentration gradient is in the wrong direction.

EXAM HINT

Remember that active transport moves substances against the concentration gradient and requires energy from ATP (metabolic energy) to do it. You must be able to discuss the role of ATP in changing the shape of the carrier protein to answer questions on this topic well.

In active transport, the movement of a substance is often linked with that of another particle such as a sodium ion. One of the best known examples of active transport is the sodium pump that actively moves potassium ions into the cell and sodium ions out. This pump is vital for the working of the nervous system – each nerve impulse depends on an influx of sodium ions through the axon membrane. These ions must be actively pumped out of the neurone again so that another impulse can pass.

The combination of diffusion, facilitated diffusion and active transport means that the cell surface membrane provides control over what moves into or out of the cell. The concentration of ions and molecules within the cell can be maintained at very different levels from those of the external fluids. In a similar way, the membranes around the organelles and in the cytoplasm provide a range of microenvironments within the cell itself, each suited to different functions such as the protein-packaging systems in the Golgi apparatus which you will learn about in **Section 3A.3**.

DID YOU KNOW?

Evidence for active transport

Active transport requires energy in the form of ATP produced during cellular respiration. Much of the evidence for active transport comes from linking these two processes together, showing that without ATP, active transport cannot take place.

- **Active transport takes place only in living, respiring cells.**
- **The rate of active transport depends on temperature and oxygen concentration. These affect the rate of respiration and so the rate of production of ATP.**
- **Many cells that are known to carry out a lot of active transport contain very large numbers of mitochondria – the site of aerobic cellular respiration and ATP production.**
- **Poisons that stop respiration or prevent ATPase from working also stop active transport. For example, cyanide prevents the synthesis of ATP during cellular respiration. It also stops active transport. However, if ATP is added artificially, active transport starts again.**

ENDOCYTOSIS AND EXOCYTOSIS

Diffusion and active transport allow the movement of small particles across membranes. However, there are times when larger particles need to enter or leave a cell. An example of this is when white blood cells ingest bacteria or gland cells secrete large steroid hormones. Membrane transport systems cannot do these jobs, but the membrane has properties that make it possible to move larger particles into or out of the cell.

Membrane-bound vesicles can surround and take up materials in a process known as endocytosis (see **fig B**). This can occur on a relatively large scale, for example the ingestion of bacteria during **phagocytosis** (cell eating). It also happens at a microscopic level, when tiny amounts of the surrounding fluid are taken into minute vacuoles. This is known as **pinocytosis** (cell drinking). Electron microscope studies have shown that pinocytosis is very common as cells take in the extracellular fluid as a source of minerals and nutrients (see **fig C**). Exocytosis is the term for the emptying of a membrane-bound vesicle at the surface of the cell or elsewhere (see **fig B**). For example, in cells producing hormones, vesicles containing the hormone fuse with the cell surface membrane to release their contents. These processes are made possible by the fluid mosaic nature of the membrane. The formation of vesicles and the fusing of vesicles with the surface cell membrane are both active processes, requiring energy supplied by ATP.

EXAM HINT

Remember to describe exocytosis as a vesicle *fusing* with the cell surface membrane - joining it or binding to it are not the same thing.

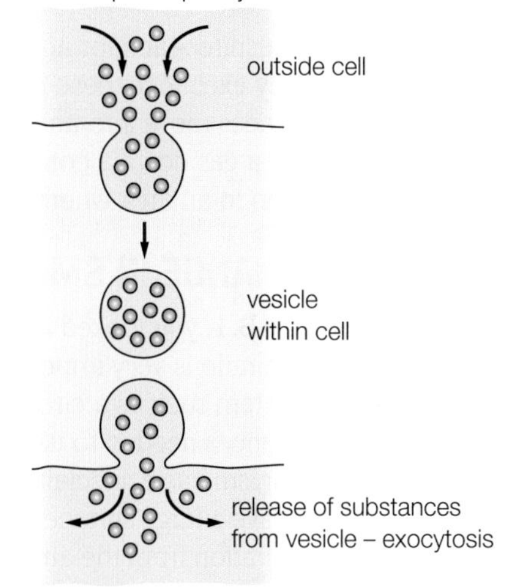

▲ **fig B** The properties of the cell membrane allow cells to take in large particles or release secretions.

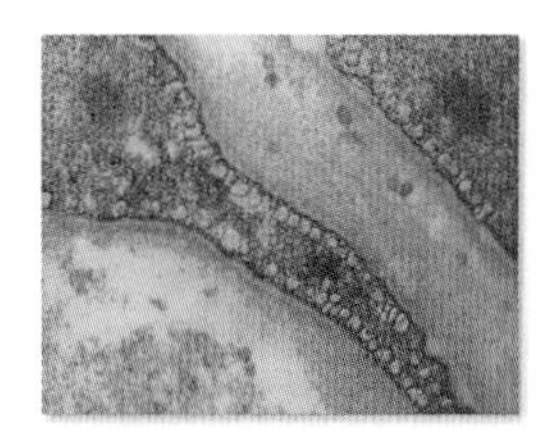

▲ **fig C** The mass of tiny vesicles along these cell membranes show pinocytosis.

CHECKPOINT

1. Explain the importance of active transport in cells.
2. Discuss the role of ATP in active transport in the cell.

SKILLS ADAPTIVE LEARNING

3. Suggest how endocytosis and exocytosis provide evidence for the fluid mosaic model of membranes.

SUBJECT VOCABULARY

ATPase an enzyme that catalyses the hydrolysis of ATP, releasing energy to move carrier systems and drive metabolic reactions
cyanide a metabolic poison that stops mitochondria working
phagocytosis the active process when a cell engulfs something relatively large such as a bacterium and encloses it in a vesicle
pinocytosis the active process by which cells take in tiny amounts of extracellular fluid by tiny vesicles

5 THE NEED FOR GAS EXCHANGE SURFACES

SPECIFICATION REFERENCE 2.1(i) 2.1(ii)

LEARNING OBJECTIVES

- Know the properties of gas exchange surfaces in living organisms.
- Understand that the rate of diffusion is dependent on these properties and can be calculated using Fick's Law of Diffusion.

Organisms respire – and for aerobic respiration they need oxygen and produce waste carbon dioxide. They exchange these gases with the environment in which they live. One of the main ways substances move into and out of cells is by diffusion. This is the free movement of particles in a liquid or a gas down a concentration gradient from an area where they are at a relatively high concentration to an area where they are at a relatively low concentration.

LEARNING TIP

Do not confuse surface area with surface area : volume ratio.

A hydra has a small surface area but a large surface area : volume ratio. A horse has a large surface area but a small surface area : volume ratio.

GAS EXCHANGE IN SMALL ORGANISMS

In **Section 1B.1**, you looked at the principles of surface area : volume ratio. The surface area : volume ratio is very important in determining whether or not an organism needs a mass transport system such as a circulatory system to overcome the limitations of diffusion and supply all the substances needed to its cells. The ratio is also the key factor that determines whether or not the organism has a specialised gas exchange system. Single-celled and very small multicellular organisms have a large surface area : volume ratio. This means they can get the oxygen they need for cellular respiration from the air or water they live in through their outer body surface. As organisms get bigger, the surface area : volume ratio gets smaller and they begin to need specialised surfaces and systems to get the oxygen they require.

GAS EXCHANGE IN LARGE ORGANISMS

In contrast to unicellular organisms such as *Amoeba*, larger organisms consist of billions of cells, often organised into specialised tissues and organs. Substances need to travel long distances from the outside to reach the cytoplasm of all the cells. Nutrients and oxygen would eventually reach the inner cells of the body by simple diffusion, but not fast enough to keep the organism alive (see **fig A**).

The metabolic rate of larger organisms, especially larger animals, tends to be higher than that in smaller animals. Mammals and birds have very high metabolic rates, an adaptation which allows them to be very active and to control their own body temperature. The demands of each individual cell for oxygen and nutrients, and the amount of carbon dioxide and other wastes produced, is much higher in each individual mammal cell than in unicellular organisms.

Complex organisms have evolved specialised systems to exchange the gases they need, taking oxygen in and removing carbon dioxide. In humans and many other large land animals, gas exchange takes place in the lungs; in fish, it is the gills; in insects, it is the tracheal system; and in plants, most gas exchange takes place in the leaves.

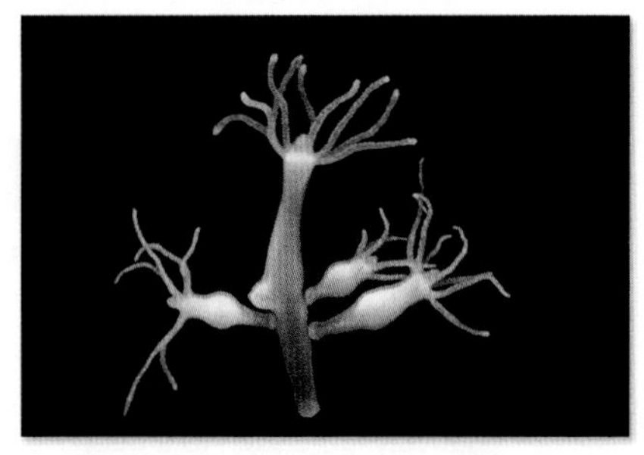

fig A This *Hydra* is less than 15 mm long, and has only two layers of cells. The Arabian horse is around 155 cm tall at the shoulder, and contains billions of cells. The surface area : volume ratio of the *Hydra* means it can get the oxygen it needs by simple diffusion. The surface area : volume ratio of the horse is very small – it needs lungs, organs specially adapted for gas exchange.

KEY PROPERTIES OF GAS EXCHANGE SYSTEMS

Gas exchange systems are specialised for the exchange of oxygen and carbon dioxide between the body of the organism and the environment. These gases are exchanged by simple diffusion. The rate of diffusion across a membrane is controlled by several factors.

- The surface area – the bigger the surface area, the more particles can be exchanged at the same time.
- The concentration gradient of the particles diffusing – particles diffuse from an area where they are at a relatively high concentration to an area where they are at a relatively low concentration. This means that the more particles there are on one side of a membrane compared with the other, the faster they move across. Maintaining the gradient (e.g. by transporting substances away once they have diffused) makes diffusion faster.
- The thickness of the exchange surfaces – the shorter the diffusion distance, the faster diffusion can take place.

You can use this information to calculate the rate at which substances of a given size will diffuse at a known temperature. This relationship is known as Fick's Law of Diffusion:

$$\text{rate of diffusion} = \frac{\text{surface area} \times \text{concentration difference}}{\text{thickness of exchange membrane or barriers}}$$

Any factor which makes the top number large, such as a big surface area or a high concentration gradient, or makes the bottom number small, such as a thin exchange membrane, will increase the rate of diffusion. In any gas exchange system, look for the factors which will affect the rate of diffusion and make it as rapid as possible. You will look at how Fick's Law applies to the human gas exchange system in the next section.

EXAM HINT

When you look at gas exchange systems in different organisms, take note of the adaptations which increase the rate of diffusion.

CHECKPOINT

SKILLS CRITICAL THINKING

1. Explain why large animals cannot take in all the substances they need from outside the body through their skin.
2. Here are three facts about gas exchange in humans.
   - Oxygen enters the body and carbon dioxide leaves it through the lungs.
   - The lungs are made of thousands of tiny air sacs with very thin walls surrounded by blood vessels which also have very thin walls.
   - The surface area of the lungs is approximately $50\,m^2$.

   Explain how this helps the two gases to diffuse quickly into and out of the blood.

2A

6 THE MAMMALIAN GAS EXCHANGE SYSTEM

SPECIFICATION REFERENCE 2.1(ii) 2.1(iii)

LEARNING OBJECTIVES

- Understand how the rate of diffusion is dependent on surface area : volume ratio, thickness of the exchange surfaces and difference in concentration, and can be calculated using Fick's Law of Diffusion.
- Understand how the structure of the mammalian lung is adapted for rapid gas exchange.

For organisms that live on land there is a continual conflict between the need for oxygen and the need for water. The conditions that favour the diffusion of oxygen into an organism also favour the diffusion of water out of that organism. This conflict of needs illustrates the way in which many respiratory systems have evolved. Animals and plants need a large, moist surface area for successful gas exchange, but they also need to limit the water loss from this same surface as much as possible. The evolution of lungs has solved the problem for air-breathing vertebrates. Some lungs are quite simple and merely add a little extra to the area for gas exchange which is already provided by the body's surface – frogs are an example of this. Other animals cannot use their outer surface for gas exchange at all and so are much more dependent on efficient lungs; this includes all the birds and mammals.

EFFECTIVE GAS EXCHANGE

Effective gas exchange systems have several features in common.

- A large surface area giving sufficient gas exchange to supply all the needs of the organism – it has to compensate for the relatively small surface area : volume ratio of the whole organism.
- Thin layers to minimise the diffusion distances from one side to the other.
- In animals, a rich blood supply to the respiratory surfaces. The blood is involved in the transport of the respiratory gases to and from the site of gas exchange and helps to maintain a steep concentration gradient.
- Moist surfaces because diffusion takes place with the gases in solution.
- Permeable surfaces that allow free passage of the respiratory gases.

Mammals, including humans, have very efficient gas exchange systems. They are well adapted to enable the maximum volume of oxygen to move into the body, and the maximum volume of carbon dioxide to be removed. You will be looking at the human gas exchange system as a mammalian example.

THE HUMAN GAS EXCHANGE SYSTEM

Most of the human gas exchange system is found within the chest. It is linked with the outside world through the mouth and nose (see **fig A**). Almost all of the breathing system has evolved to make gas exchange in the alveoli as rapid and efficient as possible. The passages of the nasal cavity have a relatively large surface area, but no gas exchange takes place here. The passages have a good blood supply, and the lining secretes mucus and is covered in hairs. This means that the external air is prepared before entering the rest of the system. The hairs and mucus filter out and remove much of the dust, small particles and pathogens such as bacteria that you breathe in. This protects your lungs from damage and infection. The moist surfaces increase the level of water vapour in the air and the rich blood supply raises the temperature of the air, if this is necessary. This means that the air entering the lungs has as little effect as possible on the internal environment.

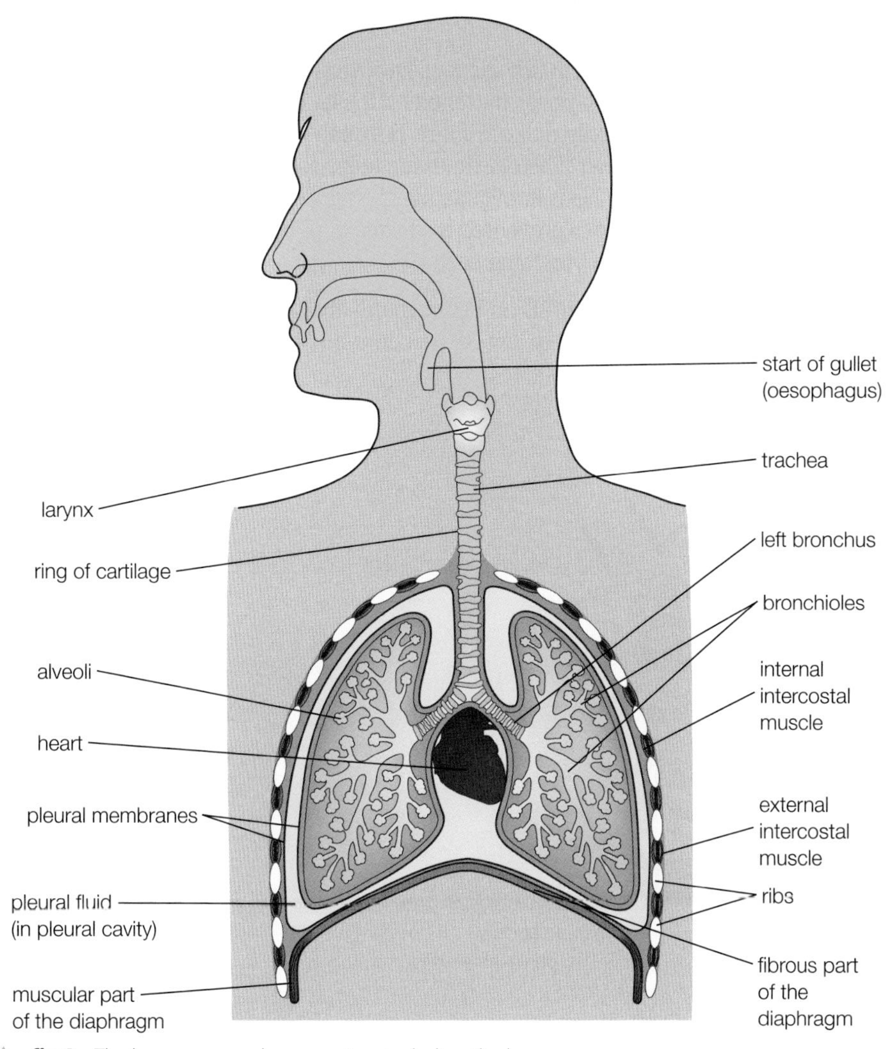

▲ **fig A** The human gas exchange system including the lungs

HOW DOES IT WORK?

The different parts of the breathing system of a mammal, represented here by a person, are all adapted to their roles. The whole breathing system has evolved to make sure gas exchange takes place as rapidly as possible in the lungs.

- **Nasal cavity**: this is the main route by which air enters the gas exchange system.
- **Mouth**: air can enter the respiratory system here, but misses out on the cleaning, warming and moistening effects of the nasal route.
- **Epiglottis**: a flap of tissue that closes over the glottis in a reflex action when food is swallowed. This prevents food from entering the gas exchange system.
- **Larynx**: the voice box, which uses the flow of air across it to produce sounds.
- **Trachea**: the major airway to the bronchi, lined with cells including mucus-secreting goblet cells. Cilia on the surface of the trachea move mucus and any trapped microorganisms and dust away from the lungs.
- **Incomplete rings of cartilage**: these prevent the trachea and bronchi from collapsing but allow food to be swallowed and moved down the oesophagus.
- **Left and right bronchi**: these tubes lead to the lungs and are similar in structure to the trachea but narrower. They divide to form bronchioles.
- **Bronchioles**: small tubes that spread through the lungs and end in alveoli. Their main function is still as an airway, but some gas exchange may occur.
- **Alveoli**: the main site of gas exchange in the lungs.
- **Ribs**: protective bony cage around the gas exchange system.
- **Intercostal muscles**: found between the ribs and important in breathing, which moves air into and out of the lungs to maintain a steep concentration gradient for rapid gas exchange.
- **Pleural membranes**: surround the lungs and line the chest cavity forming a sterile, sealed unit.
- **Pleural cavity**: space between the pleural membranes, usually filled with a thin layer of lubricating fluid that allows the membranes to slide easily with breathing movements.
- **Diaphragm**: broad sheet of tissue made of tendon and muscle that forms the floor of the chest cavity, also important in breathing movements.

EXAM HINT

Ensure your description states the alveoli have a wall that is one cell thick and not that the alveoli have a cell wall that is one cell thick.

Remember - animal cells do not have cell walls.

GAS EXCHANGE IN THE ALVEOLI

In the lungs most of the gas exchange occurs in tiny air sacs known as alveoli (singular: alveolus) (see **fig B**). An alveolus is made of a single layer of flattened epithelial cells. The capillaries that run close to the alveoli also have a wall that is only one cell thick. Between the two is a layer of elastic connective tissue holding everything together. The elastic tissue helps to force air out of the lungs, which are stretched when you breathe in. This is known as the elastic recoil of the lungs. The alveoli have a natural tendency to collapse, but this is prevented by a special phospholipid known as **lung surfactant** that coats the alveoli and makes breathing easier.

Gas exchange occurs by a process of simple diffusion between the alveolar air and the deoxygenated blood in the capillaries. This blood has a relatively low oxygen content and a relatively high carbon dioxide content.

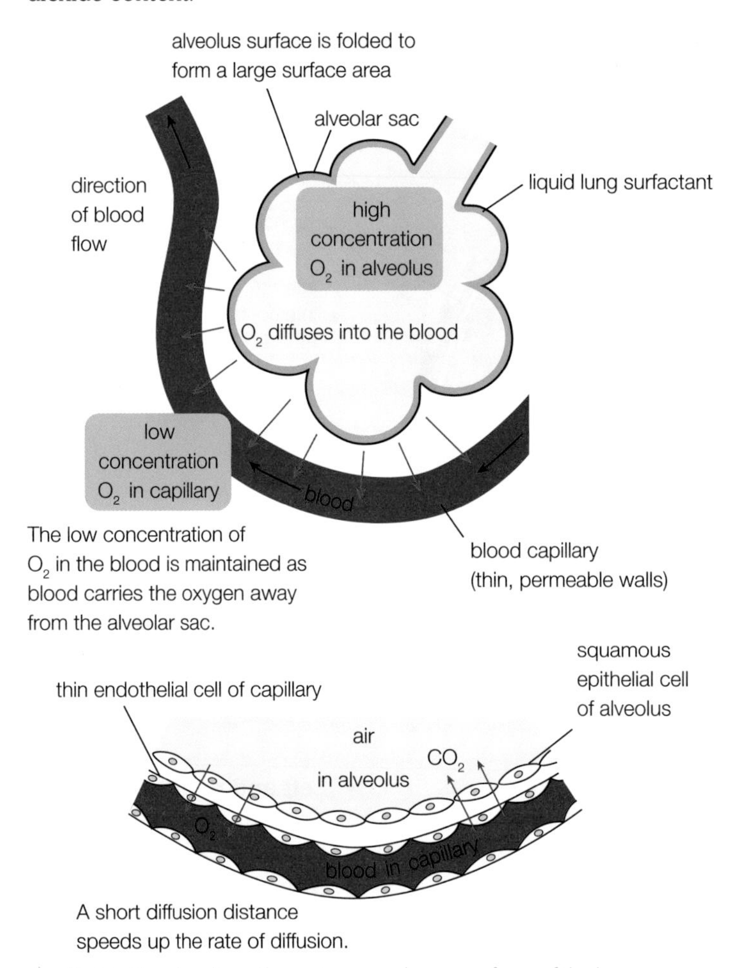

▲ **fig B** The alveoli are the main gas exchange surfaces of the lungs.

Understanding Fick's Law can help us explain how the adaptations of the human respiratory system make it possible to exchange gases as efficiently as possible in the lungs by diffusion. Remember:

$$\text{rate of diffusion} = \frac{\text{surface area} \times \text{concentration difference}}{\text{thickness of exchange membrane or barriers}}$$

LARGE SURFACE AREA

The alveoli provide an enormous surface area for the exchange of gases in the human body. Recent calculations have shown that an average adult human has around 480–500 million alveoli in their lungs. This means the surface area for gas exchange is around 40–75 m^2 packed into your chest – that is the surface area of between 10 and 18 table tennis tables.

STEEP CONCENTRATION GRADIENT

Blood is continuously flowing through the capillaries past the alveoli, exchanging gases. The continuous flow of the blood maintains the concentration gradient on the capillary side. The air within the alveoli is constantly being refreshed with air from outside by breathing (see **table A**). Movement of gases into and out of the alveoli is mainly by diffusion, but movement of air into and out of the lungs is by a mass transport system (see **fig C**).

GAS	PERCENTAGE OF GAS IN:		
	INSPIRED AIR	ALVEOLAR AIR	EXPIRED AIR
oxygen	20.70	13.20	14.50
carbon dioxide	0.04	5.00	3.90
nitrogen	78.00	75.60	75.40
water vapour	1.24	6.20	6.20

table A The composition of the gases in the human gas exchange system

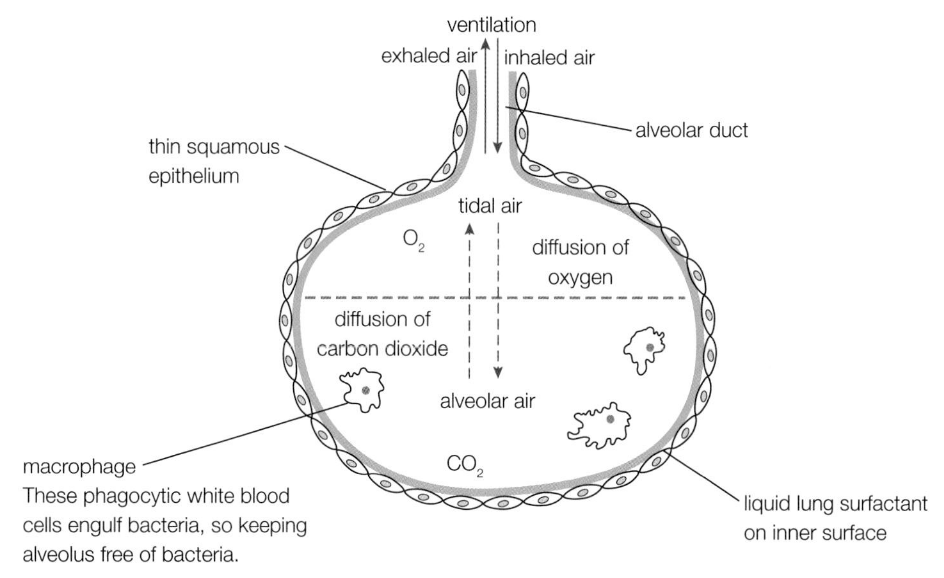

▲ **fig C** Diffusion across the alveolar surfaces provides the blood with oxygen and disposes of carbon dioxide.

SHORT DIFFUSION DISTANCE

The walls of the alveoli are only one cell thick, as are the walls of the capillaries that run beside them. This means the distance that diffusing gas molecules need to travel between them is only around 0.5–1.5 μm (micrometres, microns, 10^{-6} m). This makes it as easy as possible for diffusion to take place rapidly and effectively.

BREATHING (VENTILATION)

The exchange of gases at the alveolar surfaces in the lungs happens by passive diffusion alone, but moving air between the lungs and the external environment is an active process known as **breathing** or **ventilation**. This is key to rapid gas exchange taking place in the alveoli. By delivering air rich in oxygen, and removing air loaded with waste carbon dioxide, breathing maintains a steep concentration gradient for diffusion between the blood in the capillaries and the air in the lungs. There are two parts to the process of breathing: taking air into the chest, called **inhalation**, and breathing air out again, called **exhalation**. The chest cavity is effectively a sealed unit for air, with only one way in or out – through the trachea. Breathing involves a series of pressure changes in the chest cavity that in turn bring about movements of the air.

Inhalation is an active, energy-using process. The muscles around the diaphragm contract and, as a result, it is lowered and flattened. The intercostal muscles between the ribs also contract, raising the rib cage upwards and outwards. These movements result in the volume of the chest cavity increasing, which reduces the pressure in the cavity. The pressure within the chest cavity is now lower than the

LEARNING TIP

Remember that movement of carbon dioxide towards the lungs and oxygen away from the lungs in the blood is an important part of maintaining the concentration gradient across the alveolar walls.

EXAM HINT

The correct name for breathing is ventilation because ventilation refreshes the air in the lungs.

Be careful not to confuse it with respiration, which is a series of chemical reactions in the cells and is correctly called cellular respiration.

pressure of the atmospheric air outside, so air moves through the trachea, bronchi and bronchioles into the lungs to equalise the pressure inside and out.

EXAM HINT

An explanation of how ventilation works will require this simple level of physics to explain why the volume and pressure in the lungs rise and fall.

Normal exhalation is a passive process. The muscles surrounding the diaphragm relax so that it moves up into its resting domed shape. The intercostal muscles also relax so that the ribs move down and in, and the elastic fibres around the alveoli of the lungs return to their normal length. As a result, the volume of the chest cavity decreases, causing an increase in pressure. The pressure in the chest cavity is now greater than that of the outside air, so air moves out of the lungs, through the bronchioles, bronchi and trachea to the outside air (see **fig D**).

LEARNING TIP

The process of ventilating the lungs is an important adaptation for maintaining the steep concentration differences needed for rapid gas exchange.

If you need to, you can force air out of your lungs more rapidly than passive exhalation allows. The internal intercostal muscles contract, pulling the ribs down and in, and the abdominal muscles contract forcing the diaphragm upwards. This increases the pressure in the chest cavity, causing exhalation. This is known as forced exhalation. Singers use this to achieve a powerful voice and to maintain long notes, and free divers do it before a dive, so they can fill their lungs with as much air as possible. Coughing is an exaggerated form of forced exhalation which is used to force mucus out from the respiratory system.

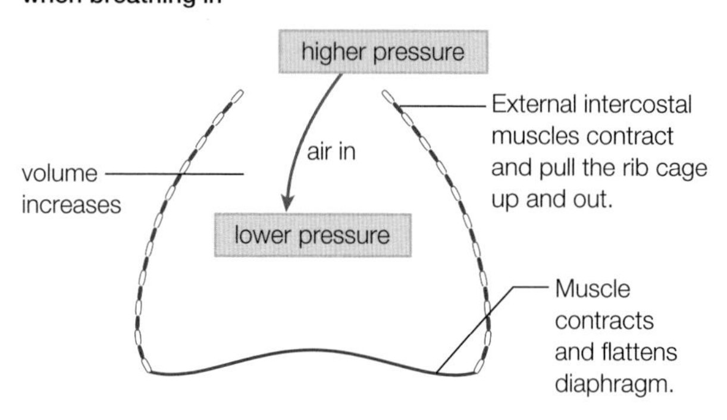

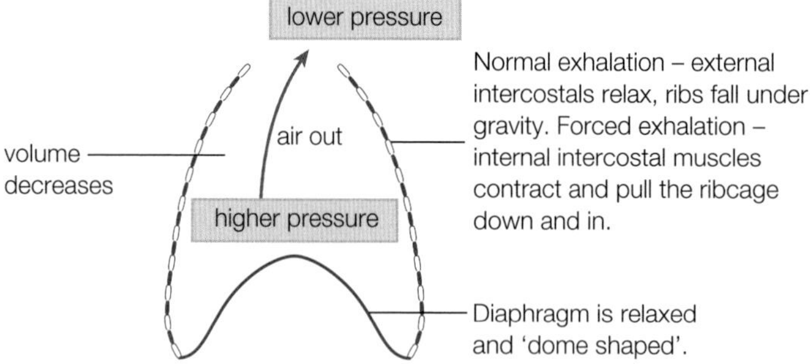

▲ **fig D** You can feel the movements of your ribs during inhalation and exhalation, but the movements of your diaphragm are less obvious.

PROTECTING THE LUNGS

Your gas exchange system carries out the exchange of oxygen and carbon dioxide. As well as gases, the air you breathe in also carries lots of tiny particles such as dust, pollen grains and smoke particles which could block the tiny alveoli. It also carries microscopic organisms like bacteria and viruses. Some of these microscopic organisms are **pathogens** – they cause diseases – and the respiratory system provides a potential route for them to get into the body. To reduce the chances of damage happening to your lungs and of infection, your respiratory system produces lots of mucus that lines your airways and traps these tiny particles and organisms. The mucus is usually very runny, so that it is easily moved up the airways by cilia that sweep upwards to the back of your throat (see **fig E**). Here most of the mucus is swallowed without you even noticing it. The acid in your stomach and your digestive enzymes digest the mucus and everything carried with it.

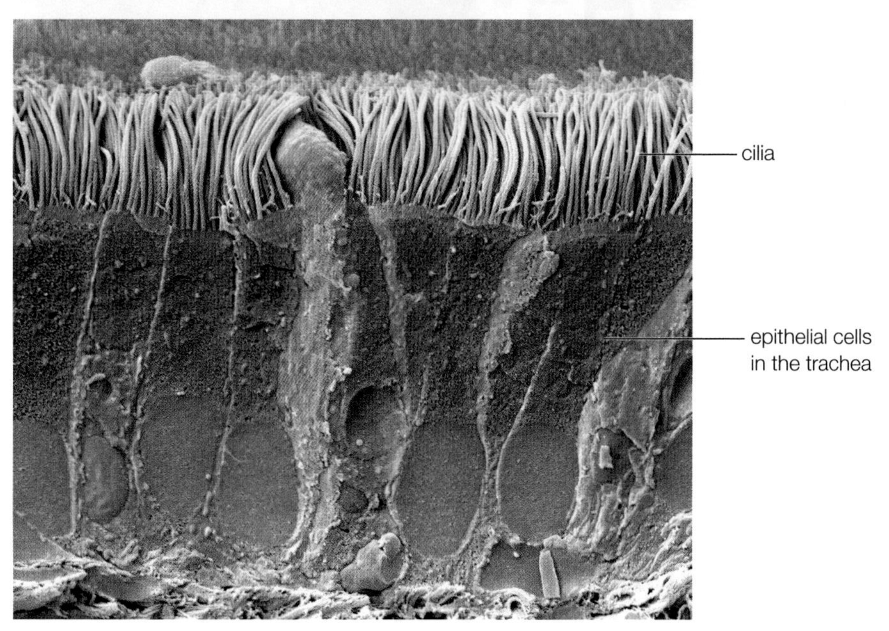

▲ **fig E** The cilia in your trachea and bronchi beat constantly to move mucus with its load of pathogens and dirt out of your gas exchange system.

CHECKPOINT

1. Explain carefully why humans need a complex internal gas exchange system.
2. ▶ Why is breathing through your nose better for your body than breathing through your mouth?
3. ▶ Explain why breathing is important for rapid gas exchange in the alveoli.
4. Compare the last two columns in **table A**. Explain the difference in oxygen and carbon dioxide percentages in expired air compared with alveolar air.

SKILLS ▶ PROBLEM SOLVING

SUBJECT VOCABULARY

lung surfactant a special phospholipid that coats the alveoli and prevents them from collapsing
breathing (ventilation) the process in which physical movements of the chest change the pressure so that air is moved in or out
inhalation breathing in
exhalation breathing out
pathogens microorganisms that cause disease

2A THINKING BIGGER

ASTHMA AND AIR POLLUTION: HOW STRONG IS THE LINK?

CRITICAL THINKING, ANALYSIS, REASONING, INTERPRETATION, CREATIVITY, INNOVATION, PERSONAL AND SOCIAL RESPONSIBILITY, CONTINUOUS LEARNING

Beijing authorities are familiar with declaring red alerts for air quality. But what effect is the air quality having on the public? Asthma is a growing problem in many larger cities with hospital admissions rising and people migrating to get away from the problem.

MEDICAL JOURNAL ARTICLE

LINK BETWEEN ENVIRONMENTAL AIR POLLUTION AND ALLERGIC ASTHMA: EAST MEETS WEST

Abstract

There is increasing evidence of the negative health impact resulting from environmental air pollution, in particular that associated with respiratory diseases and allergy. The increasing prevalence of respiratory diseases and allergy such as asthma has drawn attention to the potential role of air pollution in causing this. While this has been first noticed and reported in Europe and North America, this is now being seen in many of the rapidly-growing economies of South East Asia, particularly in China, as a result of the fast pace of urbanization and increased energy consumption that occurs with rapid industrialization and the increasing number of vehicles. This is having a significant impact on mortality and health of Asian populations and air pollution is one of the major factors that affects the health of Asians. Recent data published by the Health Effects Institute indicate that a 10 μg/m^3 increase in PM_{10}, the coarse particulate fraction of air pollution, is associated with an increase in mortality of 0.6% in daily all natural cause mortality in major cities in India and China. The health effects of air pollution particularly on the common lung diseases such as asthma and COPD are also being felt particularly in Asia. The low levels of allergy and asthma that have been seen previously is now rising to match those levels observed in Western countries, and both epidemiological cohort and experimental exposure studies provide evidence to implicate a harmful impact of traffic air pollution on both the development of allergic diseases and asthma and the increase in asthma symptoms and exacerbations. Experimental exposure studies also indicate a causative relationship between air pollution and allergic airways disorders through the induction of inflammation and oxidative stress in the lungs leading to a preferential T-helper type 2 lineage. In this review, we will examine this evidence implicating the deleterious effect of environmental pollution. One of the major issues of interest will be whether the much higher levels of environmental pollution in Asia will lead to a greater impact on lung diseases particularly asthma and allergic diseases.

From: Qingling Zhang, Zhiming Qiu, Kian Fan Chung, Shau-Ku Huang. *Journal of Thoracic Disease*. 2015 Jan; 7(1): 14–22
https://www.ncbi.nlm.nih.gov/pmc/articles/PMC4311080/#r3

SKILLS

CRITICAL THINKING, ANALYSIS, INTERPRETATION, REASONING

SCIENCE COMMUNICATION

The extract is an introduction to a scientific article. The intended audience for this article includes medical professionals and research scientists.

1 Below are three phrases from the article. In each case, look up what is meant by the phrase and explain how it is being used in this article.

(a) increasing prevalence of respiratory diseases and allergy

(b) a 10 μg/m^3 increase in PM_{10}

(c) both epidemiological cohort and experimental exposure studies

BIOLOGY IN DETAIL

Now apply your biological knowledge to the information on the effects of asthma. You should be able to answer all of these questions now, although you may wish to return to the final question if you go on to study hormones at A level.

You may like to visit a public access site about asthma. The Association of the British Pharmaceutical Industry (ABPI) has a public access site providing information. Visit the page on breathing and asthma at: www.abpischools.org.uk/topic/breathingandasthma
This resource includes a number of animations including one showing the effects of asthma and how asthma-relieving medication works.

SKILLS CONTINUOUS LEARNING, INTELLECTUAL INTEREST AND CURIOSITY

2 One important part of modern medicine is education – informing people about the problems caused by the environment or by their lifestyle.

(a) How do animations help to develop understanding of complex processes such as breathing, asthma and the action of asthma-relieving drugs?

SKILLS COMMUNICATION

(b) Compare the advantages and disadvantages of an online resource for delivering biological knowledge.

3 (a) How does an asthma attack interfere with normal breathing?

(b) How does this interfere with efficient gas exchange?

4 The effect of asthma is usually measured using FEV1. This is the forced expiratory volume in 1 second. The graph below shows the effect of asthma on FEV1.

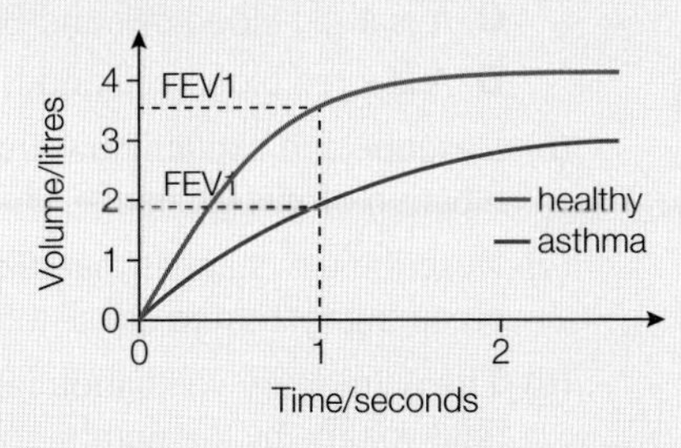

(a) What proportion of the air in the lungs has been expelled in 1 second:

(i) in a healthy person?

(ii) in a person with asthma?

(b) What is the percentage reduction in FEV1 of a person with asthma compared to a healthy person?

(c) What is the percentage reduction in the total volume of air expelled?

(d) Evaluate the reasons why FEV1 is used to determine the severity of an asthma attack rather than total volume expelled.

TIP

Measure (b) after 1 second of expiration. Measure (c) after 2.5 seconds of expiration.

THINKING BIGGER TIP

Evaluate means to review the information available and consider the strengths and weaknesses of the evidence.

SKILLS CRITICAL THINKING, REASONING, DECISION MAKING, PERSONAL AND SOCIAL RESPONSIBILITY, ETHICS, INTEGRITY

5 Research into conditions such as asthma is often funded by the pharmaceutical companies that manufacture the medicines. Their reports may not give a completely independent view. What is independent research? What are its advantages and disadvantages?

SKILLS CONTINUOUS LEARNING, INTELLECTUAL INTEREST AND CURIOSITY

6 The prevalence of asthma in many Chinese cities is 11% and rising. Many people take relieving medications rather than preventative medications. These medications affect the cells lining the bronchi and bronchioles. Find out more about the action of one common medication used to relieve asthma and one used to prevent asthma. What effect does it have on the cells and how does it help to reduce the symptoms of an asthma attack? Produce clear diagrams explaining the action at a cellular level, showing how it returns the body to normal function and helps to maintain normal function.

2A EXAM PRACTICE

1 Substances are moved in and out of cells by diffusion, facilitated diffusion, active transport, exocytosis and endocytosis.

(a) Which method(s) of transport require energy from cell metabolism?

A active transport, facilitated diffusion, exocytosis and endocytosis

B active transport, exocytosis and endocytosis

C diffusion, facilitated diffusion, exocytosis and endocytosis

D only active transport [1]

(b) Which statement best describes diffusion?

A random movement of particles due to their kinetic energy

B movement of particles from a less concentrated area to a more concentrated area

C movement of particles down a concentration gradient

D movement of particles across a membrane using metabolic energy [1]

(c) (i) The graph below shows the changes in concentration of substance A on the inside and outside of a partially permeable membrane, during a 50-minute period.

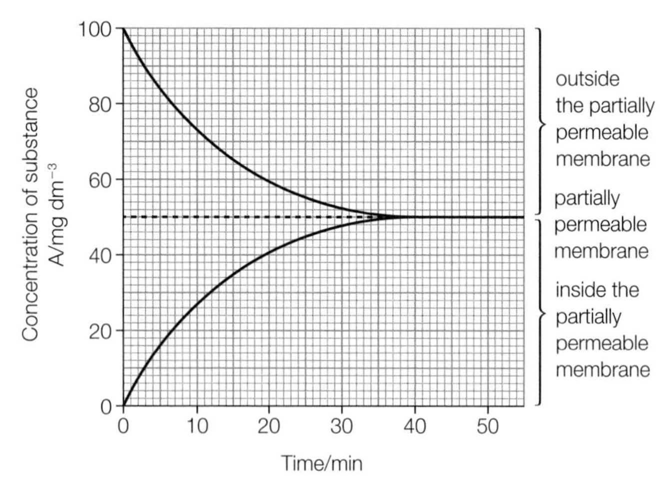

'Substance A crosses from one side of this membrane to the other by diffusion.'

Justify this conclusion about the data provided in the graph. [3]

(ii) Suggest how the changes in concentration of substance A would be different if substance A was being transferred across the membrane using active transport. [2]

(Total for Question 1 = 7 marks)

2 The diagram below shows a section through an alveolus and the surrounding tissue.

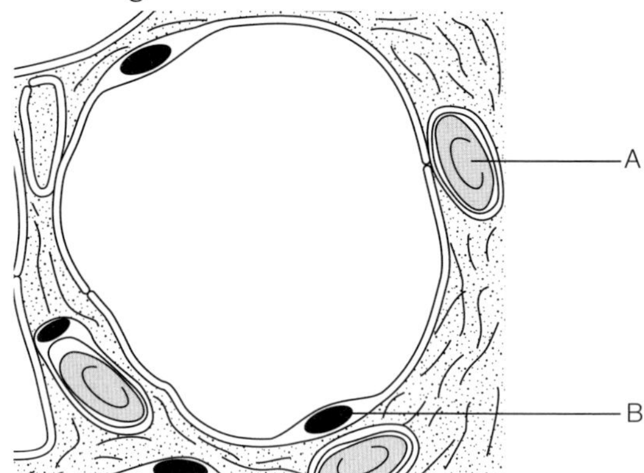

(a) What are the correct names for structures A and B?

A A is the alveolus and B is a bacterium

B A is a capillary and B is an epithelium cell

C A is a red blood cell and B is an epithelium cell

D A is a red blood cell and B is a nucleus [1]

(b) Describe and explain how alveoli are adapted for the function of gas exchange. [4]

(c) Describe the changes that take place in the lungs and breathing system to cause inhalation. [4]

(d) The pulmonary ventilation rate is found by multiplying the tidal volume by the number of breaths taken per minute. Calculate the pulmonary ventilation rate for a person breathing a tidal volume of $0.45\,dm^3$, 18 times per minute. [1]

(Total for Question 2 = 10 marks)

3 (a) What is the correct description of the structure of a phospholipid molecule?

A one glycerophosphate and two fatty acids attached by condensation bonds

B one glycerol and two fatty acids attached by condensation bonds

C one glycerol attached to three fatty acids by ester bonds

D one glycerol attached to two fatty acids and a phosphate group by ester bonds [1]

(b) The presence of a phosphate makes part of the phospholipid molecule hydrophilic. Explain what is meant by the term hydrophilic. [1]

(c) Describe the role of phospholipids in the cell surface membrane. [2]

(d) Draw a diagram representing the fluid mosaic model of cell membrane structure. [5]

(Total for Question 3 = 9 marks)

4 The diagram below represents the structure of the cell surface membrane.

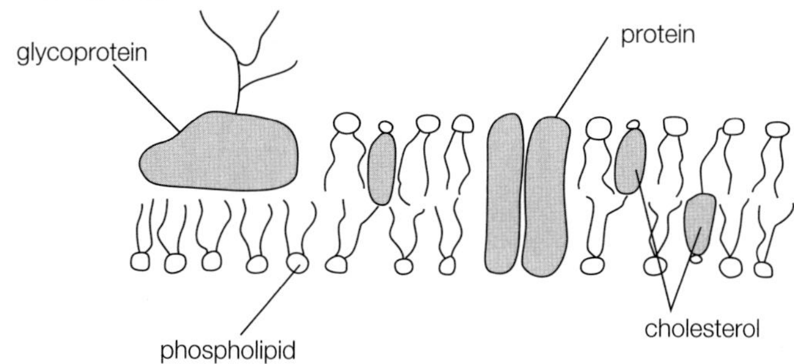

(a) Explain why the phospholipid molecules form a bilayer. [2]

(b) The ability of a substance to pass into a cell depends on its solubility in oil and water. The oil–water partition coefficient is a measure of the solubility of a substance in oil compared to water. The equation below shows how it is calculated.

$$\text{oil water partition coefficient} = \frac{\text{solubility in oil}}{\text{solubility in water}}$$

The graph below shows the relationship between membrane permeability and the oil–water partition coefficient for four different substances, A, B, C and D.

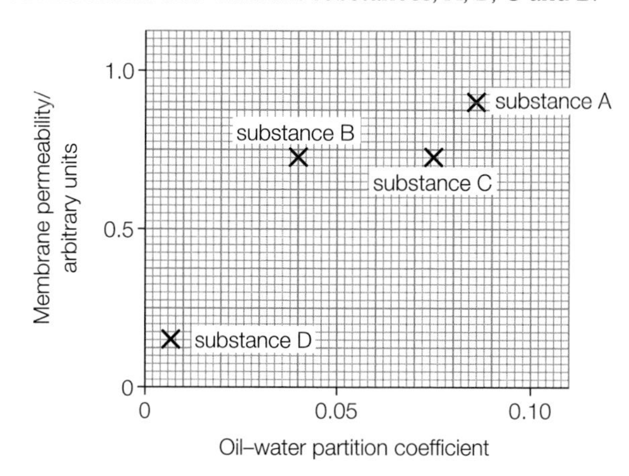

(i) Compare the ability of substances A and D to cross a cell surface membrane. [3]

(ii) Using the information shown in the graph and your knowledge of the cell surface membrane, suggest how substance A crosses a membrane. Justify your answer. [3]

(Total for Question 4 = 8 marks)

5 (a) A student carried out an experiment to investigate the effect of alcohol concentration on the permeability of beetroot membranes. Beetroots are root vegetables that appear red because the vacuoles in their cells contain a water-soluble red pigment. This pigment cannot pass through membranes. Eight pieces of beetroot were cut. One piece of beetroot was placed into a tube containing 15 cm^3 of water and left for 15 minutes. The procedure was repeated for seven different concentrations of ethanol. After 15 minutes, each piece of beetroot was removed from the tubes and a sample of the fluid removed and placed in a colorimeter. The colorimeter was used to determine the intensity of red coloration of the fluid.

The graph below shows the results of the investigation.

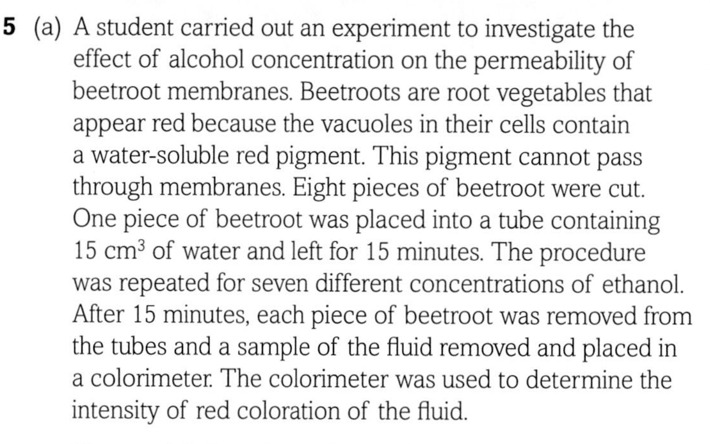

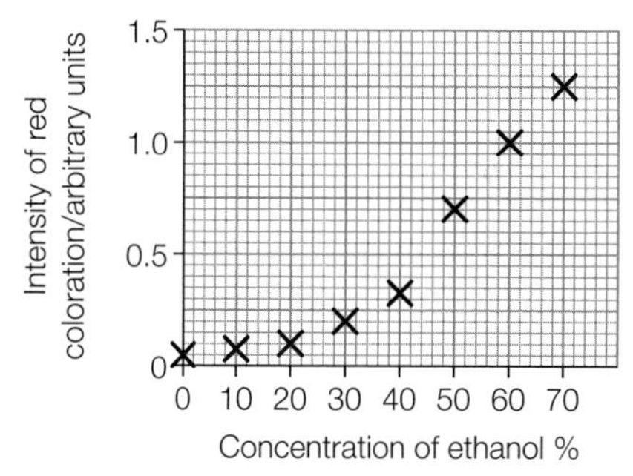

(i) Suggest **two** variables, other than those stated above, which should be kept constant during this investigation. [2]

(ii) There was some red coloration in the tube containing only water. Suggest an explanation for this. [2]

(iii) Describe what the student should have done to reduce the red coloration in the tube containing only water. [2]

(b) The graph above shows that ethanol has an effect on the permeability of beetroot.

(i) State the effect that changing the ethanol concentration has on the intensity of the red coloration. [1]

(ii) Suggest an explanation for this effect. [2]

(Total for Question 5 = 9 marks)

TOPIC 2 MEMBRANES, PROTEINS, DNA AND GENE EXPRESSION

CHAPTER 2B PROTEINS AND DNA

A small child with a swollen belly sits listlessly in the Caribbean sun. Like many millions of infants, she is suffering from kwashiorkor. She is a 'sugar baby'; she has enough calories but lacks protein. The enzymes that control metabolism in cells are made of protein, so if the diet is severely lacking in protein, the health of the child (or adult) will fail.

In this chapter, you will study the importance of proteins as enzymes and how enzymes work to speed up chemical reactions inside and outside cells.

You will also study the nucleotides including DNA and RNA. These molecules carry the genetic code, which is used to manufacture proteins and produce our phenotype. You will discover the structure of deoxyribonucleic acid (DNA) and crack the code it carries with the information needed to build an entire new organism. You will learn about the different types of ribonucleic acid (RNA) and how they work together with the DNA to translate the genetic code into the phenotype of the cell or organism through protein synthesis.

MATHS SKILLS FOR THIS CHAPTER

- **Recognise and make appropriate use of units in calculations** (*e.g. mmol s^{-1}*)
- **Recognise and use expressions in decimal and standard form** (*e.g. considering the number of base pairs in a genome*)
- **Identify uncertainties in measurements and use simple techniques to determine uncertainty when data are combined** (*e.g. when measuring the rate of enzyme-controlled reactions*)
- **Translate information between graphical, numerical and algebraic forms** (*e.g. describe the effect of changing conditions on enzyme activity from a graph*)
- **Construct and interpret frequency tables and diagrams, bar charts and histograms** (*e.g. plot enzyme activity over time represented on a graph*)
- **Substitute numerical values into algebraic equations using appropriate units for physical quantities** (*e.g. calculate the Q_{10} value for an enzyme-controlled reaction*)
- **Plot two variables from experimental or other data** (*e.g. draw graphs from a table of results*)
- **Understand that $y = mx + c$ represents a linear relationship** (*e.g. predict/sketch the shape of a graph with a linear relationship such as the effect of substrate concentration on the rate of an enzyme-controlled reaction*)
- **Determine the intercept of a graph** (e.g. *state the optimum conditions for an enzyme-controlled reaction*)
- **Calculate rate of change from a graph showing a linear relationship** (*e.g. calculate a rate from a graph of an enzyme-controlled reaction*)
- **Draw and use the slope of a tangent to a curve as a measure of rate of change** (*e.g. amount of product formed plotted against time when the concentration of enzyme is fixed*)

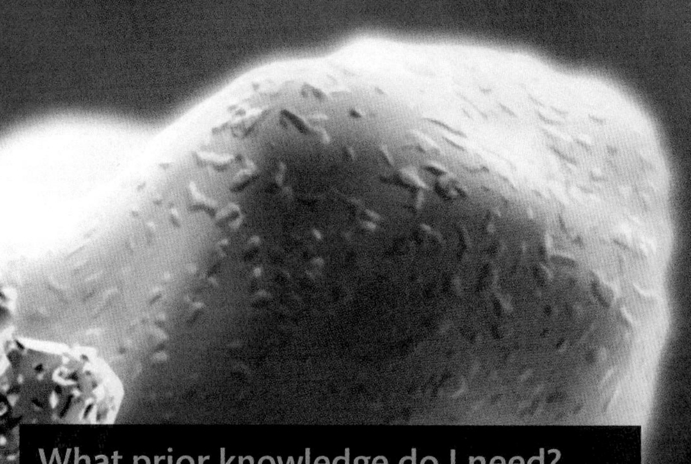

What prior knowledge do I need?

- DNA is a polymer made up of two strands that form a double helix
- DNA is the genetic material of the cell
- The genome is the entire genetic material of the cell

Chapter 1A

- Proteins are long chains of amino acids
- Enzymes are made of proteins
- The importance of inorganic ions in biological systems

What will I study in this chapter?

- The activity of enzymes inside and outside the cell
- The structure of nucleotides
- The structure of the DNA molecule including the double helix
- How the DNA code works and the experimental evidence used to prove it
- How proteins are synthesised on the surface of the ribosomes and the roles of the different types of RNA in the process

What will I study later?

Chapter 2C

- What happens to the DNA in chromosome mutations and how they can affect the phenotype of the resulting individual

Chapter 3A

- The site of the genetic material in the nucleus of the cell

Chapter 3B

- What happens to the genetic material during mitotic and meiotic cell division

Chapter 4B

- The impact of DNA sequencing on classification

Chapter 4C

- The importance of mutation and genetic variation in evolution by natural selection

Topic 7 (Book 2: IAL)

- How proteins and lipids act as hormones
- How carbohydrates and proteins act as signalling molecules in and on cell membranes
- The importance of the enzymes formed during protein synthesis in processes such as cellular respiration and photosynthesis

Chapter 8C (Book 2: IAL)

- Gene technology and genetic engineering

2B 1 ENZYMES

SPECIFICATION REFERENCE 2.7(ii) 2.7(iii)

LEARNING OBJECTIVES

- Understand that enzymes are biological catalysts.
- Know that there are intracellular enzymes catalysing reactions inside cells and extracellular enzymes catalysing reactions outside cells.

In **Section 1B.5**, you studied the biochemistry of the most important molecules in living things. You saw there that proteins are large molecules made of long chains of amino acids linked by peptide bonds. They often have secondary, tertiary and even quaternary structures which determine the shape of the molecule. These complex protein shapes are held together by a combination of hydrogen bonds, disulfide bonds and ionic bonds. Proteins carry out a large number of functions in living organisms, and one of their key roles in cells is to act as enzymes.

WHAT IS AN ENZYME?

A **catalyst** is a substance that changes the rate of a reaction without changing the substances produced. The catalyst is unaffected at the end of the reaction and can be used again. **Enzymes** are biological catalysts, which control the rate of the reactions that occur in individual cells and in whole organisms. In some cases – for example, when organisms such as fungi and flies digest their food externally – they control the rate of reactions happening outside the bodies of organisms. Most of the reactions that provide cells with energy and produce new biological material would take place very slowly at the temperature and pH of living things – too slowly for life to exist. Enzymes make life possible by speeding up the chemical reactions in cells without changing the conditions in the cytoplasm.

Enzymes are globular proteins (see **Section 1B.5**) which are produced during protein synthesis as mRNA transcribed from the DNA molecule is translated (see **Section 2B.6**). They have a very specific shape as a result of their primary, secondary, tertiary and quaternary structures (see **Section 1B.5**). This means that each enzyme will only catalyse a specific reaction or group of reactions. We say enzymes show great **specificity**. Changes in temperature and pH affect the efficiency of an enzyme because they affect the intramolecular bonds within the protein that are responsible for the shape of the molecule.

LEARNING TIP

Remember that enzymes are proteins - so they show all the same properties as proteins and these properties can be explained by their primary, secondary, tertiary and quaternary structure.

Within any cell, many chemical reactions are going on at the same time. Those reactions that build up new chemicals are known as **anabolic reactions** ('ana' means up, as in 'build up'). Those that break substances down are **catabolic reactions** ('cata' means down, as in 'break down'). The combination of these two processes is **metabolism**. Most of the reactions of metabolism occur as part of a sequence of reactions known as a **metabolic chain** or **metabolic pathway**. Although we usually think of enzymes speeding up reactions, they sometimes act to slow them down, or stop them completely. **Fig A** is an electron micrograph of part of a cell. Hundreds of reactions would be going on in this cell, each controlled by specific enzymes.

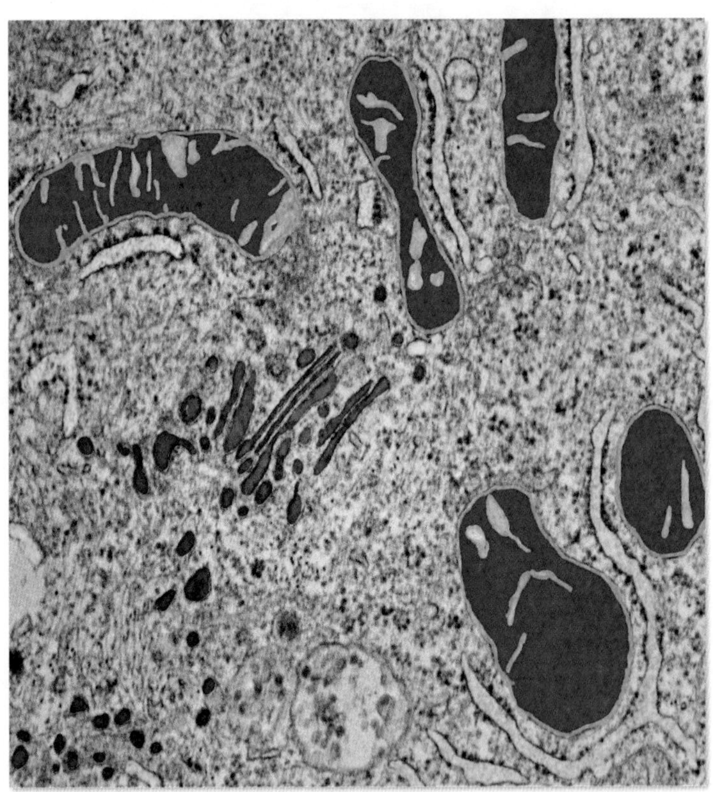

fig A Each cell contains several hundred different enzymes to control all the reactions going on inside.

NAMING ENZYMES

In the study of biology, in medicine, in cellular and genetic research and in industries that use biotechnology, it is important to be able to refer to the action of specific enzymes. To do this, we need to understand how enzymes are named.

Many of the enzymes found in animals and plants work inside the cells. The enzymes that catalyse reactions inside cells are known as **intracellular enzymes**; examples include DNA polymerase and DNA ligase (see **Section 2B.4**). Cells secrete other enzymes that work outside the cell membrane. The enzymes that catalyse reactions outside the cells are called **extracellular enzymes**. The digestive enzymes and lysozyme, the enzyme in your tears, are good examples of these.

Most enzymes – both intracellular and extracellular – have several names including:

- a relatively short recommended name, which is often the name of the molecule that the enzyme works on (the substrate) with '-ase' on the end, or the substrate with an indication of what it does (e.g. creatine kinase)
- a longer systematic name describing the type of reaction being catalysed (e.g. ATP : creatine phosphotransferase)
- a classification number (e.g. EC 2.7.3.2).

Some enzymes, such as urease, ribonuclease and lipase, are known by their recommended names. There are still some enzymes that are known by common, but uninformative names – trypsin and pepsin for example. However, the names of most enzymes give you useful information about the role of the enzyme in the cell or the body.

DID YOU KNOW?

The discovery of enzymes

In 1835 people noticed that malt (sprouting barley) breaks down starch into sugars even more effectively than sulfuric acid does.

People also suspected there were 'ferments' in yeast (a single-celled fungus) used for making bread and turning sugar to alcohol, and in 1877 the name *enzyme* was introduced. In 1897, Eduard Buchner (1860–1917) extracted the enzyme responsible for fermenting sugar from yeast cells, and showed it worked outside a living cell.

In 1926, James B. Sumner (1887–1955) extracted the first pure, crystalline enzyme from jack beans (see **fig B**). It was urease, the enzyme that catalyses the breakdown of urea. Sumner found the crystals were protein and concluded that enzymes must therefore be proteins. Unfortunately, no one believed the young researcher at the time, because many established scientists had been trying and failing to isolate enzymes for years. However, 20 years later Sumner was awarded a Nobel Prize for his ground-breaking work.

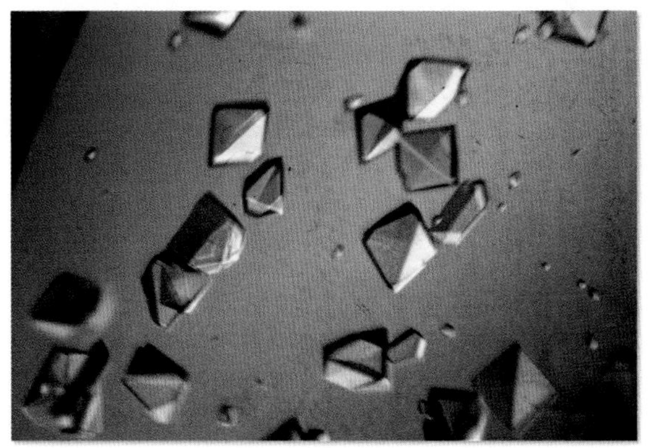

fig B Pure urease does not look very exciting, but the ability to isolate and extract enzymes has revolutionised our understanding of biology and the way we can use enzymes in industry.

LEARNING TIP

Remember that most substances with a name ending in *ase* are enzymes.

CHECKPOINT

1. From which organisms were the first enzymes isolated?

SKILLS INTERPRETATION

2. What is the difference between an intracellular enzyme and an extracellular enzyme?
3. Investigate Sumner's work and discover which scientists were particularly against his ideas and why.

SUBJECT VOCABULARY

catalyst a substance that speeds up a reaction without changing the substances produced or being changed itself
enzymes proteins that act as biological catalysts for a specific reaction or group of reactions
specificity the characteristic of enzymes that means that each enzyme will catalyse only a specific reaction or group of reactions; this is due to the very specific shapes which come from the tertiary and quaternary structures
anabolic reaction a reaction that builds up (synthesises) new molecules in a cell
catabolic reaction a reaction which breaks down substances within a cell
metabolism the sum of the anabolic and catabolic processes in a cell
metabolic chain (metabolic pathway) a series of linked reactions in the metabolism of a cell
intracellular enzymes enzymes that catalyse reactions within the cell
extracellular enzymes enzymes that catalyse reactions outside of the cell in which they were made

2B 2 HOW ENZYMES WORK

SPECIFICATION REFERENCE 2.7(i) 2.7(ii) CP4

LEARNING OBJECTIVES

- Understand that enzymes are biological catalysts that reduce activation energy.
- Understand the mechanism of action and the specificity of enzymes in terms of their three-dimensional structure.

For a chemical reaction to occur, the reacting molecules must have enough energy to break the chemical bonds that hold them together. A simple model is that the reaction must get over an 'energy hill', known as the **activation energy**, before it can get started.

Raising the temperature increases the rate of a chemical reaction by giving more molecules sufficient energy to react. However, living cells could not survive the temperatures which are needed to make many cellular reactions fast enough – and the energy demands to produce the heat would be enormous. Enzymes solve the problem by lowering the activation energy needed for a reaction to take place (see **fig A**).

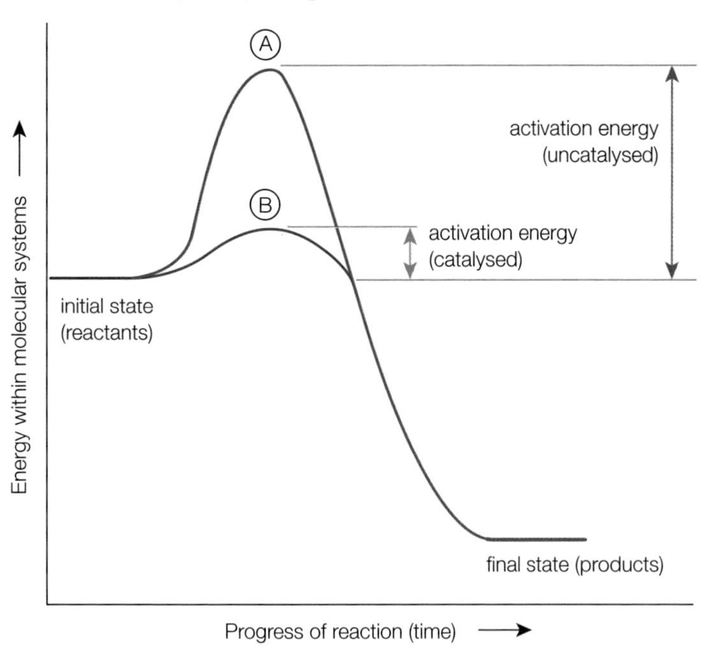

▲ **fig A** Energy diagram showing the difference between an uncatalysed and a catalysed reaction.

HOW DO ENZYMES WORK?

To lower the activation energy and catalyse a reaction, enzymes form a complex with the **substrate** or substrates of the reaction. A simple model of enzyme action in a catabolic reaction is:

substrate + enzyme ⇆ enzyme–substrate complex
⇆ enzyme + products

Once the products of the reaction are made, they are released and the enzyme is free to form a new complex with more substrate. How does this relate to the structure of the enzyme? The **lock-and-key hypothesis** gives us a simple model that helps us understand what happens (see **fig B**). Within the globular protein structure of each enzyme is an area known as the **active site** that has a very specific shape. Only one substrate or type of substrate will fit the shape of the active site, and it is this that gives each enzyme its specificity. The enzyme and substrate slot together to make a complex in the same way as a key fits into a lock.

The formation of the enzyme–substrate complex lowers the activation energy of the reaction. The active site affects the bonds in the substrate, making it easier for them to break. The reacting substances are brought close together, making it easier for bonds to form between them. Once the reaction is complete, the products are not the right shape to stay in the active site and the complex breaks up. This releases the products and frees the enzyme for further catalytic activity.

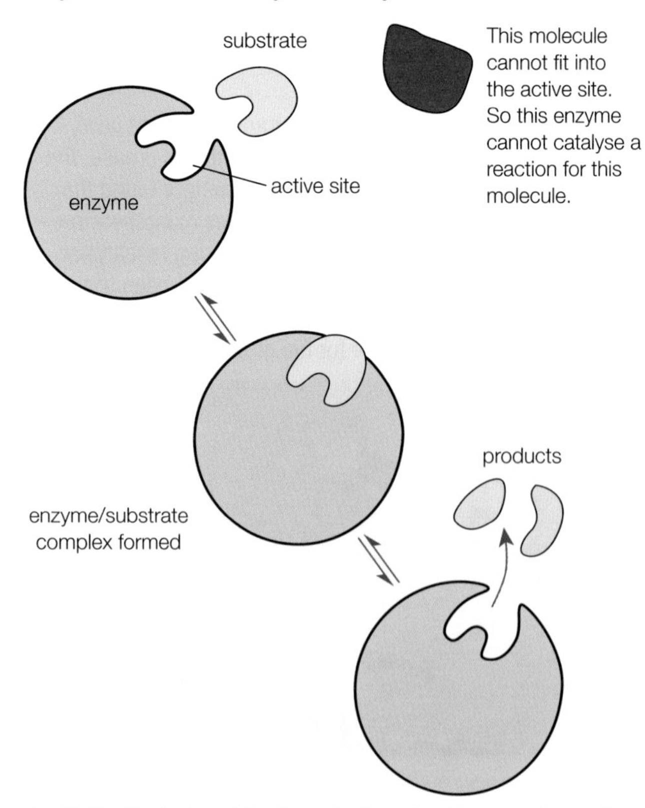

▲ **fig B** The lock-and-key hypothesis underpins our understanding of how enzymes work.

The lock-and-key hypothesis fits most of our evidence about enzyme characteristics. However, it is now thought to be an over-simplification. Evidence from X-ray crystallography, chemical analysis of active sites and other techniques suggests that the active site of an enzyme is not simply a rigid shape. The **induced-fit hypothesis** is generally accepted as the best current model of enzyme action. In this, the active site still has a distinctive shape and arrangement, but it is a flexible one. After the substrate enters the active site, the shape of the site is modified around it to form the active complex. Once the products have left the complex, the enzyme returns to its inactive, relaxed form until another substrate molecule binds (see **fig C**).

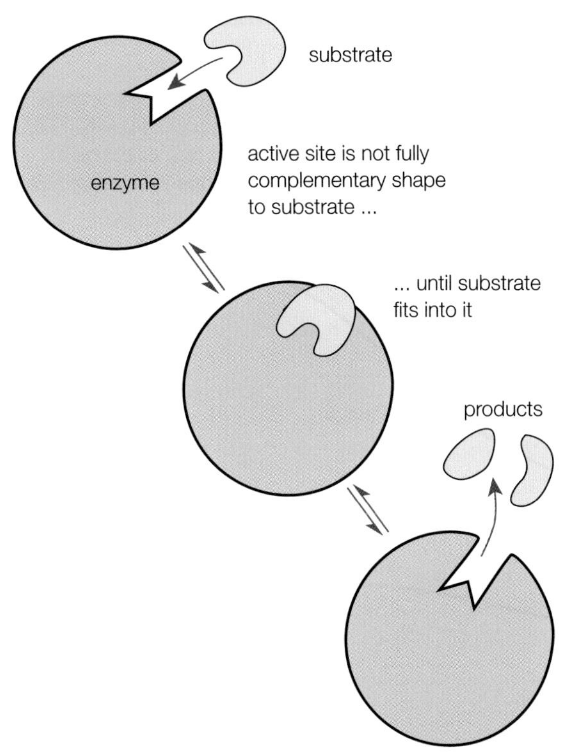

▲ **fig C** The induced-fit theory of enzyme action proposes that the catalytic groups of the active site are not brought into their most active positions until a substrate is bound to the site, inducing a change in shape.

WHAT DO WE KNOW ABOUT ENZYMES?

Our current model of enzymes is that they are globular proteins (see **Section 2B.1**). They contain an active site, which is essential to the functioning of the enzyme. The active site is a small depression on the surface of the molecule that has a specific shape because of the way the whole large protein molecule is folded. Anything affecting the shape of the protein molecule affects its ability to do its job. This indicates that the three-dimensional nature of the molecule is important to the way it works. A change in shape changes the shape of the active site as well – and so the enzyme can no longer function.

Enzymes change *only* the rate of a reaction. They do not change or contribute to the end products and they do not affect the equilibrium of the reaction. They act only as catalysts and do not modify the reaction in any other way.

EVIDENCE FOR THE RELATIONSHIP BETWEEN THE STRUCTURE AND FUNCTIONS OF ENZYMES

Observing the factors that affect the rate of enzyme activity helps us to understand the relationship between the structure of an enzyme and the way it functions.

Enzymes speed up reactions so much that only very small amounts of them are needed to catalyse the reaction of many substrate molecules into products. This is described by the **molecular activity** or **turnover number** of an enzyme, which measures the number of substrate molecules transformed per minute by a single enzyme molecule. Most enzymes can catalyse thousands of molecules per minute. Some can achieve many more: the number of molecules of hydrogen peroxide catalysed by the enzyme catalase extracted from liver cells is 6×10^6 in 1 minute. If every enzyme molecule is involved in a reaction, it will not go any faster unless there is an increase in the enzyme concentration. In other words, enzyme-controlled reactions are affected by the concentration of the enzyme.

Enzymes are very specific to the reaction that they catalyse. Inorganic catalysts such as platinum frequently catalyse many different reactions, often only at extremes of temperature and pressure. In comparison, some enzymes are so specific that they will catalyse only one particular reaction. Others are specific to a group of molecules that are all of similar shape, or to a type of reaction that always involves the same groups. This suggests that there is a physical site within the enzyme with a specific shape, which a specific substrate fits into.

The number of substrate molecules present (the concentration of the substrate) affects the rate of an enzyme-catalysed reaction. Take a simple reaction where substrate A is converted to product Z. If the concentration of A increases, the rate of the enzyme-catalysed reaction $A \rightarrow Z$ increases – but only for a limited period. Then the enzyme becomes saturated – all of the active sites are occupied by substrate molecules – and a further increase in substrate concentration will not increase the rate of the reaction further (see **fig D**). At this point, only an increase in enzyme concentration will increase the rate of the reaction.

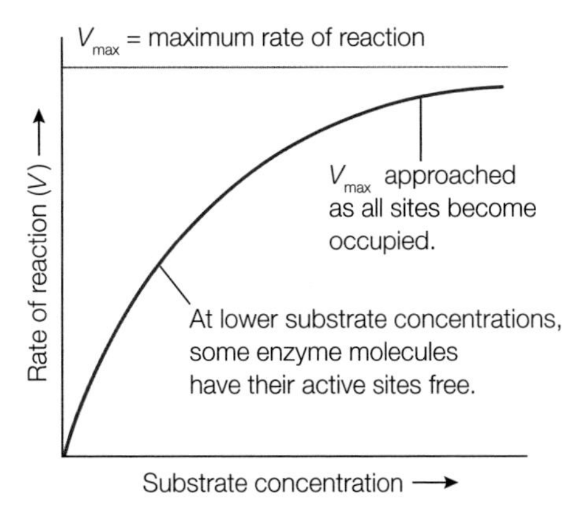

▲ **fig D** The effect of substrate concentration on an enzyme-catalysed reaction, showing how the enzyme becomes saturated with substrate molecules.

Temperature affects the rate of an enzyme-catalysed reaction in a characteristic way (see **fig E**). Temperature affects all reactions because the number of successful collisions leading to a reaction increases at higher temperatures. The effect of temperature on the rate of any reaction can be expressed as the **temperature coefficient, Q_{10}**. This is expressed as:

$$Q_{10} = \frac{\text{rate of reaction at } (x + 10)\,°\text{C}}{\text{rate of reaction at } x\,°\text{C}}$$

Between about 0 °C and 40 °C, Q_{10} for any reaction is 2 – the rate of the reaction doubles for every 10 °C rise in temperature. However, outside this range, Q_{10} for enzyme-catalysed reactions in human beings decreases markedly, while Q_{10} for other reactions changes only slowly. The rate of enzyme-catalysed reactions in human beings falls as the temperature rises, and at about 60 °C the reaction stops completely in most cases. At temperatures over 40 °C most proteins, including most enzymes, start to lose their tertiary and quaternary structures – they **denature**. When enzymes denature, the shape of the active site changes and so they lose their ability to catalyse reactions. There are some exceptions to this rule. For example, the enzymes of thermophilic bacteria, which live in hot springs at temperatures of up to 85 °C, can work at very high temperatures. They are made of temperature-resistant proteins that contain a very high density of hydrogen bonds and disulfide bonds, which hold them together even at high temperatures (see **Section 1B.5**). However, the optimum temperature of the enzymes of many organisms, including cold water fish and many plants, is much lower than 40 °C.

LEARNING TIP

Remember that not all enzymes have the same optimum conditions. Enzymes in humans have an optimum temperature of about 37 °C because that is our body temperature. Enzymes in other organisms may have an optimum set of conditions that is very different.

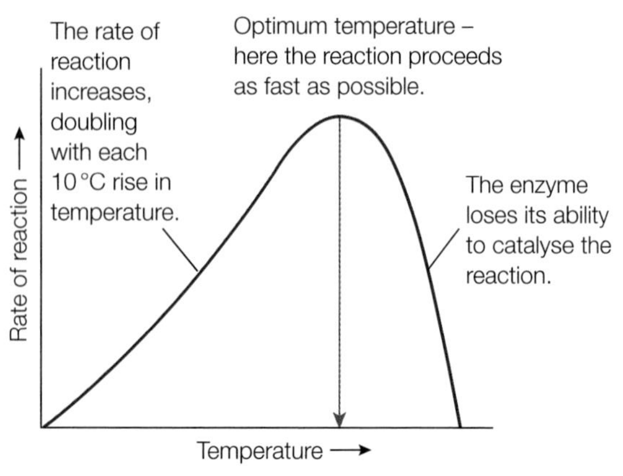

▲ **fig E** The effect of temperature on the rate of a typical enzyme-catalysed reaction. All other factors must be kept constant.

EXAM HINT

When you describe the effect of changing temperature on the rate of an enzyme-controlled reaction, ensure you describe the effect between 0 and 40 °C and what happens above 40 °C.

pH also has a major effect on enzyme activity by affecting the shape of protein molecules. Different enzymes work in different ranges of pH (see **fig F**). This is because changes in pH affect the formation of the hydrogen bonds and disulfide bonds that hold the three-dimensional structure of the protein together. The optimum pH for an enzyme is not always the same as the pH of its normal surroundings. This seems to be one way in which cells control the effects of their intracellular enzymes, increasing or decreasing their activity by very small changes in the pH.

EXAM HINT

When you describe the effect of changing temperature or pH on an enzyme molecule, you should always refer to changes in the shape of the active site.

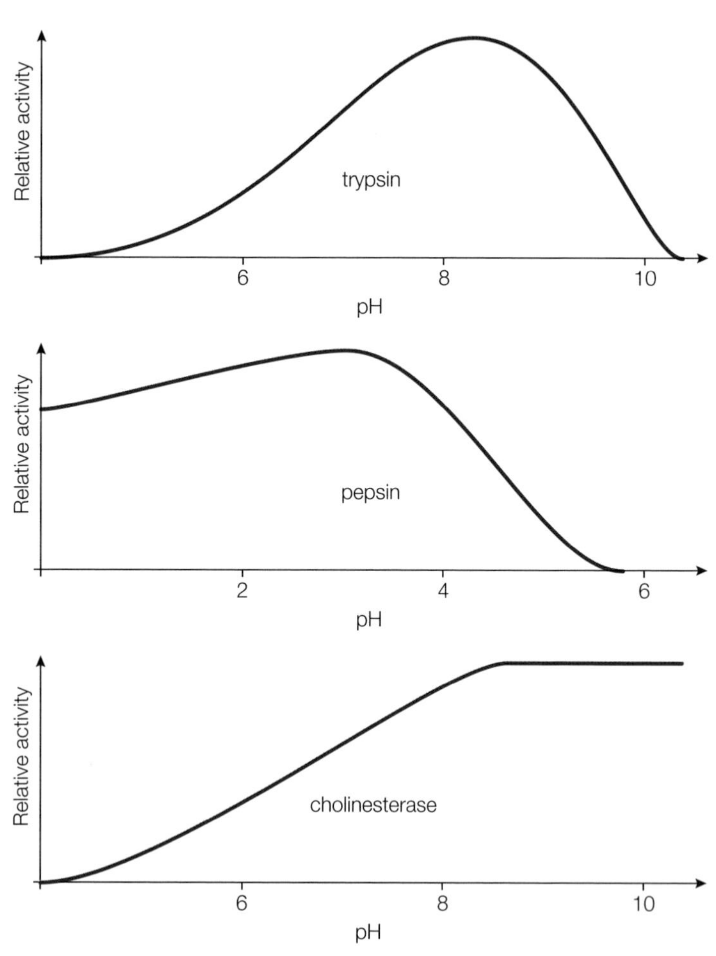

▲ **fig F** Different enzymes work best at different pH levels. All other factors must be kept constant.

PRACTICAL SKILLS CP4

Measuring reaction rate

When scientists are investigating enzymes and how they act as catalysts, they frequently measure the reaction rate. For example, one practical way of demonstrating the effect of an enzyme on a reaction is to measure the rate of the reaction with and without the enzyme. Using this method, it has been shown that when urea breakdown is catalysed by urease extracted from the jack bean, the rate of the reaction increases by a factor of 1014. Enzymes are such efficient catalysts that they generally increase reaction rates by factors from 108 to 1026. This is why only tiny amounts of most enzymes are needed.

Much of the evidence for the structure of enzymes and the way this relates to their functions comes from practical investigations into the effect of different factors on the rate of enzyme-catalysed reactions. In the laboratory, you may be asked to investigate the effect of a number of factors on enzyme activity. Such factors include temperature, pH, enzyme concentration and substrate concentration. To investigate the way a factor affects the rate of reaction, biologists measure the **initial rate of reaction** each time the independent variable is changed. Every other factor must be kept the same so that any changes are the result of changing the one variable.

It is important to provide a large excess of substrate in enzyme experiments, unless the effect of substrate concentration is under investigation. The enzyme–substrate complexes develop quickly and the reaction rapidly takes place at a steady rate. As soon as this point is reached, the reaction rate is recorded accurately by measuring the amount of product over a short period of time. Measuring the initial rate only, with an excess of substrate, means other factors such as build-up of products, lack of substrate and changes in pH will not have time to influence the rate. You can calculate the initial reaction rate of an enzyme by drawing a line which is a tangent to the curve at the beginning of the reaction (see **fig G**).

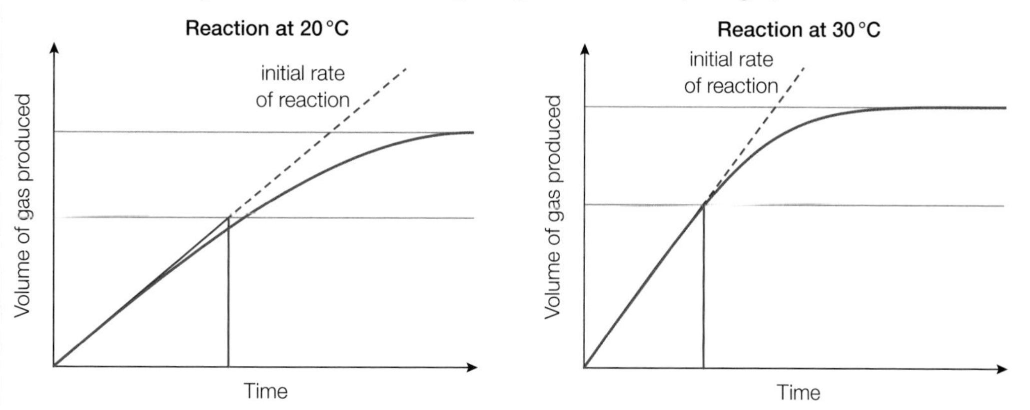

▲ **fig G** The initial reaction rate is clearly higher when the temperature increases from 20 °C to 30 °C.

CHECKPOINT

1. (a) Summarise the characteristics of enzymes.
   (b) Explain how each characteristic of enzymes provides evidence for the induced-fit hypothesis.

SKILLS EXECUTIVE FUNCTION

2. Plan a practical investigation into the effect of temperature on enzyme activity.

SUBJECT VOCABULARY

activation energy the energy needed for a chemical reaction to get started
substrate the molecule or molecules on which an enzyme acts
lock-and-key hypothesis a model that explains enzyme action by an active site in the protein structure that has a very specific shape; the enzyme and substrate slot together to form a complex in the same way as a key fits in a lock
active site the area of an enzyme that has a specific shape into which the substrate(s) of a reaction fit
induced-fit hypothesis a modified version of the lock-and-key model of enzyme action where the active site is considered to have a more flexible shape; after the substrate enters the active site, the shape of that site changes around it to form the active complex; after the products have left the complex, the enzyme returns to its inactive, relaxed form
molecular activity (turnover number) the number of substrate molecules transformed per minute by a single enzyme molecule
temperature coefficient (Q_{10}) the measure of the effect of temperature on the rate of a reaction
denaturation the loss of the three-dimensional shape of a protein (e.g. caused by changes in temperature or pH)
initial rate of reaction the measure taken to compare the rates of enzyme-controlled reactions under different conditions

2B 3 THE STRUCTURE OF DNA AND RNA

SPECIFICATION REFERENCE 2.9(i) 2.9(ii)

LEARNING OBJECTIVES

- Know the basic structure of mononucleotides and the structures of DNA and RNA as polynucleotides composed of mononucleotides linked by condensation reactions which form phosphodiester bonds.
- Know how complementary base pairing and the hydrogen bonding between two complementary strands are involved in the formation of the DNA double helix.

MONONUCLEOTIDES

Mononucleotides are key molecules in biology. They provide the energy currency of cells in the form of **adenosine triphosphate (ATP)**. They also provide the building blocks for the mechanism of inheritance in the form of **deoxyribonucleic acid (DNA)** and **ribonucleic acid (RNA)**.

Each nucleotide has three parts:

- a 5-carbon pentose sugar
- a nitrogen-containing base
- a phosphate group.

The pentose sugar in RNA is **ribose**, and in DNA is **deoxyribose**. Deoxyribose, as its name suggests, contains one fewer oxygen atom than ribose (see **fig A**).

LEARNING TIP

Remember that a nucleotide contains a pentose sugar, an organic base and a single phosphate.

The most common types of nucleotide have either a **purine base**, which has two nitrogen-containing rings, or a **pyrimidine base**, which has only one. The most common purines are **adenine** (A) and **guanine** (G) and the most common pyrimidines are **cytosine** (C), **thymine** (T) and **uracil** (U).

A phosphate group ($-PO_4^{3-}$) is the third component of a nucleotide. Inorganic phosphate ions are present in the cytoplasm of every cell (see **Section 1A.1** for more about inorganic ions). This phosphate group means that the nucleotides are acidic molecules and carry a negative charge.

The sugar, the base and the phosphate group are joined together by condensation reactions, with the elimination of two water molecules, to form a mononucleotide (see **fig A**).

POLYNUCLEOTIDES

Reproduction is one of seven key processes in living organisms. If the individuals in a species do not reproduce, then that species will die out. Multicellular organisms also need to grow, and to replace worn-out cells. Within every cell is a set of instructions for the assembling of new cells. These can be used to produce both offspring and identical cells for growth. Over the last 75 years or so, scientists have made enormous advances towards understanding how the genetic code works. In unravelling the secrets of the genetic code, people have come closer than ever before to understanding the mystery of life itself.

Nucleic acids, also known as **polynucleotides**, are the information molecules of the cell. They carry all the information needed to make new cells. They are polymers, consisting of many mononucleotide monomer units. The chromosomes in the nucleus of eukaryotic cells like our own store the genetic information; but in prokaryotes (e.g. bacteria), the DNA is found floating freely in the cytoplasm of the cells. The information is stored as a code in the molecules of DNA. Parts of the code are used as a template to make a complementary strand of messenger RNA (mRNA) and direct the production of the proteins that build the cell and control its actions.

BUILDING THE POLYNUCLEOTIDES

Nucleic acids are chains of nucleotides linked together by condensation reactions that produce **phosphodiester bonds** between the sugar on one nucleotide and the phosphate group of the next nucleotide (see **fig B**). These nucleic acids can be millions of nucleotide units long. Both DNA and RNA have this sugar–phosphate backbone. The sugar of one nucleotide bonds to the phosphate group of the next nucleotide and so polynucleotides always have

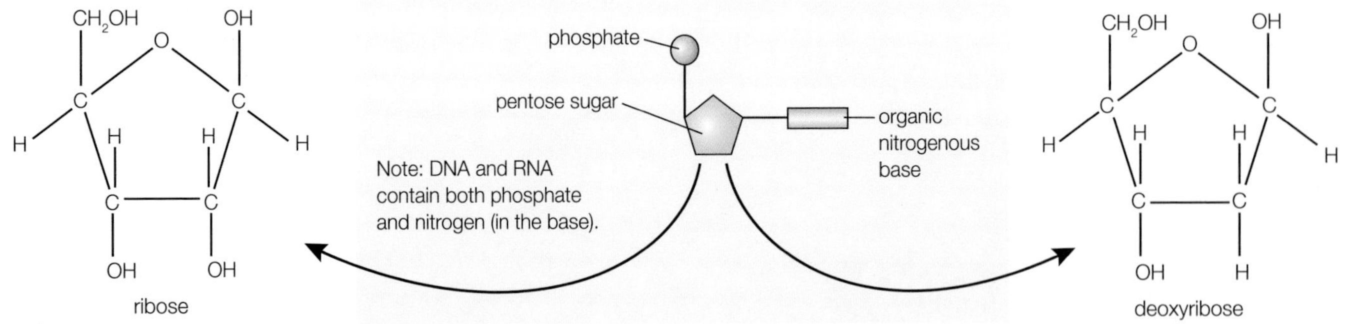

▲ **fig A** The structure of a mononucleotide. The properties of mononucleotide molecules are crucial to the roles of ATP, DNA and RNA.

a hydroxyl group at one end and a phosphate group at the other. This structural feature is important in the role of the nucleic acids in the cell. Long chains of nucleotides containing the bases C, G, A and T join to form DNA. Chains of nucleotides containing C, G, A and U make RNA. Knowledge of how these units join together, and the three-dimensional structures in DNA in particular, is the basis of our understanding of molecular genetics.

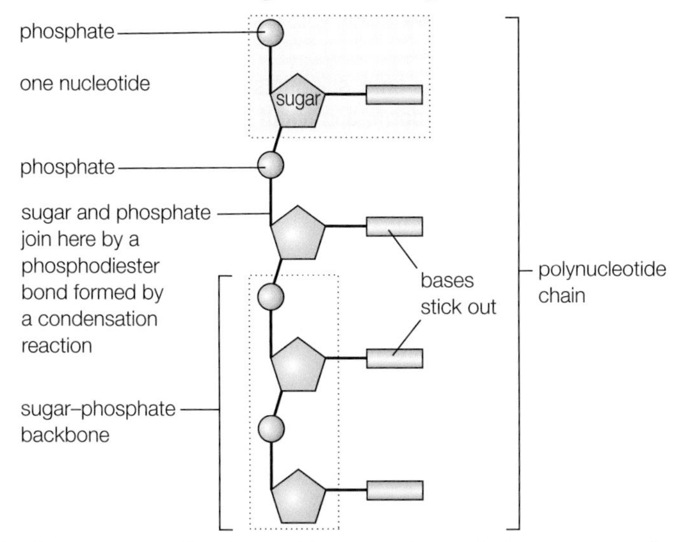

fig B A polynucleotide strand like this makes up the basic structure of both DNA and RNA.

RNA molecules form single polynucleotide strands. These can either fold into complex shapes, held in place by hydrogen bonds, or remain as long thread-like molecules. DNA molecules consist of two polynucleotide strands twisted around each other. The sugars and phosphates form the backbone of the molecule and, pointing inwards from the two sugar–phosphate backbones, are the bases which make pairs in specific ways. A purine base always pairs with a pyrimidine base. In DNA, adenine pairs with thymine, and cytosine pairs with guanine. This results in the famous DNA *double helix*, a massive molecule that is the same shape as a spiral staircase (see **fig C**).

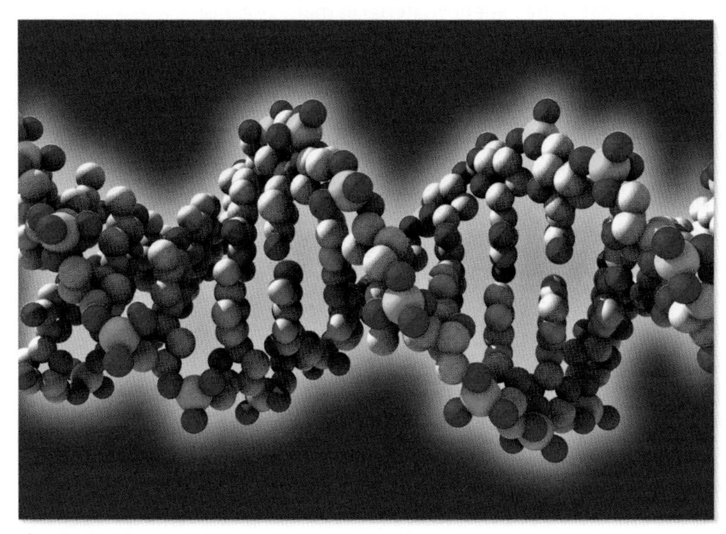

fig C The double helix structure of a DNA molecule is not just an iconic image of science – it is vital to the role of DNA in cells.

EXAM HINT

Remember the differences between DNA and RNA:

- they have a different pentose sugar
- in RNA, the base uracil replaces the thymine found in DNA
- RNA is single-stranded.

The two strands of the DNA double helix are held together by hydrogen bonds between the **complementary base pairs** (see **fig D**). These hydrogen bonds form between the amino and the carbonyl groups of the purine and pyrimidine bases on the opposite strands. There are three hydrogen bonds between C and G but only two between A and T. There are 10 base pairs for each complete twist of the helix.

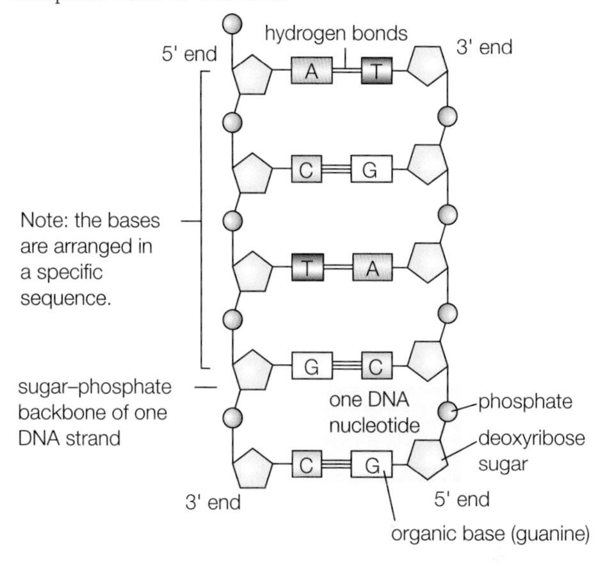

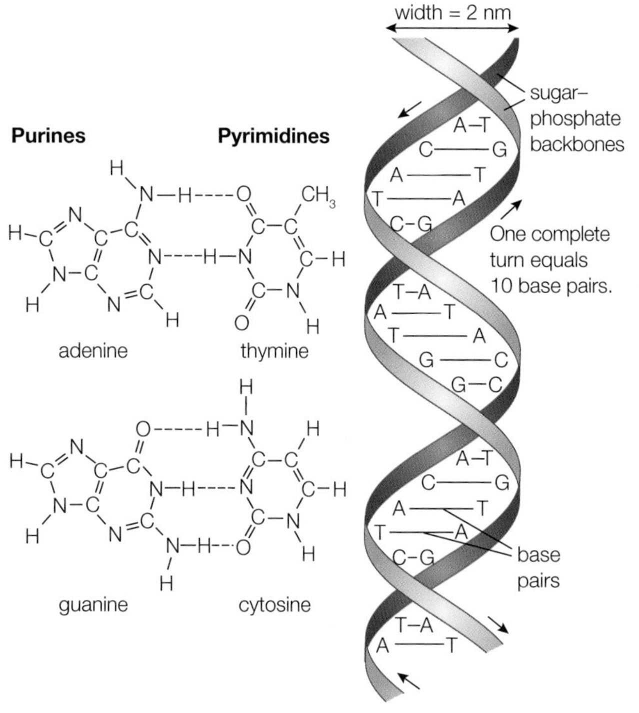

fig D The double helix structure of DNA depends on the hydrogen bonds that form between the base pairs.

EXAM HINT

Make sure you use the terms mononucleotide and polynucleotide correctly. Do not get them confused.

DID YOU KNOW?

Sequencing the genome

From the late 20th century onwards, scientists from around the world collaborated in the Human Genome Project. This was an ambitious project that set out to identify the human **genome** – that is all of the sequences of the 3 billion base pairs which make up the human DNA. The scientists worked on DNA from anonymous donors and showed that every individual has at least 99.9% of their DNA the same. Advances in technology meant the project finished ahead of the expected date, but it still took almost 13 years.

In 2008 a new project began – the 1000 Genomes Project. This time, scientists analysed the DNA of 1092 people from all around the world. The aim was to gain information about differences in our DNA that can, among many things, have an impact on the diseases that may affect us. The 1000 Genomes Project took 6 years.

The 100 000 Genomes Project is now sequencing the genomes of 100 000 people from around the world with rare genetic diseases and cancers. It passed the 10 000 genomes sequenced milestone in 2016. Scientists are also attempting to sequence at least one genome from every species of vertebrate animal. Projects like these are only possible as a result of huge improvements in sequencing technology. This means that processes that once took weeks and months, now take hours and days. It should greatly increase our understanding, diagnosis and even treatment of rare genetic conditions and give us a much greater insight into evolutionary relationships.

CHECKPOINT

SKILLS PROBLEM SOLVING

1. What is a mononucleotide and which mononucleotides are found in both DNA and RNA?
2. (a) Explain how complementary base pairing and hydrogen bonding contribute to the structure of DNA.

   (b) Look carefully at the structural formulae of the purine bases and the pyrimidine bases of the DNA molecule. Suggest reasons why the pairs of bases always involve one purine and one pyrimidine base, never two purines or two pyrimidines.

SUBJECT VOCABULARY

mononucleotides molecules with three parts - a 5-carbon pentose sugar, a nitrogen-containing base and a phosphate group - joined by condensation reactions
adenosine triphosphate (ATP) a molecule that acts as the universal energy supply molecule in cells; it is made up of the base adenine, the pentose sugar ribose and three phosphate groups
deoxyribonucleic acid (DNA) a nucleic acid that is the genetic material in many organisms
ribonucleic acid (RNA) a nucleic acid which is the genetic material in some organisms and is involved in protein synthesis
ribose a pentose sugar that is part of the structure of RNA
deoxyribose a pentose sugar that is part of the structure of DNA
purine base a base found in nucleotides that has two nitrogen-containing rings
pyrimidine base a base found in nucleotides that has one nitrogen-containing ring
adenine a purine base found in DNA and RNA
guanine a purine base found in DNA and RNA
cytosine a pyrimidine base found in DNA and RNA
thymine a pyrimidine base found in DNA
uracil a pyrimidine base found in RNA
nucleic acids/polynucleotides polymers made up of many nucleotide monomer units that carry all the information needed to form new cells
phosphodiester bond bond formed between the phosphate group of one nucleotide and the sugar of the next nucleotide in a condensation reaction
complementary base pairs complementary purine and pyrimidine bases which align in a DNA helix, with hydrogen bonds holding them together (C–G, A–T)
genome the entire genetic material of an organism

2B 4 HOW DNA WORKS

SPECIFICATION REFERENCE 2.10(i) 2.10(ii)

LEARNING OBJECTIVES

- Understand the process of DNA replication, including the role of DNA polymerase.
- Understand how Meselson and Stahl's classic experiment provided new data that supported the accepted theory of DNA replication and refuted competing theories.

One of the most important features of the DNA molecule is that it can replicate (copy) itself exactly. This, together with the fact that it can carry a complex code, means it can pass on genetic information from one cell or generation to another.

UNCOVERING THE MECHANISM OF REPLICATION

After Watson and Crick had produced their double helix model for the structure of DNA, it took scientists some years to work out exactly how the molecule replicates itself.

There were two main ideas about how replication happens: **conservative replication** and **semiconservative replication**. In the conservative replication model, the original double helix remained intact and in some way instructed the formation of a new, identical double helix, made up entirely of new material.

The semiconservative replication model assumed that the DNA 'unzipped' and new nucleotides aligned along each strand. Each new double helix contained one strand of the original DNA and one strand made up of new material. This was the Watson and Crick hypothesis – they felt the double helix would unzip along the hydrogen bonds in their structural model, allowing semiconservative replication to occur. It took a classic piece of practical investigation to settle the argument.

LEARNING TIP

Remember that *semi* means half: half the molecule (or one strand) is conserved in each new molecule.

EXPERIMENTAL EVIDENCE

Matthew Meselson (1930–) and Franklin Stahl (1929–) carried out an elegant set of experiments in the late 1950s at the California Institute of Technology, USA. These led to semiconservative replication becoming the accepted model of DNA replication.

They grew several generations of the gut bacteria *Escherichia coli* (*E. coli*) in a medium where their only source of nitrogen was the radioactive **isotope** ^{15}N from $^{15}NH_4Cl$. Atoms of ^{15}N are denser than those of the usual isotope, ^{14}N. The bacteria grown on this medium took up the radioactive isotope to make the cell chemicals, including proteins and DNA. After several generations, the entire bacterial DNA was made using ^{15}N ('heavy' nitrogen).

They moved the bacteria to a medium containing normal $^{14}NH_4Cl$ as the only nitrogen source, and measured the density of the DNA as the cells reproduced.

Meselson and Stahl predicted that if DNA reproduced by conservative replication, some of the DNA would have the density expected if it contained only ^{15}N (the original strands), and some of it would have the density expected if it contained only ^{14}N (the new strands). However, if DNA reproduced by semiconservative replication, then after one replication cycle all of the DNA would have the same density, half-way between that of ^{15}N- and ^{14}N-containing DNA.

After one replication cycle, they found that all the DNA did have the same density, half-way between that of ^{15}N- and ^{14}N-containing DNA – and so they concluded that DNA must replicate semiconservatively (see **fig A**, on the next page).

Conservative replication, where the double helix remains intact and new strands form on the outside, would give:

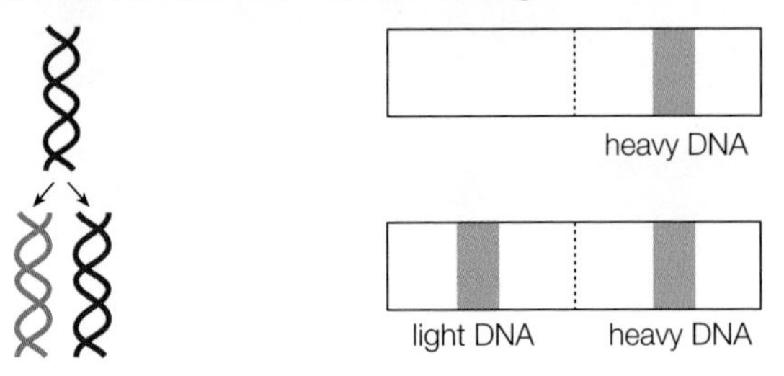

Replicates in medium containing only light nitrogen.

Half of the DNA molecules have two light strands and half have two heavy strands.

Semiconservative replication, where the double helix unzips and each strand replicates to produce a second, new strand, would give:

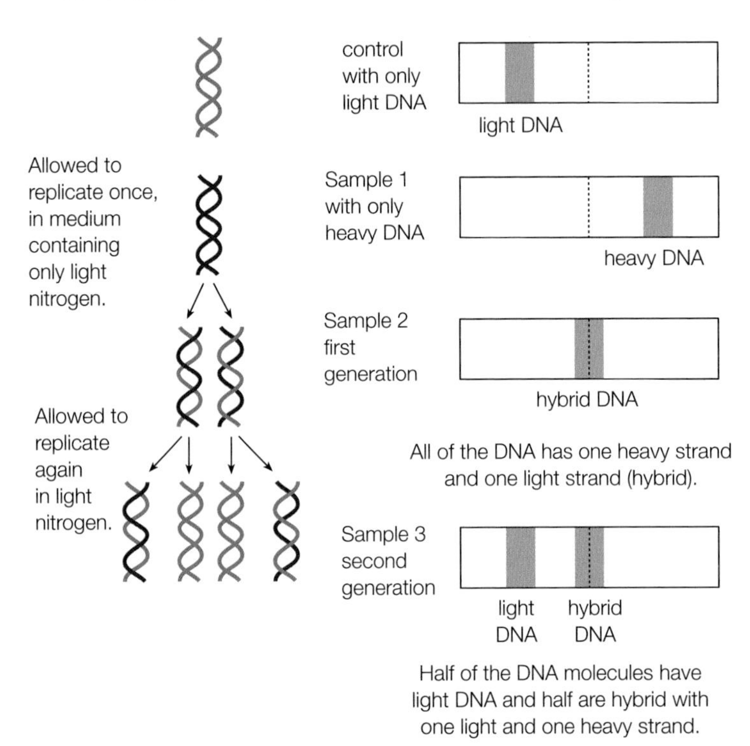

▲ **fig A** The results of these experiments by Meselson and Stahl confirmed the model of semiconservative replication of DNA – and at the same time put an end to the theory of conservative replication.

EXAM HINT

This experiment is a true classic because it is very simple and yet it managed to answer one of the great questions about DNA. Make sure you understand the principles involved as exam questions may supply evidence from experiments and ask you to interpret the evidence.

HOW DNA MAKES COPIES OF ITSELF

A careful look at the process of the semiconservative replication of DNA shows clearly how important the structure and properties of the DNA molecule are to its role as the genetic material of the cell. The complete process depends on three enzymes. **DNA helicase** unzips the two strands of the DNA. **DNA polymerase** lines up the new nucleotides along the DNA template strands, and **DNA ligase** catalyses the formation of the phosphodiester bonds between the new nucleotides. This is shown in **fig B**.

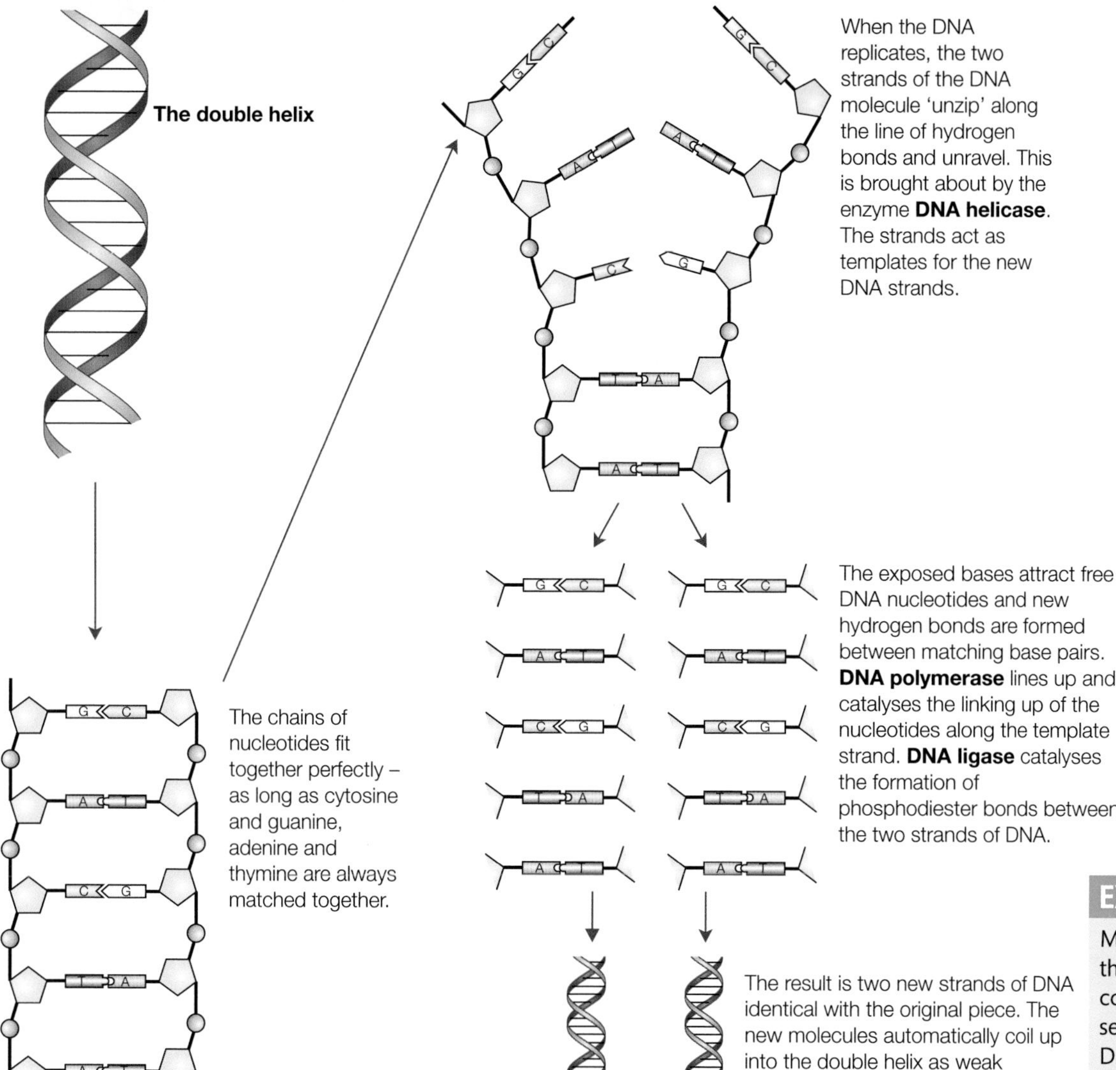

▲ **fig B** The semiconservative replication of DNA

EXAM HINT

Make sure you are clear about the difference between conservative and semiconservative models of DNA replication and can explain how the evidence supports the second model.

CHECKPOINT

1. Make a flow diagram to describe the replication of DNA.
2. Explain how the work of Meselson and Stahl destroyed support for the model of the conservative replication of DNA.

SKILLS INTERPRETATION

SUBJECT VOCABULARY

conservative replication a model of DNA replication which suggests that the original double helix remains intact and in some way instructs the formation of a new, identical double helix made up entirely of new material
semiconservative replication the accepted model of DNA replication in which the DNA 'unzips' and new nucleotides align along each strand; each new double helix contains one strand of the original DNA and one strand made up of new material
isotopes different atoms of the same element, with the same number of protons but a different number of neutrons; isotopes have the same chemical properties
DNA helicase an enzyme involved in DNA replication that 'unzips' the two strands of the DNA molecules
DNA polymerase an enzyme involved in DNA replication that lines up the new nucleotides along the DNA template strands
DNA ligase an enzyme involved in DNA replication that catalyses the formation of phosphodiester bonds between the nucleotides

2B 5 THE GENETIC CODE

SPECIFICATION REFERENCE 2.11 2.12

LEARNING OBJECTIVES

- Understand the nature of the genetic code as a triplet code which is non-overlapping and degenerate.
- Know that a gene is a sequence of bases on a DNA molecule that codes for a sequence of amino acids in a polypeptide chain.

We know that DNA has a double helix structure and can replicate itself exactly. But how does DNA act as the genetic material and carry the information needed to make new cells and whole new organisms? The key is the link between DNA and proteins. DNA controls protein synthesis and so the DNA instructions control not only how the cell is built, but also how it works.

Proteins are made up of amino acids. Using the DNA code, the 20 naturally occurring amino acids are joined together in countless combinations to make an almost infinite variety of proteins. This process of **translation** happens on the surface of the **ribosomes** (see protein synthesis in **Section 2B.6**).

WHAT IS THE GENETIC CODE?

In the DNA double helix, the components that vary are the bases. So scientists guessed that it was the arrangement of the bases that carries the genetic code – but how? There are only four bases, so if one base coded for one amino acid there could be only four amino acids. Even a combination of two bases does not give enough amino acids – the possible arrangements of four bases into groups of two is $4 \times 4 = 16$. However, a **triplet code** of three bases gives $4 \times 4 \times 4 = 64$ possible combinations – more than enough for the 20 amino acids that are coded for.

CRACKING THE CODE

The genetic code is based on genes. We can define a **gene** as a sequence of bases on a DNA molecule which codes for a sequence of amino acids in a polypeptide chain. This polypeptide affects a characteristic in the phenotype of the organism. By the early 1960s, it had been proved that a triplet code of bases was the cornerstone of the genetic code. Each sequence of three bases along a strand of DNA codes for something very specific. Most code for a particular amino acid, but some triplets signal the beginning or the end of a particular amino acid sequence.

A sequence of three bases on the DNA or RNA is called a **codon**. The codons of the DNA are difficult to work out because the molecule is so large, so most of the work was done on the codons of the smaller molecule **messenger RNA (mRNA)**. This mRNA is formed as a **complementary strand** to the DNA, so it is like a reverse image of the original base sequence. When we know the RNA sequence, we can work out the DNA sequence because of the way bases always pair: T and U with A, and G with C. Sequencing tasks like this have become much easier in the 21st century as technology has advanced.

LEARNING TIP

Remember that the base thymine is found in DNA but not in RNA. When converting the code from DNA to make mRNA, the base U pairs up with A on the DNA.

EXAM HINT

The universal nature of the genetic code is used as a strong line of evidence for the idea that all living organisms originate from one group.

The result of all this work is a sort of dictionary of the genetic code (see **tables A** and **B**). Much of the original work, done in the 1960s, used the gut bacteria *E. coli*, but all the studies done since then suggest that the genetic code is identical throughout the living world.

Large parts of the DNA do not code for proteins. Scientists think the non-coding DNA sequences are very important – 98% of the human DNA is non-coding. These DNA sequences are involved in regulating the protein-coding sequences – effectively turning genes on or off. Many organisms have similar non-coding sequences, which suggests they are useful, but in many cases we still do not know exactly what they do.

In the 2% of the human DNA that codes for proteins, some codons code for a particular amino acid, while others code for the beginning or the ending of a particular amino acid sequence. So, for example, the codon to start a polypeptide chain is TAC. This is also the codon for methionine – so the first amino acid in a polypeptide chain is always methionine. We now know that the genetic code is not only a triplet code, it is also a **non-overlapping code** and **degenerate code**.

DID YOU KNOW?

DNA code-breakers

The first breakthrough in decoding the genetic code came in 1961 when M.W. Nirenberg (1927–2010) and J.H. Matthaei (1929–) in the United States prepared artificial mRNA where all the bases were uracil. They added their polyU – chains reading UUUUUUUUUUU ... – to all the other ingredients needed for protein synthesis (ribosomes, tRNAs, amino acids, etc). When they analysed the polypeptides made, they found chains of a single type of amino acid, phenylalanine. UUU appeared to be the mRNA codon for phenylalanine. So the DNA codon would be AAA. The scientists soon showed that CCC codes for proline and AAA for lysine. Evidence for the triplet code – three non-overlapping bases with some degeneracy – built up swiftly from this early work. It was also shown that the minimum length of artificial mRNA that would bind to a ribosome was three bases long – a single codon. This would then bind with the corresponding tRNA. After this, it was a case of careful and precise work to identify all of the codons and their corresponding amino acids.

first letter of the codon	second letter of the codon: A	G	T	C	third letter of the codon
A	AAA phenylalanine	AGA serine	ATA tyrosine	ACA cysteine	A
A	AAG phenylalanine	AGG serine	ATG tyrosine	ACG cysteine	G
A	AAT leucine	AGT serine	**ATT** stop codon	**ACT** stop codon	T
A	AAC leucine	AGC serine	**ATC** stop codon	ACC tryptophan	C
G	GAA leucine	GGA proline	GTA histidine	GCA arginine	A
G	GAG leucine	GGG proline	GTG histidine	GCG arginine	G
G	GAT leucine	GGT proline	GTT glutamine	GCT arginine	T
G	GAC leucine	GGC proline	GTC glutamine	GCC arginine	C
T	TAA isoleucine	TGA threonine	TTA asparagine	TCA serine	A
T	TAG isoleucine	TGG threonine	TTG asparagine	TCG serine	G
T	TAT isoleucine	TGT threonine	TTT lysine	TCT arginine	T
T	**TAC** methionine; start codon	TGC threonine	TTC lysine	TCC arginine	C
C	CAA valine	CGA alanine	CTA aspartic acid	CCA glycine	A
C	CAG valine	CGG alanine	CTG aspartic acid	CCG glycine	G
C	CAT valine	CGT alanine	CTT glutamic acid	CCT glycine	T
C	CAC valine	CGC alanine	CTC glutamic acid	CCC glycine	C

table A The triplet code that underpins all work on genetics. This shows the antisense or template strand sequence of the DNA code.

first letter of the codon	second letter of the codon: U	C	A	G	third letter of the codon
U	UUU phenylalanine	UCU serine	UAU tyrosine	UGU cysteine	U
U	UUC phenylalanine	UCC serine	UAC tyrosine	UGC cysteine	C
U	UUA leucine	UCA serine	**UAA** stop codon	**UGA** stop codon	A
U	UUG leucine	UCG serine	**UAG** stop codon	UGG tryptophan	G
C	CUU leucine	CCU proline	CAU histidine	CGU arginine	U
C	CUC leucine	CCC proline	CAC histidine	CGC arginine	C
C	CUA leucine	CCA proline	CAA glutamine	CGA arginine	A
C	CUG leucine	CCG proline	CAG glutamine	CGG arginine	G
A	AUU isoleucine	ACU threonine	AAU asparagine	AGU serine	U
A	AUC isoleucine	ACC threonine	AAC asparagine	AGC serine	C
A	AUA isoleucine	ACA threonine	AAA lysine	AGA arginine	A
A	**AUG** methionine; start codon	ACG threonine	AAG lysine	AGG arginine	G
G	GUU valine	GCU alanine	GAU aspartic acid	GGU glycine	U
G	GUC valine	GCC alanine	GAC aspartic acid	GGC glycine	C
G	GUA valine	GCA alanine	GAA glutamic acid	GGA glycine	A
G	GUG valine	GCG alanine	GAG glutamic acid	GGG glycine	G

table B This second table shows the RNA code for the same amino acids as the DNA code in **table A**.

A NON-OVERLAPPING CODE ...

After scientists had worked out that the genetic code was based on triplets of DNA bases, they wanted to find out how the code was read. Do the triplets of bases follow each other along the DNA strand like beads on a necklace, or do they overlap? For example, the mRNA sequence UUUAGC could code for two amino acids, phenylalanine (UUU) and serine (AGC). Alternatively, if the code overlaps, it could code for four: phenylalanine (UUU), leucine (UUA), a nonsense or stop codon (UAG) and serine (AGC).

An overlapping code would be very economical – relatively short lengths of DNA could carry the instructions for many different proteins. However, it would also be very limiting, because the amino acids that could be coded for side by side would be limited. In the example given, only leucine out of the 20 available amino acids could ever follow phenylalanine, because only leucine has an mRNA codon starting with UU–.

Scientists rely on experimental observations to help decide whether the genetic code is overlapping or not. If a codon consists of three nucleotides and is completely overlapping, and a single nucleotide is altered by a **point mutation**, then three amino acids will be affected by that single change. If the code is only partly overlapping, then a single point mutation would result in two affected amino acids. But if the codons do not overlap at all, then a change in a single nucleotide mutation would affect only one amino acid, which is what has been observed, for example in sickle cell disease. All the evidence available suggests that the code is not overlapping and this is generally accepted among scientists.

LEARNING TIP

Remember that for many amino acids only the first two bases are important in the code. The third base in a triplet may vary but does not change the code.

... AND A DEGENERATE CODE

When you look at the genetic code, it seems that the code is degenerate, also known as redundant. In other words, it contains more information than it needs. If you look carefully at **table B**, you will see that often only the first two of the three nucleotides in a codon seem to determine which amino acid results. This may not seem useful, but if each amino acid was produced by only one codon, then any error or mutation could cause chaos. With a degenerate code, if the final base in the triplet is changed, this mutation could still produce the same amino acid and have no effect on the organism. Only methionine and tryptophan are represented by a single codon. Remember, methionine is always the first amino acid in an amino acid chain.

Mutations can happen any time the DNA is copied – the degenerate code at least partly protects living organisms from their effects.

CHECKPOINT

SKILLS INTERPRETATION

1. Explain what is meant by the genetic code.
2. What is non-coding DNA?
3. What are the benefits to an organism of having:
   (a) a non-overlapping code?
   (b) a degenerate code?

SUBJECT VOCABULARY

translation the process by which proteins are produced, via RNA, using the genetic code found in the DNA; it takes place on the ribosomes
ribosomes the site of protein synthesis in the cell
triplet code the code of three bases that is the basis of the genetic information in the DNA
gene a sequence of bases on a DNA molecule; it contains coding for a sequence of amino acids in a polypeptide chain that affects a characteristic in the phenotype of the organism
codon a sequence of three bases in DNA or mRNA
messenger RNA (mRNA) the RNA formed in the nucleus that carries the genetic code out into the cytoplasm
complementary strand the strand of RNA formed that complements the DNA acting as the coding strand
non-overlapping code a code where each codon codes for only one thing with no overlap between codons
degenerate code a code containing more information than is needed
point mutation a change in a single base of the DNA code

2B 6 DNA AND PROTEIN SYNTHESIS

SPECIFICATION REFERENCE 2.13(i) 2.13(ii)

LEARNING OBJECTIVES

- Understand the process of protein synthesis (transcription and translation) including the role of RNA polymerase, translation, messenger RNA, transfer RNA, ribosomes and the role of start and stop codons.
- Understand the roles of the DNA template (antisense) strand in transcription, codons on messenger RNA and anticodons on transfer RNA.

In eukaryotes, the DNA that codes for the individual proteins is in the nucleus of the cell. The proteins are synthesised at the ribosomes, which are in the cytoplasm. DNA from the nucleus has never been detected in the cytoplasm, so the message cannot be carried directly. RNAs (ribonucleic acids) carry the information from the nuclear DNA to the ribosomes.

DIFFERENT TYPES OF RNA

RNA is closely related to DNA (see **Section 2B.4**). However, it contains a different sugar (ribose) and a different base (uracil instead of thymine). It consists of a single helix and does not form enormous and complex molecules like DNA. The sequence of bases along a strand of RNA relates to the sequence of bases on a small part of the DNA in the nucleus. RNA enables DNA to act as the genetic material. It carries out three main functions in the process of protein synthesis:

- it carries the instructions for a polypeptide from the DNA in the nucleus to the ribosomes where proteins are made
- it picks up specific amino acids from the protoplasm and carries them to the surface of the ribosomes
- it makes up the bulk of the ribosomes themselves.

To perform these three very different functions, there are three different types of RNA.

MESSENGER RNA

Messenger RNA (mRNA) is formed in the nucleus. A piece of mRNA usually has instructions for one polypeptide. This is different from the double helix of DNA which carries information about a vast array of proteins. The **sense strand** of the DNA strand carries the code for the protein to be formed, but the messenger RNA forms on the **antisense strand** of the DNA, also called the **template strand**. Look at **fig A** to see how this works – it allows the code of the mRNA to reflect the sense code of the DNA. The mRNA which is formed codes for a polypeptide.

Part of the DNA unravels and unzips, exposing the bases which act as a template. Beginning at a **start codon**, which is also the code for methionine (see the codons in **tables A** and **B**, **Section 2B.5**), RNA nucleotides align along the exposed sequence of DNA bases in the normal complementary fashion. Then **RNA polymerase** joins the chain of RNA nucleotides together. The process ends when the chain reaches a **stop codon** and the mRNA chain separates from the DNA template, allowing the DNA chains of the double helix to re-join. Hydrogen bonds maintain the helical structure of the RNA molecule. In the same way as in the DNA, the bases of the mRNA form a triplet code and each triplet of bases is a codon. The relatively small mRNA molecules pass easily through the pores in the nuclear membrane, carrying the instructions from the genes in the nucleus to the cytoplasm. They then move to the surface of the ribosomes, where protein synthesis takes place (see **fig A**).

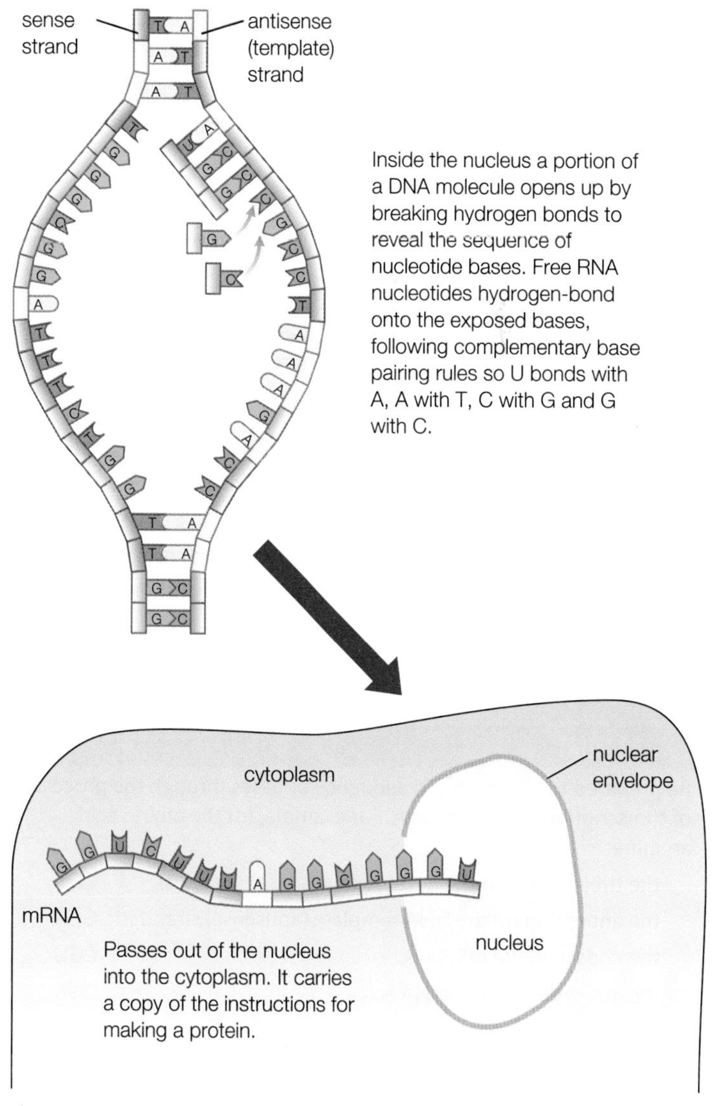

fig A The transcription of the DNA message. Any mistakes in this process can have fatal consequences for the cell or even the whole organism if the wrong protein is made.

LEARNING TIP

Remember that mRNA is made on the antisense (template) strand of the DNA, not the sense strand.

TRANSFER RNA

Transfer RNA (tRNA) is found in the cytoplasm. It has a complex shape that enables it to carry out its function (see **fig B**). This shape is the result of hydrogen bonding between different bases. One part of the tRNA molecule has a sequence of three bases that matches the genetic code of the DNA and corresponds to one specific amino acid. This sequence of three bases is called the **anticodon**. Each tRNA molecule also has a binding site with which it picks up one specific amino acid from the vast numbers always free in the cytoplasm.

The tRNA molecules, each carrying a specific amino acid, align beside the mRNA on the surface of the ribosome. The anticodons of the tRNA align with the codons of the mRNA on the surface of the ribosome, held in place by hydrogen bonds between the corresponding bases. The correct sequence of amino acids is assembled because the anticodon has a sequence of bases that align with the corresponding bases in the mRNA on the ribosomal surface. Once the amino acids are lined up together, peptide bonds form between them, building up a long chain of amino acids.

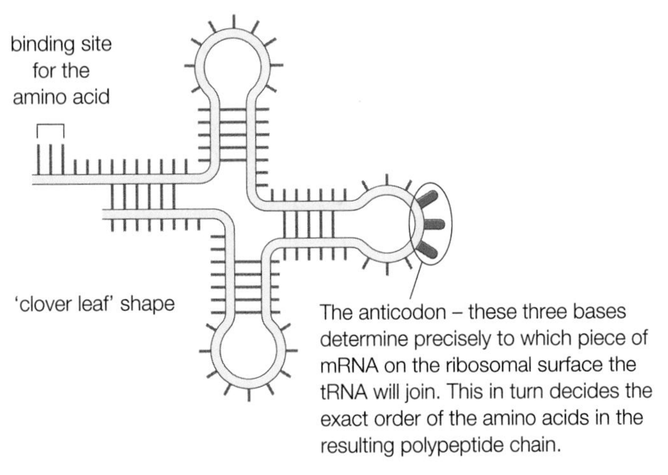

▲ **fig B** There are 61 types of tRNA molecule available to carry all the necessary amino acids to the surface of the ribosomes ready for synthesis into protein molecules.

EXAM HINT

Be prepared to follow a short sequence of bases through the process of transcription and translation. For example, for the amino acid arginine:

the triplet code on the DNA coding (sense) strand is:	CGT
the anticodon on the DNA template (antisense) strand is:	GCA
the codon on the mRNA is:	CGU
the anticodon on the tRNA is:	GCA

RIBOSOMES

Ribosomes are made up of a combination of ribosomal RNA (rRNA) and proteins. They consist of a large and a small subunit. Ribosomes surround and bind to the parts of the mRNA that are being actively **translated**, and then move along to the next codon. Their job is to hold together the mRNA and tRNA and act as enzymes controlling the process of protein synthesis.

PROTEIN SYNTHESIS

In the process of protein synthesis, the genetic code of the DNA of the nucleus is **transcribed** onto messenger RNA. This mRNA moves out of the nucleus into the cytoplasm and becomes attached to a ribosome. Molecules of transfer RNA carry individual amino acids to the surface of the ribosome. The tRNA anticodon lines up beside a complementary codon in the mRNA, held in place by hydrogen bonds, while enzymes link the amino acids together. The tRNA then breaks away and returns to the cytoplasm to pick up another amino acid. The ribosome moves along the molecule of mRNA until it reaches the stop codon at the end, leaving a completed polypeptide chain. The message may be read again and again.

Protein synthesis is a continual process, like many other events in living things. However, it is simpler to understand if we look at the two main aspects of it separately. The events in the nucleus involve the transcription of the DNA message (see **fig A**). In the cytoplasm, that message is translated into polypeptide molecules and then into proteins (see **fig C**).

LEARNING TIP

Remember that the DNA code determines the mRNA code which determines the amino acid sequence.

EXAM HINT

Make sure you understand and can explain the difference between transcription and translation.

MASS PRODUCTION

The cytoplasm of cells contains many **polysomes**. These are groups of ribosomes joined by a thread of mRNA and they are used to mass-produce specific proteins. Ribosomes attach in a steady stream to the mRNA and move along one after the other producing lots of identical polypeptides.

This is how the genetic code carried on the DNA is translated into living material by the synthesis of proteins.

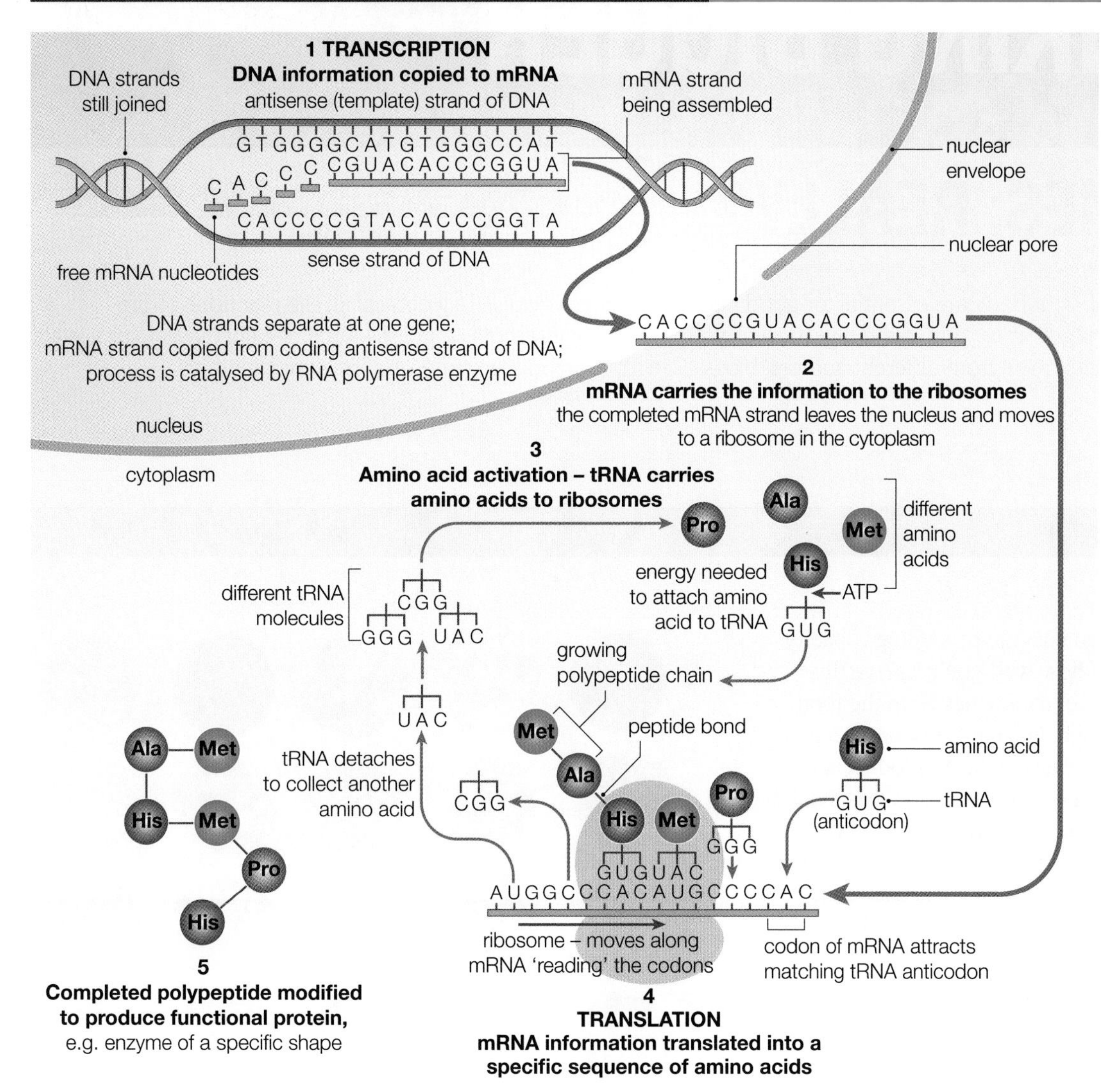

fig C A simplified diagram to show how the information held in the DNA sequence in the nucleus is translated into a sequence of amino acids in a polypeptide chain. The mRNA strand always begins with the code for methionine and ends with a stop codon. Both the mRNA strand and the amino acid chain may be thousands of units long.

CHECKPOINT

SKILLS INTERPRETATION, CREATIVITY

1. DNA and RNA are the information molecules of the cell. Explain the differences in the basic structures of these two molecules.
2. In many organisms, the DNA is in the nucleus of the cells and the proteins for which it codes are in the cytoplasm. Explain carefully the roles of the following in translating the genetic code into an active enzyme in the cytoplasm of a cell:
   (a) DNA
   (b) messenger RNA
   (a) transfer RNA
   (b) ribosomal RNA.

SUBJECT VOCABULARY

sense strand the DNA strand that carries the code for the protein to be produced
antisense strand (template strand) the DNA strand which acts as a template for an mRNA molecule
start codon the sequence of bases which indicates the start of an amino acid chain - TAC; this is the code for the amino acid methionine
RNA polymerase the enzyme that polymerises nucleotide units to form RNA in a sequence determined by the antisense strand of DNA
stop codon one of three sequences of bases which indicate the end of an amino acid chain
transfer RNA (tRNA) small units of RNA that pick up specific amino acids from the cytoplasm and transport them to the surface of the ribosome to align with the mRNA
anticodon a sequence of three bases on tRNA that are complementary to the bases in the mRNA codon
translation the process by which the DNA code is converted into a protein from the mRNA strand made in the nucleus
transcription the process by which the DNA sequence is used to make a strand of mRNA in the nucleus
polysomes groups of ribosomes, joined by a thread of mRNA, that can produce large quantities of a particular protein

2B THINKING BIGGER

RAW ENZYMES: REALLY?

SKILLS CRITICAL THINKING, ANALYSIS, DECISION MAKING, INTERPRETATION, CREATIVITY, INNOVATION, SELF DIRECTION

The enzymes made by the cells of your body are essential for good health. Inside your cells, they control all the reactions of life. Outside your cells, they are particularly important in the digestion of your food. The internet is a great source of information but not all of it is reliable. Read the following extracts from different authors' blogs. The topic is food, enzymes and healthy eating.

BLOG EXTRACTS

Author 1

Each person is born with a limited enzyme-producing capacity. Your life expectancy depends on how well you preserve this enzyme potential. You need to take in enzymes from the food you eat. If you don't take in enough enzymes, it imposes a great strain on your digestive system because it has to produce all the enzymes you need. This reduces the number of enzymes available for the metabolic reactions taking place in your cells – and this is the cause of many chronic health problems. The solution is simple: eat at least 75% of your food raw to make use of the enzymes in the food, eat less, chew your food well and don't chew gum!

Author 2

When food is cooked, enzymes are destroyed by the heat. Enzymes help us digest our food. Enzymes are proteins, and they work because they have a very specific 3D structure in space. Once they are heated much above 118 degrees, this structure can be changed so they no longer work. Cooked foods contribute to chronic illness, because their enzyme content is damaged and so we have to make our own enzymes to process the food. This uses up valuable metabolic enzymes. It takes a lot more energy to digest cooked food than raw food – the evidence being that raw food passes through the digestive tract about 50% faster than cooked food. Eating enzyme-dead (cooked!) foods overworks and eventually exhausts your pancreas and other organs. Many people progressively lose the ability to digest their food after years of eating cooked and processed food.

The cells of raw fruit and vegetables are full of enzymes. How much use are they to you?

Author 3

Enzymes are an essential part of a healthy diet. As an expert explains, 'Science cannot duplicate enzymes. Only raw food has functional living enzymes. The chain reaction generated by enzymes helps to send fats to where they are needed in our body, instead of being stored.'

Based on a number of different websites promoting good health.

SCIENCE COMMUNICATION

This information comes from blogs written by people promoting healthy eating. Think about the way they are using scientific information as you try and answer the following questions.

1 (a) Who do you think these web resources are aimed at?

(b) Do you think that the people producing these resources are writing objectively? Explain your answer.

(c) What tactics are used to try to persuade people that eating raw food provides you with useful enzymes and that cooking food is bad for you?

INTERPRETATION NOTE

If the word *explain* is used in a question, you should include a clear description and give reasons. It would be helpful to give examples to support your point.

SKILLS CRITICAL THINKING, ANALYSIS, INTERPRETATION, DECISION MAKING, COMMUNICATION

BIOLOGY IN DETAIL

Your knowledge of biochemistry now allows you to read the blog with a scientific mind.

2 Make a table to separate the information in the blog that is biologically correct from that which you think is not biologically correct.

3 Do you think the people writing this web resource are real biologists or doctors? Explain your opinion.

SKILLS INTERPRETATION, DECISION MAKING

4 Write a blog post describing the dangers of articles like these and putting right all of the biological misconceptions you found in question 2.

SKILLS ASSERTIVE COMMUNICATION, RESPONSIBILITY

ACTIVITY

Enzymes are essential for life. A healthy diet provides your body with the materials it needs to make enzymes. You do not directly use the enzymes contained in the food that you eat.

You are going to prepare a three-minute talk for a debate titled 'Raw food – the only healthy way to support your enzymes'.

Choose whether you want to support this idea or oppose it.

Focus on the biology of enzymes and of the compounds that make up your food. Whichever side you choose, your argument must be backed up by good scientific evidence.

THINKING BIGGER TIP

Consider what you have learned about enzymes and their roles in the cells and in the digestive systems of organisms, including people. You can also do more research, but make sure that your sources are reliable.

2B EXAM PRACTICE

1 (a) The three-dimensional structure of a protein is held together by:

A peptide, hydrogen and ionic bonds
B hydrogen, ester and ionic bonds
C disulfide bridges and ester bonds
D disulfide bridges, hydrogen and ionic bonds [1]

(b) Amylase is an enzyme that can digest starch to produce maltose sugars. Which test could be used to show that starch has been digested?

A biuret test **C** Benedict's test
B iodine **D** emulsion test [1]

(c) A student investigated the initial rate of reaction using amylase and different concentrations of starch. She did this first with copper ions present and then with no copper ions present. The results are shown in the graph below.

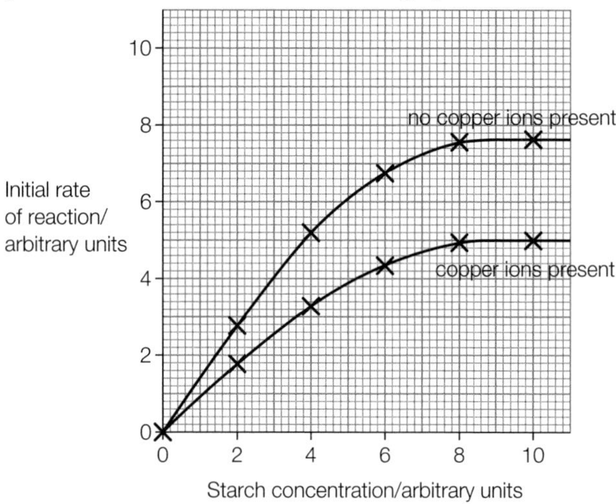

(i) Name the limiting factor in this reaction at starch concentrations below 8 au. Justify your answer. [2]

(ii) Explain why the initial rate of reaction was measured in this investigation. [2]

(iii) Describe the effect of increasing the concentration of copper ions on the rate of reaction. [2]

(iv) Calculate the percentage change in the rate of reaction caused by adding copper ions to a starch concentration of 4 au. [3]

(v) Suggest how copper ions affect amylase. [3]

(Total for Question 1 = 14 marks)

2 (a) The graph below shows the change in energy that takes place during a chemical reaction.

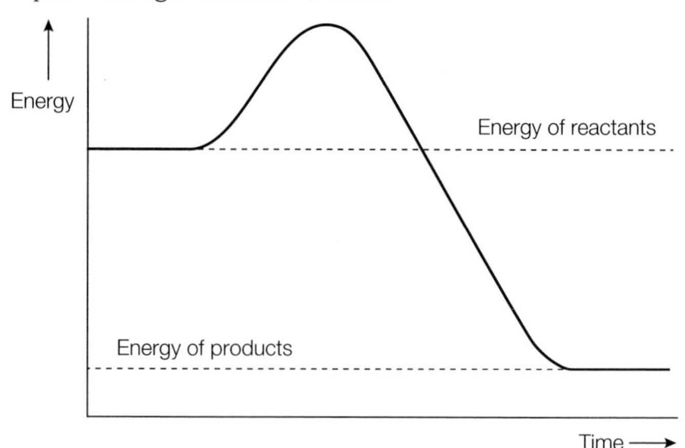

(i) With reference to enzyme activity, explain the meaning of the term *activation energy.* [2]

(ii) On the graph above, draw the energy changes that would take place if the same chemical reaction was catalysed by an enzyme. [2]

(b) An experiment was carried out to determine the effect of temperature on the activity of a protein-digesting enzyme (a protease). Solutions of the protease were incubated with a protein called gelatine at three temperatures: 20 °C, 30 °C and 40 °C. The concentrations of amino acids were measured over a 48-hour period. The results are shown in the graph below.

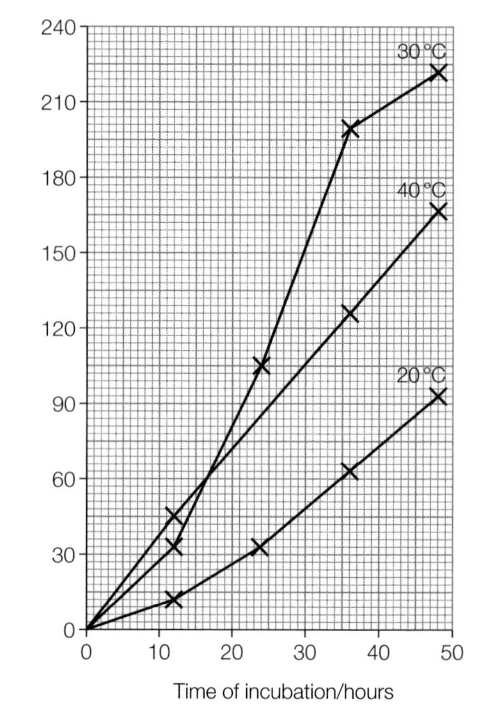

(i) Name the type of reaction catalysed by this protease and the bond broken. [2]

(ii) Calculate the mean rate of production of amino acids at 40 °C during the first 36 hours of incubation. Show your working and give your answer in arbitrary units hr^{-1}. [2]

(iii) Evaluate the conclusion that the optimum temperature for this reaction is 30 °C. [2]

(Total for Question 2 = 10 marks)

3 (a) Which mRNA sequence would be produced from the DNA sequence **CCGAAACGACTC**?

A CCGUUUCGUCAC
B GGCUUUGCUGAG
C CCGAAACGACUC
D GGCAAAGCAGTG [1]

(b) The strand of DNA that is transcribed is called the:

A sense strand
B antisense strand
C same sense strand
D missense strand. [1]

(c) The diagram below shows the structure of a mononucleotide from a DNA molecule.

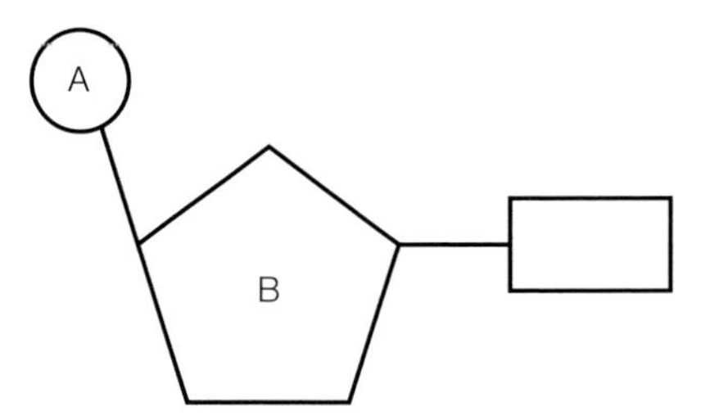

Name the parts of the mononucleotide labelled A and B. [1]

(d) The table below shows the percentage of different bases present in the DNA from a cow.

Percentage of each base present			
Adenine	**Guanine**	**Thymine**	**Cytosine**
		29	

Complete the table to show the percentage of adenine, guanine and cytosine in the DNA of the cow. [1]

(e) (i) State where transcription takes place in the cells of a cow. [1]

(ii) During transcription, part of a DNA molecule unwinds and the DNA strands separate.

Describe the events that follow to produce a messenger RNA (mRNA) molecule. [3]

(f) Oligonucleotides are short chains of nucleotides. Some of these are synthetic and have been used as drugs to treat many different diseases. They work by binding to mRNA or DNA and inhibiting protein synthesis. The drugs are described as antisense drugs when they bond to mRNA and triplex drugs when they bind to DNA.

State which stage of protein synthesis will be inhibited by each of the following:

(i) antisense drugs [1]

(ii) triplex drugs [1]

(Total for Question 3 = 10 marks)

4 Meselson and Stahl's experiment showed that DNA replication is semiconservative. Bacteria grown on a medium containing ^{15}N over many generations would have DNA containing this one isotope of nitrogen. They were then transferred to grow on a medium that contained ^{14}N.

(a) Describe how the following enzymes are used in DNA replication.

(i) DNA helicase [2]

(ii) DNA polymerase [3]

(iii) DNA ligase [2]

(b) Calculate the proportion of bacterial DNA that contains only ^{14}N after the third round of replication. [2]

(c) The strands of DNA were separated according to their density. Explain why it was necessary that ^{15}N is a stable isotope of nitrogen. [2]

(d) If DNA replicated conservatively then the results after the first round of replication would have differed from the observed result with semiconservative replication.

Describe how the DNA strands formed by one round of semiconservative replication are different from those that might have formed from one round of conservative replication. [2]

(Total for Question 4 = 13 marks)

TOPIC 2 MEMBRANES, PROTEINS, DNA AND GENE EXPRESSION

CHAPTER 2C GENE EXPRESSION AND GENETICS

With many experimental organisms, we can control the individuals that are allowed to breed and manipulate the environment in which they grow and develop. However, we cannot choose human partners or deliberately manipulate human environments for genetic experiments. This means that we need to understand how genes are passed from generation to generation without conducting breeding experiments.

In this chapter, you will consider how errors during DNA replication can produce mutations in genes. These mutations affect protein synthesis and they may affect the phenotype. Mutations may be lethal, they may have some effect on the phenotype or they may have no visible effect at all. Some mutations cause diseases such as cystic fibrosis, which have huge effects on those people who possess the mutation. You will learn how genetic information is transferred from parents to their offspring and how such diseases pass from generation to generation. You will be able to construct genetic crosses and pedigree diagrams for different organisms and be able to explain the inheritance of single factor (monohybrid) characteristics. Further, you will learn about sex linkage and the importance of genes that are missing from the X chromosome and how this explains the inheritance of certain conditions, such as colour blindness, in humans. As sexual reproduction involves random fertilisation, the results of a genetic cross may not be as expected.

Finally, you will see that we can test parents to see if they are likely to have children with a genetic disease. We can even test unborn children to see if they have a genetic disorder in a process called genetic screening.

MATHS SKILLS FOR THIS CHAPTER

- **Use ratios, fractions and percentages** (*e.g. representing monohybrid crosses*)
- **Estimate results** (*e.g. estimate ratios for genetic crosses*)
- **Understand simple probability** (*e.g. probabilities associated with genetic inheritance*)
- **Understand the terms mean, median and mode** (*e.g. mean of a set of data referring to particular phenotypic characteristics*)

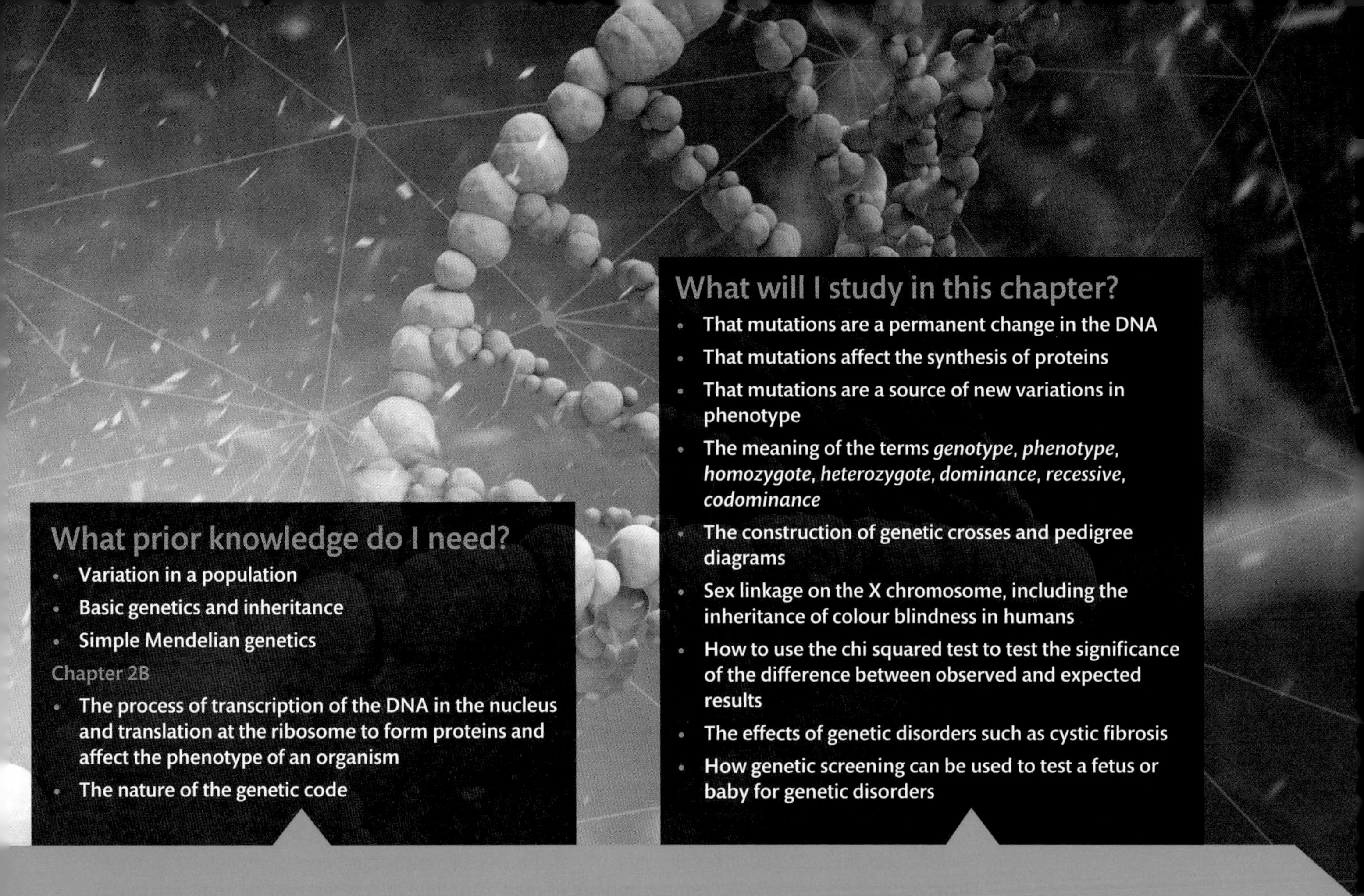

What prior knowledge do I need?

- Variation in a population
- Basic genetics and inheritance
- Simple Mendelian genetics

Chapter 2B

- The process of transcription of the DNA in the nucleus and translation at the ribosome to form proteins and affect the phenotype of an organism
- The nature of the genetic code

What will I study in this chapter?

- That mutations are a permanent change in the DNA
- That mutations affect the synthesis of proteins
- That mutations are a source of new variations in phenotype
- The meaning of the terms *genotype*, *phenotype*, *homozygote*, *heterozygote*, *dominance*, *recessive*, *codominance*
- The construction of genetic crosses and pedigree diagrams
- Sex linkage on the X chromosome, including the inheritance of colour blindness in humans
- How to use the chi squared test to test the significance of the difference between observed and expected results
- The effects of genetic disorders such as cystic fibrosis
- How genetic screening can be used to test a fetus or baby for genetic disorders

What will I study later?

Chapter 3C

- How gene expression can be controlled
- How gene expression can be modified by the environment
- How therapeutic use of stem cells has potential to cure genetic disorders

Chapter 4C

- The concept of the gene pool as a set of genes to which all members of a population have access
- How changes in allele frequencies may be the result of chance and not selection, including genetic drift, genetic bottlenecks and the founder effect
- How the Hardy-Weinberg equation can be used to monitor changes in the allele frequencies in a population

Chapter 5C (Book 2: IAL)

- That selection pressures acting on the gene pool change allele frequencies in the population
- How stabilising selection maintains continuity in a population and disruptive selection leads to changes or speciation

Chapter 8C (Book 2: IAL)

- How gene technology can be used to modify genes and the genome

1 GENE MUTATION

LEARNING OBJECTIVES

- Understand how errors in DNA replication can give rise to mutations (substitution, insertion and deletion of bases).
- Know that some mutations will give rise to cancer or genetic disorders, but that many mutations will have no observable effect.

The genetic code carried on the DNA is translated into living cellular material through protein synthesis. If a single codon is changed, then the amino acid for which it codes may be different from the original. Consequently, the whole polypeptide chain and the final protein may be altered. A change like this is known as a **mutation**. A mutation is a permanent change in the DNA of an organism. A mutation can happen when the **gametes** (sex cells) form, although they also occur during the division of body cells.

A tiny alteration at this molecular level may not affect any part of the organism, or it may affect the whole organism in devastating ways.

DIFFERENT TYPES OF MUTATION

Gene mutations involve changes in the bases making up the codons. The chance of a mutation occurring during DNA replication is around 2.5×10^{-8} per base, but it is very difficult to measure so estimates can be very different. Fortunately, the body has its own DNA repair systems. Specific enzymes cut out or repair any parts of the DNA strands that become broken or damaged. However, some mutations do remain and are copied from the DNA when new proteins are made.

Some mutations occur when just one, or a small number of nucleotides, are miscopied during replication (see **fig A**). These are **point mutations** or **gene mutations**. If you think of the amino acids produced from the codons as similar to letters of the alphabet, the result of a point mutation is like changing a letter in one word. It may still make an acceptable word, but the meaning will probably be different. These gene mutations include the following types:

- **substitution**, where one base substitutes for another
- **deletion**, where a base is completely lost from the sequence
- **insertion**, when an extra base is added, which may be a repetition of one of the bases already there or a different base entirely.

You can see examples of these different types of mutation in **fig A**.

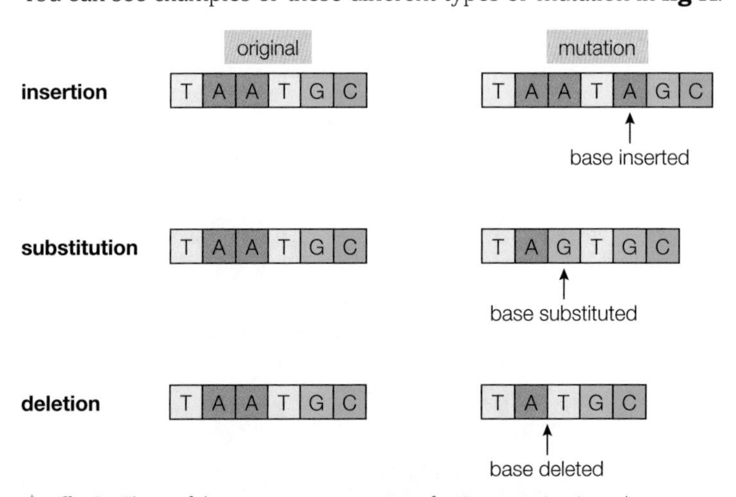

▲ **fig A** Three of the most common types of point mutation in a chromosome

Chromosomal mutations involve changes in the positions of whole genes within the chromosomes. Think of this as rearranging the words within a sentence – if you are lucky, the rearranged words will still make sense, but it will not mean the same as the original sentence. Finally, there are **whole-chromosome mutations**, where an entire chromosome is either lost during meiosis (cell division to form the sex cells) or duplicated in one cell by errors in the process. This is like losing or repeating a whole sentence. For example, Down syndrome is caused by a whole-chromosome mutation at chromosome 21 – individuals with this syndrome have three copies of this chromosome instead of the usual two.

HOW GENE MUTATIONS CAN AFFECT THE PHENOTYPE

Mutations can produce **variation** within an organism. If the different arrangements of nucleotides code for the same amino acid (see **Section 2B.6**), a point mutation will have no effect. Very occasionally, a mutation occurs that produces a new and superior protein. This may help the organism gain a reproductive advantage so that it leaves more offspring than other individuals of that species, particularly if environmental conditions change. Most mutations are neutral, which means that they neither improve nor worsen the chances of survival. Some mutations cause much damage, disrupting the biochemistry of the whole organism. If a harmful mutation is in a protein that is important to the function of a cell – for example, the active site of an enzyme – the effect can be catastrophic (see **fig B** on p130).

EXAM HINT

You may be asked to describe the effect of a mutation on protein synthesis. Remember to describe the whole sequence of events giving detail of how the mutation in DNA affects each stage to bring about formation of a different protein.

Random mutations in the genetic material of the gametes are the cause of many human genetic diseases. Examples include:

- thalassaemia, in which the blood proteins are not manufactured correctly
- cystic fibrosis, in which a membrane protein does not function properly (see **Section 2C.4**).

Mutations in the somatic cells of the body as they divide result in many different types of cancer, depending on where in the body they occur.

However, most mutations will have no observable effects on the organism. This may be because:

- the mutations occur in part of the non-coding DNA which does not affect the way the genetic code is read
- the code is degenerate (see **Section 2B.5**), and one small change in the code may not alter the amino acid coded for.

SICKLE CELL DISEASE: WHEN THE CODE GOES WRONG

Sickle cell disease is a genetic disease that affects the protein chains of the haemoglobin in the red blood cells. It is the result of a point mutation. A change of one base in one codon changes a single amino acid in a chain of 147 amino acids – but that change alters the nature of the protein (see **table A**). As a result, the haemoglobin molecules stick together to form rigid rods that give the red blood cells a sickle shape. They do not carry oxygen efficiently and can prevent the blood flowing in the capillaries. This single tiny change in one nucleotide is enough to cause severe pain and even death to the people affected.

SEQUENCE FOR HEALTHY HAEMOGLOBIN								
ATG	GTG	CAC	CTG	ACT	CCT	GAG	GAG	TCT
Start	Val	His	Leu	Thr	Pro	Glu	Glu	Ser
SEQUENCE FOR SICKLE CELL HAEMOGLOBIN								
ATG	GTG	CAC	CTG	ACT	CCT	GTG	GAG	TCT
Start	Val	His	Leu	Thr	Pro	Val	Glu	Ser

table A This table shows the change in the single codon that causes sickle cell disease (only the first nine codons are shown). This shows the 'sense' or 'coding' strand sequence.

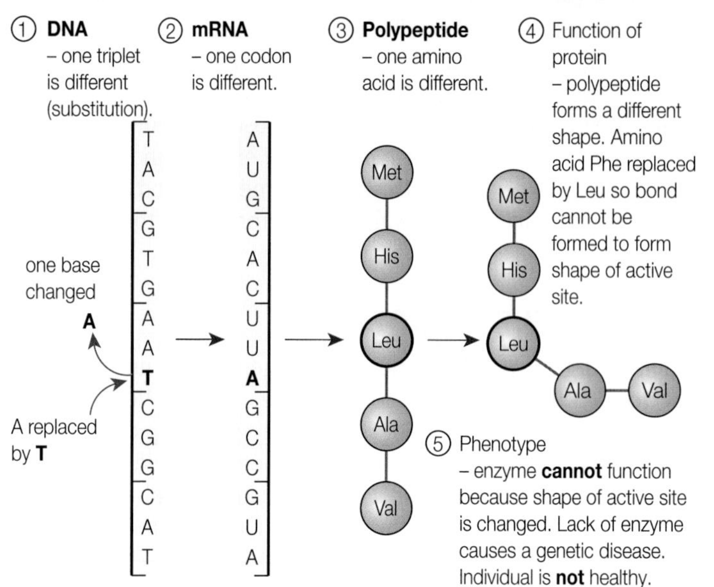

fig B This example shows how a change in a single nucleotide can affect the structure and function of the protein formed.

EXAM HINT

Be prepared to search a set of base sequences to find an error: here the error is in the sixth triplet where an adenine is replaced by a thymine to produce a triplet that codes for a different amino acid.

Mutations can happen to any cell at any time, though they usually occur during the copying of DNA for cell division. Mutations in the body cells can cause problems such as cancer. The most damaging mutations occur in the gametes because they will be passed on to future offspring. These are the mutations that lead to genetic diseases. X-rays, ionising radiation and certain chemicals are called **mutagens** because exposure to them increases the rate at which mutations occur. It is best to avoid exposure to mutagens whenever possible.

CHECKPOINT

SKILLS INTERPRETATION, CREATIVITY, CRITICAL THINKING

1. Some point mutations will have as big an impact on the way the body works as any chromosomal or whole-chromosome mutation. Others have no effect at all on the organism. Explain.
2. Explain how a change in a single base in the sickle cell mutation has such a dramatic effect on affected individuals.

SUBJECT VOCABULARY

mutation a permanent change in the DNA of an organism
gametes haploid sex cells that fuse to form a new diploid cell (zygote) in sexual reproduction
point mutation (gene mutation) a change in a single base of the DNA code
substitution a type of point mutation in which one base in a gene is substituted for another
deletion a type of point mutation in which a base is completely lost
insertion a type of point mutation in which an extra base is added into a gene, which may be a repeat or a different base
chromosomal mutations changes in the position of entire genes within a chromosome
whole-chromosome mutations the loss or duplication of a whole chromosome
variation differences between organisms which may be the result of different genes or the environment they live in
mutagen anything that increases the rate of mutation

2 PATTERNS OF INHERITANCE

LEARNING OBJECTIVES

- Understand what is meant by the terms *gene*, *allele*, *genotype*, *phenotype*, *recessive*, *dominant*, *incomplete dominance*, *homozygote* and *heterozygote*.
- Understand patterns of inheritance, including the interpretation of genetic pedigree diagrams, in the context of monohybrid inheritance.

For centuries, people bred the plants we grow for food and for our gardens, and the animals we keep for food and as companion animals. They did this to get more desirable crops and animals yet did not understand the selection process. Now we understand many of the underlying mechanisms of inheritance, and the science of genetics has emerged.

GENETICS: THE BASIS OF INHERITANCE

The physical and chemical characteristics that make up the appearance of an organism are known as its **phenotype**. Examples include the size of an olive, the colour of a flower, the shape of a nose. The phenotype is partly the result of the **genotype** (the combination of alleles) passed from parents to their offspring and partly the effects of the environment in which the organism lives. For example, the size of an olive will depend on levels of soil nutrients and sunlight as well as on the genetic make-up of the individual olive plant.

Differences in the genotype between individuals of a species are due to:

- rearrangement of genes during meiosis (see **Section 3B.3**)
- inheritance of genes from two different individuals in sexual reproduction.

The cells of every individual organism within a species contain the same number of chromosomes. So, for example, humans have 46 chromosomes. Half the chromosomes are inherited from the female parent and the other half come from the male parent. The two sets can be arranged as matching pairs, called **homologous pairs**.

Along each chromosome are hundreds of genes. Each gene is a different segment of DNA coding for a particular protein or polypeptide. The chromosomes in a homologous pair carry the same genes – except for the sex chromosomes. The gene for any particular characteristic is always found in the same position or **locus**, which means that you usually carry two genes for each characteristic.

EXAM HINT

Ensure you know the meanings of all the terms used here. You may be asked to explain the difference between certain pairs of terms such as *genotype* and *phenotype*, *homozygous* and *heterozygous*, *dominant* and *recessive*, *gene* and *allele*.

Each gene exists in slightly different versions called **alleles**. For example, at the locus for the gene for the height of a pea plant, the allele may code for a tall plant or for a dwarf plant. As the pea plant has two homologous chromosomes carrying this gene, it may have two alleles for the tall characteristic, two for the dwarf or one of each (see **fig A**).

If both alleles that code for a particular characteristic are identical, then the individual is homozygous for that characteristic – it is a **homozygote** ('homo' means 'the same'). If the two alleles coding for a characteristic are different, the individual is heterozygous for that characteristic and is called a **heterozygote** ('hetero' means 'different').

Some phenotypes are **dominant**: their effect is always expressed (shown) whether the individual is homozygous or heterozygous for the allele. If one allele for the dominant phenotype is present

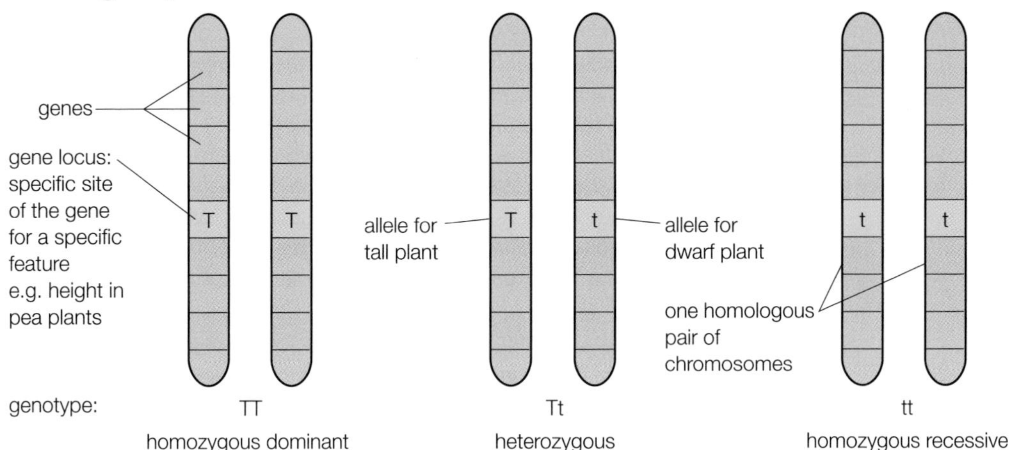

fig A In sexual reproduction, the variants of the genes passed on determine the genotype and, therefore, eventually an aspect of the phenotype of the offspring.

it will be expressed even if an allele for the **recessive** phenotype is there. Recessive phenotypes are only expressed when there are two alleles coding for the recessive feature, in other words, when the individual is homozygous recessive. In genetic diagrams, the alleles coding for dominant phenotypes are usually represented by a capital letter and those for recessive phenotypes by the lower-case version of the same letter.

MONOGENIC (MONOHYBRID) CROSSES

Homozygotes are referred to as **true breeding**, because if you cross two individuals that are homozygous for the same characteristic, all the offspring of all the generations that follow will show this same characteristic in their phenotype (unless a mutation occurs). Heterozygotes are not true breeding. If two heterozygotes are crossed, the offspring will include homozygous dominant, homozygous recessive and heterozygous types and at least two different phenotypes.

When genes are considered individually in a genetic cross, it is called a **monohybrid cross**. We can represent these crosses using simple diagrams called Punnett squares. A Punnett square shows you the potential alleles inherited from both parents, and the potential offspring that result. For example, **fig C** shows a cross between a pea plant homozygous for the dominant round pea seed shape and a pea plant homozygous for the recessive wrinkled pea (see **fig B**). The first generation of this cross is called the F_1 (first filial generation) and you can see that they all have the same genotype for the characteristic and they are heterozygous. They also all have the same phenotype – round pea shape because the round allele is dominant. There is no sign of the wrinkled pea allele. If we cross individuals from the F_1 generation we call the next generation the F_2 (second filial generation). In **fig C** you can see that theory predicts the ratio of the genotypes to be 1 homozygous dominant : 2 heterozygous dominant : 1 homozygous recessive. In terms of the phenotypes that would result from these genotypes, you would expect to see three round peas for every wrinkled one. The recessive trait of wrinkled peas has become visible again, after being 'hidden' in the F_1 generation.

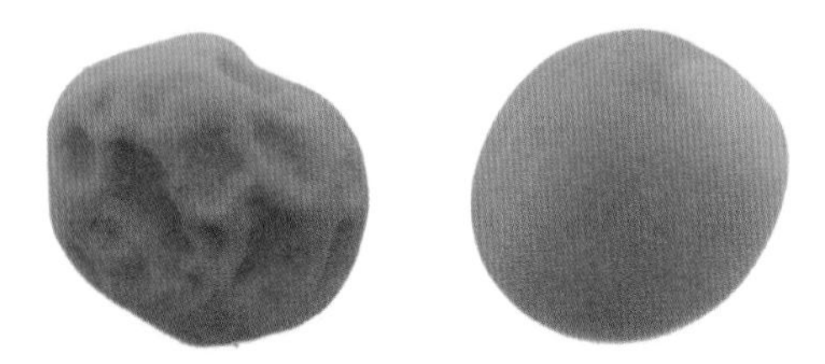

▲ **fig B** Round or wrinkled peas may not seem very exciting, but they show genetic characteristics that are easy to identify and have been studied since the earliest days of genetics.

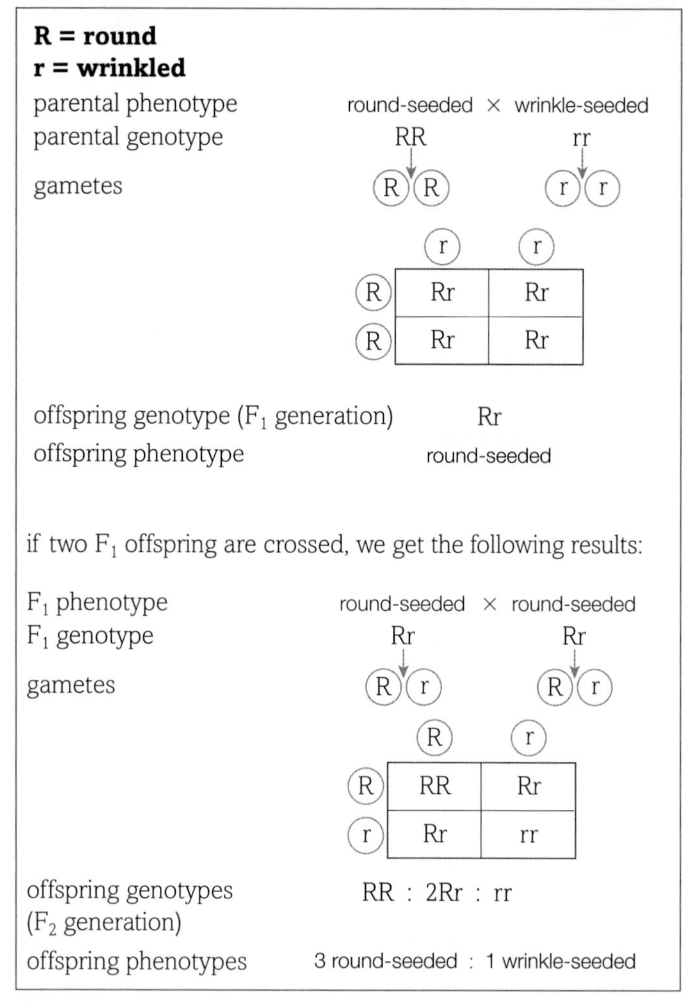

▲ **fig C** Using a Punnett square makes it easy to predict the results of crosses between different parent pea plants.

LEARNING TIP

Remember to put a circle around gametes as this helps to distinguish them from individuals.

TEST CROSS

As you can see from **fig C**, individuals that are homozygous dominant or heterozygous have identical dominant phenotypes. For a plant or animal breeder this can cause many difficulties. A breeder often needs to know that the stock will breed true, in other words, that it is homozygous for the desired feature. If the feature is a recessive phenotype, then any plant showing the feature in the phenotype must be homozygous. However, if the required feature is a dominant phenotype the physical appearance does not show whether the individual is homo- or heterozygous. To find out which it is, the individual must be crossed with a homozygous recessive individual (see **fig D**). This type of cross, known as a test cross, reveals the parental genotype.

LEARNING TIP

When you draw a genetic cross, always show the parent genotypes and phenotypes. Don't just draw the gametes into a Punnett square.

Y = yellow
y = green

if a homozygous yellow parent is crossed:

parental phenotype yellow seeds × green seeds
parental genotype YY yy
gametes (Y)(Y) (y)(y)

	(y)	(y)
(Y)	Yy	Yy
(Y)	Yy	Yy

offspring genotype (F_1 generation) Yy
offspring phenotype yellow seeds

if a heterozygous yellow parent is crossed:

parental phenotype yellow seeds × green seeds
parental genotype Yy yy
gametes (Y)(y) (y)(y)

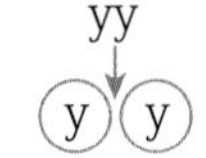

	(y)	(y)
(Y)	Yy	Yy
(y)	yy	yy

offspring genotypes Yy : yy
offspring phenotypes 1 yellow : 1 green

▲ **fig D** A test cross reveals the genotype for seed colour of a parent pea plant.

CODOMINANCE

In the human ABO blood group system there are three possible alleles – A, B and O. These are usually shown as I^O, I^A and I^B. The different alleles code for the presence or absence of antigens on the surface of the erythrocytes (red blood cells). I^O codes for no antigens, I^A codes for antigen A and I^B codes for antigen B (see **fig E**).

I^O is recessive. A homozygote with I^O will have no antigens on their erythrocytes and have blood group O. Both I^A and I^B are dominant to I^O so heterozygotes have the blood group of the dominant phenotype. An individual who inherits I^OI^A or I^AI^A will have A antigens on the erythrocytes and have blood group A, while someone who inherits I^OI^B or I^BI^B will have B antigens and be blood group B.

I^A and I^B are **codominant**. This means both alleles are expressed and produce their proteins, which act together without mixing. So an individual who inherits I^A and I^B (I^AI^B) will have both antigen A and antigen B on the surface of their erythrocytes and they will have the blood group AB. The A and B antigens will act in just the same way as if they were there individually – there is no blending in the phenotype. This is the key feature in codominance. Humans are not the only species that have the ABO multiple allele blood groups with codominance. They are seen in bonobos, chimpanzees and gorillas, and also in cows and sheep.

♀ I^AI^A, ♂ I^OI^O

♂ \ ♀	I^A	I^A
I^O	I^AI^O	I^AI^O
I^O	I^AI^O	I^AI^O

parental phenotypes: mother group A
father group O

offspring genotypes: all I^AI^O
offspring phenotypes: all group A

♀ I^BI^O, ♂ I^AI^A

♂ \ ♀	I^B	I^O
I^A	I^BI^A	I^OI^A
I^A	I^BI^A	I^OI^A

parental phenotypes: mother group B
father group A

offspring genotypes: I^BI^A : I^OI^A
offspring phenotypes: 1AB : 1A

♀ I^AI^O, ♂ I^BI^O

♂ \ ♀	I^A	I^O
I^B	I^AI^B	I^OI^B
I^O	I^AI^O	I^OI^O

parental phenotypes: mother group A
father group B

offspring genotypes: I^AI^B : I^OI^B : I^AI^O : I^OI^O
offspring phenotypes: 1AB : 1B : 1A : 1O

▲ **fig E** The genetics of the ABO blood group system

DID YOU KNOW?

Carrying out genetic experiments

It is unethical to use people as experimental organisms in genetics. Also, people usually have only one baby at a time, after nine months gestation; families are small and many phenotypic characteristics do not show up until adulthood. Much of our understanding of human genetics comes from experiments on other organisms. An organism for a genetic experiment should, ideally, have the following features:

- be relatively easy and cheap to grow, to maximise the chance of successful breeding and minimise experimental costs
- have a short life cycle so that the results of crosses and/or mutations can be seen quickly
- produce large numbers of offspring so that the results of any crosses are statistically relevant
- have clear, easily distinguished characteristics, such as tall or dwarf plants, or the wing length in insects.

Organisms commonly used by scientists for genetic investigations include the fruit fly *Drosophila melanogaster*, pea plants, fungi such as *Aspergillus nidulans* and a variety of bacteria including *Escherichia coli*. Using organisms such as plants, bacteria and fruit flies also raises fewer ethical concerns than experimenting with larger organisms such as mammals.

EXAM HINT

When asked to explain why a certain organism is used for scientific experiments, students often forget to include the obvious reasons: that the organism may be cheap and easy to keep, it may have a very short life cycle or it may produce many offspring.

SAMPLING ERRORS

The theoretical ratios of phenotypes that are predicted by a genetic cross are usually seen (approximately) in real genetic experiments. However, the numbers are never precise. There are several reasons for this.

- Reproduction is a result of chance. The combination of alleles in each gamete is completely random and so is the joining of particular gametes. However, the theoretical diagrams that we draw do not show this.
- Some offspring die before they can be sampled. For example, some seeds do not germinate and some embryos miscarry.
- Inefficient sampling techniques. For example, it is very easy to allow a few *Drosophila* to escape.

So the sampling error must be taken into account when you look at a real genetic cross, and the smaller the sample, the larger the potential sampling error. If a 3 : 1 ratio of phenotypes is expected in the offspring, it is unlikely to show itself if only four offspring are produced. Looking at 400 offspring increases the likelihood of the expected ratio emerging, and 4000 offspring is even better. This is why organisms such as fast-growing plants, *Drosophila,* and certain fungi and bacteria are so useful – they all produce large numbers of offspring in a short time.

GENETIC PEDIGREE DIAGRAMS

Genetic diagrams, such as the Punnett squares in **figs C**, **D** and **E** show how a trait can be theoretically passed on and the probability that different offspring are produced. The use of a family tree or genetic pedigree diagram (see **fig G**) can show us what happens in reality over a long period of time. A pedigree diagram includes all the members of a family, or breeding programme, indicating their sex and whether or not they have a particular characteristic or a disease. Humans are not available as experimental animals and so much of our direct understanding of human genes has come from the analysis of genetic pedigree diagrams. Genetic pedigree diagrams highlight carriers of a recessive phenotype. These can be useful for families affected by conditions such as thalassaemia and cystic fibrosis as they help to predict which family members may be carriers of the genetic mutation. This allows people to consider their options before they conceive a child. Genetic pedigree diagrams are also extremely useful for identifying **sex-linked traits** (see **Section 2C.3**), because a family tree indicates the males and females in the family.

LEARNING TIP

Always remember to look for the individuals that show the recessive phenotype because these are the only ones where you can be sure of the genotype - they are double recessive.

DID YOU KNOW?

The albino trait

The formation of the pigments that colour the hair and skin of animals is controlled by multiple genes. Albinism is a condition in which the natural melanin pigment of the skin, eyes and hair does not form. It is seen in many species (see **fig F**). There are several different forms of albinism, as mutations in several different genes can give similar results in the phenotype.

fig F Some albino organisms lack an enzyme which makes the pigment melanin.

One of the most common types of albinism is due to a mutant allele that prevents the formation of a normal enzyme in the cells. The enzyme tyrosinase is normally active in the melanocytes or pigment-forming cells. However, in this condition it is not formed correctly so the reactions that make melanin cannot take place. Albinism of this type is a recessive phenotype. The parents may appear normal, but both carry the recessive albino allele. Or, one or both of the parents may be an albino themselves. In the general human population, about 1 person in 30 000 is affected by albinism. People who inherit this condition not only lack pigment in the cells of their skin, hair and eyes – their vision is often poor and they are at higher risk of developing skin cancers because they do not have the natural protective pigment melanin. However, albinism is not life-threatening and both difficulties can usually be overcome. Genetic pedigrees can show how the recessive allele is passed on through a family (see **fig G**).

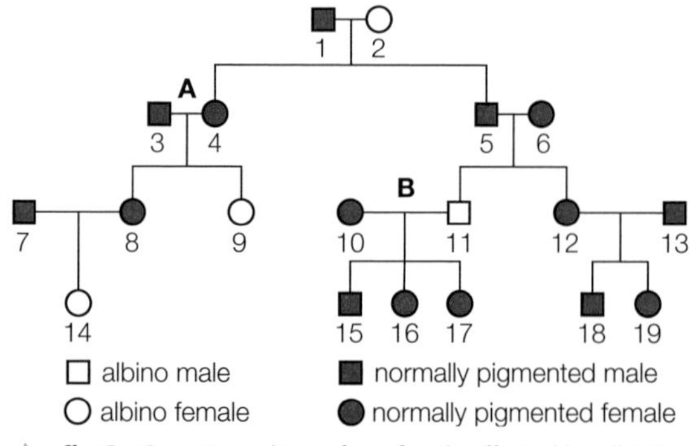

fig G Genetic pedigree for a family affected by albinism

Genetic pedigree diagrams are also widely used wherever people selectively breed animals. For example, pedigrees are extremely important in the breeding of thoroughbred racehorses, and in specific breeds of animals such as dogs, cats, cattle and sheep. They can also track mutations in rare animals such as white Bengal tigers. Normal tigers have black stripes on a golden orange coat. White tigers have black stripes on a white coat. This is the result of a mutation in a single gene. The white phenotype is recessive, so we know that any white tiger is homozygous recessive. You can see how a genetic pedigree works by looking at the family tree for a group of tigers shown in **fig H**.

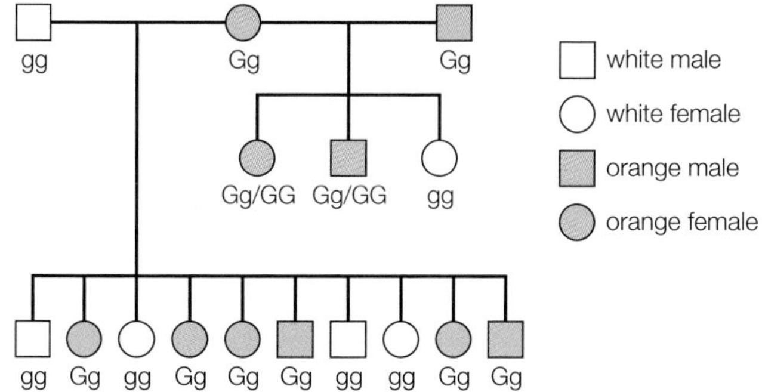

▲ **fig H** This genetic pedigree shows you the results of two crosses between a golden female and two different male tigers. You can see the phenotypes of the cubs, and work out possible genotypes. The dominant allele for golden/orange coat is shown as **G**; the recessive allele for white coat is shown as **g**.

CHECKPOINT

1. Define the following terms: *homozygous*; *heterozygous*; *dominant* and *recessive*.
2. In pea plants, the gene for plant height has two alleles. The tall phenotype is dominant to the dwarf phenotype. Choose suitable letters for the two alleles and draw genetic diagrams to show the following:
   (a) a cross between two homozygous tall parents
   (b) a cross between heterozygous parents
   (c) the F_1 and F_2 generations of a cross between a homozygous tall parent and homozygous dwarf parent.
3. Explain the importance to plant breeders of a test cross.
4. Why are organisms such as *Drosophila*, *Aspergillus* and *E. coli* bacteria suitable as experimental organisms in genetics?
5. In **fig H**, the genotypes of all the cubs in the cross between the white male tiger and the golden orange female tiger are known. In the cross between the two golden orange tigers, only the genotype of the white cub is known for certain.
   (a) Explain why this is the case.
   (b) How could you try to determine the genotypes of the two golden orange cubs resulting from the cross between the two golden orange parents and what are the limitations of this?

SUBJECT VOCABULARY

phenotype the physical traits, including biochemical characteristics, expressed as a result of the interactions of the genotype with the environment
genotype the genetic make-up of an organism with respect to a particular feature
homologous pairs matching pairs of chromosomes in an individual which both carry the same genes, although they may have different alleles
locus the site of a gene on a chromosome
alleles versions of a gene, variants
homozygote an individual where both alleles coding for a particular characteristic are identical
heterozygote an individual where the two alleles coding for a particular characteristic are different
dominant a characteristic which is expressed in the phenotype whether the individual is homozygous or heterozygous for that allele
recessive a characteristic which is only expressed when both alleles code for it; in other words, the individual is homozygous for the recessive trait
true breeding a homozygous organism which will always produce the same offspring when crossed with another true-breeding organism for the same characteristic
monohybrid cross a genetic cross where only one gene for one characteristic is considered
codominance in heterozygotes, where both alleles at a gene locus are fully expressed in the phenotype
sex-linked traits characteristics which are inherited on the sex chromosomes

2C 3 SEX LINKAGE

SPECIFICATION REFERENCE 2.15(iii)

LEARNING OBJECTIVES

- Understand sex linkage on the X chromosome, including red–green colour blindness in humans.

DID YOU KNOW?

KARYOTYPES

When the cells of the body divide, the chromosomes shorten and thicken and at this stage they will take up stains. You will learn more about this in **Section 3B.1**. Scientists can take photographs of the chromosomes and then match up the individual chromosome pairs. This is called a karyotype and is a picture of the chromosomes of an individual (see **fig A**). Karyotypes have a number of uses, from helping discover chromosome problems in a developing fetus to checking the sex of athletes in international competitions.

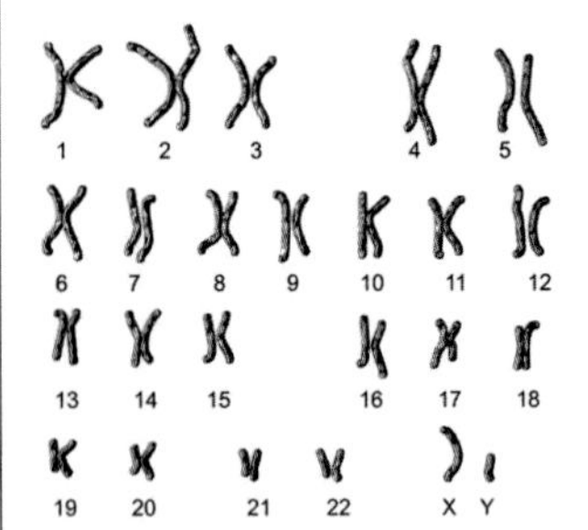

▲ **fig A** This human male karyotype shows clearly the difference in size between the X and the Y chromosomes.

SEX DETERMINATION

The chromosomes in diploid organisms occur as matching or homologous pairs. In organisms such as mammals, where there are clear differences between the males and females, sex is determined by the chromosomes. The **autosomes** are all except one of the homologous pairs of chromosomes. They carry information about the general body cells and their biochemistry. The remaining pair of chromosomes are not a homologous pair. They are the X chromosome and the Y chromosome, known together as the sex chromosomes. They carry information about the sex of the individual. In mammals, the female has two large X chromosomes. Therefore, all of her eggs contain an X chromosome and she is **homogametic**. The male has one X chromosome and one much smaller Y chromosome. Half of his sperm will each contain an X chromosome and the other half will each contain a Y chromosome. Males are **heterogametic**. In humans, there are 22 pairs of autosomes and one pair of sex chromosomes. In some organisms, such as birds and many reptiles, it is the male that is homogametic.

The small human Y chromosome has about 23 million base pairs and only 78 protein-coding genes. It mainly carries male sex information, including the crucial SRY gene (sex-determining region Y). This gene starts the development of the testes, which then make male hormones to produce a male fetus. The X chromosome is much bigger, with over 150 million base pairs. Estimates for the number of protein-coding genes on the X chromosome vary from around 800 to around 1200. The X chromosome codes for all the female characteristics. It also carries other genes coding for traits including the clotting factors in the blood and the ability to distinguish between certain colours.

Despite their obvious differences, there is enough similarity between the X and Y chromosomes to allow them to pair up during cell division.

SEX LINKAGE

Genes that are carried on the X chromosome are said to be sex-linked. Recessive or mutant alleles on the X chromosome passed from a female parent to her male offspring will be expressed in the phenotype because there is no corresponding allele on the homologous Y chromosome. They are shown on the genetic diagrams because of the importance of the X and Y chromosomes in sex linkage. Sex linkage was first discovered in *Drosophila* by Thomas Morgan in the beginning of the 20th century and we now know it occurs in many different organisms.

In *Drosophila*, eye colour is sex-linked and is the result of multiple alleles. The most common eye colour is red and is inherited through a dominant allele on the X chromosome. However, there is a wide range of mutant alleles resulting in flies with different eye colours including white, apricot and purple (see **fig B**).

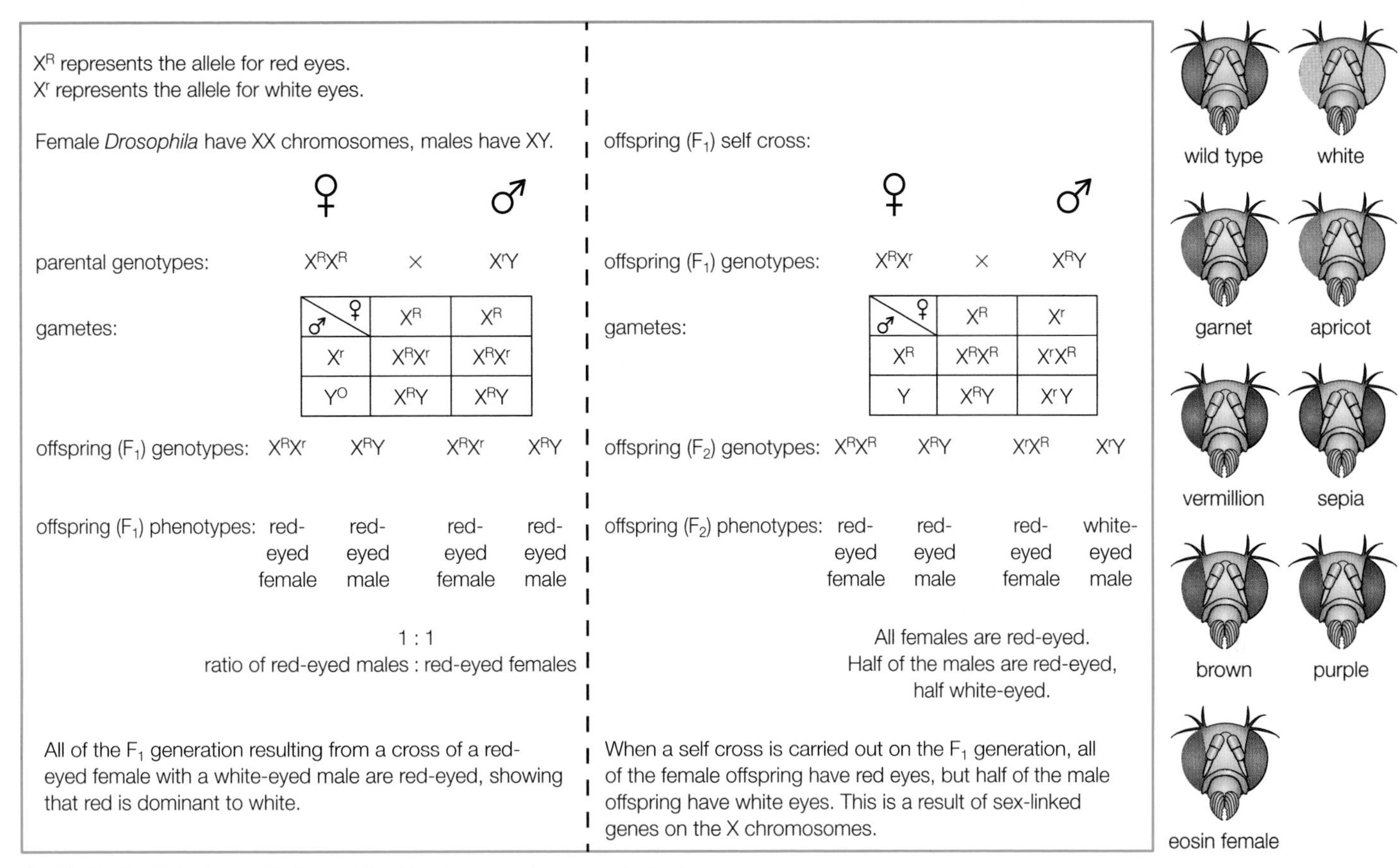

▲ **fig B** Sex-linked genes in *Drosophila* affect the eye colours seen in the insects.

SEX LINKAGE IN HUMANS

Human genetics work in the same way as the genetics of the peas and fruit flies that you have been studying. In people, as in most organisms, few characteristics are the result of single genes. Almost every aspect of your phenotype results from interactions between variants of multiple genes, together with **transcription factors** and **epigenetics**. Studying the inheritance of some single human genes can still be helpful, especially in understanding some of the more common inherited diseases that affect people around the world. For example, multiple genes affect the structure of the haemoglobin molecule that carries oxygen in the blood (see **Section 1A.5**). A mutation in any one of them can affect the final structure of the haemoglobin molecule and, therefore, its ability to carry oxygen. However, looking at one of the multiple alleles alone can help our understanding of inherited conditions.

SEX-LINKED DISEASES IN HUMANS

Sex-linked genes occur in humans just as they do in other organisms. A mother always donates an X chromosome to her sons. The father always donates the Y chromosome. Any mutations in a gene on the X chromosome will affect the phenotype of the offspring, even if the characteristic it codes for is recessive. This is because the Y chromosome is small and carries only genes which code for traits associated with maleness. It follows that sex linkage in humans leads to a variety of conditions known as **sex-linked diseases**. Some of these are relatively minor. Some are life-threatening or even fatal.

LEARNING TIP

Always remember that it is the mother who passes sex-linked characters to her son on the X chromosome.

RED–GREEN COLOUR BLINDNESS

Human colour vision is the result of three different types of light sensitive cell found in the retina of the eye (you will learn more about how you see in **Book 2 Chapter 8A**). The ability to see in colour is the result of multiple genes coding for different aspects of the process. Many of these genes are found on the X chromosome. Mutations in these genes can affect our ability to see in colour, causing different types of colour blindness. **Red–green colour blindness** is the result of one of these mutations. People affected still see red and green colours, but can have difficulty seeing the difference between some tones of colour (see **fig C**).

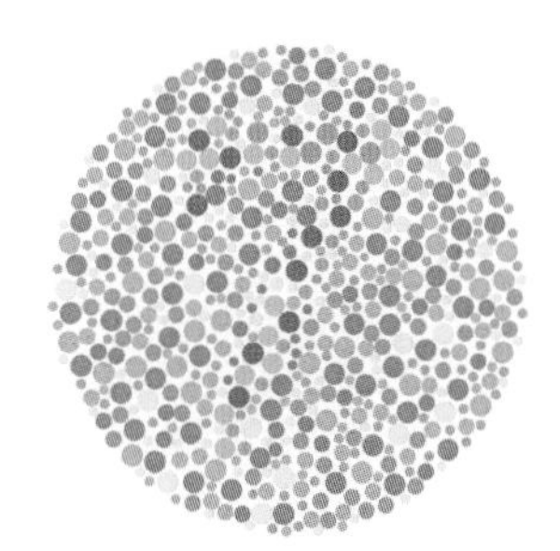

▲ **fig C** Tests like this can demonstrate colour blindness – if you have normal colour vision you will see a 7 in this image.

Red–green colour blindness is usually an inconvenience but nothing more. It is caused by a recessive mutation of a gene on the X chromosome. It is much more common in men than in women because the condition is sex linked. However, colour blindness does occasionally occur in women (see **fig D**) because the gene does not markedly affect the chances of survival of an individual, and because the homozygous form is not lethal. In many populations, around 7–8% of males are affected by red–green colour blindness, but fewer than 1% of females. In the genetic pedigree in **fig D**, symptom-free carriers are shown as half-shaded circles.

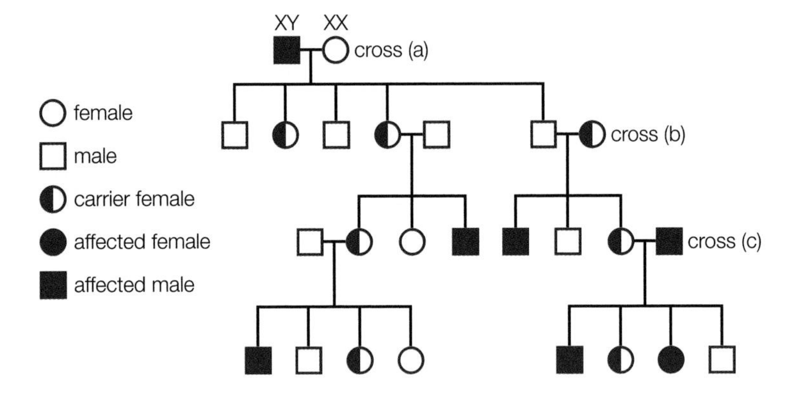

X^C represents the allele for normal vision.
X^c represents the allele for colour blindness.

cross (a)

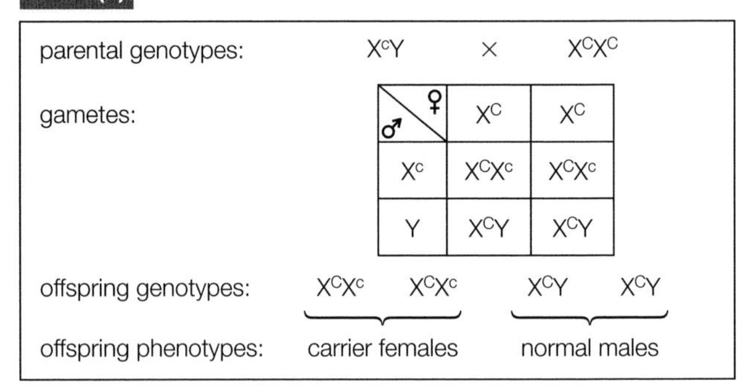

cross (b)

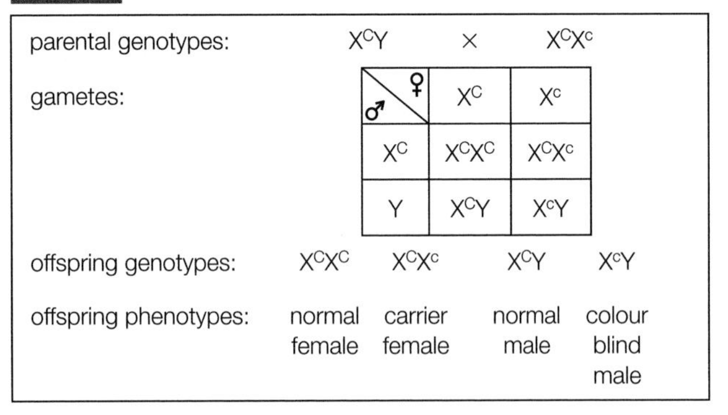

cross (c)

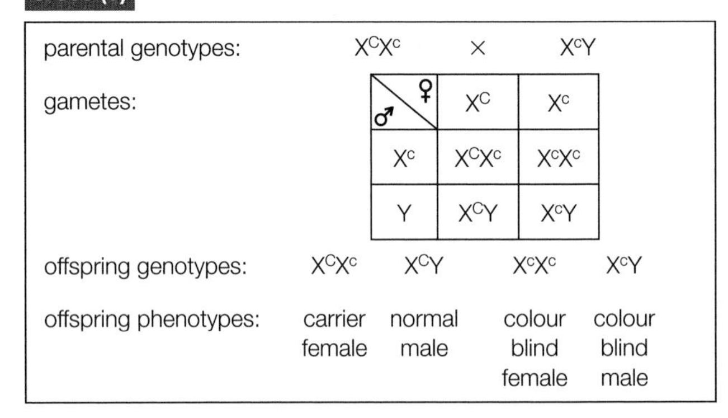

▲ **fig D** These genetic diagrams show how the colour blindness genes are transmitted through a family tree.

EXAM HINT

Always read the key to a pedigree chart closely. Males and females are represented by squares and circles. If you confuse them, your explanation will be incorrect.

HAEMOPHILIA

Colour blindness is usually only an inconvenience. **Haemophilia** is a much more severe sex-linked trait in which one of the proteins needed for blood to clot is missing. The components of the blood-clotting process (see **Section 1B.2**) are coded for by multiple genes. Many of these genes are carried on the X chromosome, so problems with blood clotting are often sex-linked diseases.

One of the most common and best understood forms of haemophilia is haemophilia A (see **fig E**). This is a sex-linked condition in which clotting factor VIII is missing. It is also known as factor VIII deficiency. Globally, the condition affects between 1 in 4000 and 1 in 5000 live male births. The severity of the disease varies, but it can be fatal if not treated. The homozygous form is rare and very severe and most female fetuses that are affected will not survive birth. In an untreated male with haemophilia, the slightest injury can lead to death through excessive bleeding – even exercise can result in internal bleeding of the joints. In the past, people with haemophilia were treated by blood transfusions or with factor VIII extracted from donated blood. Now recombinant (genetically engineered) bacteria produce pure factor VIII in large quantities (see **Chapter 8C**). Regular treatment with pure human factor VIII made by bacteria has given people with haemophilia A an almost normal life expectancy and life quality. For this to happen, it is important to diagnose and treat people with the disease as soon after birth as possible – or even before birth.

X^H represents the normal allele.
X^h represents the haemophiliac allele.

parental genotypes: X^HX^h × X^HY

gametes:

♂ \ ♀	X^H	X^h
X^H	X^HX^H	X^hX^H
Y	X^HY	X^hY

offspring genotypes:	X^HX^H	X^HY	X^hX^H	X^hY
offspring phenotypes:	normal female	normal male	carrier female	haemophiliac male

fig E It only takes one recessive allele for haemophilia to affect a family with this sex-linked disease.

CHECKPOINT

SKILLS PROBLEM SOLVING

1. Globally, about 8% of men have red–green colour blindness and around 0.5% of women are affected. Using genetic diagrams to help your explanations, answer the following questions.
   (a) Why are there more red–green colour-blind men than women?
   (b) Does every red–green colour-blind man have a colour-blind parent?
   (c) Does every red–green colour-blind woman have a colour-blind parent?
2. (a) Explain the difference between genetic diagrams and genetic pedigree diagrams.
   (b) Why are genetic pedigree diagrams so useful in human genetics?
3. (a) What is haemophilia A?
   (b) Why are there almost no girls with haemophilia A?
   (c) A couple with no family history of ill health have four children. Their two daughters and one son appear healthy, but their eldest son has haemophilia A. Suggest two different ways in which this might have happened. Use genetic diagrams or pedigree diagrams in your explanation.

SUBJECT VOCABULARY

autosomes chromosomes which carry information about the body but do not determine the sex of an individual
homogametic an individual who produces gametes that contain only one type of sex chromosome – in humans this is the female
heterogametic an individual who produces two types of gamete each containing different types of sex chromosome – in humans this is the male
transcription factors proteins that bind to the DNA in the nucleus and affect the process of converting, or transcribing, DNA into RNA
epigenetics the study of changes in gene expression (active versus inactive genes) that does not involve changes to the underlying DNA sequence but affects how cells read genes
sex-linked diseases genetic diseases that result from a mutated gene carried on the sex chromosomes – in human beings, on the X chromosome
red–green colour blindness a sex-linked genetic condition which affects the ability to distinguish tones of red and green
haemophilia a sex-linked genetic disease in which one of the factors needed for blood to clot is not made in the body

4 CYSTIC FIBROSIS: A GENETIC DISEASE

SPECIFICATION REFERENCE 2.16

LEARNING OBJECTIVES

- Understand how the expression of a mutation in people with cystic fibrosis impairs the functioning of the gaseous exchange, digestive and reproductive systems.

CYSTIC FIBROSIS

Cystic fibrosis (CF) is a serious genetic disease which affects people all over the world. It is a life-threatening condition that causes severe respiratory and digestive problems as well as very salty sweat. It also often causes infertility. The chloride transport systems of the **exocrine glands** don't function properly leading to production of a thick sticky mucus. This leads to blockages of the airways of the lungs; it also affects the digestive system and the reproductive system. The sweat glands are also impaired and produce very salty sweat.

Cystic fibrosis is the most common genetic disease in the UK, affecting over 10 000 people a year. Every newborn baby is tested for the disease. Around the world, people are becoming more aware of this disease, which can be a particular risk factor when spouses are distantly related. Organisations such as the Middle East Cystic Fibrosis Association are working hard to raise awareness of this condition, and to help people get the early treatment which can improve and extend lives.

In **Sections 2A.2–2A.4**, you discovered how substances are moved across cell membranes in your body. In cystic fibrosis, one of these important membrane transport systems goes wrong because of a faulty recessive allele. In **fig A** you can see how the cystic fibrosis transmembrane regulatory (CFTR) channel protein lines the channels through which chloride ions leave the epithelial cells and move into the fluid outside the cells in the lungs. CFTR is an enormous protein, containing 1480 amino acids. The gene that codes for it is also large, and it is found on chromosome 7. A mutation in any part of the gene can affect the CFTR protein and so cause cystic fibrosis. Around 1000 different mutations in this gene have been discovered, all of them coding for a faulty CFTR protein, although some are very rare. All of the mutations are recessive. The most common mutation on the cystic fibrosis gene is known as DF508. People who inherit two copies of the faulty allele lack effective CFTR proteins. This means that chloride ions build up in their cells instead of moving out through the channels. This means that water does not move out of their cells to dilute the mucus on the surfaces of their membranes. In fact, water moves *into* the cells by osmosis from the fluid surrounding the cells, making the mucus even more thick and sticky (**fig A**).

LEARNING TIP

To explain the effects of the mutation and the non-functioning CFTR protein, you will need to refer to earlier work on how substances move across membranes.

Cystic fibrosis has many effects on people affected. It makes diffusion of gases much slower because there is a thick layer of mucus between the blood and the air, as well as the normal cell layers. It also reduces the air flow into and out of the lungs, making the diffusion gradients less steep.

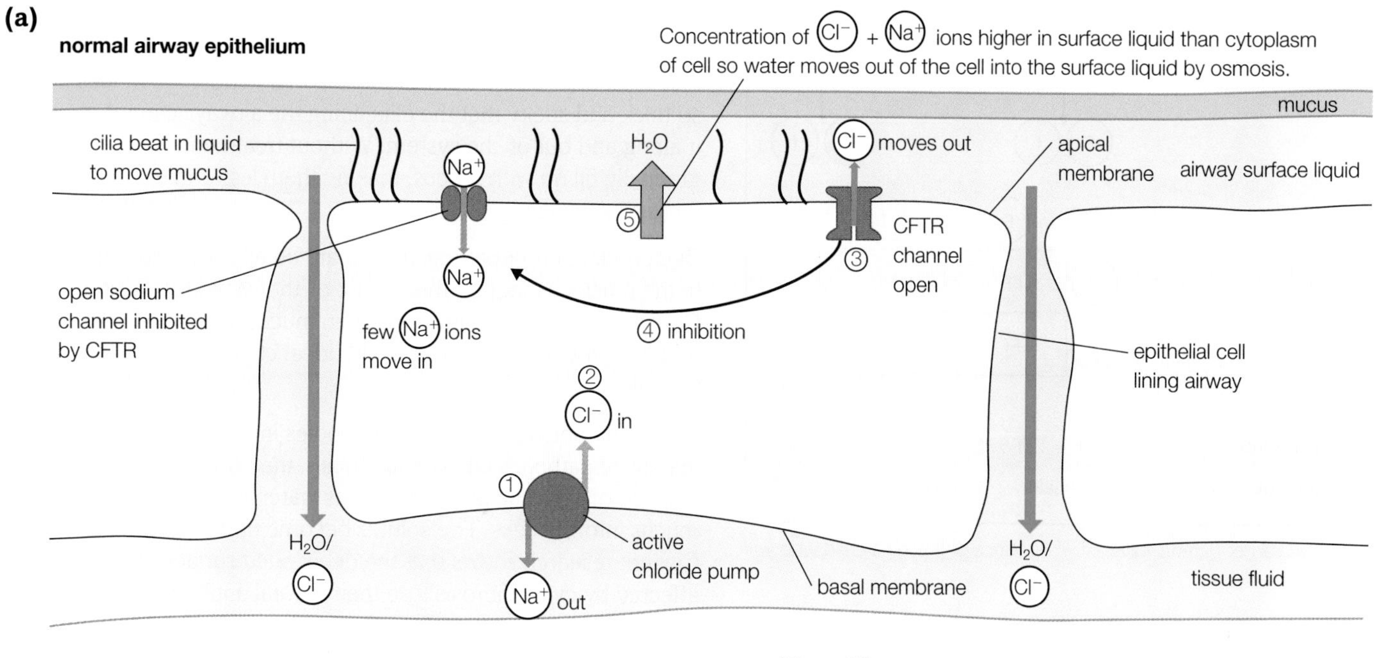

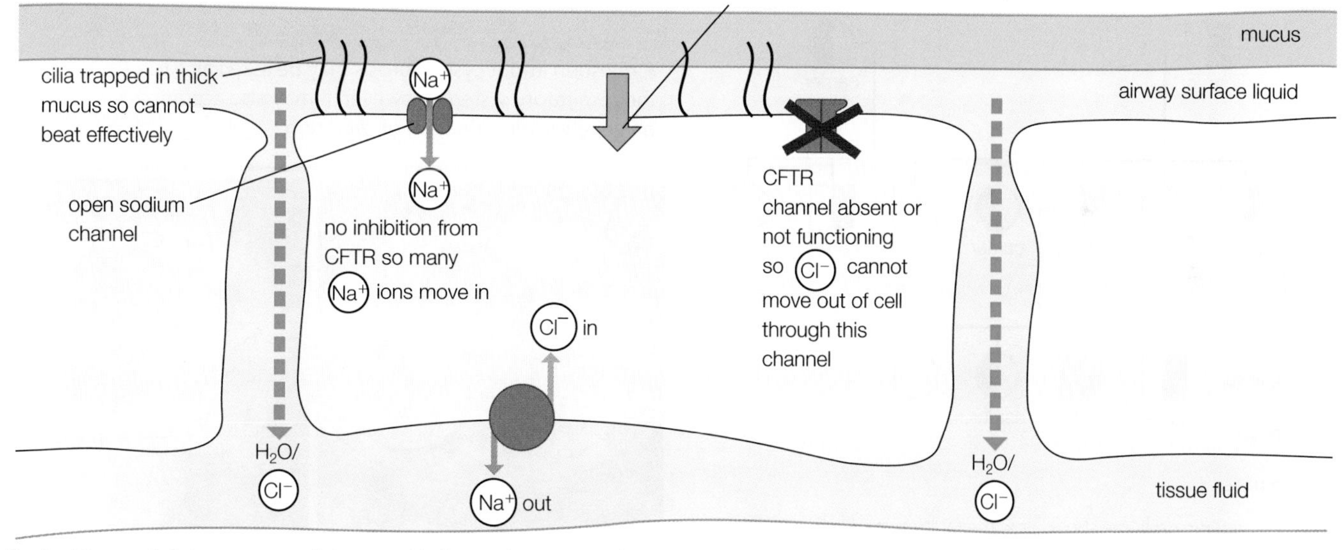

▲ **fig A** Diagram **(a)** shows a normal airway epithelium - the mucus is kept runny which helps the body prevent infection and keeps the airways from getting blocked. Diagram **(b)** shows the epithelium in someone with cystic fibrosis. Here the CFTR channel does not work. This has a dramatic effect on the water balance of the cell and on the liquid and mucus lining the airways.

THE INHERITANCE OF CYSTIC FIBROSIS

Cystic fibrosis is very common in Europe – for example, in the UK about 1 person in 25 carries a faulty CF allele – that's between 2 and 3 million people – and cystic fibrosis occurs in about 1 in 2500 babies born to white Europeans. It is much less common in other ethnic groups – for example in 2017, only 62 patients were diagnosed in the UAE outside of Dubai – but it occurs everywhere in the world and it is often under-diagnosed. Early diagnosis is vital to take advantage of the various treatments now available to affected children and adults. Cystic fibrosis is caused by a recessive allele, which means that many people carry the mutation without knowing it. These carriers are phenotypically normal and usually have no idea that they are carrying the cystic fibrosis mutation. It is only if two carriers have children together that the problems show (see **fig B**). Even then, because the allele is recessive, there is only a 1 in 4 chance that any child of these parents will develop cystic fibrosis.

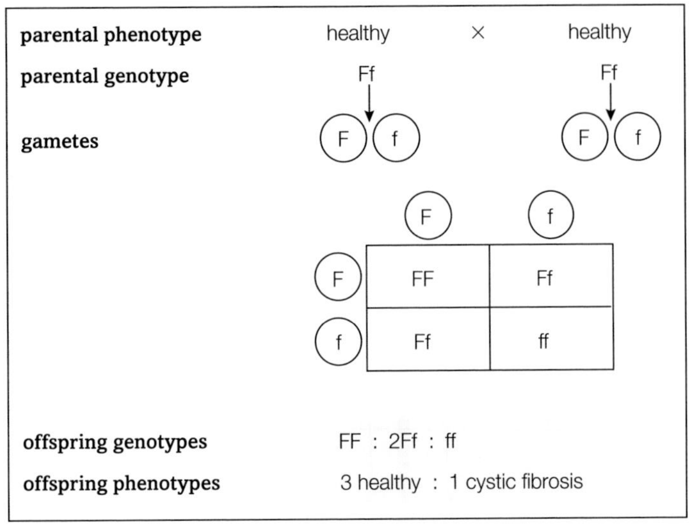

▲ **fig B** Two carrier parents have a 1 in 4 chance of having a child who will have cystic fibrosis.

Once the disease has appeared in a family, other family members become aware they may carry the faulty allele (see **fig C**). They will often be offered genetic counselling before they have children.

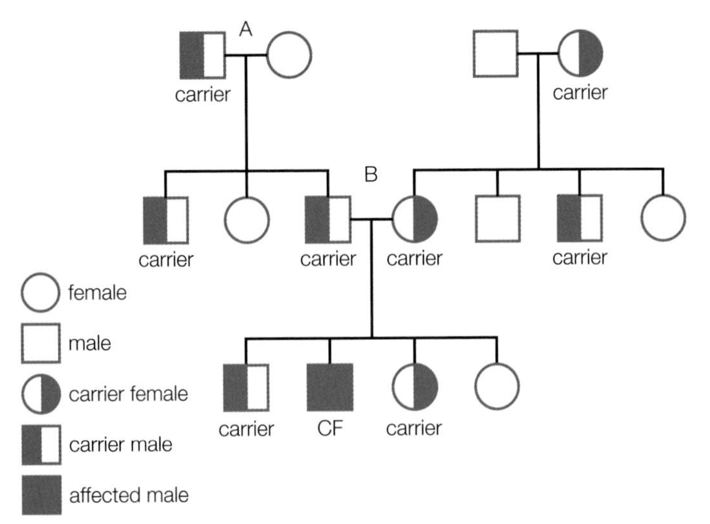

▲ **fig C** This genetic pedigree shows how the faulty cystic fibrosis allele can be passed on through families for generations until, by chance, two carriers have an affected child.

SYMPTOMS OF CYSTIC FIBROSIS

If someone inherits two alleles for cystic fibrosis it will affect many body systems. This is because it affects the mucus that lines the tubes in these systems.

THE RESPIRATORY SYSTEM

The thick, sticky mucus which is typical of cystic fibrosis builds up in the tiny airways of the lungs and reduces the flow of air into the alveoli. It often obstructs the smaller bronchioles completely, preventing air flow in the bronchioles. The reduced airflow means there is a smaller concentration gradient between the air and the blood in the lungs which reduces gas exchange. The thick sticky mucus also greatly reduces the surface area available in the lungs for gas exchange. This means that affected individuals often have severe coughing fits as their body attempts to expel the mucus. They also feel breathless and are often short of oxygen making them feel tired and lacking in energy. In addition, the mucus is so thick and sticky that the cilia lining the airways cannot move it along and out of the system. Without treatment, the lungs gradually fill up with mucus, making them less and less effective for gas exchange.

Bacteria and other pathogens that are breathed in and trapped in the mucus cannot be moved out of the respiratory system. Consequently, thick, pathogen-laden mucus builds up in the lungs, and this provides the bacteria with ideal conditions in which to grow.

Your body normally secretes antibodies into the mucus which inactivate pathogens. In cystic fibrosis, the normal chemical balance of the mucus is changed as water is moved out of the mucus into the cells. The solutes become more concentrated. Current research shows that the dehydrated surfaces of cells affected by cystic fibrosis lose their natural antibacterial properties, because the white blood cells and their antibodies cannot function effectively in the thickened mucus. However, these can be restored if the surface of the cell can be rehydrated.

EXAM HINT

A question about cystic fibrosis may be asked as part of a question on the respiratory system. However, it may also accompany a question on genetics, reproduction or digestion.

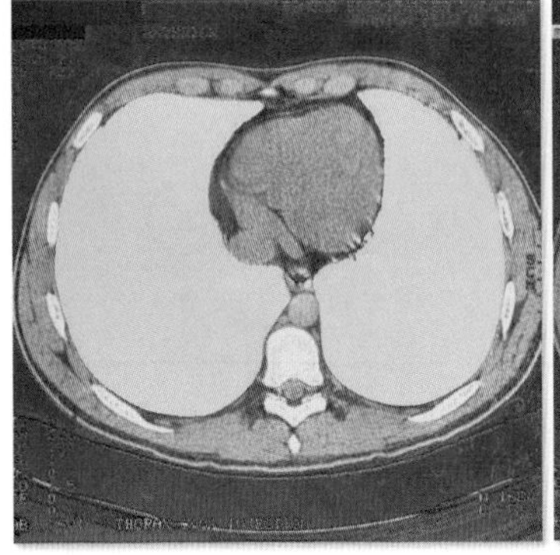
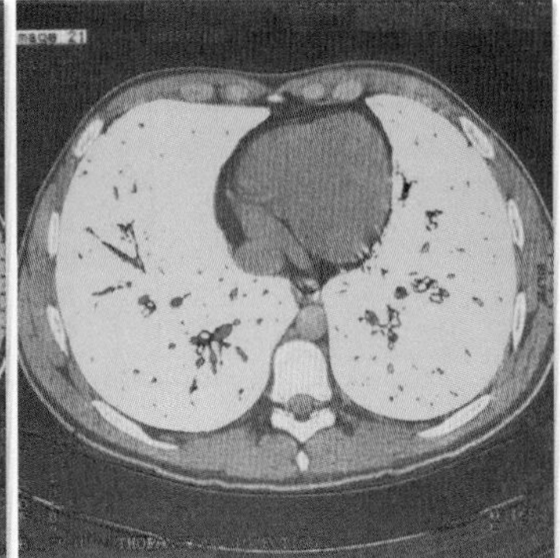

▲ **fig D** The lung tissue on the left is clear and healthy. The lung tissue on the right shows blocked, damaged, mucus-filled airways from someone affected by cystic fibrosis.

THE DIGESTIVE SYSTEM

The gut is badly affected in 85–90% of people with cystic fibrosis. Your digestive system makes enzymes that break down the large complex food molecules (carbohydrates, proteins and fats) into smaller molecules. These can then be absorbed into your blood through the lining of your small intestine, which is covered in finger-like projections called **villi**. These provide a large surface area for absorption to take place. Glands in the lining of your gut make some digestive enzymes; associated organs such as the pancreas make others.

The enzymes from the pancreas are very important in the breakdown of carbohydrates, proteins and fats in the top part of your small intestine (the **duodenum**). The enzymes pass from the pancreas into your duodenum along a tube known as the **pancreatic duct**. Thin mucus is produced by the cells lining this tube in the same way that the airways of the lungs produce it. A faulty CFTR protein means that the mucus produced in the pancreatic duct is also very thick and sticky. It often blocks the pancreatic duct, so that the enzymes do not reach the duodenum. This has two damaging effects. If the digestive enzymes do not reach your gut, you cannot digest your food properly. This means you do not get enough nutrients from the food. Also, the digestive enzymes trapped in the pancreas may start to digest and damage the cells of the pancreas. If they affect the cells which make the hormone **insulin**, then the person may also develop diabetes.

Not only does the thick mucus stop enzymes getting to the gut to digest food, it also makes it more difficult for any digested food to be absorbed into the blood. The gut secretes mucus to prevent the digestive enzymes damaging the delicate lining of the gut and to act as a lubricant. But, when this mucus is very thick and sticky, it forms a barrier between the contents of the gut and the lining of the intestine and clogs up the villi, reducing the surface area for absorption. These two effects put people with cystic fibrosis at severe risk of malnutrition and they often struggle to maintain their body mass. In fact, one of the symptoms that can suggest a baby or small child may be affected by cystic fibrosis is 'failure to thrive' – in other words, the baby fails to gain weight and grow as expected.

THE REPRODUCTIVE SYSTEM

The thick, sticky mucus produced in cystic fibrosis can have a damaging effect on the reproductive system. In women, the mucus in the reproductive system normally changes through the menstrual cycle. When the woman is fertile it becomes thinner to help the sperm get through the cervix and along the oviducts. Women with cystic fibrosis usually produce fertile eggs, but the thick mucus can block the cervix so sperm cannot reach them. It can also block the oviducts, making fertilisation even less likely. In men, the secretions of the reproductive system carry the sperm. Men with cystic fibrosis are often infertile. They may lack the tube that carries sperm out from the testis into the semen (the vas deferens). If the vas deferens is present, it may be partly or completely blocked by thick, sticky mucus so that only a reduced number of sperm (or no sperm at all) can leave the testis.

THE SWEAT GLANDS

The faulty CFTR protein means that people with cystic fibrosis usually have sweat that is more concentrated and salty than normal. Sweat is mainly salty water that is produced in your sweat glands. Normally, as the sweat passes along the duct of a sweat gland, salt (sodium chloride) is reabsorbed, largely as a result of the CFTR protein moving chloride ions *into* the cells. Sodium ions follow along a concentration gradient. This reabsorption of salt prevents you losing too much salt in the sweat. So, in sweat glands the chloride pump works in the opposite direction to that in the mucus-producing glands, where chloride ions are moved *out of* the epithelial cells.

Without functioning CFTR proteins, the chloride ions remain in the sweat, and so do the sodium ions. This means that the sweat is very salty. The loss of sodium and chloride ions causes health problems linked to the balance of ions in the body. Levels of sodium and chloride ions are very important for the proper functioning of many body systems, including the nervous system and the heart. If too much salt is lost in the sweat, the concentration of the body fluids changes, which can affect the heart.

Cystic fibrosis is often first noticed when parents comment that their baby tastes salty when they kiss it. Along with a 'failure to thrive' as a result of the gut complications that come with cystic fibrosis, salty sweat is an early warning sign to health professionals that something may be wrong.

CHECKPOINT

1. Nuala and her parents are both apparently healthy, but Nuala has a child with cystic fibrosis. Is it possible to decide the chances that Nuala's brother and sister might be carriers of the disease? Explain your answer, using genetic diagrams.
2. Using **fig C**, show genetic diagrams for couple A and couple B, giving all possible genotypes and phenotypes of the offspring. Comment on the ratio in the offspring that were actually born.

SKILLS CREATIVITY, CRITICAL THINKING

3. Someone with cystic fibrosis may often feel tired and lacking in energy. Explain why, with reference to the effect of cystic fibrosis on:
   (a) the respiratory system
   (b) the digestive system.
4. Explain why the loss of extra sodium and chloride ions in the sweat of someone with cystic fibrosis can affect all the other cells in the body.

SUBJECT VOCABULARY

cystic fibrosis (CF) a serious genetic disease caused by a recessive allele which affects the production of mucus by epithelial cells
exocrine glands glands which produce substances and secrete them to where they are needed through a small tube called a duct
villi finger-like projections of the lining of the duodenum and small intestine which increase the surface area for the absorption of digested food
duodenum the first part of the gut after the stomach
pancreatic duct the duct from the pancreas which carries digestive enzymes made in the pancreas into the duodenum
insulin a hormone made in the pancreas involved in the regulation of blood sugar levels

2C 5 GENETIC SCREENING

SPECIFICATION REFERENCE 2.17 2.18

LEARNING OBJECTIVES

- Understand the uses of genetic screening, including the identification of carriers, preimplantation genetic diagnosis (PGD) and prenatal testing including amniocentesis and chorionic villus sampling.
- Understand the implications of prenatal genetic screening.
- Identify and discuss ethical and social issues relating to genetic screening from a range of ethical viewpoints, including religious, moral and social implications.

SCREENING BEFORE AND AFTER BIRTH

In the future, it may be possible to cure genetic diseases such as cystic fibrosis. At the moment, it isn't. For individuals born with a genetic disease, it is important to diagnose the condition as early as possible to improve their chances of survival and their general state of health. Sometimes, screening may be carried out during pregnancy to find out whether the fetus is affected by a severe genetic disorder. For some conditions, screening may be carried out on newborn babies to identify problems and give the best treatment as early as possible.

For some genetic diseases, whole populations are tested. This is known as **genetic screening**. Globally, about 7 in 100 000 babies are born the genetic condition **phenylketonuria (PKU)**. It is one of the more common genetic diseases in the Middle East, with an estimated incidence of 10 cases of PKU per 100 000 live births in Bahrain, 8 per 100 000 births in Qatar and around 5 per 100 000 in the UAE. PKU is a recessive autosomal disease, so an affected child can be born to normal carrier parents. With PKU, the body lacks the enzyme that is needed to digest the amino acid phenylalanine. If the condition is left untreated, the amino acid builds up in the blood of the affected baby and this can cause serious, irreversible damage to the brain and nervous system. However, if the condition is diagnosed soon after birth, the affected baby can be given a diet free of phenylalanine and they will grow up healthy.

Phenylketonuria can be easily diagnosed using a simple blood test carried out on babies in the first few days after birth. Affected babies can be given the specialised diet they need, avoiding long-term permanent damage. Although this screening costs money, it is much cheaper than caring for severely affected children for the rest of their lives. Many countries, including Qatar, the UAE, the USA and the UK, have introduced neonatal screening programmes for PKU, and are preventing the development of serious disabilities.

As you saw in **Section 2C.4**, cystic fibrosis is another serious genetic disease. It can also be diagnosed with a simple test given to newborn babies. In the 1980s, in the UK, doctors, scientists and politicians decided it was not economic to test all newborn babies for cystic fibrosis. Those with the disease had a relatively short life expectancy and it would have cost more to screen than to treat affected children. But in the 21st century, people with cystic fibrosis can live far longer, well into adulthood, so it is much more expensive to treat them for a lifetime. As a result, screening of all newborn infants for cystic fibrosis has been introduced in the UK. This is a big advance, because the sooner treatment is started, the better the chances of avoiding serious lung damage. The cost of screening is outweighed by the health benefits and lower costs of early treatment.

For families where the pedigree shows a history of an inherited disease, different screening tests can be offered to couples before they consider having a child. These tests offer help and hope – but also lead to some very difficult choices.

▲ **fig A** This baby is being tested for PKU and other genetic diseases at University Hospital, Sharjah.

IDENTIFYING CARRIERS

If one member of a family is born with a genetic disease such as PKU or cystic fibrosis, other members of the family will be offered genetic testing. For example, it is possible to detect the cystic fibrosis allele in a carrier who has no symptoms. A sample of blood, or some cells from the inside of the mouth, can be used to carry out a simple test to identify the allele. If one partner in a couple knows they are a carrier, the other partner is advised to be tested as well, because if two carriers have a baby there is a 1 in 4 risk that it will be affected by cystic fibrosis. The same principle can be applied to many other genetic diseases.

PRENATAL SCREENING

Couples who find they are at risk of having children with a serious genetic condition have several options open to them. They can go ahead and have a family as usual, hoping that they are lucky and that their children inherit healthy genes. They can be prepared to support and take care of them if they don't. Or they may decide not to have children at all, to prevent passing on a faulty gene even in a carrier.

The third option is to go ahead with pregnancies but to have screening during each pregnancy (**prenatal screening**).

Ideally, prenatal screening is used to try and discover if a fetus is affected by a serious condition early in the pregnancy. The information gained from prenatal screening is used differently in different countries. In some parts of the world it is used to prepare parents for the fact that their child has a genetic disorder which is not compatible with life, so their child will die before or at birth, or that the child has a serious genetic defect which will mean it has serious health problems. In some countries, if prenatal testing shows that a fetus has a genetic defect that is incompatible with life, such as Meckel syndrome or Edwards syndrome, the parents may choose to have the pregnancy ended by doctors. In countries such as the UK and the US, pregnancies may be ended if the fetus has a serious but non-fatal genetic condition such as Down syndrome.

Prenatal screening tests are carried out on fetal cells collected from the mother. The fetal tissue can be obtained by amniocentesis or by chorionic villus sampling (see **fig B**). Whichever method is used to obtain the fetal cells, one of the tests which is then carried out is a karyotype (see **Section 2C.3**).

Amniocentesis involves removing about 20 cm³ of the amniotic fluid which surrounds the fetus using a needle and syringe. This is done at about the 16th week of pregnancy. Fetal epithelial cells and blood cells can be recovered from the fluid after spinning it in a centrifuge. The cells are cultured for 2–3 weeks and then a number of genetic defects and the sex of the baby can be determined from examination of the chromosomes.

Amniocentesis has the following disadvantages.

- It can only be carried out relatively late in the pregnancy making it very difficult for the parents if termination of the pregnancy is necessary.
- The results are not available until 2–3 weeks after the test.
- It carries a 0.5–1% risk of spontaneous abortion after the procedure, regardless of the genetic status of the fetus.

In **chorionic villus sampling**, a small sample of embryonic tissue is taken from the developing placenta. This makes a much bigger sample of fetal tissue available for examination. The cells can be tested for a wide range of genetic abnormalities. This diagnostic technique can be carried out much earlier in the pregnancy, so that if a termination is necessary it is physically less traumatic for the mother. The results are also available more rapidly than for amniocentesis.

There are two disadvantages to chorionic villus sampling.

- There is a 0.5–1% risk that the embryo may spontaneously abort after the tissue sample is taken, though the risk of miscarriage at this stage of pregnancy is high anyway.
- All paternal X chromosomes are inactivated in fetal placental cells so any problems in the genes on that chromosome cannot be detected by this technique.

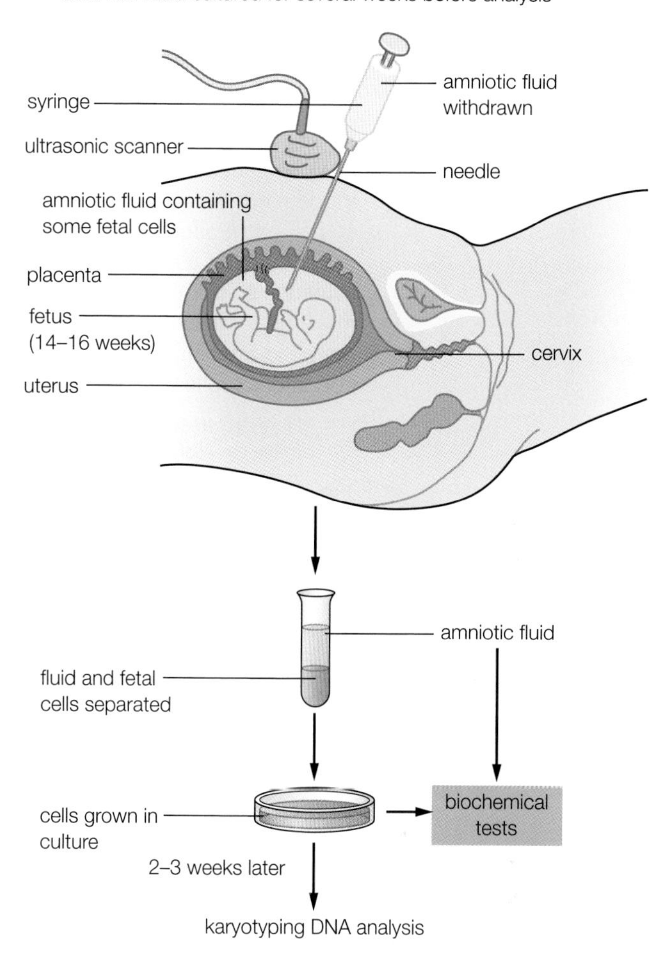

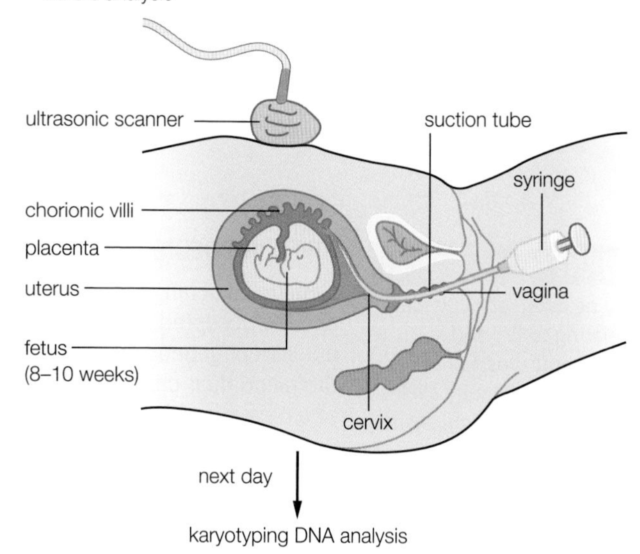

▲ **fig B** Prenatal diagnostic techniques such as amniocentesis and chorionic villus sampling make an accurate diagnosis of serious genetic disorders possible before birth.

PREIMPLANTATION GENETIC DIAGNOSIS

If parents already have a child affected by a genetic disease such as cystic fibrosis, they already know that they are both carriers. If someone else in the family has a child affected by cystic fibrosis, then a couple may choose to be tested and find out if one or other or both of them are carriers. Can parents who know they are carriers avoid having a child with cystic fibrosis without going through amniocentesis or chorionic villus sampling?

As a result of major developments in human infertility treatments over the last 30 years, sophisticated ways of genetically screening an early embryo before it is implanted in the uterus have been introduced. This technique, called **preimplantation genetic diagnosis**, is based on the technique of IVF (*in vitro* fertilisation) (see **fig C**). In this technique, the egg and sperm are fertilised outside the body. After a few cell divisions, a single cell is removed from each embryo. Amazingly, all the evidence so far suggests that this causes no harm to the development of the embryo. The genetic make-up is checked and only those embryos free of the problem alleles are placed in the mother's uterus to implant and grow. This removes the faulty allele from the gene pool. In the case of genetic conditions found only in boys (e.g. haemophilia) only female embryos would be implanted.

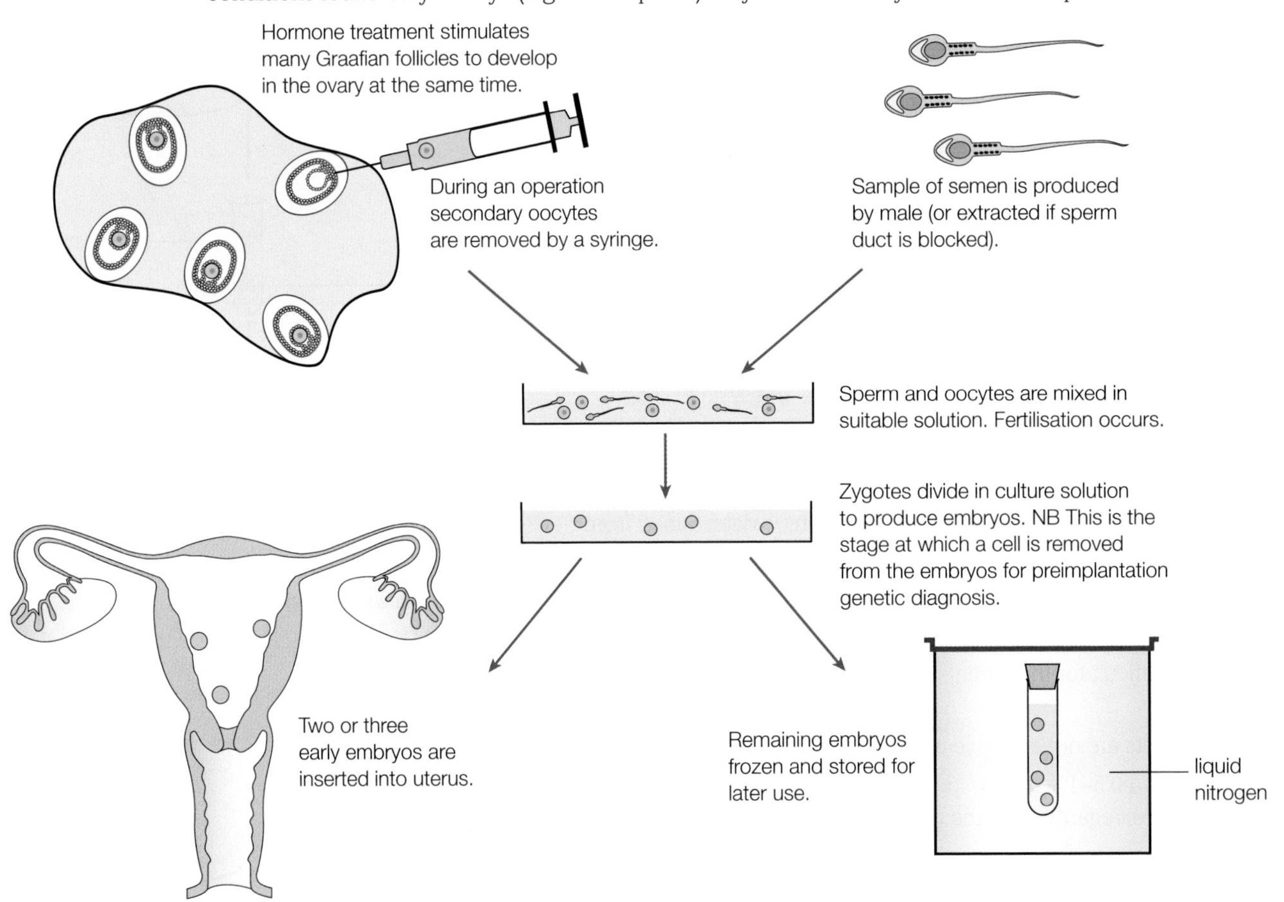

▲ **fig C** The main steps in the process of *in vitro* fertilisation (IVF) which makes preimplantation genetic diagnosis possible.

EXAM HINT

You may be asked about wider issues relating to IVF and selection of embryos. For example, removing faulty alleles from the gene pool might seem a good idea, but is this always the case?

DIFFICULT DECISIONS

Genetic screening raises many ethical issues. Depending on both the personal beliefs of the parents and the accepted moral codes and laws of a society, there are many implications. Some people will not accept genetic screening because their religious beliefs mean they accept whatever happens to them and their children. Other people have religious beliefs which mean they see scientific advances as a blessing which may help them and their unborn children.

Ethical issues also arise from the possibility of false positives (when a healthy child is wrongly shown to have a genetic defect) and false negatives (when as child with damaged chromosomes appears normal in the tests). No test is 100% reliable, and parents have to decide what their attitude is to the possibility that a test may be wrong. There is also the dilemma that sometimes tests such as amniocentesis and chorionic villus sampling can cause a miscarriage, whether or not the fetus has genetic problems.

If genetic screening shows that a fetus has genetic problems, it can be very difficult for parents to come to terms with the fact that the child they are expecting will not be the normal, healthy baby they had hoped for. In some cultures, doctors may end a pregnancy when the fetus has a condition such as anencephaly or Edwards syndrome which is not compatible with life. In other cultures, parents who find that their fetus is affected by a serious genetic disease may decide to end the pregnancy. Their reasons for this may include worries about the quality of life the child would have and/or their ability to care for the child. Other parents may choose to carry their baby to term with the knowledge of its genetic condition. For all couples, the testing can be valuable. It allows time to grieve for the loss of the healthy baby they hoped to have and, if they decide to go ahead, to come to terms with and welcome the child they have. It also gives time to put emotional and practical support in place for when the baby arrives.

The nature of the amniocentesis test means that knowledge of a genetic condition such as cystic fibrosis often comes around half-way through a pregnancy. However, it is still an option although most people now use either chorionic villus sampling or preimplantation diagnosis. Some people may feel there is social pressure towards the view that the only acceptable baby is a perfect baby; they may also feel this is a sad indictment of modern society.

EXAM HINT

When asked about the ethical or social issues surrounding preimplantation and prenatal testing, it is important to present a balanced view. If you have strong personal feelings try not to push these too forcibly without giving other views.

GENETIC COUNSELLING

For most people, finding out about genetic diseases in their family is very traumatic. All the issues discussed above – decisions concerned with having children or not, who to tell – are suddenly of immediate and personal relevance. Genetic counsellors are trained to help people to understand and come to terms with the situation of carrying a faulty allele that can cause a genetic disease. They will assess the statistical risk of a couple producing an affected child and help couples recognise the options they have. They will then work with the parents to choose what they believe is right for them within their own framework of moral, family, religious and social beliefs and traditions.

CHECKPOINT

SKILLS PERSONAL AND SOCIAL RESPONSIBILITY

1. What are the advantages and disadvantages of testing newborn babies for genetic diseases such as PKU?
2. Why is chorionic villus sampling becoming more popular than amniocentesis? Why is amniocentesis still necessary as an option?
3. What are the advantages of preimplantation genetic diagnosis over prenatal testing techniques such as amniocentesis?
4. Summarise the ethical issues raised by the new techniques for prenatal and preimplantation testing.

SUBJECT VOCABULARY

genetic screening when whole populations are tested for a genetic disease
phenylketonuria (PKU) a recessive genetic disorder where those affected lack the enzyme needed to digest the amino acid phenylalanine; the amino acid builds up in the blood and causes severe brain damage
prenatal screening screening of an embryo or fetus before birth
amniocentesis a type of prenatal screening which involves removing a sample of amniotic fluid at around 16 weeks of pregnancy, culturing the fetal cells found and analysing them for genetic diseases
chorionic villus sampling a type of prenatal screening where a small sample of embryonic tissue is taken from the developing placenta and the cells tested for genetic diseases
preimplantation genetic diagnosis testing the cells of an embryo produced by IVF to check for genetic diseases before it is implanted into the uterus of the mother

2C EXAM PRACTICE

1 (a) A young couple who are both healthy have a child who is diagnosed with cystic fibrosis. What is the probability that their next child will also have cystic fibrosis?

A 25% **C** 50%

B 33% **D** 0% [1]

(b) What is the best definition of the term *mutation*?

A a permanent change in the protein

B a permanent change in the phenotype

C a permanent change in the RNA

D a permanent change in the DNA [1]

(c) Explain how a mutation to the gene for the CFTR protein can lead to a non-functioning protein being expressed. [4]

(d) Explain how a non-functioning CFTR protein can lead to the symptoms of cystic fibrosis. [3]

(e) Genetic screening is available to families who have children with genetic conditions such as cystic fibrosis. Describe how and when genetic screening can be carried out. [3]

(Total for Question 1 = 12 marks)

2 (a) Suggest why most mutations arise during replication of the DNA. [2]

(b) Work by Nachman and Crowell (*Genetics*, September 1, 2000 vol. 156 no. 297–304) using human DNA has estimated an average mutation rate of 2.5×10^{-8} per base. Assuming there are 7×10^{9} base pairs in a human diploid cell, calculate the average number of mutations formed each time the cell divides. [2]

(c) If a cell divides 50 times, calculate the total number of mutations accumulated by an average cell. [1]

(d) Explain why most cells still produce proteins that work properly despite this number of mutations. [3]

(Total for Question 2 = 8 marks)

3 The genetics of *Drosophila* fruit flies have been studied for many years. 'Goggle-eye' is a sex-linked trait whose allele is found on the X chromosome of *Drosophila* flies. This allele is recessive to that for 'standard eye'. In addition, the allele for white eyes is recessive to red eyes, and the gene for eye colour is on an autosomal chromosome.

Using this information, answer these questions.

(a) If a heterozygous standard-eyed female mated with a goggle-eyed male, what would the proportion of standard-eyed male flies produced be?

A 0% **C** 50%

B 25% **D** 75% [1]

(b) What proportion of *Drosophila* flies would be white-eyed if two homozygous red-eyed flies mated?

A 0% **C** 75%

B 25% **D** 50% [1]

(c) *Drosophila* are commonly used for genetics investigations. Give three reasons why *Drosophila* are often used. [3]

(d) Name one condition that is sex-linked in humans and explain why the condition is found in one sex more frequently than the other. [3]

(Total for Question 3 = 8 marks)

4 In maize, the seeds can be yellow or white in colour. In addition, the seeds may have a smooth surface or a wrinkled surface. Each of these characteristics of maize seeds is an example of monohybrid inheritance. Yellow and smooth are the dominant phenotypes.

(a) (i) Draw suitable symbols that could be used in a genetic diagram for the alleles involved in these characteristics.

- allele for yellow colour
- allele for white colour
- allele for smooth seed
- allele for wrinkled seed [1]

(ii) Using these symbols, give the genotype of both the pure-breeding varieties:

- variety producing yellow, smooth seeds
- variety producing white, wrinkled seeds. [2]

(b) (i) What is meant by the term *heterozygote*? An individual that possesses:

A two identical dominant alleles

B three different dominant alleles

C two identical recessive alleles

D two different alleles. [1]

(ii) Draw a genetic diagram to show the offspring resulting from a cross between two heterozygous yellow maize plants. [3]

(c) Which is the best description of a recessive allele?

A it is always expressed

B it is never expressed

C it is only expressed when no dominant allele present

D it is only expressed when on its own [1]

(Total for Question 4 = 8 marks)

5 (a) Explain what is meant by the term *incomplete dominance.* [2]

(b) Sickle cell anaemia is caused by the allele of a single gene. People who are homozygous for this allele have sickle cell anaemia and are severely anaemic. People who are heterozygous have a less severe form of the disease.

The photograph below shows red blood cells taken from a person with sickle cell anaemia.

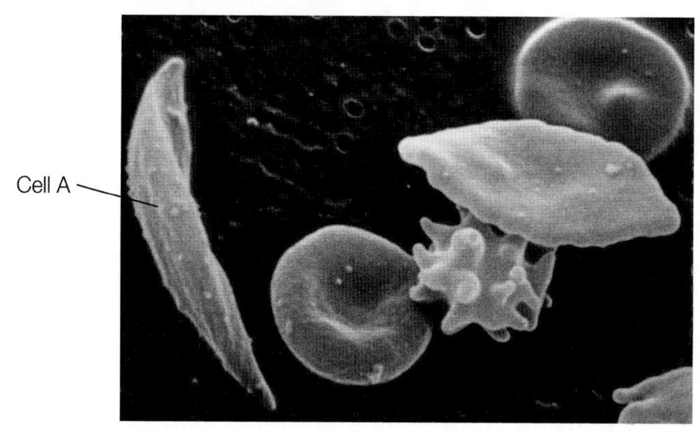

(i) State **two** reasons why oxygen transport by cell A may be less efficient than a normal red blood cell. [2]

(ii) A couple have several children. Only one of the children has sickle cell anaemia. Use a genetic diagram to show how this child inherited sickle cell anaemia. Use the symbols Hb^A to represent the normal allele and Hb^S to represent the sickle cell allele. [3]

(c) A person with sickle cell anaemia has a change in one of the amino acids in one of the polypeptide chains in their haemoglobin. Explain how a change in DNA can lead to a change in a single amino acid in a polypeptide chain. [4]

(Total for Question 5 = 11 marks)

6 (a) What is meant by the term *genetic carrier*?

A a gene carrying a mutation

B a person who has a mutated gene that does not affect the phenotype

C non-coding DNA that is faulty

D a person with an extra chromosome [1]

(b) Red–green colour blindness is a sex-linked character in humans. The diagram below shows the genetic pedigree of a family in which some members are red–green colour blind.

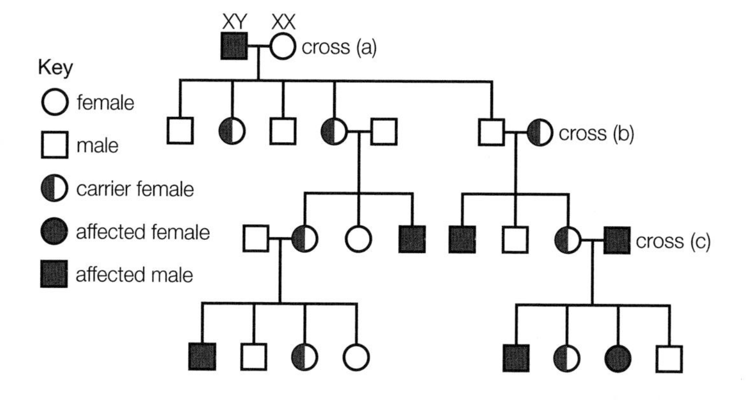

(i) What are the genotypes of the parents in cross (a)?

A X^nY and X^NX^N

B X^nY^n and X^NY^N

C XY^n and X^NX^N

D X^nY and X^NX^n [2]

(ii) Explain why there are no affected male children resulting from cross (a). [2]

(iii) Draw a genetic diagram to represent cross (b). [3]

(iv) Cross (c) has resulted in four children. What is the probability that a fifth child would be an affected female? Show your working. [4]

(Total for Question 6 = 12 marks)

7 (a) (i) Give a definition of the term *allele.* [1]

(ii) Explain what is meant by the term *codominance.* [1]

(b) Certain breeds of cattle have a coat colour known as roan. This is caused by a combination of red and white hairs. When two roan animals are bred the offspring can be red, white or roan.

(i) Explain how two roan cattle can produce offspring with a range of coloured coats. [4]

(ii) Use a genetic diagram to show what is the probability of the first offspring being roan. [3]

(Total for Question 7 = 9 marks)

TOPIC 3 CELL STRUCTURE, REPRODUCTION AND DEVELOPMENT

CHAPTER 3A CELL STRUCTURE

In October 1951, Henrietta Lacks, a 31-year-old American black woman, died of an aggressive cancer of the cervix. Before she died, doctors took a sample of her cells, which became the first human cells to grow successfully in culture. Her cells are still alive and well and are called HeLa cells (for Henrietta Lacks). Ever since they were first cultured, HeLa cells have played an important role in cell biology and medical research. The cells have been infected with viruses, and were used to help develop the first polio vaccine. HeLa cells reproduce themselves quickly and reliably and they have travelled all over the world. They have been used to investigate radiation damage and they have even gone into space.

All animals, plants and fungi are eukaryotic. In this chapter, you will discover more about eukaryotic cells like Henrietta's and the sub-cellular membrane-bound organelles they contain. Without microscopes, we could not see most cells, so you will learn about how both light and electron microscopes work.

Membranes are key to understanding how the cells are divided into compartments or organelles that each perform a specific function. By looking at the many different types of organelle within a cell, you will discover their structures, their functions and how they work together. Mitochondria are very important organelles inside cells – so important they have their own DNA that enables them to divide independently. This is where the reactions of cellular respiration take place. But other organelles are also important and the way that organelles work together is shown by the way that cells manufacture, transport, modify and release proteins such as enzymes.

The chapter covers both animal and bacterial cells. But a single cell can only become a very small and simple organism, so cells are organised into tissues, organs and systems to produce larger and more complex multicellular organisms.

MATHS SKILLS FOR THIS CHAPTER

- **Carry out calculations using numbers in standard and ordinary form** (*e.g. use of magnification*)
- **Use expressions in standard form** (*e.g. when applied to areas such as the size of organelles*)
- **Use scales for measuring** (*e.g. graticule to measure size of cells*)
- **Make order of magnitude calculations** (*e.g. use and manipulate the magnification formula: magnification = size of image/size of real object*)
- **Use and manipulate equations, including changing the subject of an equation** (*e.g. magnification*)
- **Calculate the circumference, surface areas and volumes of regular shapes** (*e.g. calculate the surface area or volume of a cell*)

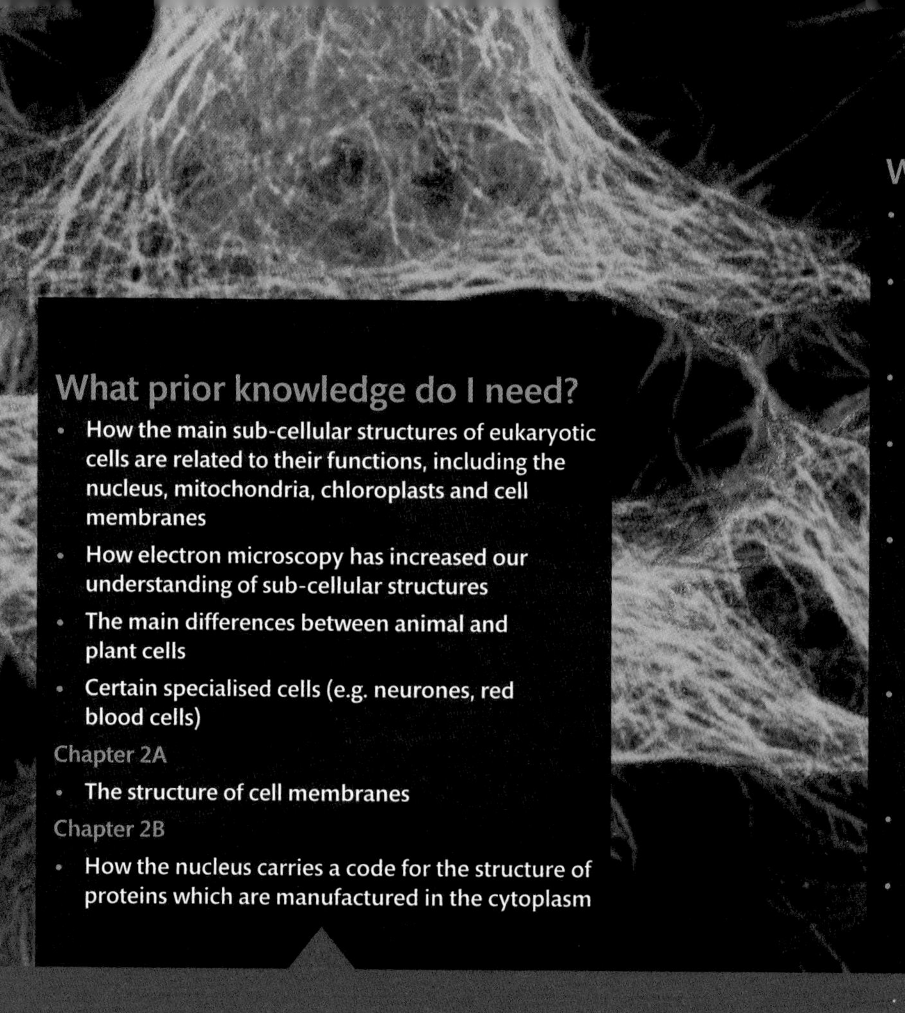

What prior knowledge do I need?

- How the main sub-cellular structures of eukaryotic cells are related to their functions, including the nucleus, mitochondria, chloroplasts and cell membranes
- How electron microscopy has increased our understanding of sub-cellular structures
- The main differences between animal and plant cells
- Certain specialised cells (e.g. neurones, red blood cells)

Chapter 2A

- The structure of cell membranes

Chapter 2B

- How the nucleus carries a code for the structure of proteins which are manufactured in the cytoplasm

What will I study in this chapter?

- The way specimens are magnified by the light microscope and the electron microscope
- The way specimens are prepared for the light microscope (including staining) and the electron microscope
- The difference between magnification and resolution
- The details of the ultrastructure of eukaryotic cells related to the functions of the membrane-bound organelles
- The main membrane-bound organelles of animal eukaryotic cells, including the nucleus, nucleolus, 80S ribosomes, rough and smooth endoplasmic reticulum, mitochondria, centrioles, lysosomes and Golgi apparatus
- The main membrane-bound organelles of bacterial prokaryotic cells, including the cell wall, pili and flagella, nucleoid, plasmids and 70S ribosomes
- The way that the organelles work together to produce and secrete proteins from the cell
- The way that cells are organised into tissues, organs and organ systems

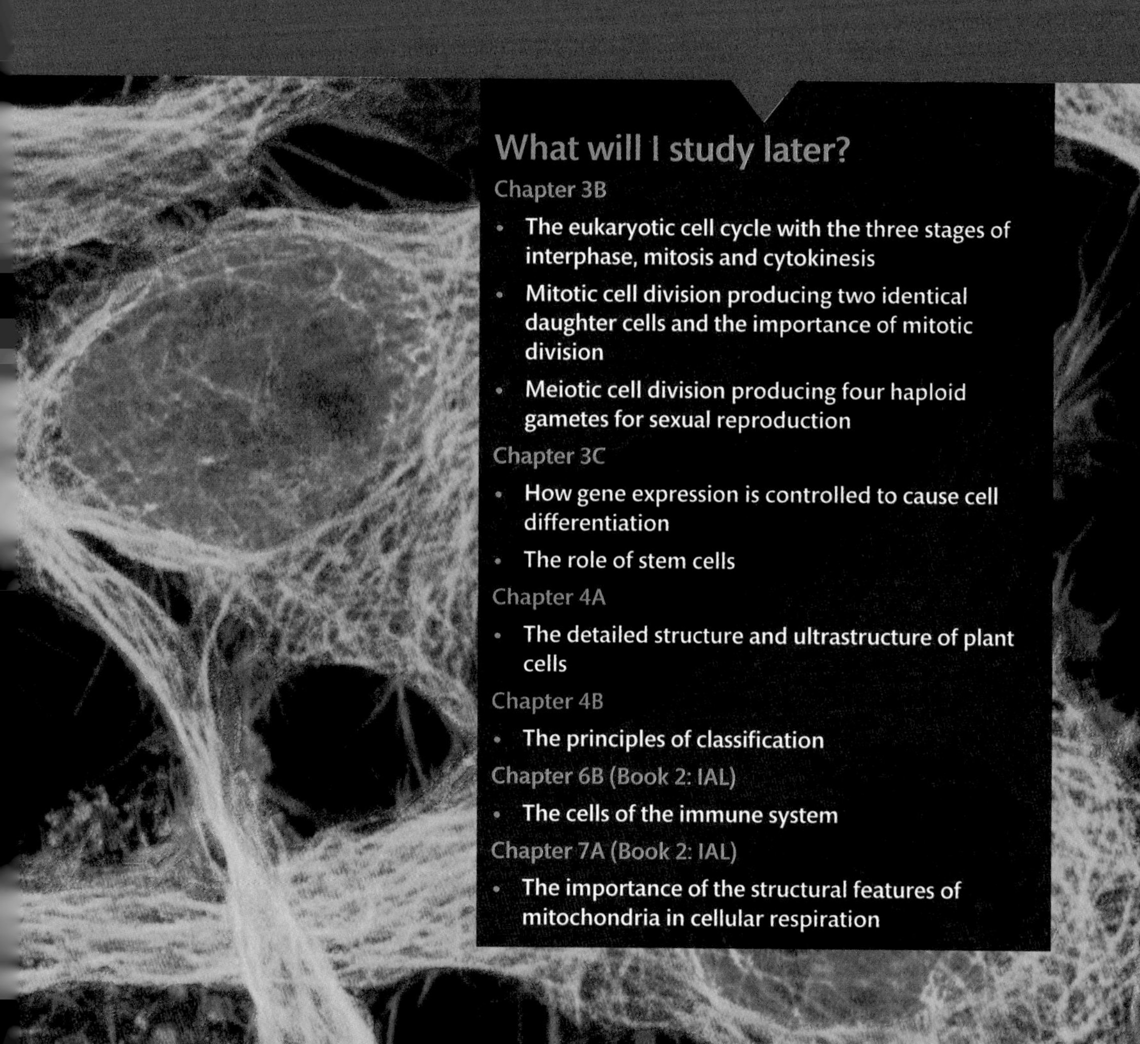

What will I study later?

Chapter 3B

- The eukaryotic cell cycle with the three stages of interphase, mitosis and cytokinesis
- Mitotic cell division producing two identical daughter cells and the importance of mitotic division
- Meiotic cell division producing four haploid gametes for sexual reproduction

Chapter 3C

- How gene expression is controlled to cause cell differentiation
- The role of stem cells

Chapter 4A

- The detailed structure and ultrastructure of plant cells

Chapter 4B

- The principles of classification

Chapter 6B (Book 2: IAL)

- The cells of the immune system

Chapter 7A (Book 2: IAL)

- The importance of the structural features of mitochondria in cellular respiration

3A 1 OBSERVING CELLS

SPECIFICATION REFERENCE

3.1 | 3.7(i) | 3.7(ii) | 3.8(ii) | CP5

LEARNING OBJECTIVES

- Know that all living organisms are made of cells, sharing some common features.
- Know how magnification and resolution can be achieved using light and electron microscopy.
- Understand the importance of staining specimens in microscopy.

Cells are discussed in the media almost daily in relation to topics such as cancer, stem cells and DNA testing. Although we have known about cells for over 300 years, most people have only a vague idea about what they are and how they function.

DISCOVERING CELLS

In 1665, Robert Hooke (1635–1703), an English architect and natural philosopher, designed and made one of the first working optical microscopes. He examined many objects, some of which were thin sections of cork, made up of tiny, regular compartments. He called these cells, as they reminded him of the monks' cells in a monastery. Anton van Leeuwenhoek (1632–1723) was a Dutch draper who ground lenses in his spare time in order to check the weave of his fabrics. In 1676, he used his lenses to observe a wide variety of living unicellular organisms in drops of water; he called these organisms 'animalcules'. At the same time, the English plant scientist Nehemiah Grew (1641–1712) was one of the first scientists to publish accurate drawings of 'tissues'. Matthias Schleiden (1804–1881) and Theodor Schwann (1810–1882) introduced their cell theory in 1839. It proposed the idea that cells are the basic units of life. By the 1840s it was widely accepted that cells are the basic units of life. Cell theory is now accepted as a unifying concept in biology. It states that cells are a fundamental unit of structure, function and organisation in all living organisms. Improved quality of lenses, new staining techniques and modern technologies such as electron and confocal microscopes have allowed us to see cells in increasing detail. This has developed our understanding of both the structure and function of cells.

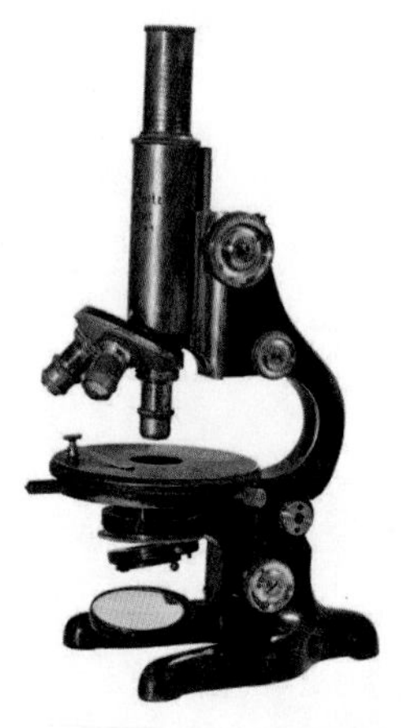

▲ **fig A** As microscopes have developed, more and more has been revealed about cells, the key to understanding biology.

MICROSCOPES

We can see some cells easily with the naked eye; for example, the ovum in an unfertilised bird's egg is a single cell. But we need a method of magnification to enable us to see most cells.

The **light microscope** or **optical microscope** was the only tool for observing cells for many years and is still widely used today (see **fig A**). A good light microscope can magnify to 1500 times and still give a clear image. At this **magnification**, an average person would appear to be 2.5 km tall.

Since the mid-20th century, the **electron microscope** has enabled scientists to understand the inner workings of cells. An electron microscope can give a magnification of up to 500 000 times, making an average person appear over 830 km tall!

DID YOU KNOW?

Magnification and **resolution (resolving power)** are the two features of any microscope that determine how clear the image is.

- Magnification is a measure of how much bigger the image you see is when compared to the real object (e.g. ×40, ×1000 or ×500 000).
- Resolution is a measure of how far apart two objects must be before we see them separately. The resolution of the naked eye is around 0.1 mm (100 000 nm), so two objects closer together than 0.1 mm cannot be seen as separate objects. The resolution of a light microscope is around 0.2 μm (200 nm) and the resolution of an electron microscope is around 0.1–1 nm.

THE LIGHT MICROSCOPE

A specimen or thin slice of biological material is placed on the stage of a light microscope (see **fig B**). This is illuminated from underneath, either by sunlight reflected with a mirror or by a built-in light source. The objective lens produces a magnified and inverted image. The eyepiece lens focuses this image at the eye.

The total magnification of the specimen is calculated as follows:

magnification of objective lens × magnification of eyepiece lens = total magnification

e.g. ×10 × ×10 = ×100

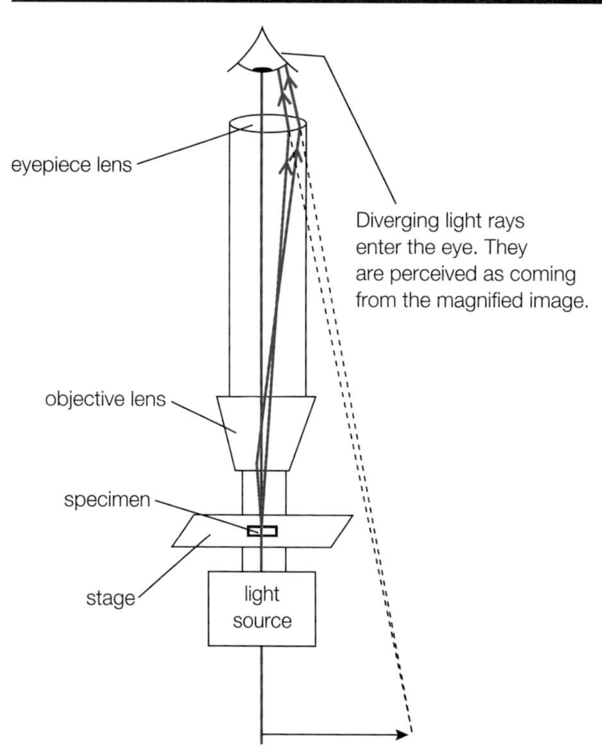

fig B Light passes through the specimen and on through the lenses to give an image that is magnified and upside down.

Providing you always record the magnification you are using, you can work out the actual size of a specimen by measuring it under the microscope:

$$\text{actual size} = \frac{\text{image size}}{\text{magnification}}$$

$$\text{or } A = \frac{I}{M}$$

WORKED EXAMPLE

For example, if the diameter of a cell measured under the light microscope at magnification ×400 is 1.0 cm (10 mm):

$$\text{actual size} = \frac{\text{image size}}{\text{magnification}}$$

$$= 10/400$$

$$= 0.025\text{ mm } (25\ \mu\text{m})$$

LEARNING TIP

Using the magnification triangle can help you remember the correct calculation:

A = actual size of specimen
I = image size
M = magnification

I = A × M
or: A = 1/M
or: M = 1/A

You can look at living organisms, tissues and cells under the light microscope. However, most of the specimens will be dead, stained, specially preserved and sectioned (very thinly sliced) before they are mounted on a slide. The staining is used to make it easier to identify types of cell or parts of cells under the microscope (see **fig C**). Stains you may come across include:

- haematoxylin – stains the nuclei of plant and animal cells purple, blue or brown
- methylene blue – stains the nuclei of animal cells blue
- acetocarmine – stains the chromosomes in dividing nuclei in both plant and animal cells
- iodine – stains starch-containing material in plant cells blue-black.

There are big advantages to using light microscopes, but there are some disadvantages too (see **table A**).

ADVANTAGES OF THE LIGHT MICROSCOPE	DISADVANTAGES OF THE LIGHT MICROSCOPE
• Can see living plants and animals, or parts of them, directly. This is useful and allows you to compare prepared slides with living tissue.	• Preserving and staining tissue can produce **artefacts**. These artefacts are *not* part of the living tissue. They are a result of the process of preserving and staining, but it is easy to mistake them for part of the tissue.
• Relatively cheap so are available in schools and universities, hospitals, industrial labs and research labs.	• Limited powers of resolution (and magnification).
• Relatively light and portable so we can use them almost anywhere, e.g. identifying malaria in the field.	

table A The advantages and disadvantages of the light microscope

Developments, including a technology called confocal microscopy, mean the information we can get from light microscopes continues to increase.

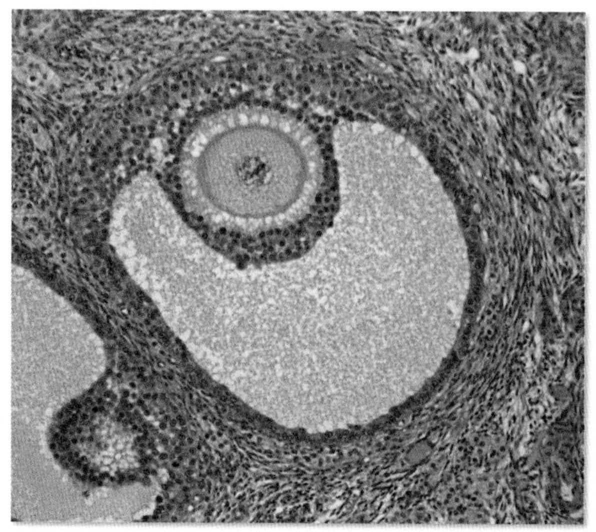

fig C A good light micrograph of tissue with staining can provide a lot of information. This is a section through ovarian tissue. Magnification ×100

PRACTICAL SKILLS CP5

Using a graticule

When you are looking at specimens under the light microscope, you can use a **graticule** to measure the size of individual cells or organelles. By using a stage micrometer - a slide with a very accurate scale in micrometres (μm) etched into it - with an eyepiece graticule, you can calibrate your graticule for the lens of the microscope you are using and then make extremely accurate measurements.

THE ELECTRON MICROSCOPE

The electron microscope uses a beam of electrons to form an image. The electrons are scattered by the specimen in much the same way as light is scattered in the light microscope. In an electron microscope, the electrons behave in the same way as light waves but with a very tiny wavelength. Electromagnetic or electrostatic lenses focus the electron beam to form an image. Resolving power increases as the wavelength gets smaller, so the electron microscope can resolve detail down to less than 1 nm or 0.0001 μm, about 1000 times better than the light microscope (see **fig E**).

For the electron microscope to work, the specimens must be in a vacuum, so they are always dead. The preparation of a specimen for the electron microscope is a very complex process that may involve chemical preservation, freeze-drying, freeze-fracturing, removing the water (dehydration), embedding, sectioning and mounting on a metal grid. Specimens for electron microscopy are often stained using heavy metal ions such as lead and uranium. This is not to identify tissues, but to improve the scattering of the electrons. This produces an image with more contrast which is clearer and easier to interpret. The image is displayed on a monitor or computer screen.

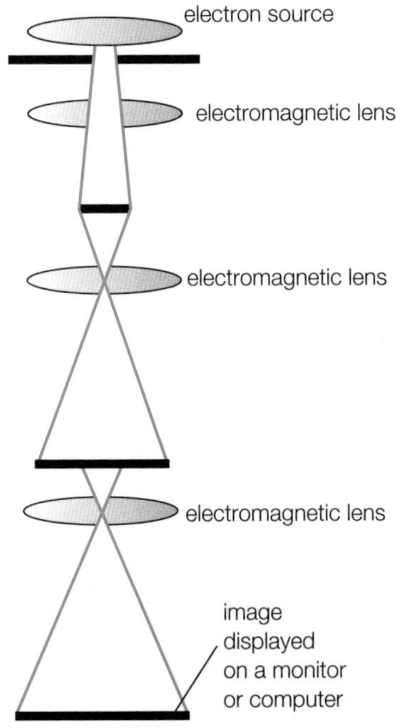

▲ **fig D** In an electron microscope, a beam of electrons passes through the specimen and on through a series of electromagnetic or electrostatic lenses. This gives a greatly magnified image.

EXAM HINT

Make sure you understand and can clearly explain the difference between magnification (how much bigger the object appears compared with the real object) and resolution (how far apart two points must be before we see them as separate objects).

There are two main types of electron micrograph. **Transmission electron micrographs (TEMs)** are two-dimensional (2D) images similar to those from a light microscope (see **fig E**). **Scanning electron micrographs (SEMs)** have a lower magnification, but are three-dimensional (3D) and can be very striking (see **fig F**). Sometimes electron micrographs are given false colours to make it easier to identify the different cells, but these are not stains. They are added digitally after the image has been taken.

There are big advantages to using electron microscopes, but there are some disadvantages too (see **table B**).

ADVANTAGES OF THE ELECTRON MICROSCOPE	DISADVANTAGES OF THE ELECTRON MICROSCOPE
• Huge powers of magnification and resolution. Many details of cell structure have been seen for the first time using an electron microscope.	• All specimens are examined in a vacuum - air would scatter the electrons and produce a blurred image of the tissue - so it is impossible to look at living material.
	• Specimens undergo severe treatment that is likely to result in artefacts. Preparing specimens for the electron microscope is very skilled work.
	• Extremely expensive.
	• The instrument is very large, must be kept at a constant temperature and pressure, and with an internal vacuum. Relatively few scientists outside research laboratories have easy access to this equipment.

table B The advantages and disadvantages of the electron microscope

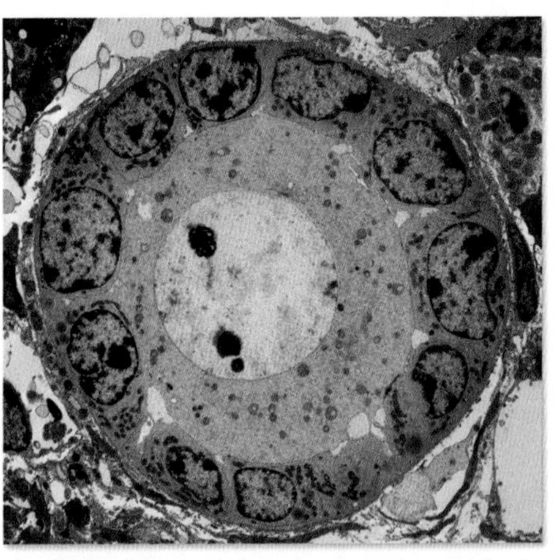

▲ **fig E** A transmission electron micrograph of a cell gives you much more detailed information than a light micrograph (see **fig C**).

fig F Scanning electron micrographs introduce a 3D world of biology on a small scale.

CHECKPOINT

SKILLS PROBLEM SOLVING

SKILLS ANALYSIS

1. Why is high magnification alone not enough to give us biological details of cells?
2. Both light and electron micrographs can be brightly coloured. Explain the differences and similarities between the way colour is used in light and electron microscopy.
3. A student measured the diameter of three cells of the same type under the light microscope. Measurement 1 was taken with a magnification of ×40, and measurements 2 and 3 with a magnification of ×100. Work out the mean diameter of the cells.

   Measurement 1 = 5 mm   Measurement 2 = 12 mm   Measurement 3 = 1 mm

SUBJECT VOCABULARY

light microscope (optical microscope) a tool that uses a beam of light and optical lenses to magnify specimens up to 1500 times life size
magnification a measure of how much bigger the image you see is than the real object
electron microscope a tool that uses a beam of electrons and magnetic lenses to magnify specimens up to 500 000 times life size
resolution (resolving power) a measure of how close together two objects must be before they are seen as one
artefacts things observed in a scientific investigation that are not naturally present; they occur as a result of the preparation or investigation
graticule a series of lines in the eyepiece of a microscope which help you measure specimens accurately
transmission electron micrographs (TEMs) micrographs produced by the electron microscope that give 2D images like those from a light microscope, but magnified up to 500 000 times
scanning electron micrographs (SEMs) micrographs produced by the electron microscope that have a lower magnification than TEMs, but produce a 3D image

2 EUKARYOTIC CELLS 1: COMMON CELLULAR STRUCTURES

SPECIFICATION REFERENCE
3.3(i) 3.3(ii)
3.6 3.8
CP5

LEARNING OBJECTIVES

- Know the ultrastructure of eukaryotic cells, including the nucleus, nucleolus, ribosomes, mitochondria, centrioles, and lysosomes.
- Understand the function of these organelles.
- Recognise these organelles in electron micrographs.

Cells appear flat and two-dimensional (2D) in most microscope images, except those of living material or from a scanning electron microscope. However, cells are spheres, cylinders or asymmetrical three-dimensional (3D) shapes – so when you look at cells, try to imagine them in three dimensions.

THE CHARACTERISTICS OF EUKARYOTIC CELLS

In eukaryotic organisms such as animals, plants and fungi, there is a very wide range of different types of cell, each with a different function. There are certain cell features that are common and which we can put together to represent a typical plant or animal cell. Remember that this typical cell does not really exist, but acts as a useful guide to what to look for in any eukaryotic cell (see **fig A**).

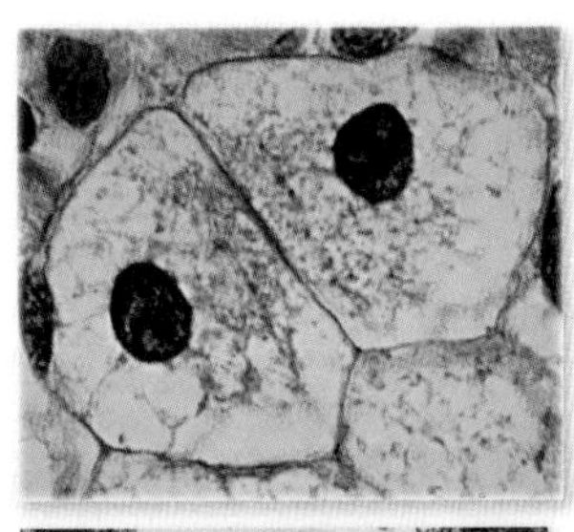

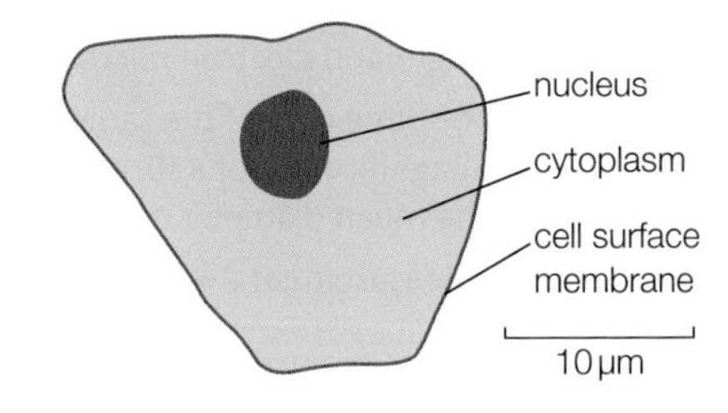

LEARNING TIP

Remember to draw large and clear diagrams with complete lines drawn using a sharp pencil.

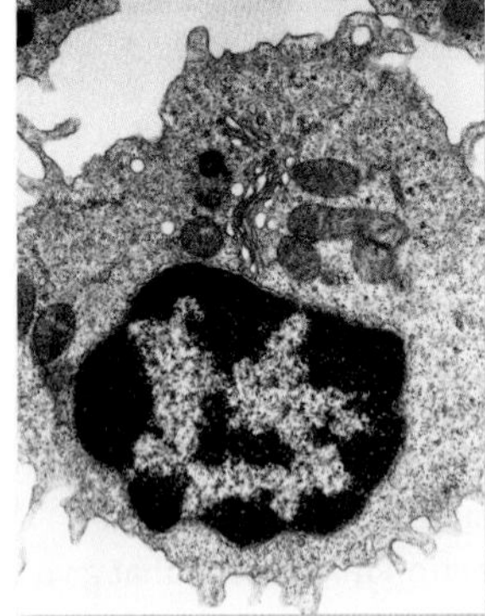

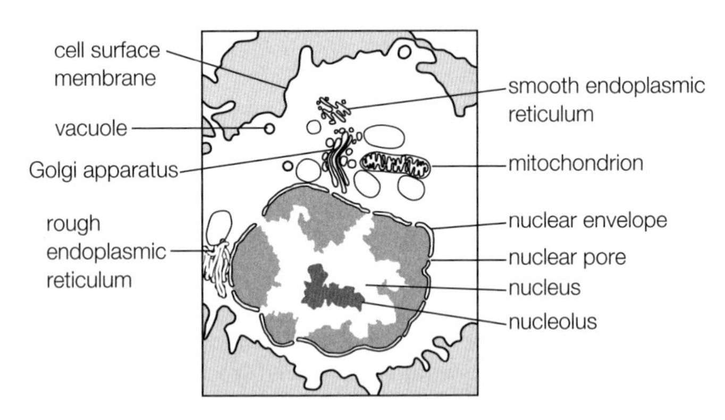

fig A These images show clearly how introducing the electron microscope increased our detailed knowledge and understanding of structures within cells.

THE TYPICAL ANIMAL CELL

A typical animal cell contains many things that are found in all eukaryotic cells, including plants and fungi. A membrane known as the cell surface membrane surrounds the cell. Inside this membrane is a jelly-like liquid called the **cytoplasm** containing a **nucleus** – the two together are known as the **protoplasm**. The cytoplasm contains the components that perform the functions of the cell. The nucleus contains the information needed to produce all the chemical substances the cell is made from. There is an enormous number of variations which adapt cells for different functions. The different parts of the cell have complex and detailed structures, which we can see more clearly when an electron microscope is used. The **ultrastructure** of the cell are those structures that can only be observed in detail using the electron microscope (see **fig B**). You will never see a cell that looks like this, but it is a useful model to show how the different organelles are arranged in the cell and how they relate to each other. The structure of each part of the cell relates closely to its function.

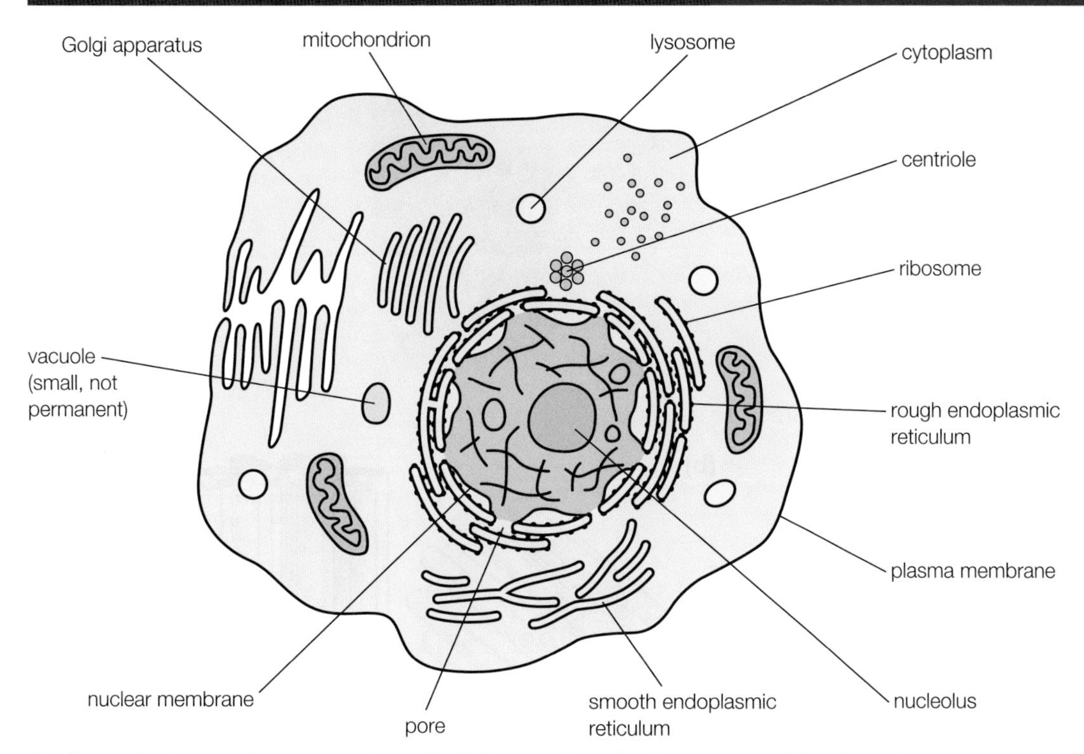

fig B This diagram shows a typical animal cell, drawn to show the ultrastructure visible with an electron microscope.

MEMBRANES

Membranes are important as an outer boundary to the cell and also as the many **intracellular** (internal) membranes. In **Chapter 2A**, you looked at the importance of cell membranes for controlling the movement of substances, but membranes inside the cell also have other functions. For example, they localise enzymes in reaction pathways (e.g. respiration in mitochondria and photosynthesis in chloroplasts). They also keep biological molecules separate (e.g. hydrolytic enzymes in lysosomes). You will find out more about membrane functions in this section as you learn about the various structures of the cell.

THE PROTOPLASM

When the light microscope was the only tool that biologists had to observe cells, they thought that the cytoplasm was a relatively structureless clear jelly. But the electron microscope revealed the cytoplasm to be full of many structures, known as organelles, some of which are described below.

THE NUCLEUS

The nucleus is usually the largest organelle in the cell (1–20 μm) and it can be seen with the light microscope. Electron micrographs show that the nucleus, which is usually spherical in shape, is surrounded by a double nuclear membrane with holes or pores. Chemical substances can pass in and out of the nucleus through these pores so that the nucleus can control events in the cytoplasm. Inside the nuclear membrane, sometimes called the nuclear envelope, are two main substances: nucleic acids and proteins. The nucleic acids are deoxyribonucleic acid (DNA) and ribonucleic acid (RNA) (see **Section 2B.4**).

When the cell is not actively dividing, the DNA is bonded to the protein to form **chromatin**, which looks like tiny granules. Also in the nucleus is at least one **nucleolus** – an extra-dense area of DNA, RNA and protein. The nucleolus is involved in the production of ribosomes. Recent research also suggests that the nucleolus plays a part in the control of cell growth and division.

MITOCHONDRIA

The name **mitochondria** (singular mitochondrion) means 'thread granules' and describes the tiny rod-like structures that are 1 μm wide by up to 10 μm long. They can be seen in the cytoplasm of almost all cells under a light microscope. In recent years, by using the electron microscope, we have been able to understand not only their complex structure, but also their vital functions.

The mitochondria are the 'powerhouses' of the cell. In a series of complicated biochemical

PRACTICAL SKILLS CP5

Practice makes perfect

When you first look at cells under a light microscope, it can be hard to identify the different structures. It takes practice to use the higher power lenses successfully.

Looking at the same type of cells using slides made with different stains is useful because the different stains show up different features of the cells and make it easier to identify them.

Drawing what you see helps to make sense of the images and gives you a record of your time with the microscope. First, you can practise by copying drawings of the micrographs in your textbook, where the cell structures are clear and labelled. Then try drawing for yourself the different cells and cell organelles that you can see under a microscope.

reactions, simple molecules are oxidised in the process of cellular respiration. This produces ATP, which can be used to drive the other functions of the cell and the whole organism. The number of mitochondria present can give you useful information about the functions of a cell. Cells that require very little energy (e.g. white fat storage cells) have very few mitochondria. Any cell with an energy-demanding function will contain many mitochondria (e.g. muscle cells or cells such as liver cells that carry out a lot of active transport).

Mitochondria have an outer and inner membrane. They also contain their own genetic material, so that when a cell divides, the mitochondria replicate themselves controlled by the nucleus. Mitochondrial DNA is part of the whole genome of the organism.

Mitochondria have an internal arrangement adapted for their function (see **fig C**). The inner membrane is folded to form **cristae**, which give a very large surface area, surrounded by a fluid matrix. This structure is closely integrated with the events in cellular respiration that occur in the mitochondrion (see **Book 2 Chapter 7A**). The fact that mitochondria (and chloroplasts) have their own DNA leads scientists to think that these organelles originated as symbiotic **eubacteria** living inside early eukaryotic cells. During millions of years, they have become an integral part of the eukaryotic cell.

LEARNING TIP

Remember to relate structure to function. If a membrane is highly folded, it is always to create a larger surface area. In mitochondria, the surface area of the inner membrane provides a site of attachment for enzymes to carry out cellular aerobic respiration.

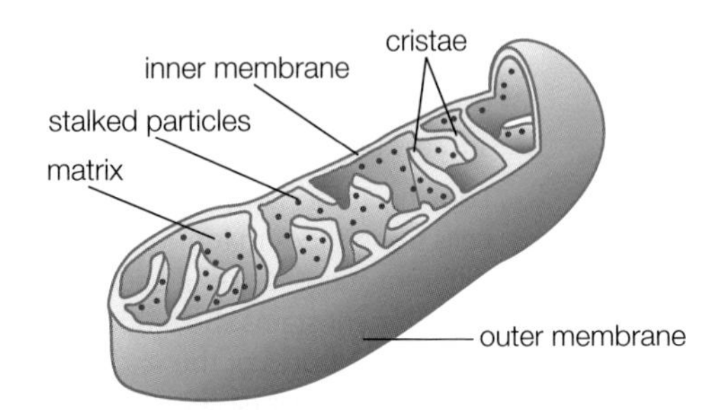

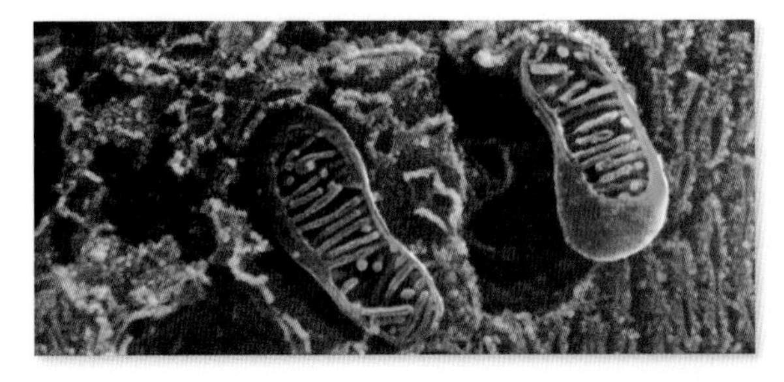

▲ **fig C** The 3D structure of the mitochondria (shown here in blue) is closely related to their functions in cellular respiration.

THE CENTRIOLES

In each cell, there is usually a pair of **centrioles** near the nucleus (see **fig D**). Each centriole consists of a bundle of nine sets of tubules and is about 0.5 µm long by 0.2 µm wide. The centrioles are involved in cell division. When a cell divides, the centrioles pull apart to produce a **spindle** of microtubules that are involved in the movement of the chromosomes, as you will see later in this chapter.

(a)

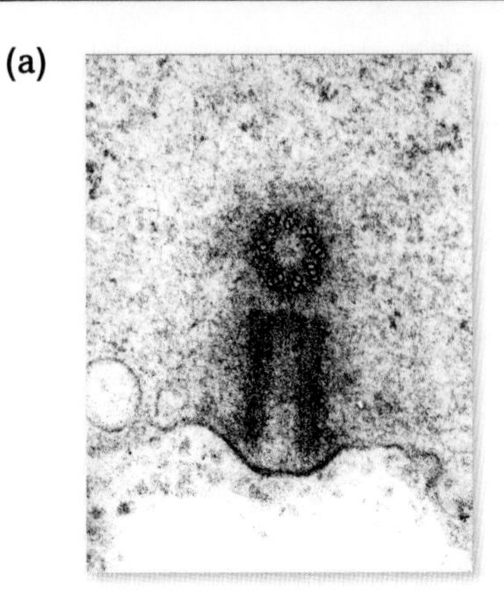

(b)

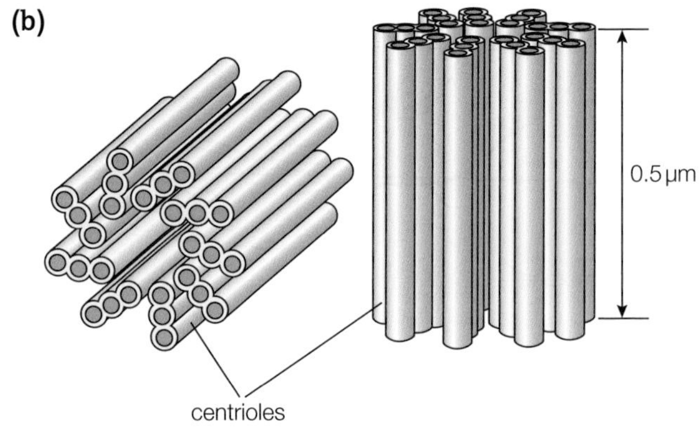

▲ **fig D** **(a)** Transmission electron micrograph of centrioles and **(b)** diagram of centrioles

80S AND 70S RIBOSOMES

In **Section 2B.6** you encountered ribosomes. Protein synthesis occurs on these organelles in the cytoplasm of the cell. Ribosomes are made from ribosomal RNA and protein, and consist of a large subunit and a small subunit. The main type of ribosomes in eukaryotic cells are **80S ribosomes**. The S stands for Svedberg, a unit used to measure how quickly particles fall to the bottom of the tube (settle) in a centrifuge. The rate of settling depends on the size and shape of the particle. When 80S ribosomes are broken into their two units, they are made up of a 40S small subunit and a 60S large subunit. The ratio of RNA : protein in 80S ribosomes is 1 : 1.

Eukaryotic cells also contain another type of ribosome. Scientists have discovered **70S ribosomes** in the mitochondria and in the chloroplasts of plant cells. These ribosomes are usually found in prokaryotic cells (bacteria and cyanobacteria). They consist of a small 30S subunit and a larger 50S subunit and the ratio of RNA : protein in 70S ribosomes is 2 : 1.

The 70S ribosomes are reproduced in the mitochondria and chloroplasts independently when a cell divides. This provides good evidence for the **endosymbiotic theory**.

LYSOSOMES

Food that is taken into the cell of single-celled protists such as *Amoeba* must be broken down into simple chemical substances that can then be used. Organelles in the cells of your body that are worn out need to be destroyed. These jobs are the function of the **lysosomes** (see **fig E**). The word 'lysis' means 'breaking down'.

Lysosomes appear as dark, spherical bodies in the cytoplasm of most cells and they contain a powerful mix of digestive enzymes. They frequently fuse with each other and with a membrane-bound vacuole which contains either food or an obsolete organelle. Their enzymes then break down the contents into molecules that can be reused. A lysosome may fuse with the outer cell membrane to release its enzymes outside the cell as extracellular enzymes, perhaps to destroy bacteria or in digestion.

Lysosomes can also self-destruct. When an entire cell is too old, needs to be removed during development, has a mutation or is under stress, then its lysosomes may rupture. They release their enzymes which then destroy the entire contents of the cell. This **programmed cell death** is known as **apoptosis**.

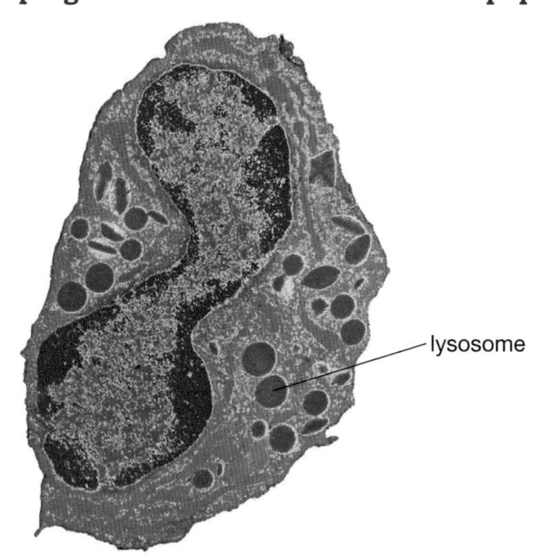

▲ **fig E** Good microscopic evidence of lysosomes, like those seen in this false colour electron micrograph, have helped scientists work out what they do in the cell.

DID YOU KNOW?

APOPTOSIS AND DISEASE

Apoptosis or programmed cell death is vital to the maintenance of a healthy body. Lysosomes rupture and their enzymes are released and kill the cells. This will include cells that are old and coming to the end of their healthy life, or cells that need to be removed during development, for example the webbing between the fingers and toes of a fetus in the uterus. Lysosomes may also destroy cells in which the DNA replication system is not functioning properly. If apoptosis stops working properly – if too many cells are destroyed or not enough lysosomes rupture so that cell death no longer takes place – this can have serious consequences for your health. For example, cancer is often thought of as a disease of uncontrolled cell growth. However, scientists are increasingly convinced that uncontrolled growth is not the whole story. Cancer cells also fail to die by apoptosis. As a result, they pass on the genetic mutations that allow them to reproduce uncontrollably. Excessive apoptosis also causes problems. It leads to the damage seen in the heart after a heart attack, and is linked to the death of T killer cells in HIV/AIDS. This is covered in more detail in **Book 2 Topic 6**. The excessive rupturing of lysosomes may also be involved in autoimmune diseases such as rheumatoid arthritis, when the body attacks and destroys cartilage tissue in the joints, and possibly in other conditions such as osteoporosis and retinitis pigmentosa.

LEARNING TIP

Remember that different types of electron microscopy provide very different types of information.
- The scanning EM can show intact organelles, allowing detailed measurements of the outer dimensions to be taken or it can take 3D images along fracture lines.
- The transmission EM provides clear images of sections of the internal structures of the organelles.
- Together the information helps us produce a detailed image of the ultrastructure of a cell.

CHECKPOINT

1. Explain the importance of organelles in eukaryotic cells.
2. Look at the images that result from transmission and scanning electron microscopes in **Sections 3A.1** and **3A.2** and describe them. Suggest the advantages of each type of image and give examples where each would be most appropriate to use.
3. Why is it important that apoptosis does not occur more or less than it should? Investigate examples of diseases that are caused at least in part by apoptosis.

SUBJECT VOCABULARY

cytoplasm a jelly-like liquid that makes up the bulk of the cell and contains the organelles
nucleus an organelle containing the nucleic acids DNA (the genetic material) and RNA, as well as protein, surrounded by a double nuclear membrane with pores
protoplasm the cytoplasm and nucleus combined
ultrastructure the detailed organisation of the cell, only visible using the electron microscope
intracellular inside the cell
chromatin the granular combination of DNA bonded to protein found in the nucleus when the cell is not actively dividing
nucleolus an extra-dense region of almost pure DNA and protein found in the nucleus; it is involved in the production of ribosomes and control of growth and division
mitochondria rod-like structures with inner and outer membranes that are the site of aerobic respiration
cristae the infoldings of the inner membrane of the mitochondria which provide a large surface area for the reactions of aerobic respiration
eubacteria true bacteria (prokaryotic organisms)
centrioles bundles of tubules found near the nucleus and involved in cell division by the production of a spindle of microtubules that move the chromosomes to the ends of the cell
spindle a set of overlapping protein microtubules running the length of the cell, formed as the centrioles pull apart in mitosis and meiosis
80S ribosomes the main type of ribosome found in eukaryotic cells, consisting of ribosomal RNA and protein, made up of a 60S and 40S subunit; they are the site of protein synthesis
70S ribosomes the ribosomes found in the mitochondria and chloroplasts of eukaryotic cells and in prokaryotic organisms
endosymbiotic theory a theory that suggests mitochondria and chloroplasts originated as independent prokaryotic organisms that began living symbiotically inside other cells as endosymbionts
lysosomes organelles full of digestive enzymes used to break down worn-out cells or organelles or digest food in simple organisms
apoptosis (programmed cell death) the breakdown of worn-out, damaged or diseased cells by the lysosomes

3A 3 EUKARYOTIC CELLS 2: PROTEIN TRANSPORT

SPECIFICATION REFERENCE

3.3(i) 3.3(ii) 3.4 3.6 3.8 CP5

LEARNING OBJECTIVES

- Know the ultrastructure of eukaryotic cells, including the rough and smooth endoplasmic reticulum and the Golgi apparatus.
- Understand the function of these organelles.
- Recognise these organelles in electron micrographs.

The cytoplasm of the cell contains the **endoplasmic reticulum (ER)**, a three-dimensional (3D) network of cavities surrounded by membranes. The electron microscope shows that some of the cavities are sac-like and some are tubular, and that the ER spreads through the cytoplasm. The ER network links with the membrane around the nucleus, and is a large part of the transport system within a cell as well as being the site of synthesis of many important chemical substances. It has been calculated that 1 cm^3 of liver tissue contains about 11 m^2 of endoplasmic reticulum. Electron microscopes also helped scientists to work out the functions of the endoplasmic reticulum, by showing up the different forms – the rough and the smooth endoplasmic reticulum.

Another useful technique is to provide cells with radioactively labelled chemical substances that are building blocks for specific modules and then find out where they appear in the cell. An example of this is using labelled amino acids to observe the synthesis of proteins. The labelled products can be tracked using microscopy. Another method of locating them is to break the cells open and then spin the contents in a centrifuge. The different parts of the cell can be separated out and the regions containing the radioactively labelled substances can be identified.

ROUGH AND SMOOTH ENDOPLASMIC RETICULUM

Electron micrographs show that much of the outside of the endoplasmic reticulum membrane is covered with granules, which are 80S ribosomes, so this is known as **rough endoplasmic reticulum (RER)** (see **fig A**). The function of the ribosomes is to make proteins. The RER has a large surface area for the synthesis of all these proteins. It then stores and transports the proteins within the cell after they have been made. Cells that secrete materials have a large amount of RER. Examples include cells producing hormones or the digestive enzymes in the lining of the gut. These proteins must be secreted without interfering with the cell's own activities. This is an example of **exocytosis**.

Not all endoplasmic reticulum is covered in ribosomes (see **fig A**). **Smooth endoplasmic reticulum (SER)** is also involved in synthesis and transport, but in this case of steroids and lipids. For example, lots of SER is found in the testes, which make the steroid hormone testosterone. SER is also plentiful in the liver, which metabolises cholesterol and other lipids. The amount and type of endoplasmic reticulum in a cell give an idea of the type of job the cell does. You can see how the two types of endoplasmic reticulum look under an electron microscope by referring back to **Section 3A.2 figs A** and **B**.

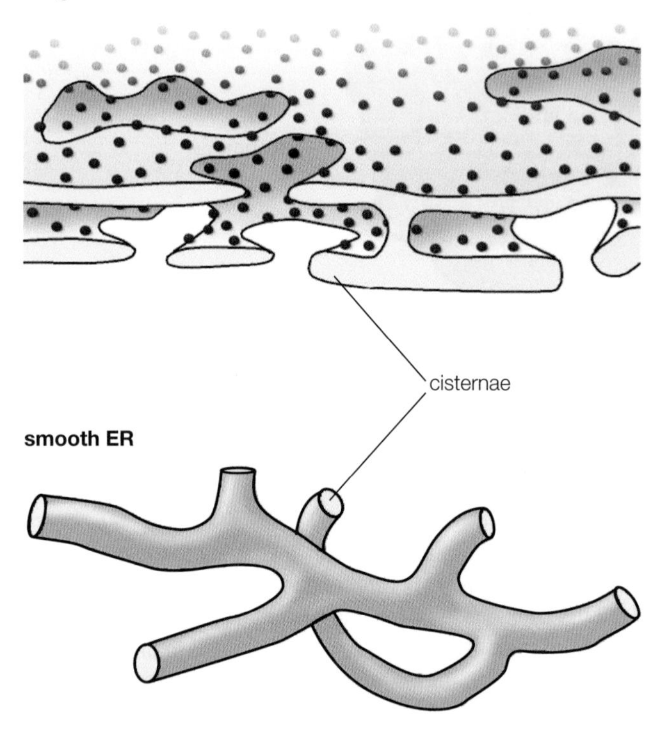

▲ **fig A** This is rough and smooth endoplasmic reticulum. Smooth ER is more tubular than rough ER and also lacks ribosomes on the surface.

THE GOLGI APPARATUS

Under the light microscope the **Golgi apparatus** looks like a rather dense area of cytoplasm. An electron microscope reveals that it is made up of stacks of parallel, flattened membrane pockets formed by vesicles from the endoplasmic reticulum fusing together.

The Golgi apparatus has a close link with, but is not joined to, the RER. It has taken scientists a long time to discover exactly what the Golgi apparatus does. Materials have been radioactively labelled and tracked through the cell to try and find out exactly what happens inside it. Proteins are brought to the Golgi apparatus in vesicles that are separate from the RER where they were made.

The vesicles fuse with the membrane sacs of the Golgi apparatus and the protein enters the Golgi stacks. The proteins are modified as they travel through the Golgi apparatus.

Carbohydrate is added to some proteins to form glycoproteins such as mucus. The Golgi apparatus also seems to be involved in producing materials for plant and fungal cell walls and insect cuticles. Some proteins in the Golgi apparatus are digestive enzymes. These may be enclosed in vesicles to form the organelles known as lysosomes. Alternatively, enzymes may be transported through the Golgi apparatus and then in vesicles to the cell surface membrane. Here the vesicles fuse with the membrane to release extracellular digestive enzymes. The process is summarised in **fig B**.

EXAM HINT

Always describe membranes as *fusing* together. This means the vesicles become part of the larger structure.

The Golgi apparatus was first reported over 100 years ago, in April 1898. An Italian scientist called Camillo Golgi (1843–1926) observed the flattened stack of membranes through a light microscope. For more than 50 years, scientists argued about what its function might be. Some thought it was an artefact from the process of fixing and staining during tissue preparation. The detailed structure of the Golgi apparatus could be seen for the first time when the electron microscope was introduced in the 1950s.

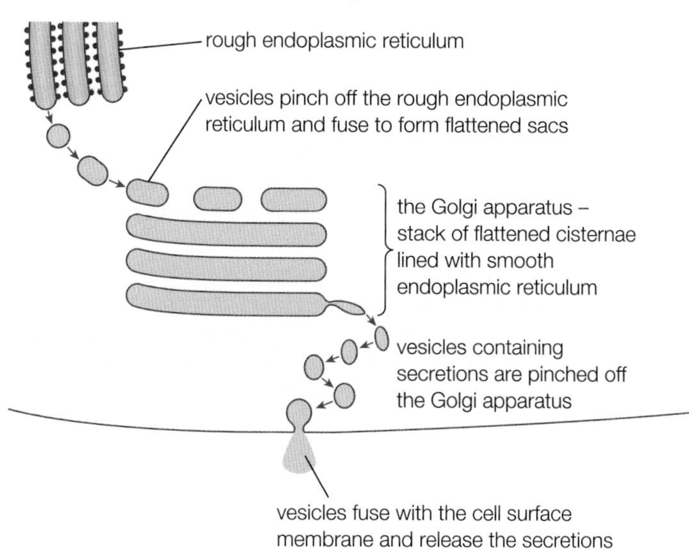

▲ **fig B** The Golgi apparatus takes proteins from the RER, assembles and packages them and then transports them to where they are needed. This may be the surface of the cell or different regions inside it.

The electron microscope has been very useful in showing details of the internal structure of the Golgi apparatus. In addition, several techniques have been developed that allow a more detailed understanding. The most important of these is the process of labelling specific enzymes so we can see them using the electron microscope. The inner areas of the Golgi apparatus, nearer to the RER, are very rich in enzymes that modify proteins in different ways. This is where most enzymes or membrane proteins are converted into the final product. In contrast, in the outer regions of the Golgi apparatus you find lots of finished protein products, but not many of the enzymes that make them. The movement of cell membrane proteins through the Golgi apparatus is very complex. The Golgi apparatus aligns the areas of the protein that need to be on the outside of the cell membrane, such as receptor binding sites. Consequently, when they arrive at the membrane they are inserted facing in the correct direction.

EXAM HINT

The secretion of proteins from the cell is a complex process. When you describe it, keep the sequence correct and use numbered points so that the sequence is clear.

CHECKPOINT

SKILLS EXECUTIVE FUNCTION

1. What type of questions would scientists ask when they set out to investigate the functions of the endoplasmic reticulum, and how might they set about finding the answers?
2. Describe the role of the RER and the Golgi apparatus in the production of both intracellular and extracellular enzymes, and explain the importance of packaging products within a cell.

SUBJECT VOCABULARY

endoplasmic reticulum (ER) a 3D network of membrane-bound cavities in the cytoplasm that links to the nuclear membrane and makes up a large part of the cellular transport system as well as playing an important role in the synthesis of many different chemical substances
rough endoplasmic reticulum (RER) endoplasmic reticulum that is covered in 80S ribosomes and which is involved in the production and transport of proteins
exocytosis the movement of large molecules out of cells by the fusing of a vesicle containing the molecules with the surface cell membrane; the process requires ATP
smooth endoplasmic reticulum (SER) a smooth tubular structure similar to RER, but without the ribosomes, which is involved in the synthesis and transport of steroids and lipids in the cell
Golgi apparatus stacks of membranes that modify proteins made elsewhere in the cell and package them into vesicles for transport, and also produce materials for plant cell walls and insect cuticles

3A 4 PROKARYOTIC CELLS

SPECIFICATION REFERENCE 3.5(ii) 3.5(iii) 3.8 CP5

LEARNING OBJECTIVES

- Know the ultrastructure of prokaryotic cells, including the cell wall, capsule, plasmid, flagellum, pili, ribosomes and circular DNA.
- Understand the function of these structures in prokaryotic cells.

Bacteria, cyanobacteria and archaebacteria are prokaryotic organisms. Bacteria alone are probably the most frequently found form of life on Earth. Some bacteria are pathogens and cause disease, but the great majority do no harm and many are beneficial to living organisms. Examples include gut bacteria and in the cycling of nutrients in the natural world. In this section, you will mainly consider the structure and functions of bacterial cells.

THE STRUCTURE OF BACTERIA

All bacterial cells have certain features in common, although these vary greatly between species (see **fig A**).

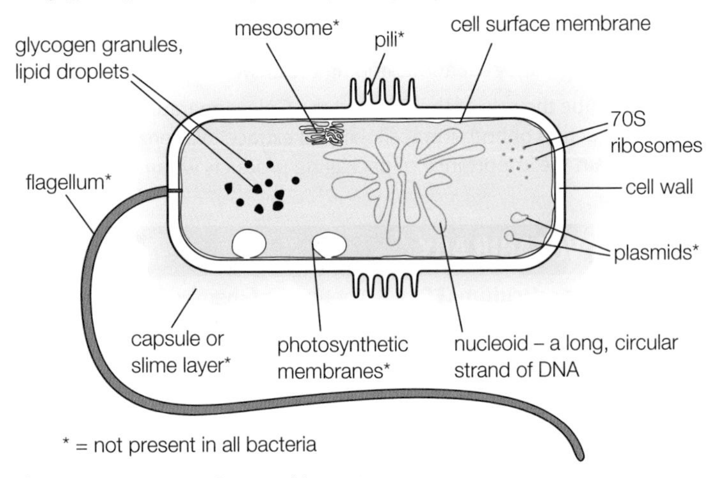

▲ **fig A** Structure of a typical bacterium

BACTERIAL CELL WALLS

All bacterial cells have a cell wall. The contents of bacterial cells are usually hypertonic to the medium around them, so water tends to move into the cells by osmosis. The cell wall prevents the cell swelling and bursting. It also maintains the shape of the bacterium, and gives support and protection to the contents of the cell. All bacterial cell walls have a layer of **peptidoglycan** that consists of many parallel polysaccharide chains with short peptide cross-linkages producing an enormous molecule with a net-like structure. Some bacteria have a **capsule** (or slime layer if it is very thin and diffuse) around their cell walls. This may be produced from starch, gelatin, protein or glycolipid, and protects the bacterium from phagocytosis by white blood cells. It also covers the cell markers on the cell membrane that identify the cell. So, a capsule can make it easier for a bacterium to be pathogenic (cause disease) because it is not so easily identified by the immune system. This is the case for the bacteria that cause pneumonia, meningitis, tuberculosis (TB) and septicaemia. However, many capsulated bacteria do not cause disease. It seems likely that capsules evolved to help the bacteria survive very dry conditions.

PILI AND FLAGELLA

Some bacteria have from one to several hundred thread-like protein projections from their surface. These are called the **pili** (singular pilus) and they are found on some well-known bacteria such as *Escherichia coli* (*E. coli*) species and *Salmonella* species. They seem to be used for attachment to a host cell and for sexual reproduction. However, they also make bacteria more vulnerable to virus infections, as a **bacteriophage** can use pili as an entry point to the cell.

Some bacteria can move themselves using **flagella** (singular flagellum). These are made of a many-stranded helix of the protein flagellin. The flagellum moves the bacterium by rapid rotations – about 100 revolutions per second.

CELL SURFACE MEMBRANE

The cell surface membrane in prokaryotes is similar in both structure and function to the membranes of eukaryotic cells. However, bacteria have no mitochondria, so the cell membrane is also the site of some of the respiratory enzymes. In some bacterial cells such as *Bacillus subtilis*, a common soil bacterium, the membrane shows infoldings known as **mesosomes**. There is still some debate about their function. Some scientists think they may be an artefact from the process of preparing the cell for an electron micrograph, others believe they are associated with enzyme activity, particularly during the separation of DNA and the formation of new cell walls when the bacteria divide. It appears that other infoldings of the bacterial cell surface membrane may be used for photosynthesis by some bacterial species.

NUCLEOID

The genetic material of prokaryotic cells consists of a single circular strand of DNA, which is *not* contained in a membrane-bound nucleus. This is an important, identifying difference between prokaryotic and eukaryotic cells. However, the DNA is folded and coiled to fit into the bacterium. The area in the bacterial cell where this DNA tangle is found is known as the **nucleoid** (see **fig B**). In an *E.coli* bacterium, it occupies about half of the cytoplasm.

LEARNING TIP

Do not confuse the nucleoid with a nucleus or nucleolus. Make sure your spellings are correct.

PLASMIDS

Some bacterial cells also contain one or more much smaller circles of DNA known as **plasmids**. A plasmid codes for a particular aspect of the bacterial phenotype in addition to the genetic information in the nucleoid. For example, plasmids can code for the production of a particular toxin or resistance to a particular antibiotic. Plasmids can reproduce themselves independently of the nucleoid. They can be transferred from one bacterium to another in a form of sexual reproduction using the pili.

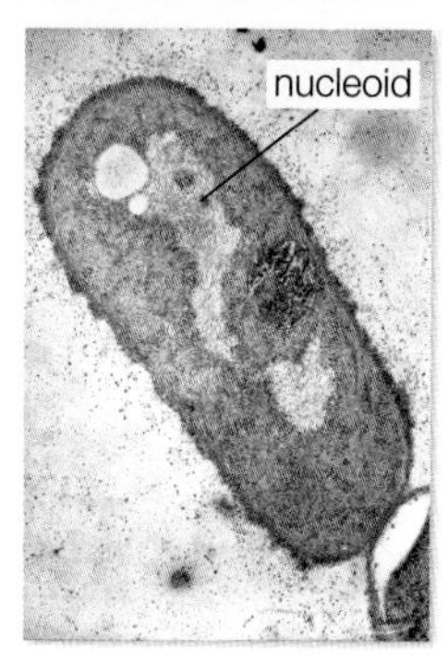

▲ **fig B** The nucleoid area of a bacterium

70S RIBOSOMES

The bacteria, cyanobacteria and archaebacteria have no membrane-bound organelles, but they do have ribosomes where protein synthesis occurs. The ribosomes in bacterial cells are 70S, smaller than the 80S ribosomes which dominate in eukaryotes. They have two subunits. The smaller is 30S and the larger is 50S (see **Section 3A.3**). They are involved in the synthesis of proteins in a similar way to eukaryotic ribosomes.

GRAM STAINING AND BACTERIAL CELL WALLS

All bacterial cell walls contain peptidoglycan but there are two types which can be distinguished by **Gram staining**. This staining technique was developed in 1884 by Christian Gram (1853–1938) and is still used today. It is valuable because different types of disease-causing bacteria are vulnerable to different types of antibiotic and the type of cell wall they have is one of the factors that affects how vulnerable they are.

Before staining, bacteria are often colourless. The cell walls of **Gram-positive bacteria** (e.g. methicillin-resistant *Staphylococcus aureus*, MRSA) have a thick layer of peptidoglycan containing chemical substances such as **teichoic acid** within the net-like structure. The crystal violet/iodine complex in the Gram stain is trapped in the thick peptidoglycan layer and resists decolouring when the bacteria are dehydrated using alcohol. As a result, the bacteria do not pick up the red safranin counterstain and appear purple/blue when viewed in a light microscope (see **fig C**).

The cell walls of **Gram-negative bacteria** have a thin layer of peptidoglycan with no teichoic acid between the two layers of membranes. The outer membrane is made up of lipopolysaccharides. This layer dissolves when the bacteria are dehydrated in ethanol. This exposes the thin peptidoglycan layer and the crystal violet/iodine complex is washed out. The peptidoglycan then takes up the red safranin counterstain. The cells appear red when viewed in a light microscope (see **fig C**).

LEARNING TIP

Remember that ribosomes are the only organelles shared with the eukaryotes. Prokaryotes have no membrane-bound organelles.

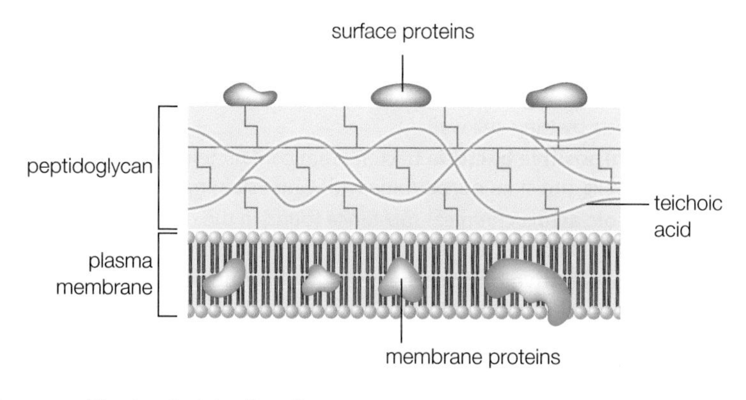

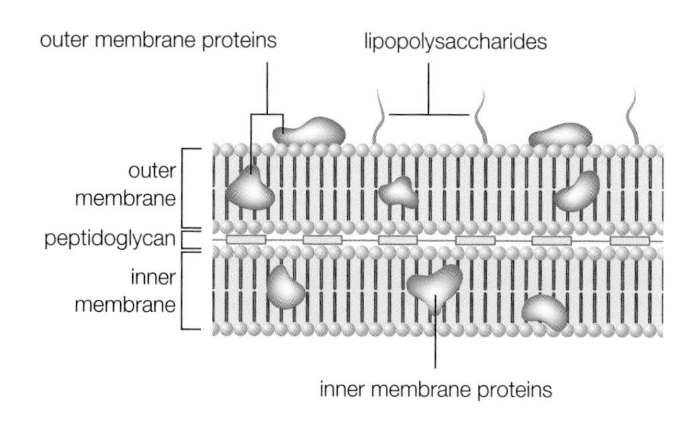

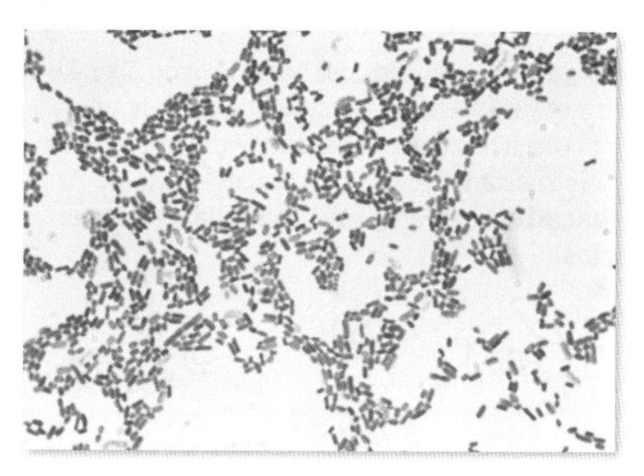

▲ **fig C** The difference in the cell wall structure of the bacteria results in the different reactions with the Gram stain.

DID YOU KNOW?

ANTIBIOTICS AND BACTERIAL CELL WALLS

Antibiotics are drugs that are used against bacterial pathogens. There are a number of different types of antibiotic, each working in a different way. They may work by affecting the bacterial cell walls, the cell membranes, the genetic material, the enzymes or the ribosomes. Antibiotics usually target features that bacterial cells have but eukaryotic cells do not, including bacterial cell walls and 70S ribosomes.

Different types of bacteria are sensitive to different types of antibiotic. Doctors need to know if a pathogenic bacterium is Gram-positive or Gram-negative as this will affect which antibiotic is chosen to treat the disease.

To pinpoint the actions of an antibiotic, first think about the difference between human cells and bacterial cells, and then about the differences between Gram-positive and Gram-negative bacteria.

Antibiotics such as beta-lactam antibiotics (penicillins and cephalosporins) inhibit the formation of the peptidoglycan layer of the bacterial cell wall. As a result, they are very effective against Gram-positive bacteria, because these have a thick peptidoglycan layer on the surface of the cell. But they are less effective against Gram-negative bacteria, because their peptidoglycan layer is hidden and less vital to the wall structure. These antibiotics don't affect human cells because they don't have a peptidoglycan cell wall.

Glycopeptide antibiotics such as vancomycin are large polar molecules that cannot penetrate the outer membrane layer of Gram-negative bacteria. However, they are very effective against Gram-positive bacteria, even ones that have developed resistance to many other antibiotics.

Polypeptide antibiotics such as polymixins are rarely used because they can have serious side-effects. They are very effective against Gram-negative bacteria because they interact with the phospholipids of the outer membrane. They do not affect Gram-positive bacteria.

Most other antibiotics affect both Gram-positive and Gram-negative bacteria because they target common processes such as protein synthesis by the ribosomes. They only target prokaryote ribosomes, not eukaryotic ones.

You will learn more about antibiotics in **Book 2**.

ALTERNATIVE WAYS OF CLASSIFYING BACTERIA

Grouping bacteria by the way their cell walls do or do not take up Gram stains is not very useful in classifying the different types. Another way in which bacteria can be identified is by their shape (see **fig D**). Some bacteria are spherical (**cocci**), some are rod-shaped (**bacilli**). Others are twisted (**spirilla**) or comma-shaped (**vibrios**).

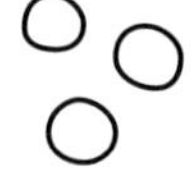

▲ **fig D** The shapes of different types of bacteria can be seen clearly under powerful microscopes.

Bacteria are also sometimes grouped by their respiratory requirements. **Obligate aerobes** need oxygen for respiration. **Facultative anaerobes** use oxygen if it is available, but can manage without it. Many human pathogens are in this group. **Obligate anaerobes** can only respire in the absence of oxygen – in fact, oxygen will kill them.

CHECKPOINT

1. Make a table to compare and contrast prokaryotic and eukaryotic cells.

2. What is the difference in the structure of the walls of Gram-positive and Gram-negative bacteria?

SUBJECT VOCABULARY

peptidoglycan a large, net-like molecule found in all bacterial cell walls made up of many parallel polysaccharide chains with short peptide cross-linkages
capsule a layer formed from starch, gelatin, protein or glycolipid, found around the outside of some bacteria
pili thread-like protein projections found on the surface of some bacteria
bacteriophage virus that attacks bacteria
flagella many-stranded helices of the contractile protein flagellin found on some bacteria; they move the bacteria by rapid rotations
mesosomes infoldings of the cell membrane of bacteria
nucleoid the area in a bacterium containing the single circular loop of coiled DNA
plasmids small, circular pieces of DNA that code for specific aspects of the bacterial phenotype
Gram staining a staining technique used to distinguish types of bacteria by their cell wall
Gram-positive bacteria bacteria that contain teichoic acid in their cell walls and stain purple/blue with Gram staining
teichoic acid a chemical substance found in the cell walls of Gram-positive bacteria
Gram-negative bacteria bacteria that have no teichoic acid in their cell walls; they stain red with Gram staining
cocci spherical bacteria
bacilli rod-shaped bacteria
spirilla bacteria with a twisted or spiral shape
vibrios comma-shaped bacteria
obligate aerobes organisms that need oxygen for respiration
facultative anaerobes organisms that use oxygen if it is available, but can respire and survive without it
obligate anaerobes organisms that can only respire in the absence of oxygen and are killed by oxygen

5 THE ORGANISATION OF CELLS

LEARNING OBJECTIVES

- Understand how the cells of multicellular organisms are organised into tissues, tissues into organs and organs into organ systems.

Multicellular organisms consist of specialised cells but these cells do not operate on their own. The specialised cells are organised into groups of cells known as **tissues**. These tissues consist of one or more types of cell all carrying out a function in the body. However, tissues do not operate in isolation. Many tissues are further organised into **organs**.

TISSUES

Tissues are groups of cells that all develop from the same kind of cell. Although there are many different specialised cells, there are only four main tissue types in the human body – epithelial tissue, connective tissue, muscle tissue and nervous tissue. Modified versions of these tissue types containing different specialised cells perform all the functions of the body. **Fig A** shows some different **epithelial tissues**, which are tissues that line the surfaces both inside and outside of the body. Although some epithelial tissues consist of more than one kind of cell, they all originate from the basement membrane. Cells in epithelial tissues usually sit tightly together and form a smooth surface that protects the cells and tissues below.

Squamous epithelium is frequently found lining the surfaces of blood vessels, and forms the walls of capillaries and the lining of the alveoli. Cuboidal and columnar cells line many other tubes in the body. Ciliated epithelia often contain goblet cells that produce mucus. These epithelia form the surfaces of tubes in the gas exchange system and the oviducts. The regular waving of the cilia from side to side moves materials along inside the tubes. Compound epithelia are found where the surface is continually scratched and abraded, such as the skin. The thickness of the tissue protects what lies beneath as new cells continue to grow from the basement membrane.

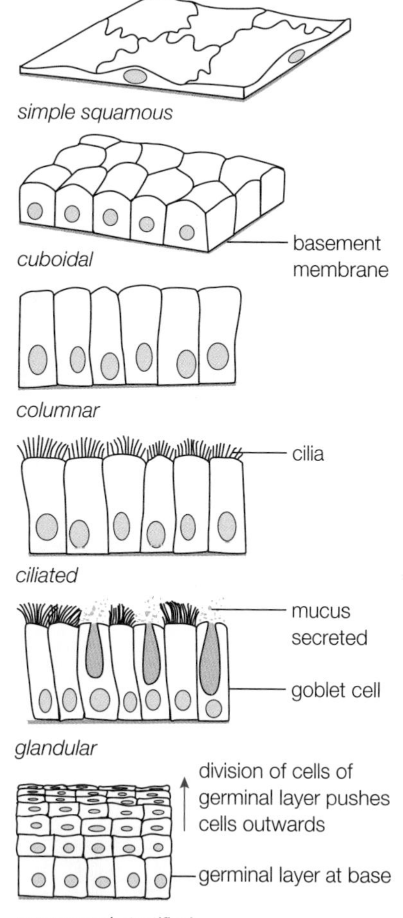

fig A There are many different kinds of epithelial tissue inside the human body.

EXAM HINT

Remember that each tissue has a particular function and that the cells will be specialised for that function. The ultrastructure is also part of that specialisation.

There are many other tissues in the body, including muscle tissue, nervous tissue, the collagen tissue and elastin tissue found in artery walls and the glandular tissue that secretes substances from inside the cells. Connective tissue is the main supporting tissue in the body, and includes bone tissue and cartilage tissue as well as packing tissue that supports and protects some of the organs. Some of these tissues are shown in **fig B**.

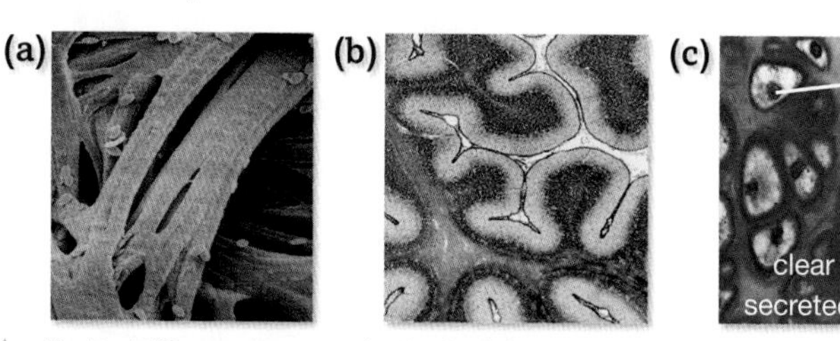

fig B Different tissues in the body: **(a)** cardiac muscle tissue; **(b)** brain tissue; **(c)** cartilage tissue

ORGANS

An organ is a structure made of several different tissues grouped into a structure so that they can work effectively together to carry out a particular function. There are many organs in the human body, some of which are shown in **fig C**.

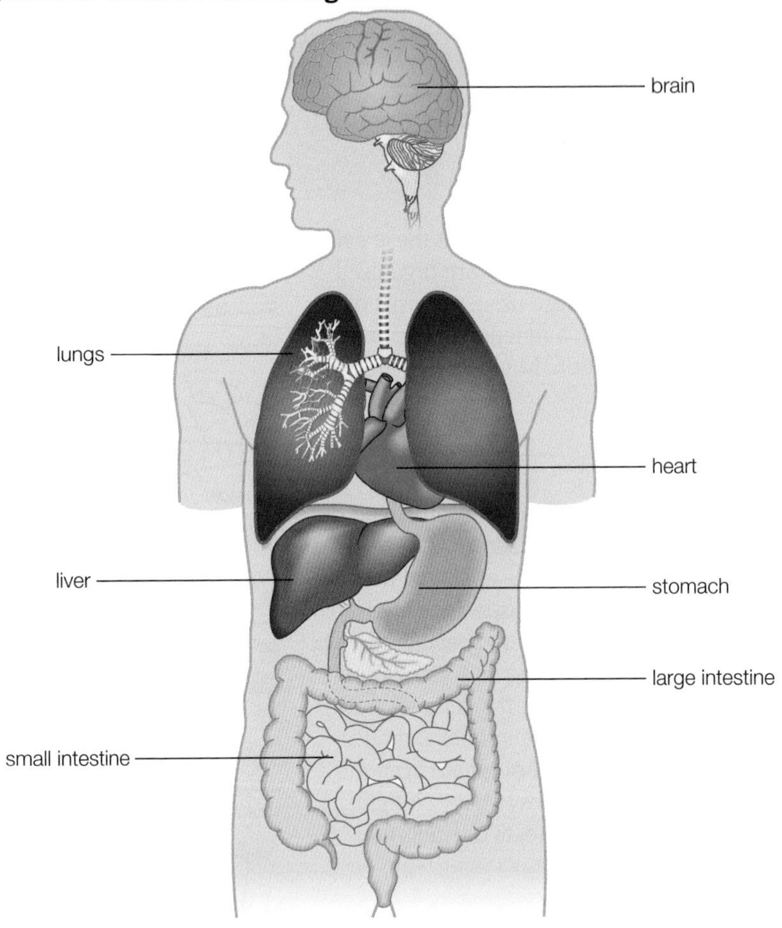

▲ **fig C** Some of the organs and organ systems of the human body.

Plants also have cells grouped into tissues and organs. For example, the leaf is an organ that is composed of vascular tissue, epithelial tissue and mesophyll tissue as shown in **fig D**.

LEARNING TIP

Other plant organs include the stem and the roots. You need to be able to recognise the pattern of tissues in transverse section.

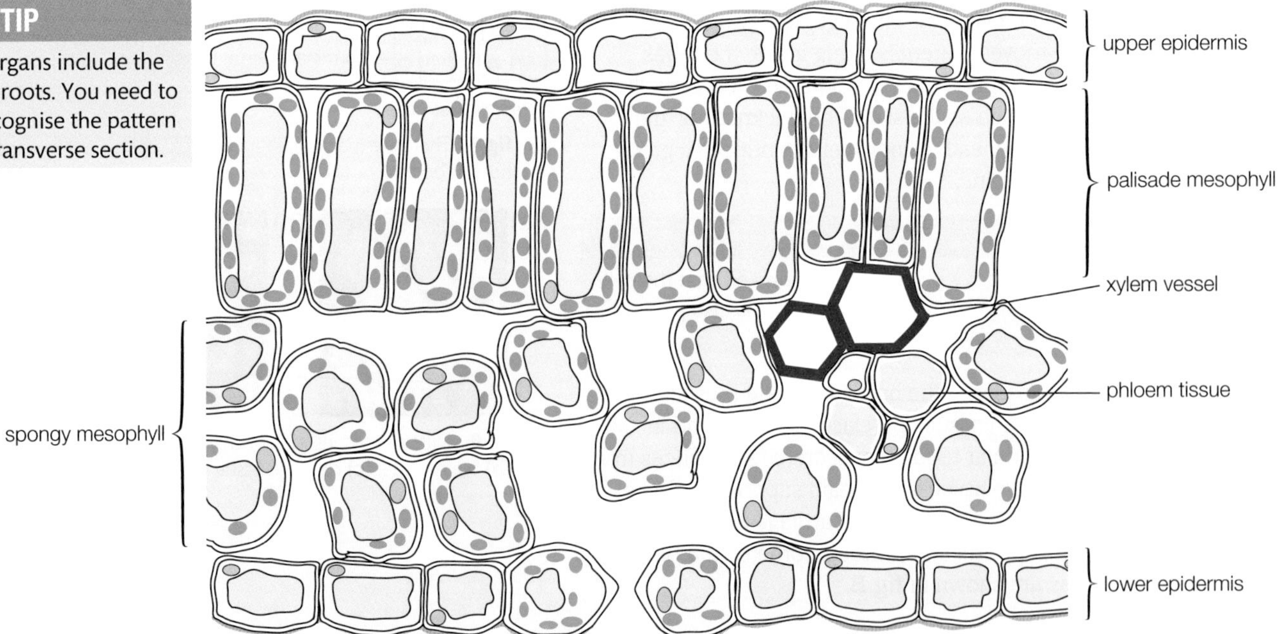

▲ **fig D** Some of the tissues found in a leaf – the photosynthetic, food-making organ of a plant.

ORGAN SYSTEMS

In animals, often several organs work together as an **organ system** to perform large-scale functions in the body. For example, the digestive system includes the stomach, pancreas, and small and large intestines. The nervous system includes the brain, spinal cord and all peripheral nerves.

Most of the cells in tissues, organs and systems have differentiated (changed to become specialised) during their development so that they can perform their specific function. You will find out more about how this process happens in **Chapter 3C**.

CHECKPOINT

SKILLS INTERPRETATION

1. Explain how the structure of the following tissues is related to their function.
   (a) squamous epithelium lining an alveolus
   (b) ciliated epithelium lining a bronchus
   (c) muscle tissue in the biceps muscle
2. (a) Choose one of the systems in the human body and describe briefly the cells, tissues and organs found within that system.
   (b) Explain why this grouping enables the system to carry out its function effectively.

SUBJECT VOCABULARY

tissues groups of specialised cells carrying out particular functions in the body
organs structures made up of several different types of tissue to carry out particular functions in the body
epithelial tissues tissues that form the lining of surfaces inside and outside the body
organ system a group of organs working together to carry out particular functions in the body

3A EXAM PRACTICE

1 The photograph below shows a mitochondrion seen through an electron microscope.

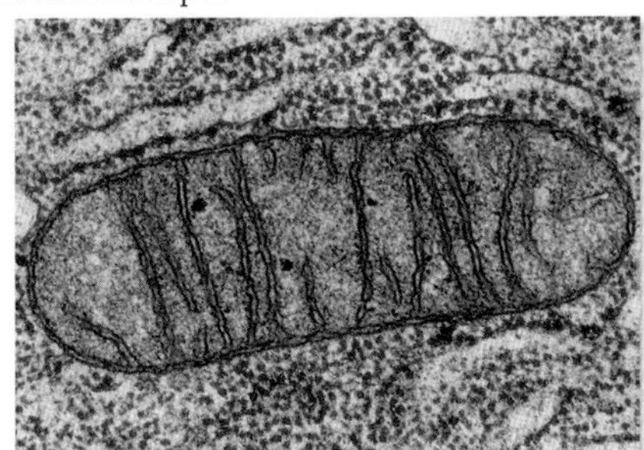

(a) What is the function of the folded inner membrane of the mitochondrion?

A to provide a larger surface area for oxygen absorption
B to divide the mitochondrion into compartments
C to provide a large surface area for chemical reactions to take place
D the folds are artefacts caused by preparation techniques [1]

(b) Which tissue would contain more mitochondria?

A muscle
B squamous epithelium
C nervous tissue
D cartilage [1]

(c) Describe the function of mitochondria in a cell. [2]

(d) Make an accurate drawing of this mitochondrion enlarged ×2. On your drawing, label the matrix and a crista. [4]

(Total for Question 1 = 8 marks)

2 The photograph shows part of a nucleus.

(a) (i) What type of microscope was used to take this image?

A transmission electron microscope
B light microscope
C scanning electron microscope
D confocal microscope [1]

(ii) Justify your answer to part (i). [2]

(b) (i) State the function of the nucleus in an animal cell. [1]

(ii) State the function of the pores that can be seen in the photograph. [2]

(c) (i) What is the equation for the surface area of a sphere? [1]

A $4/3\pi r^3$
B πr^2
C $4\pi r^2$
D $2\pi r$

(ii) The nucleus shown has a diameter of 10 μm. Calculate the surface area of the nucleus. [2]

(iii) The part of the nucleus shown in the photomicrograph is 3.5 μm along one side and there are approximately 230 nuclear pores visible. Calculate the total number of pores in this nucleus. [3]

(Total for Question 2 = 12 marks)

3 (a) Draw and label a diagram to show the structure of a nucleus, as seen using a transmission electron microscope. [4]

(b) The photograph below shows a group of mitochondria in a liver cell, as seen using an electron microscope. The magnification is ×10 000.

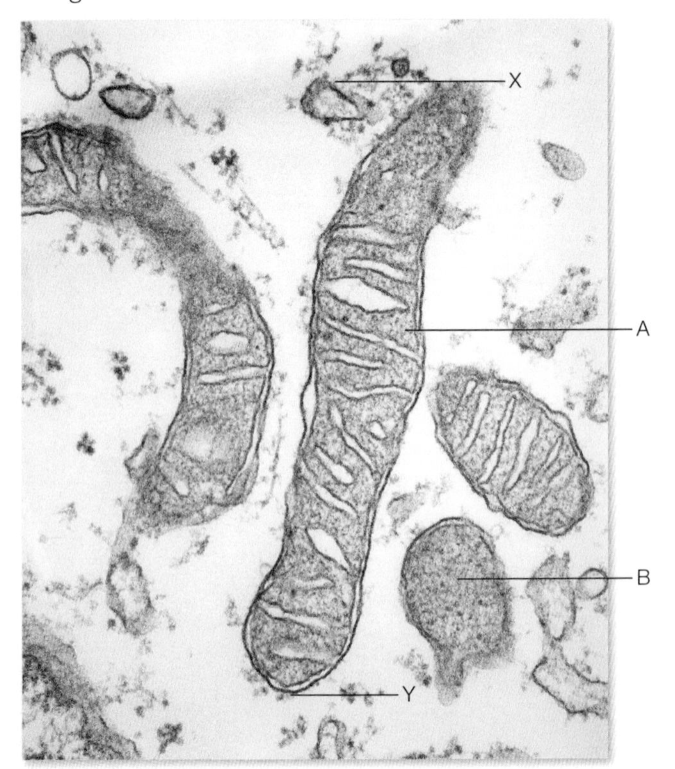

(i) Measure the length of the mitochondrion labelled A between X and Y. Calculate the actual length of this mitochondrion in μm. Show your working. [3]

(ii) Suggest one other structure that you might see in the cytoplasm of this liver cell if the magnification used were higher. [1]

(iii) Suggest one reason why the double membrane is not clearly visible all around the mitochondrion labelled A. [1]

(iv) Name structure B and give a reason for your choice. [2]

(Total for Question 3 = 11 marks)

4 (a) The table below refers to some cell structures. Complete the table by inserting the correct word, words, or diagram in the appropriate boxes. Leave the shaded grey boxes empty.

Name of cell structure	Description of cell structure	Sketch of cell structure
	darkly stained region in the nucleus where ribosomal RNA is made	
centrioles		
lysosome		
	hollow cylinders made of protein form spindle fibres	

[4]

(b) The photograph below shows some animal cells as seen using the high power of a light microscope.

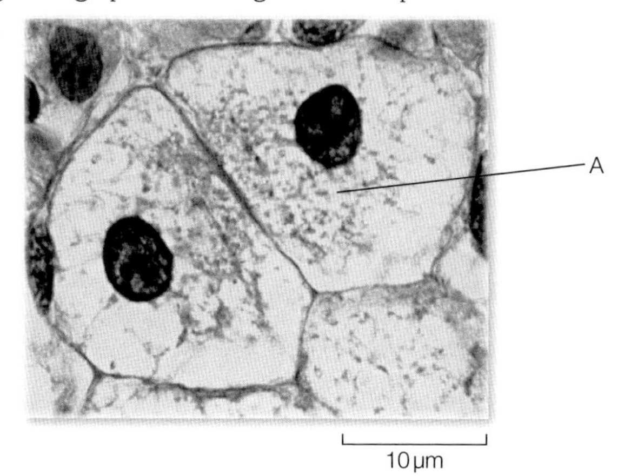

(i) Make an accurate labelled drawing of the cell labelled **A**. [4]

(ii) State the magnification of your diagram. [2]

(Total for Question 4 = 10 marks)

5 Scientists using light microscopes were unable to distinguish small organelles like ribosomes.

(a) Name **three** other organelles that were too small for these scientists to see. [3]

(b) Name **two** organelles that would have been seen using the light microscope. [2]

(c) Light microscopes can be used to watch single-celled organisms move in real time. Suggest why this is impossible in an electron microscope. [2]

(Total for Question 5 = 7 marks)

6 (a) The table below refers to some features of prokaryotic and eukaryotic cells. If the feature is present, place a tick (✓) in the appropriate box and if the feature is absent, place a cross (✗) in the appropriate box.

Feature	Prokaryotic cell	Eukaryotic cell
nuclear envelope		
cell surface membrane		
mitochondrion		
Golgi apparatus		
ribosomes		

[5]

The endosymbiotic theory suggests that mitochondria and chloroplasts originated from bacteria infecting a larger cell.

(b) Both chloroplasts and mitochondria have double membranes. When a white blood cell engulfs bacteria it surrounds them with a vacuole. Suggest how this adds support to the endosymbiotic theory. [1]

(c) List **three** things that chloroplasts and mitochondria have in common with prokaryotic organisms. [3]

(Total for Question 6 = 9 marks)

7 (a) What are the correct definitions of a tissue and an organ? [1]

	Tissue	Organ
A	a group of cells	a group of tissues
B	a group of organs that perform the same function	a group of cells working together
C	a group of cells that perform the same function	a group of tissues working together to perform a function
D	a group of organs	a group of tissues

(b) Explain why prokaryotic organisms do not have organs. [1]

(c) What are the advantages of being multicellular? [3]

(d) Draw a diagram of glandular ciliated epithelium and annotate your diagram to show how the cells are adapted to their role. [5]

(Total for Question 7 = 10 marks)

TOPIC 3 CELL STRUCTURE, REPRODUCTION AND DEVELOPMENT

CHAPTER 3B MITOSIS, MEIOSIS AND REPRODUCTION

Normal growth and division of cells occur in a cycle that allows cells to grow and replicate their DNA and all their organelles before dividing. In this chapter, you will study mitotic cell division in eukaryotic cells. The cell cycle is of great importance and you will discover the main stages of the cycle, and how the speed of the cycle varies at different ages and in different tissues.

You will learn the stages of cell division and come to understand how mitosis results in two genetically identical daughter cells: the chromosomes replicate and separate in a graceful 'dance', followed by the rest of the cytoplasm of the cell. You will consider the importance of mitosis in living organisms – it is the basis of asexual reproduction in many animals, plants and fungi. Mitosis produces offspring that are genetically identical to the single parent, and it can result in enormous numbers of offspring being produced at one time. Mitosis is also important for the repair of damaged tissues and for normal growth from infancy to adulthood.

When we consider cell division, we always talk about eukaryotic cells having two sets of chromosomes, one from each parent. In this chapter, you will look at meiotic cell division in eukaryotic cells. Meiosis is the process in which the chromosome number is halved to produce sex cells or gametes. Meiosis only takes place in the sex organs. You will learn the process of meiotic cell division and its importance in introducing genetic variation. Genetic variation is essential for survival in changing conditions, for natural selection and for evolution.

You are also going to look how the sex cells or gametes are specialised to perform their function and then go on to explore how those gametes join at fertilisation to form a new genetic individual.

MATHS SKILLS FOR THIS CHAPTER

- **Carry out calculations using numbers in standard and ordinary form** (*e.g. use of magnification*)
- **Use scales for measuring** (*e.g. measuring sizes of chromosomes during cell division*)
- **Find arithmetic means** (*e.g. measuring sizes of cells at different stages of the cell cycle*)
- **Make order of magnitude calculations** (*e.g. use and manipulate the magnification formula below*)

$$\text{magnification} = \frac{\text{size of image}}{\text{size of real object}}$$

- **Use and manipulate equations, including changing the subject of an equation** (*e.g. magnification*)
- **Use ratios, fractions and percentages** (*e.g. calculate the mitotic index*)

What prior knowledge do I need?

- Asexual reproduction in living things
- The role of meiotic cell division in halving the chromosome number to form gametes

Chapter 2B

- The way in which DNA replicates in the nucleus

Chapter 2C

- Mutations as a source of new variations

Chapter 3A

- The ultrastructure of eukaryotic cells including the nucleus, nuclear membrane, chromosomes, centriole, etc.

What will I study in this chapter?

- The cell cycle as a regulated process made up of interphase, mitosis and cytokinesis, in which cells divide to produce two identical daughter cells
- The replication and separation of the genetic material in the main stages of mitosis
- The importance of mitosis in growth, repair of damaged or ageing tissues and asexual reproduction, to produce offspring that are identical to the one parent
- The role of meiosis in the production of haploid gametes, including the stages of meiosis
- The replication and separation of the genetic material in the main stages of meiosis
- The ways in which meiosis results in genetic variation through recombination of alleles, including independent assortment and crossing over
- The development of the female and male gametes in mammals and in plants with the formation of the pollen grain and the embryo sac
- The adaptations of the gametes
- Fertilisation in mammals and double fertilisation in plants including the roles of the tube nucleus and the generative nucleus
- How random fertilisation during sexual reproduction brings about genetic variation

What will I study later?

Chapter 3C

- Cell differentiation
- The control of cell differentiation
- How post-transcriptional changes in mRNA can result in different products from the same gene
- Autosomal and sex linkage of alleles
- Stem cells in animals and plants
- Epigenetic modifications and their effect on totipotent stem cells in the embryo
- The early development of a mammalian embryo to the blastocyst stage

Chapter 6B (Book 2: IAL)

- The immune response of the body including clonal selection and expansion by rapid mitosis in the production of plasma cells and T killer cells as well as T and B memory cells

Chapter 8B (Book 2: IAL)

- Plant responses to environmental stimuli that depend on cell division and growth

3B 1 THE CELL CYCLE

SPECIFICATION REFERENCE 3.14

LEARNING OBJECTIVES

- Understand the role of the cell cycle in producing genetically identical daughter cells for growth and asexual reproduction.

One of the most awe-inspiring processes of life is the way in which organisms reproduce. Like creates like – bougainvillea produce new bougainvillea, single-celled organisms such as *Amoeba* produce more *Amoeba*, and liver cells generate more liver cells. Most new biological material results from the process of nuclear division known as **mitosis**, followed by the rest of the cell dividing. **Asexual reproduction** (the production of genetically identical offspring from a single parent cell or organism) and growth (an increase in cell numbers) are both the result of mitotic cell division. The production of offspring by **sexual reproduction** is also mostly dependent on mitosis to produce new cells after the gametes (sex cells) have fused. In mitosis, the chromosomes of a cell are duplicated and the genetic information is then equally shared out between the two daughter cells. The formation of the sex cells involves a different process of nuclear division called **meiosis** (see **Section 3B.3**).

LEARNING TIP

You could create a mnemonic to help remember: MIRGAS = Mitosis Repair, Growth, Asexual Reproduction.

WHAT ARE CHROMOSOMES?

Eukaryotic cell division involves replicating the chromosomes that carry the genetic information. A chromosome is made up of a mass of coiled threads of DNA and proteins. If a chromosome were the same length as five consecutive letters on this page, the DNA molecule it contained would stretch the length of a football pitch or more. In a cell that is not actively dividing, the chromosomes are translucent to both light and electrons, so we cannot see them easily or identify them as individual structures. When the cell starts to actively divide, the chromosomes condense – they become much shorter and denser and will take up (absorb) stains very readily. This is the basis of the name 'chromosome' (meaning 'coloured body'). As a result, at this stage of the process we can identify individual chromosomes.

When the DNA molecules condense, they need to be packaged very efficiently. This is achieved with the help of positively charged basic proteins called **histones**. The DNA winds around the histones to form dense clusters known as **nucleosomes** (see **fig A**). These interact to produce more coiling and then supercoiling to form the dense chromosome structures you can see through the microscope in the nucleus of a dividing cell. In the supercoiled areas, the genes are not available to be transcribed to make proteins.

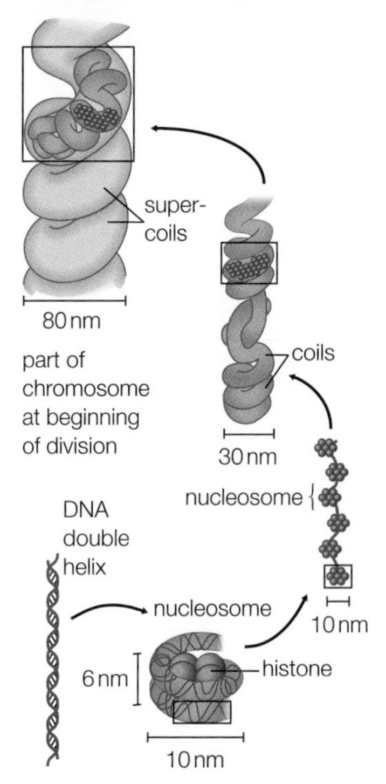

▲ **fig A** Histones play an important role in the organisation of DNA into orderly chromosomes that can be replicated.

The cells of each species possess a characteristic number of chromosomes – in humans, this is 46. These chromosomes occur in matching pairs, one of each pair originating from each parent. In mitosis, the two cells that result from the division must both receive a full set of chromosomes. So, before a cell divides, it must duplicate the original set of chromosomes. During mitosis, these chromosomes are divided equally between the two new cells so that each has a complete and identical set of genetic information. During the active phases of cell division, the chromosomes become very coiled and condensed. In this state they can be photographed to produce a **karyotype**, a special display showing all the chromosomes of the cell (see **fig B** and **Section 2C.3 fig A**).

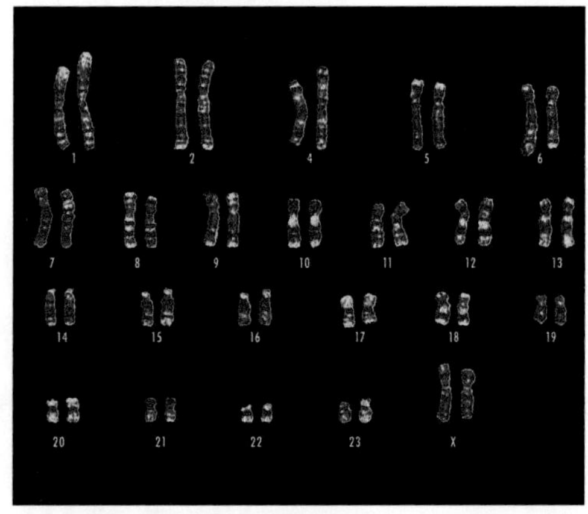

▲ **fig B** This female human karyotype shows the 22 pairs of autosomes and one pair of sex chromosomes found in every healthy human cell, except the eggs and sperm.

THE CELL CYCLE

Cells divide on a regular basis to bring about growth and asexual reproduction. They divide in a sequence of events known as the **cell cycle**, which involves several different phases, as you can see in **fig C**. **Interphase** is a period of non-division when the cells increase in mass and size, carry out normal cellular activities and replicate their DNA ready for division. This is followed by mitosis, a period of active division, and **cytokinesis** when the new cells separate. The length of the cell cycle is variable. It can be very rapid, taking 24 hours or less, or it can take a few years.

PHASES OF THE CELL CYCLE

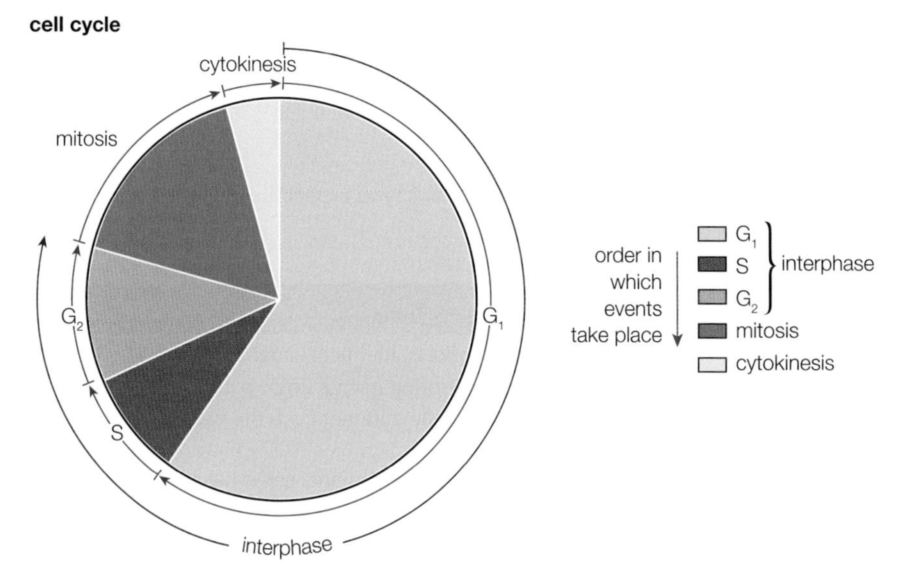

▲ **fig C** In very actively dividing tissue, the cell cycle is repeated as fast as possible; in other tissues, the time between successive divisions may be years.

G_1 (gap 1) is the time between the end of the previous round of mitotic cell division and the start of chromosome replication. The cell takes in material, grows and develops. This is the time that is most variable. In actively dividing cells, G_1 is very short – a matter of hours or days. In other cells, it can be months or even years.

S is the stage when the chromosomes replicate and become double-stranded **chromatids** ready for the next cell division.

G_2 (gap 2) is the time that the organelles and other materials needed for cell division are synthesised – before a cell can divide, it needs two of everything.

Mitosis is when the nucleus is actively dividing.

Cytokinesis is the final stage of the cell division when the new cells separate.

In multicellular organisms, the cell cycle is repeated very frequently in almost all cells during development. However, once the organism is mature, it may slow down or stop completely in some tissues. The cell cycle is controlled by chemical signals which are made in response to different genes. This control is brought about at checkpoints when the cell cycle moves from one phase to the next. The chemical substances which control this are small proteins called **cyclins**. These build up and attach to enzymes called **cyclin-dependent kinases (CDKs)**. The cyclin–CDK complex that forms phosphorylates other proteins, changing their shape and bringing about the next stage in the cell cycle. Examples include:

- the phosphorylation of the chromatin in the nucleus, which results in the chromosomes becoming denser
- the phosphorylation of some of the proteins in the nuclear membrane, which leads to the breakdown of the nuclear membrane structure during cell division.

LEARNING TIP

Chromosomes replicate to form two chromatids. These are called chromosomes again when they separate.

A chromatid is itself a double helix so, after DNA replication, the chromosome consists of two DNA molecules, or two double helices, until the chromatids separate.

DID YOU KNOW?

PERMANENT CELLS

There are some cells that do not enter the cell cycle once they have formed – they must last a lifetime. They are known as permanent cells. Examples include nerve cells, the light sensitive cells of your retina, the transparent cells of the lens of your eye and the cardiac muscle – the muscle that makes up your heart.

CHECKPOINT

SKILLS PROBLEM SOLVING

SKILLS ANALYSIS

1. Why do chromosomes only become visible as a cell goes into mitosis?
2. (a) **Fig C** shows you the relative lengths of the different stages of the cell cycle. Work out the percentage of the cell cycle taken by each stage.
   (b) Using your answer to (a), answer the following question.
   If a culture of cells is dividing every 48 hours, how long would you expect the different stages of the cycle to take?

SUBJECT VOCABULARY

mitosis the process by which a cell divides to produce two genetically identical daughter cells
asexual reproduction the production of genetically identical offspring from a single parent or organism
sexual reproduction the production of offspring that are genetically different from the parent organism or organisms by the fusing of two sex cells (gametes)
meiosis a form of cell division in which the chromosome number of the original cell is halved, leading to the formation of the gametes
histones positively charged proteins involved in the coiling of DNA to form dense chromosomes in cell division
nucleosomes dense clusters of DNA wound around histones
karyotype a way of displaying an image of the chromosomes of a cell to show the pairs of autosomes and sex chromosomes
cell cycle a regulated process of three stages (interphase, mitosis and cytokinesis) in which cells divide into two genetically identical daughter cells
interphase the period between active cell divisions when cells increase their size and mass, replicate their DNA and carry out normal metabolic activities
cytokinesis the final stage of the cell cycle before the cell enters interphase again - division of the cytoplasm at the end of mitosis to form two independent, genetically identical cells
chromatid one strand of the replicated chromosome pair that is joined to the other chromatid at the centromere
cyclins small proteins that build up during interphase and are involved in the control of the cell cycle by their attachment to cyclin-dependent kinases
cyclin-dependent kinases (CDKs) enzymes involved in the control of the cell cycle by phosphorylating other proteins, activated by attachment to cyclins

3B 2 MITOSIS

LEARNING OBJECTIVES

- Understand the role of mitosis in producing genetically identical daughter cells for growth and asexual reproduction.
- Know how to calculate mitotic indices.

A cell is in the interphase stage of the cell cycle for much of its life. This used to be called the resting phase, but this is not a good description. During interphase, the normal metabolic processes of the cell continue and new DNA is produced as the chromosomes replicate. New proteins, cytoplasm and cell organelles are also made so that the cell is prepared to produce two new cells. ATP production is stepped up at times to provide the extra energy needed as the cells divide. **Fig A** shows a cell in interphase. When everything the cell needs is present, and the parent cell is large enough, interphase ends and mitosis begins. You will mainly consider mitosis in animal cells.

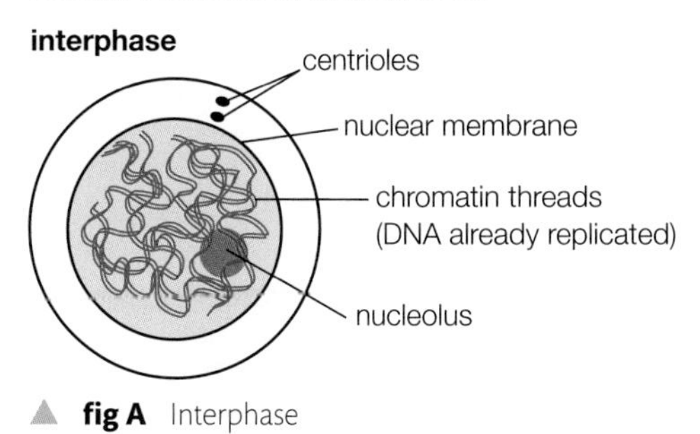

fig A Interphase

THE STAGES OF MITOSIS

During the process of cell division, the chromosomes that are duplicated during interphase are divided with the remaining contents of the cell to produce two identical daughter cells. Walther Flemming (1843–1905), a German cytologist, was the first to describe what is sometimes called the 'dance of the chromosomes'. It refers to the complex series of movements that occur during cell division as the chromosomes compete for space in the middle of the nucleus and then pull apart to opposite ends of the cell. As in many biological processes, the events of mitosis are continuous, but it is easier to describe what is happening by considering the sequence of events as a series of phases. These are known as **prophase**, **metaphase**, **anaphase** and **telophase**.

PROPHASE

Before mitosis begins, the genetic material replicates to produce exact copies of the original chromosomes. By the beginning of prophase, both the originals and the copies are referred to as chromatids. In prophase, the chromosomes coil up and can take up stains to become visible. At this point, each chromosome consists of two daughter chromatids that are attached to each other at a region known as the **centromere**. The nucleolus breaks down and the centrioles begin to pull apart to form the spindle (see **fig B**).

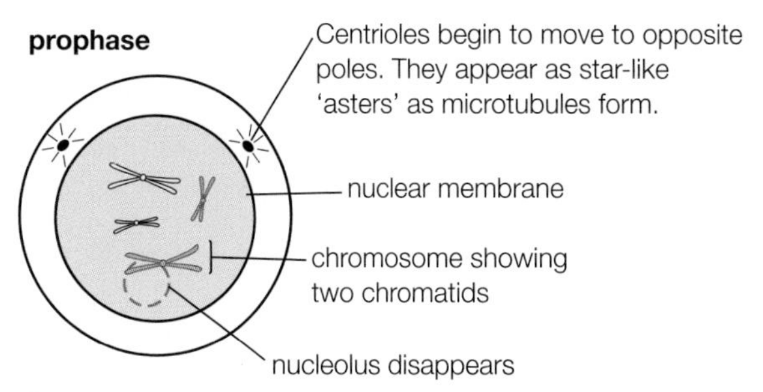

fig B Prophase

EXAM HINT

Remember that the spindle is what moves the chromosomes. It must form before the chromosomes start to move.

METAPHASE

The nuclear membrane has broken down and the centrioles have moved to opposite poles of the cell. In moving apart, the centrioles have formed between them a set of microtubules that is known as the spindle. The chromatids appear to compete for position on the **metaphase plate (equator)** of the spindle during metaphase. They eventually line up along this position. The centromere of each chromatid is associated with a separate microtubule of the spindle (see **fig C**). Plant cells also form a spindle but they do not have a centriole. Scientists still do not completely understand exactly how spindle formation takes place in plant cells.

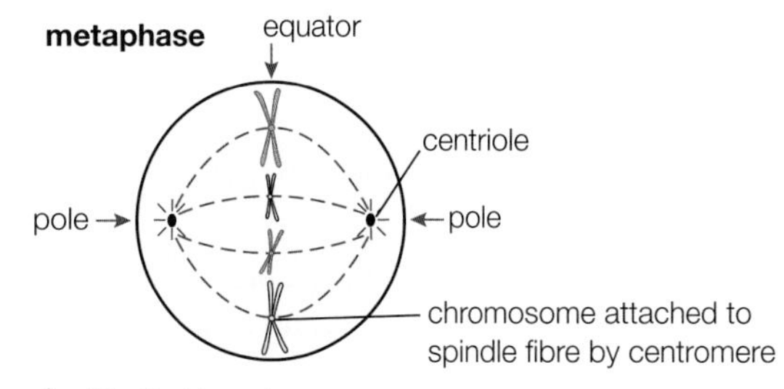

fig C Metaphase

ANAPHASE

The centromeres split so that the two identical linked chromatids become separate entities (see **fig D** overleaf). They are now new chromosomes. The chromatids from each pair are pulled, centromere first, towards opposite poles of the cell. This separation occurs quickly, taking only a matter of minutes. At the end of anaphase, the two sets of chromatids have been separated to opposite ends of the cell. The chromatids cannot move on their own. They rely on the microtubules of the spindle to move them. For many years, the spindle was thought of as a structure running from one end of the cell to the other. It is now known to consist of overlapping microtubules containing contractile fibres, which are similar to those in animal muscle cells. The overlapping fibres contract and cause the movement of the chromatids. This is an energy-using process, and the energy is supplied by ATP produced during cell respiration.

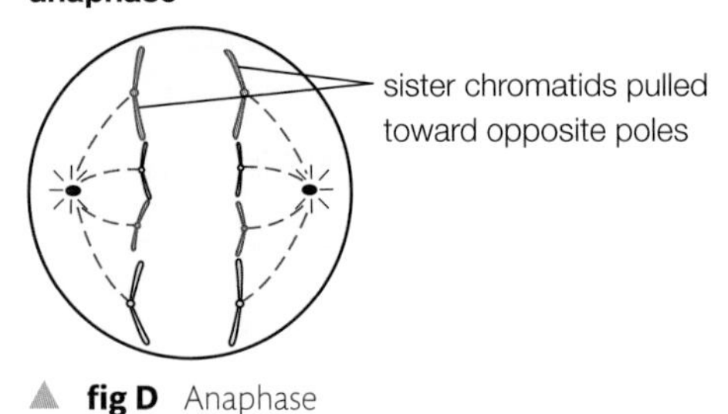

▲ **fig D** Anaphase

TELOPHASE

During telophase, the spindle fibres break down and nuclear envelopes form around the two sets of chromosomes (see **fig E**). The nucleoli and centrioles are also re-formed. The chromosomes begin to unravel and separate, becoming less dense and harder to see.

LEARNING TIP

Remember PMAT for the sequence of stages in mitosis: Prophase, Metaphase, Anaphase and Telophase.

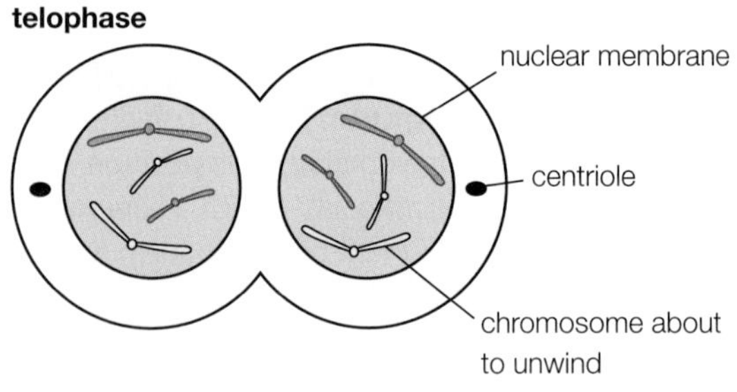

▲ **fig E** Telophase

CYTOKINESIS

The final phase of the cell cycle is cytokinesis, when the cytoplasm divides (see **fig F**).

In animal cells, a ring of contractile fibres tightens around the centre of the cell similar to a belt tightening around a sack of flour. These fibres seem to be the same as those found in animal muscle cells. They continue to contract until the two cells have been separated.

In plant cells, the division of the cell occurs differently. A cellulose cell wall builds up from the inside of the cell outwards.

In both cases, the result is that two genetically identical daughter cells are formed. These cells enter interphase and begin to prepare for the next cycle of division.

(a)

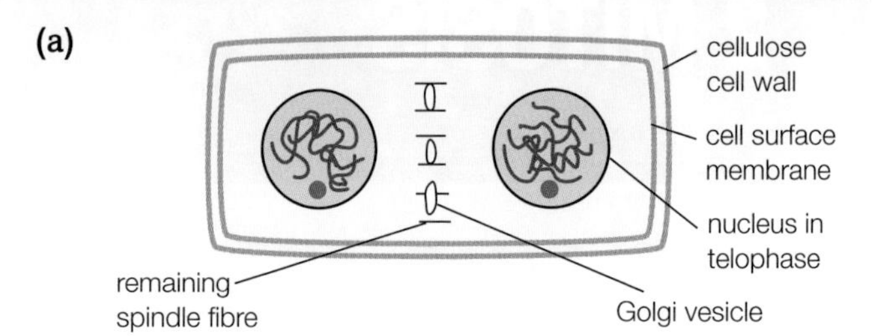

Some spindle fibres remain and guide Golgi vesicles to the equator of the cell.

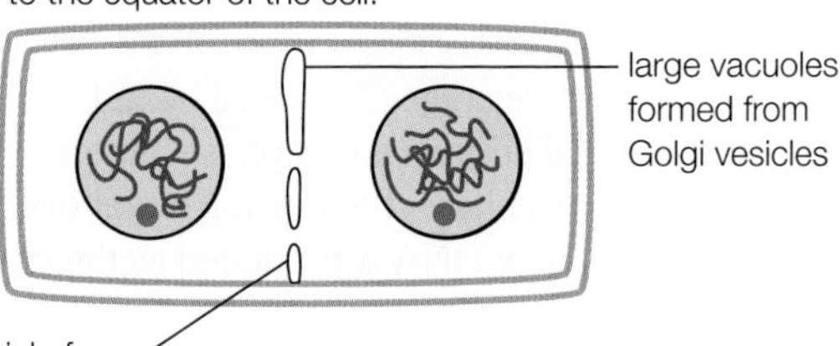

The vesicles enlarge and fuse together, forming a cell plate.

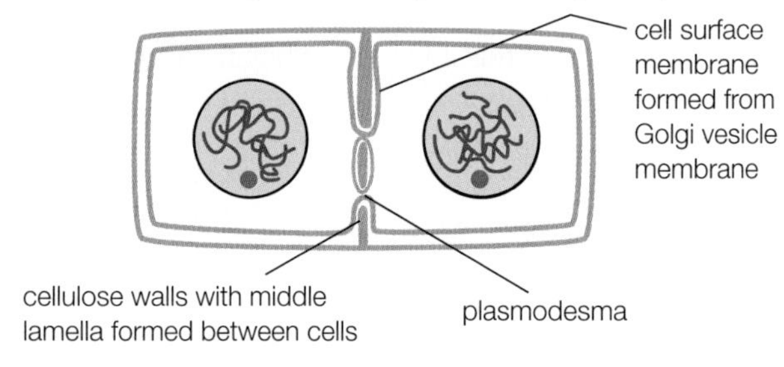

The basic structure of the cell walls forms within each vesicle, and the vesicles fuse to join the cell wall together. Small gaps left between the vesicles form plasmodesmata (see **Section 4A.1**).

(b)

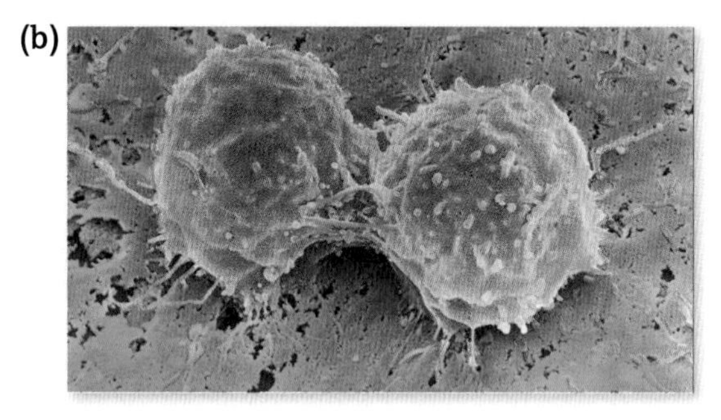

▲ **fig F** The final stages of the cell cycle in **(a)** a plant cell and **(b)** an animal cell

THE IMPORTANCE OF MITOSIS

Mitosis is how organisms grow and replace old cells. It is also the method which organisms use for asexual reproduction. Asexual reproduction involves only one parent individual and it results in genetically identical individuals or **clones**. It has many advantages for an organism. It does not rely on finding a mate and can give rise to large numbers of offspring very rapidly. However, it also has one big disadvantage – the offspring are (mostly) genetically identical to the parent organism. This becomes a problem when living conditions change in some way. The introduction of a new disease to an environment, a change in the temperature or human intervention can cause the total destruction of a group of genetically identical organisms, because if one cannot cope, neither can all the others. Many species of plants and fungi undergo both sexual and asexual reproduction as a matter of course.

PRACTICAL SKILLS CP6

Observing mitosis

You can observe mitosis quite easily in the cells of rapidly dividing tissues such as the meristem at a growing root tip (see **fig G**). Although we do not understand the details of spindle formation without centrioles in plant cells, the process of mitosis appears very similar indeed. We often use plant cells to observe mitosis because they are bigger than animal cells and readily available.

Using a dye such as acetic orcein, which stains the chromosomes, you can make a temporary tissue squash preparation showing the stages of mitosis. You can also observe mitosis in living tissue, and dramatic recordings of the activity of the chromosomes have been made using time-lapse photography. This has increased our understanding considerably. Viewing the movements of the cell contents during mitosis shows it as a dynamic process and explains why it is called the 'dance of the chromosomes'.

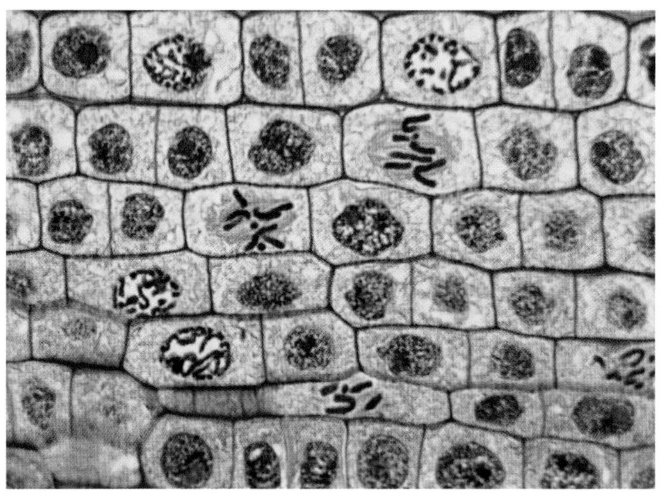

▲ **fig G** This stained section of a root tip squash shows cells in different stages of the cell cycle, including active mitosis.

MITOTIC INDEX

Some cells – for example, most nerve cells – never undergo mitosis once the organism is mature. Others – for example, skin cells – are always undergoing mitosis to replace lost and damaged cells. Some cells undergo mitosis more often than they should when the control mechanisms fail and a tissue becomes cancerous.

The **mitotic index** is a measure of how actively the cells in a tissue are dividing. It is the ratio between the number of cells in a tissue sample that are in mitosis and the total number of cells in the sample. To calculate the mitotic index, we can analyse micrographs and count the number of cells that are undergoing mitosis (actively dividing) and the total number of cells. We can identify the cells that are actively dividing by the presence of visible chromosomes (see **fig H**).

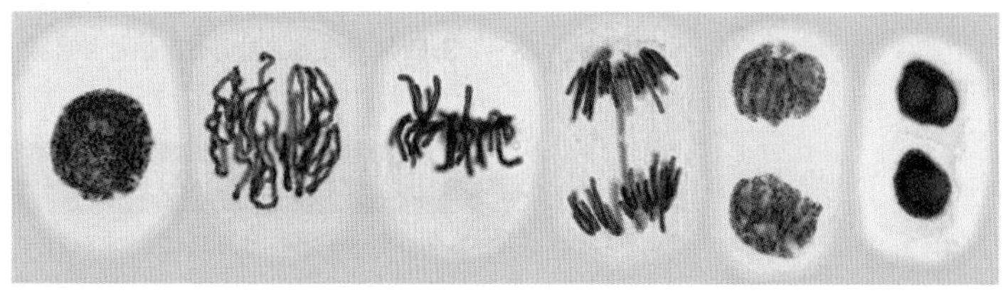

▲ **fig H** Once the cell leaves interphase and enters mitosis, the chromosomes can be seen. You can use this when counting cells to work out the mitotic index.

We can then calculate the mitotic index by dividing the number of actively dividing cells by the total number of cells (see **fig I**):

$$\text{mitotic index} = \frac{\text{cells in mitosis}}{\text{total number of cells}}$$

Cells with visible chromosomes: 12

Total number of cells: 25

$$\text{Mitotic index} = \frac{12}{25} = 0.48$$

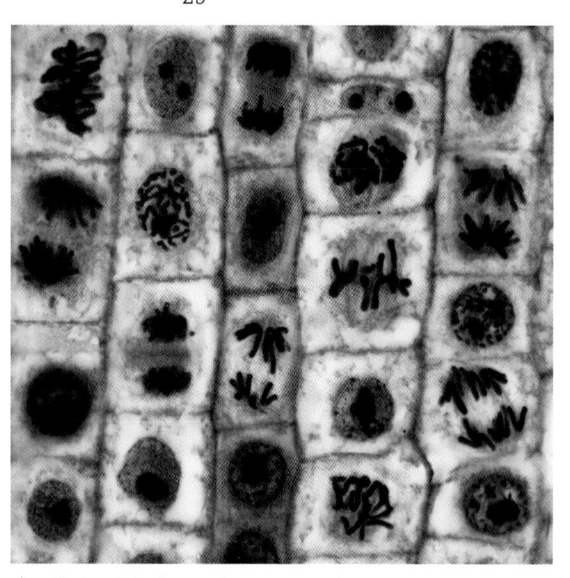

▲ **fig I** Calculating the mitotic index

We can use the mitotic index to identify actively dividing tissues, including cancerous tissue. Cancerous tissue is dividing more rapidly than it should, so we can also use the mitotic index to measure the effectiveness of treatments for cancer. If the treatment is working, the mitotic index of the tumour will fall.

EXAM HINT

Mathematical skills are an important part of science. Simple calculations like this will often appear in examination papers.

CHECKPOINT

SKILLS CRITICAL THINKING

SKILLS ANALYSIS

1. Summarise the stages of mitotic cell division in animal cells.
2. Explain why root tips are particularly suitable material to use for preparing slides to show mitosis.
3. Work out the mitotic index of the tissue shown in **fig G** to one significant figure.

SUBJECT VOCABULARY

prophase the first stage of active cell division where the chromosomes are coiled up and consist of two daughter chromatids joined by the centromere; the nucleolus breaks down
metaphase the second stage of active cell division where a spindle of overlapping protein microtubules forms and the chromatids line up on the metaphase plate
anaphase the third stage of active cell division where the centromeres split so chromatids become new chromosomes; they are moved to the opposite poles of the cell, centromere first, by contractions of the microtubules of the spindle
telophase the fourth stage of active cell division where a nuclear membrane forms around the two sets of chromosomes, the chromosomes unravel and the spindle breaks down
centromere the region where a pair of chromatids are joined and which attaches to a single strand of the spindle structure at metaphase
metaphase plate (equator) the region of the spindle in the middle of the cell along which the chromatids line up
clones genetically identical individuals resulting from asexual reproduction in a single parent
mitotic index the ratio between the number of cells in a tissue sample that are in mitosis and the total number of cells in the sample

3B 3 SEXUAL REPRODUCTION AND MEIOSIS

SPECIFICATION REFERENCE 3.10

LEARNING OBJECTIVES

- Understand the role of meiosis in ensuring genetic variation through the production of non-identical gametes, as a consequence of:
  - the independent assortment of chromosomes in metaphase 1
  - the crossing over of alleles between chromatids in prophase 1.

Asexual reproduction can be very successful at producing new individuals, but leaves the population vulnerable to changes in the environment. The offspring are mostly identical to their parents. There is very limited genetic variation, although spontaneous mutations do occur.

Not many organisms use only asexual reproduction. Most have a system of sexual reproduction, which they can use when necessary, to introduce the genetic variation that may enable the population or species to survive. In more complex organisms, particularly animals and flowering plants, sexual reproduction is the main way of producing fertile offspring.

Sexual reproduction is the production of a new individual resulting from the joining (fusion) of two specialised cells known as gametes. Sexual reproduction produces individuals that are not genetically the same as either of their parents, but contain genetic information from both (see **fig A**). Sexual reproduction relies on two gametes meeting and fusing. It is not always easy to find a mate, particularly if you are a solitary predator. It also uses more bodily resources because it usually involves special sexual organs. However, the great advantage of sexual reproduction is that it increases genetic variation because gametes from two different individuals are fused together. In a changing environment, this gives a greater chance that one or more of the offspring will have a combination of genes that improves their chance of surviving and going on to reproduce.

EXAM HINT

Remember that variation is the key to natural selection and evolution.

▲ **fig A** The genetic variation in offspring produced by sexual reproduction can be very easy to see.

WHAT ARE GAMETES?

The nucleus of a cell contains the chromosomes. In most of the cells of any individual, the chromosomes occur in pairs. A cell containing two full sets of chromosomes is called **diploid ($2n$)** and the number of chromosomes in a diploid cell is characteristic for that species. However, if two diploid cells combined to form a new individual in sexual reproduction, the offspring would have four sets of chromosomes, losing the characteristic number for the species. Each new generation would get more genetic material until eventually the cells would break down and fail to function. To avoid this, **haploid (n)** nuclei are formed with one set of chromosomes (half of the full chromosome number), usually within the specialised cells called gametes. Sexual reproduction occurs when two haploid nuclei fuse to form a new diploid cell called a **zygote** (see **fig B** overleaf). This process is called **fertilisation**.

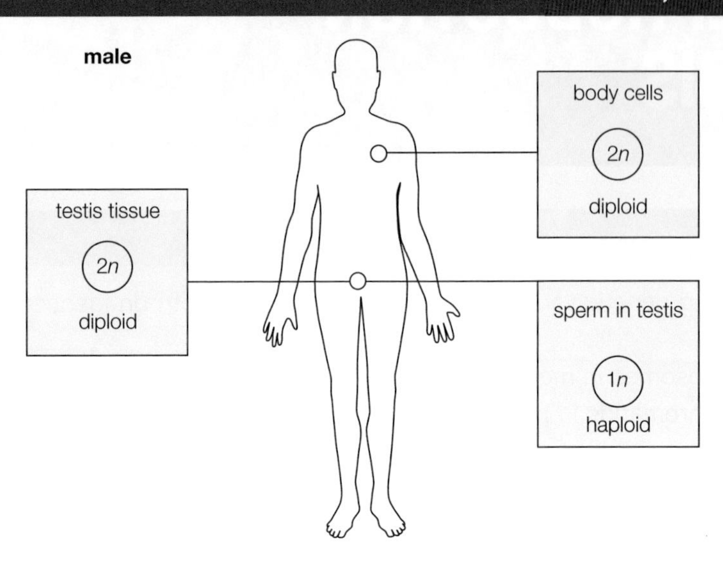

	TISSUE TYPE	DIPLOID OR HAPLOID
Male	body tissue	diploid
	testis tissue	diploid
	sperm in testis	haploid
Female	body cells	diploid
	ovary tissue	diploid
	ovum in ovary	haploid
	zygote tissue	diploid

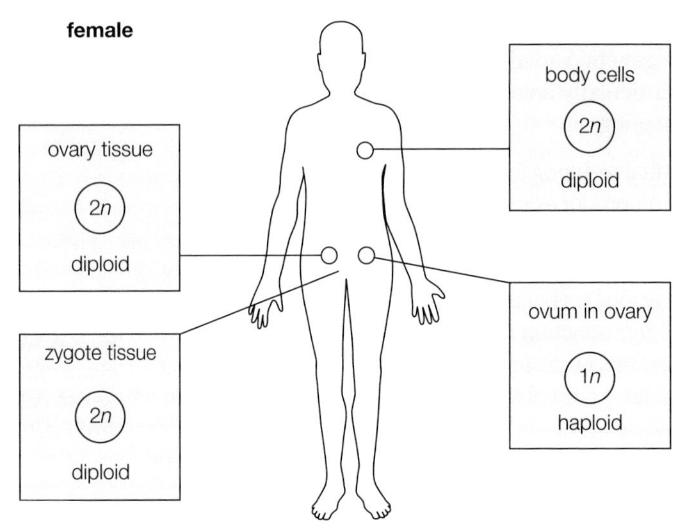

fig B The only cells in the human body that are haploid are the gametes.

DID YOU KNOW?

POLYPLOIDY

Although most eukaryotic organisms are diploid, a number have stable forms of **polyploidy**. Fish can be polyploid; for example, salmon have four sets of chromosomes and some fish have as many as 400 chromosomes in total. Some reptiles are also polyploid. Polyploidy is very common in plants, particularly ferns and flowering plants. Potatoes are polyploid and so are the members of the genus of flowering plants *Dendranthema*. The haploid number of the genus is 9, the diploid number is 18 (two sets of chromosomes) but they may have up to 198 chromosomes (22 sets).

Crop plants are often polyploid. Some wheat varieties are diploid, tetraploid or hexaploid, and the cabbage family is also hexaploid.

THE FORMATION OF GAMETES

Gametes are formed in special sex organs. In simpler animals and plants, the sex organs are often temporary, formed only when they are needed. In more complex animals, the sex organs are usually more permanent structures that we sometimes call the **gonads**. In flowering plants, the female sex organs are the **ovaries** and the male ones are the **anthers**. The female gametes, **ovules**, are made in the ovaries. The male gametes are produced in the anthers: the gamete cells are contained within a spore which we know as **pollen**. In animals, the male gonads are the **testes**, which produce the male gametes known as **spermatozoa**, or more commonly, **sperm**. The female gonads are the ovaries and they produce the female gametes known as **ova**. The male gametes are often much smaller than the female ones, but they are usually produced in much larger quantities.

LEARNING TIP

Gametes can be summarised as:

- male: many, mini, motile
- female: few, fat, fixed.

MEIOSIS

In **Section 3B.2** you saw that when cells divide by mitosis, the number of chromosomes in both daughter cells is the same as in the original parent cell. In the cell divisions that form gametes, the chromosome number needs to be halved to give the necessary haploid nuclei. To bring about this reduction in the chromosome number, gametes are formed by a different process of nuclear division known as meiosis.

Meiosis is a reduction division and it occurs only in the sex organs. In animals, the gametes are formed directly from meiosis. In flowering plants, meiosis forms special male cells called **microspores** and female cells called **megaspores**, which then develop into the gametes. Meiosis is of great biological significance – it is the basis of the variation that allows species to evolve.

THE CHROMOSOMES IN MEIOSIS

In meiosis, two nuclear divisions produce four haploid daughter cells, each with its own unique combination of genetic material (see **fig C**). The events of meiosis are continuous although we describe the stages as separate phases. As in mitosis, the contents of the cell, in particular the DNA, are replicated while the cell is in interphase. When the cell has all the materials it needs, it can enter meiosis.

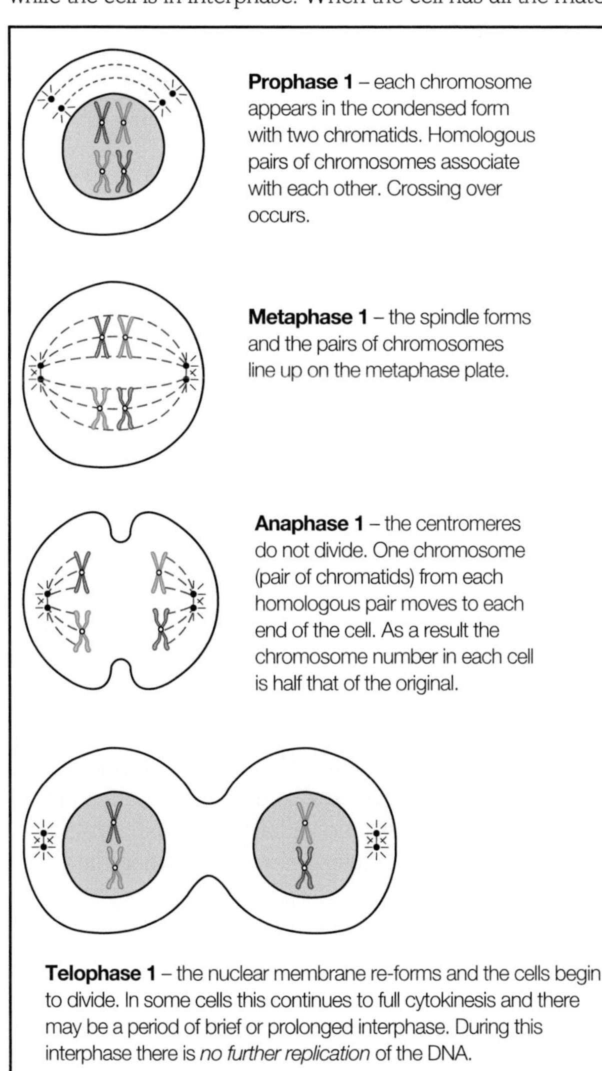

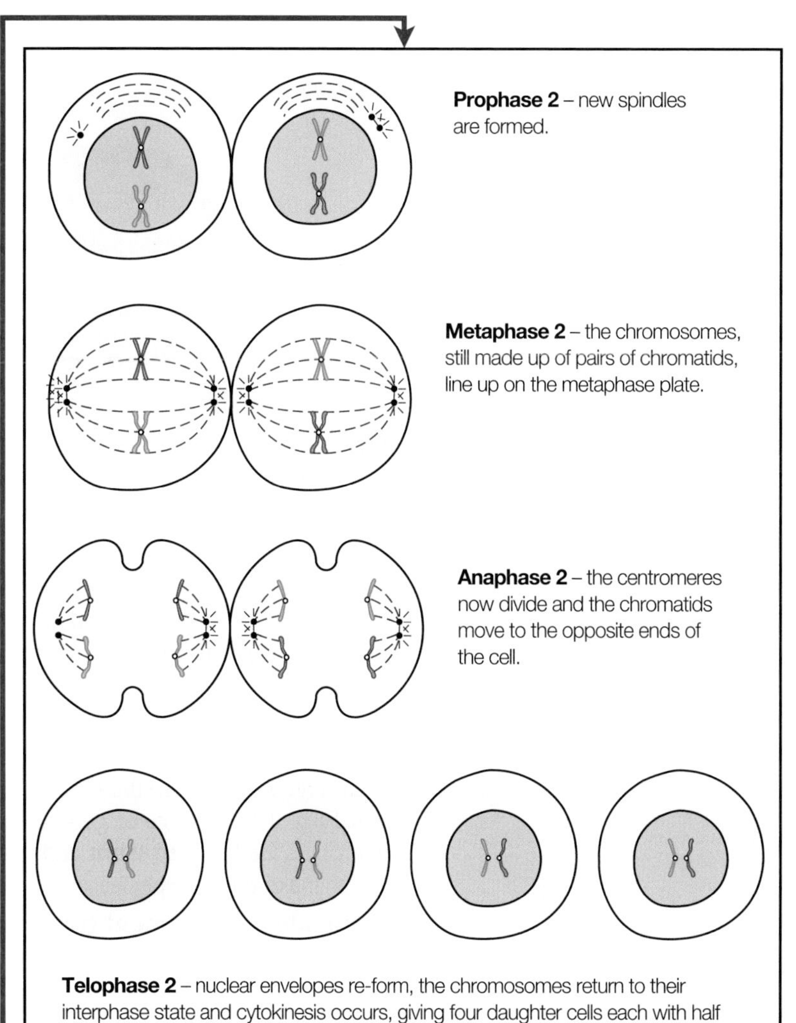

▲ **fig C** These are the main steps in the process of meiosis, which results in the formation of haploid gametes. This is a simplified version of meiosis shown in a cell with only two pairs of chromosomes, to make it easier to see what is happening.

Many of the stages of meiosis are very similar to those of mitosis, with just a few important differences. Some of the stages of meiosis are shown in **fig D** (overleaf). Before meiosis starts, the chromosomes replicate to form chromatids joined by a centromere as in mitosis. However, in

EXAM HINT

If you draw diagrams of meiosis take care with the labelling, it must be clear what part of the diagram a label line is going to.

prophase 1 of meiosis the two chromosomes of each pair, known as **homologous pairs**, stay close together. At this stage, **crossing over (recombination)** introduces genetic variation as the chromatids may break and recombine (see **fig E**). Just as in mitosis, the nuclear membrane and nucleolus break down and the centrioles pull apart to form the spindle. The centromeres do not split in the first division of meiosis, so pairs of chromatids move to the opposite ends of the cell.

The cell then immediately goes into a second division without any further replication of the chromosomes. This division is just like mitosis. The centromeres divide and chromatids move to opposite poles of the cell. Finally, the nuclear membranes re-form as the chromosomes decondense and become invisible again. Cytokinesis occurs producing four haploid daughter cells, each with half the chromosome number of the original parent cell. These daughter cells later develop into gametes.

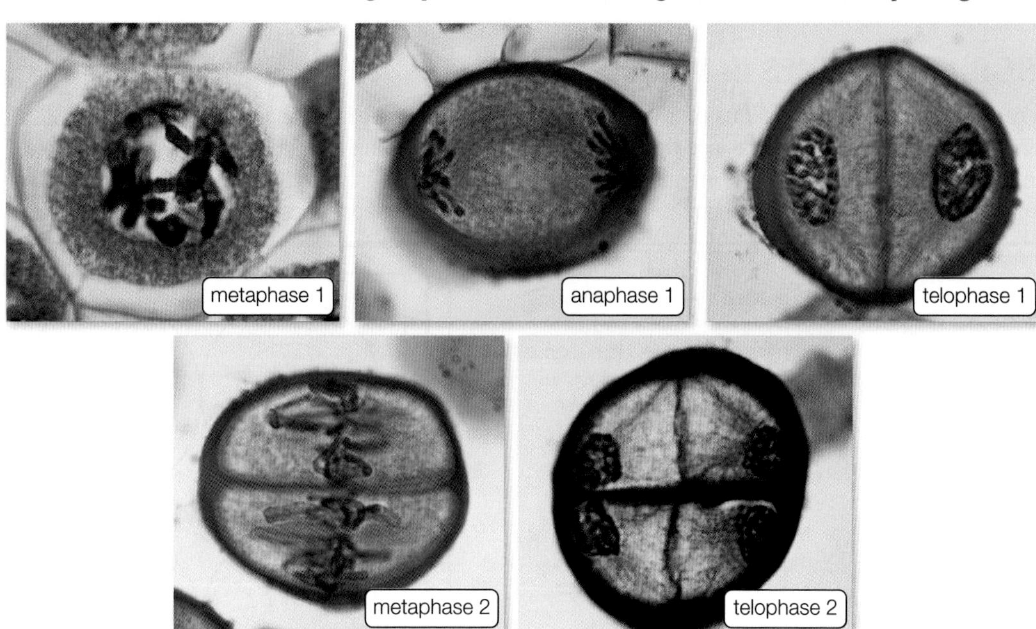

▲ **fig D** The stages of meiosis are not easy to see in cells, but these images, taken from the testis of a locust and the anther of a plant, show you some of them.

LEARNING TIP

Remember that in the first division, the homologous chromosomes (each consisting of two chromatids) separate.

In the second division, the individual chromatids separate.

THE IMPORTANCE OF MEIOSIS

Meiosis reduces the chromosome number in gametes from diploid to haploid. This means that sexual reproduction is possible without each following generation having more and more genetic material. It is also the main way in which genetic variation is introduced to a species. This variation is introduced in two main ways.

- **Crossing over (recombination)**: this process occurs in prophase 1 of meiosis when large, multi-enzyme complexes 'cut and join' bits of the maternal and paternal chromatids together (see **fig E**). The points where the chromatids break are called **chiasmata**. These are important in two ways. First, the exchange of genetic material leads to added genetic variation. Second, errors in the process lead to **mutation** and this is a further way of introducing new combinations into the genetic make-up of a species.
- **Independent assortment (random assortment)**: the maternal and paternal chromosomes are distributed into the gametes completely at random. For example, each gamete receives 23 chromosomes. In each new gamete, any number from none to all 23 can come from either the maternal or paternal chromosomes. There are more than eight million possible genetic combinations within the sperm or the egg. This guarantees huge variety in the gametes. This random assortment occurs at metaphase 1 in the process of meiosis, when the chromatids line up on the metaphase plate.

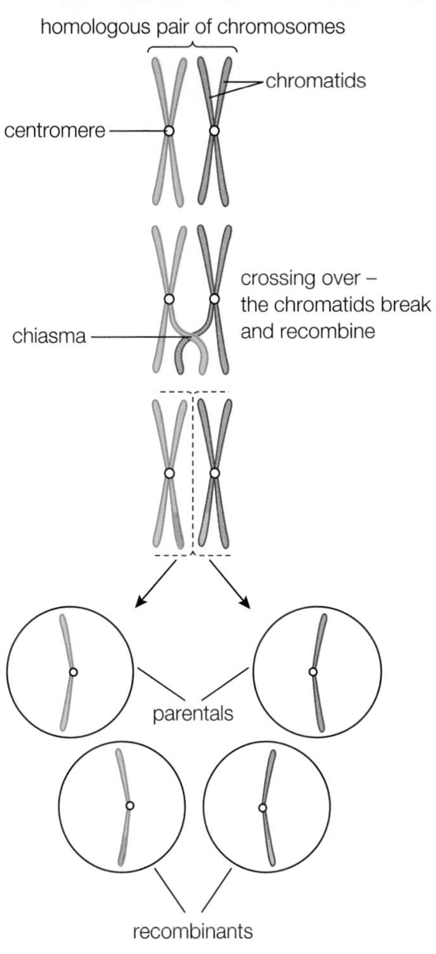

▲ **fig E** Chromosomes crossing over in meiosis. This process introduces more variation into the gametes.

EXAM HINT

Make sure you are very clear about the difference between independent assortment, recombination and mutation as sources of genetic variation.

CHECKPOINT

1. Give examples of conditions when sexual reproduction would be more advantageous in the production of offspring than asexual reproduction, and of conditions when asexual reproduction would be more advantageous than sexual reproduction. Explain your choices.
2. Make a table to summarise the main stages in the process of meiosis.

SKILLS INTERPRETATION

3. Explain how meiosis leads to variation between offspring.

SUBJECT VOCABULARY

diploid (2*n*) a cell with a nucleus containing two full sets of chromosomes
haploid (*n*) a cell with a nucleus containing one complete set of chromosomes
zygote the cell formed when two haploid gametes fuse at fertilisation
fertilisation the fusing of the haploid nuclei from two gametes to form a diploid zygote in sexual reproduction
polyploidy a cell or an organism with more than two sets of chromosomes
gonads the sex organs in animals
ovaries the female sex organs in both animals and plants; they produce the female gametes called ovules in plants and ova in animals
anthers male sex organs in plants that produce the male gametes contained in the pollen
ovules the haploid female gametes in plants
pollen the spore which contains the haploid male gametes of plants
testes the male sex organs in animals that produce the male gametes - sperm
spermatozoa (sperm) the haploid male gametes in animals
ova the haploid female gametes in animals (singular = ovum)
microspores the result of meiosis in plants that develop into the spore (pollen) containing the male gametes
megaspores the result of meiosis in plants that develop into the female gametes, ovules
homologous pairs matching pairs of chromosomes in an individual which both carry the same genes, although they may have different alleles
crossing over (recombination) the process by which large multi-enzyme complexes cut and re-join parts of the maternal and paternal chromatids at the end of prophase I, introducing genetic variation
chiasmata the points where the chromatids break during recombination
mutation a permanent change in the DNA of an organism
independent assortment (random assortment) the process by which the chromosomes derived from the male and female parent are distributed into the gametes at random

4 GAMETES: STRUCTURE AND FUNCTION

LEARNING OBJECTIVES

- Understand the role of meiosis in ensuring genetic variation through the production of non-identical gametes.
- Understand how mammalian gametes are specialised for their functions.

The gametes that make sexual reproduction possible are formed in a process called **gametogenesis**. Meiosis is just one stage in gamete formation, which produces different male and female sex cells. You are going to consider the way in which sperm and ova are made in the sex organs of mammals, using humans as an example, and also how gametes are formed in flowering plants.

GAMETE FORMATION IN MAMMALS

Many millions of sperm are released every time a male mammal ejaculates. The eggs in a sexually mature female are usually numbered in thousands and will eventually run out. Special cells (the primordial germ cells) in the gonads divide, grow, divide again and then differentiate into the gametes.

Both mitosis and meiosis have a role in gametogenesis. Mitosis provides the precursor cells. Meiosis causes the reduction of genetic material in divisions that result in gametes. In human males, the process of gametogenesis involving meiotic and mitotic cell divisions happens constantly from puberty onwards. In females, mitotic divisions occur before birth to form diploid primary **oocytes**, which remain inactive until after puberty. The second meiotic divisions are only completed if the ovum is fertilised. **Fig A** shows the structure of an ovum and a sperm – the drawings are not to scale as a sperm is much smaller than an ovum.

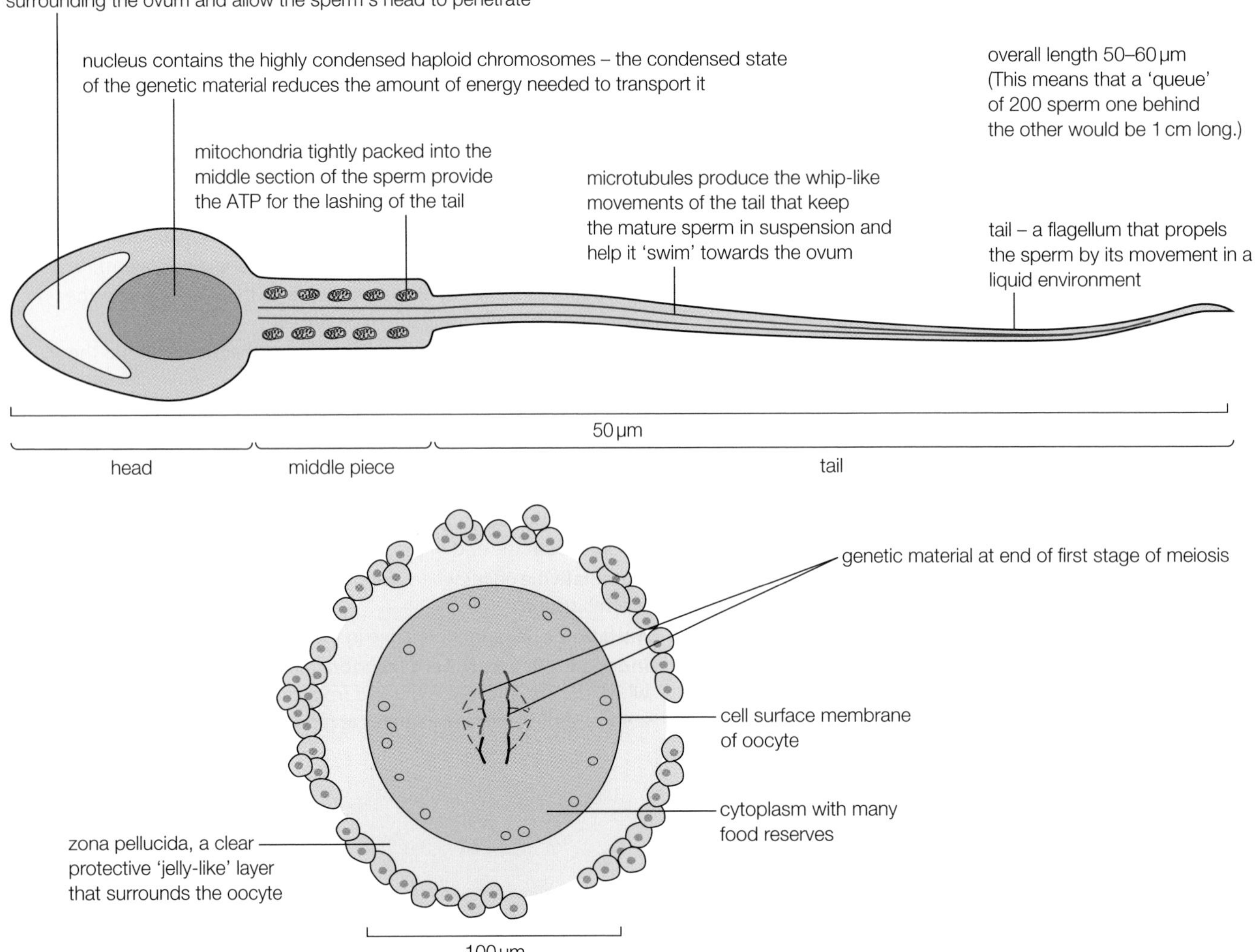

fig A Human gametes – the sperm and ova show clear specialisations that fit them for their function.

CHARACTERISTICS OF THE GAMETES

SPERMATOZOA: MANY, MINI, MOTILE

The male gametes or spermatozoa of most mammalian species, including humans, are around 50 μm long. They have several tasks to fulfil.

- They must carry the genetic information in the nucleus.
- They must remain in suspension in the semen so they can be transported through the female reproductive tract. For this, they need a long, beating tail.
- They must be able to penetrate the protective barrier around the ovum and deliver the male haploid genome safely inside. They penetrate the protective barrier of the egg using enzymes contained in the **acrosome**.

The close relationship between the structure of human spermatozoa and their functions is shown in **fig A**. Many millions of these motile gametes are produced in the lifetime of a human male. Human family sizes vary from one to around 20 children, with only one spermatozoan needed to fertilise each ovum. This gives you an idea of the scale of biological wastage.

OVA: FEW, FAT, FIXED

Although spermatozoa of most animals are very similar in size, the same is not true for ova. These vary tremendously in both their diameter and their mass. The human ovum is about 0.1 mm across, while the ovum in an ostrich egg is around 6 mm in diameter. Eggs do not move on their own, so they do not need contractile proteins, but they usually contain food for the developing embryo. They have a protective layer of jelly around them known as the **zona pellucida**. The main difference between eggs of various species is the quantity of stored food they contain. In birds and reptiles, a lot of development takes place before the animal hatches so the egg contains a large food store. In mammals, the developing fetus implants in the uterus and is then supplied with nutrients from the blood supply of the mother. This means large food stores in the egg are unnecessary.

LEARNING TIP

Do not confuse the egg with the ovum - you can remember it as: the ovum is a cell inside the egg (in birds).

THE GAMETES IN PLANTS

The formation of gametes in flowering plants is more complex because plants have two phases to their life cycles. The **sporophyte generation** is diploid and produces spores by meiosis. The resulting **gametophyte generation** is haploid and produces the gametes by mitosis. In plants such as mosses and ferns, these two phases exist as separate plants. In flowering plants, the two phases have been combined into one plant. The main body of the plant that we see is the diploid **sporophyte**. The haploid gametophytes are reduced to part of the contents of the anther and the ovary. They are produced by meiosis from spore mother cells.

POLLEN

The anthers of flowering plants are equivalent to the testes of animals. Meiosis occurs here, resulting in vast numbers of pollen grains that carry the male gametes. The male gametes in plants are known as **microgametes**. Each pollen grain contains two haploid nuclei: the **tube nucleus** and the **generative nucleus** (see **fig B**). The tube nucleus has the function of producing a **pollen tube** that penetrates through stigma, style and ovary and into the ovule. The generative nucleus then fuses with the nucleus of the ovule to form a new individual.

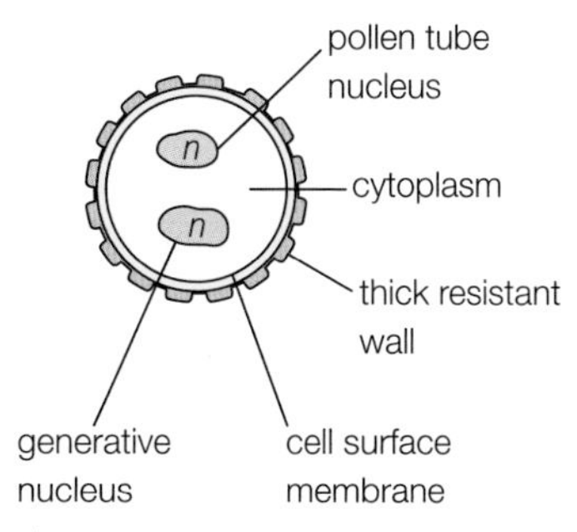

fig B Pollen grain containing two haploid nuclei

DID YOU KNOW?

THE POLLEN RECORD

The surface patterns of pollen grains are unique and specific to the species (see fig C). They are extremely tough and resistant to decay and can remain in the soil for thousands of years. Palaeobotanists can tell what plants were growing thousands of years ago and how abundant they were by analysing the pollen they find in archaeological digs.

fig C This scanning electron micrograph shows just some of the amazing shapes and sizes of pollen grains from different types of flowering plant.

OVULES

The ovary of the plant is equivalent to the animal ovary. Meiosis results in the formation of a relatively small number of ova contained within ovules inside the ovary. Some plants – an example is the peach – produce only one ovule (egg chamber), while others such as peas produce several. The ovule is attached to the wall of the ovary by a pad of special tissue called the **placenta**. Inside the ovule, the embryo sac forms the gametophyte generation (see **fig D**).

EXAM HINT

Make sure you use the correct biological terminology particularly when there are words that have similar sounds but very different meanings such as ovule and ovary.

A combination of meiotic and mitotic cell divisions results in an egg cell, known as the **megagamete**. It contains two polar nuclei and other small cells, some of which degenerate.

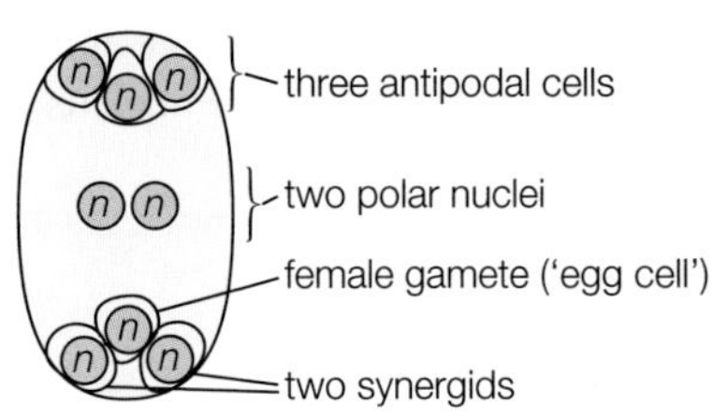

▲ **fig D** Mature embryo sac containing the egg cell, which is the female gamete in flowering plants.

CHECKPOINT

SKILLS INTERPRETATION

1. What role do gametes play in sexual reproduction?
2. Explain the role of meiosis in the production of gametes. How does this vary in animals and plants?
3. How are the human male gametes adapted to fit their role?
4. Compare the adaptations of female gametes in plants and mammals.

SUBJECT VOCABULARY

gametogenesis the formation of the gametes by meiosis in the sex organs
oocyte a cell in an ovary which may form an ovum if it undergoes meiotic division
acrosome the region at the head of the sperm that contains enzymes to break down the protective layers around the ovum
zona pellucida a layer of protective jelly around the unfertilised ovum
sporophyte generation the diploid generation in plants that produces spores by meiosis
gametophyte generation the haploid generation in plants that gives rise to the gametes by mitosis
sporophyte the diploid main body of the plant
microgametes the male gametes produced in plants, the pollen grains
tube nucleus the male nucleus that will control the production of the pollen tube in fertilisation
generative nucleus the male nucleus that will fuse with the female nucleus
pollen tube a tube that grows out of a pollen grain down the style, into the ovary and through the micropyle of the ovule to carry the generative nucleus (which divides to form two male nuclei) to the ovule
placenta (plant) the pad of special tissue that attaches the plant ovule to the ovary wall
megagamete the female gamete, the egg cell, in plants

5 FERTILISATION IN MAMMALS AND PLANTS

LEARNING OBJECTIVES

- Know the process of fertilisation in mammals, including the acrosome reaction, the cortical reaction and the fusion of the nuclei.
- Know the process of fertilisation in flowering plants, starting with the growth of a pollen tube and ending with the fusion of the nuclei.

GETTING TOGETHER

Asexual reproduction is a guaranteed method of passing on the genes from one individual into the next generation. For sexual reproduction to succeed, the gametes must meet. If these gametes come from two different individuals, the male gamete needs to be transferred to the female gamete. In plants, some flowers attract other organisms such as insects, birds or mammals. The other organism transfers the pollen from one plant to another, known as **pollination**. Other plants rely on the wind to carry their pollen from plant to plant.

Animals use a wide variety of strategies to make sure the gametes meet. They fall into two main categories.

- **External fertilisation** occurs outside the body, with the female and male gametes discharged directly into the environment where they meet and fuse. External fertilisation is usually seen in aquatic species, because spermatozoa and ova are very vulnerable to drying and are rapidly destroyed in the air. Simpler animals such as jellyfish release copious amounts of male and female gametes into the sea. It is largely a matter of chance whether fertilisation takes place. Many coral colonies have timed spawning events. On these occasions, several different coral colonies release their eggs and sperm into the water at the same time. Scientists think this synchronisation is in response to environmental cues including temperature changes and day length. More complex animals such as fish and amphibians have evolved rituals. These increase the likelihood of fertilisation by ensuring that the ova and sperm are released at the same time close to each other. Even with these strategies, many of the gametes do not meet. External fertilisation is very wasteful, and is not an option for organisms that live on land.
- **Internal fertilisation** involves the transfer of the male gametes directly to the female. This does not guarantee fertilisation, but makes it much more likely. The way in which the sperm are transferred varies greatly. In many species, the male produces packages of sperm for the female to pick up and transfer to her body. More complex animals such as insects and some of the vertebrates have evolved a system whereby the male gametes are released directly into the body of the female during **mating**. This makes sure that the ova and sperm are kept in a moist environment and are placed as close together as possible, which maximises the chances of successful fertilisation.

FERTILISATION IN HUMANS

For sexual reproduction to be successful in humans, as in any other species, the gametes must meet and fuse. **Fig A** is a scanning electron micrograph of an ovum surrounded by sperm, which gives you a good idea of the scale of the male and female gametes. The ovum is fully viable and able to receive the male gamete for only a few hours. The sperm will survive a day or two in the female reproductive tract. There is little evidence to suggest that the sperm are attracted to the egg in any way – their meeting seems to be entirely a matter of chance. Nevertheless, they frequently do meet and fuse. As sperm move through the female reproductive tract, the acrosome region matures so it can release enzymes and penetrate the ovum.

EXAM HINT

Remember that many areas of biology are linked. A question could test these links; for example, the acrosome is a specialised lysosome.

The ovum released at ovulation has not fully completed meiosis. It is surrounded by a protective jelly-like layer known as the zona pellucida and some of the follicle cells. Many sperm cluster around the ovum and, as soon as the heads of the sperm touch the surface of the ovum, the **acrosome reaction** is triggered (see **fig B**). Enzymes are released from the acrosome, which digest the follicle cells and the zona pellucida. One sperm alone does not produce enough enzyme to penetrate the protective layers around the ovum. This seems to be one reason for the very large number of sperm released in ejaculation, providing enough in the oviduct to surround the ovum and digest its defences.

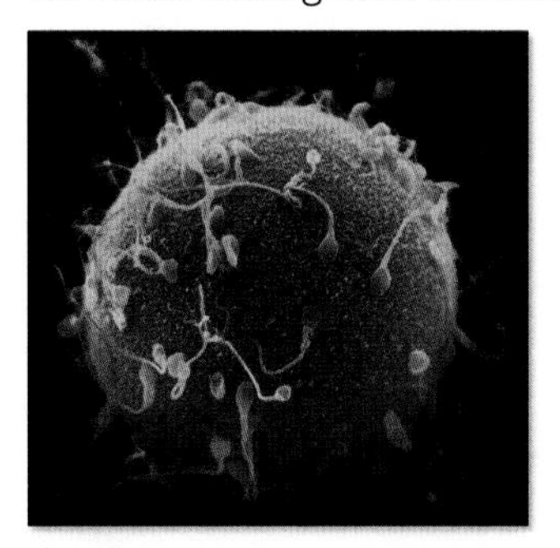

▲ **fig A** The fertilisation of a human ovum

Eventually, one sperm will wriggle through the weakened protective barriers and touch the surface membrane of the oocyte. This has several almost instantaneous effects. The second meiotic division takes place providing a haploid ovum nucleus to fuse with the haploid male nucleus. It is essential that no other sperm enter now, as this would result in **polyspermy** (fertilisation by more than one sperm) and would produce a nucleus containing too many sets of chromosomes.

The events that follow fertilisation prevent polyspermy. Ion channels in the cell membrane of the ovum open and close so that the inside of the cell, instead of being electrically negative with respect to the outside, becomes positive. This alteration in charge blocks the entry of any further sperm. It is a temporary measure

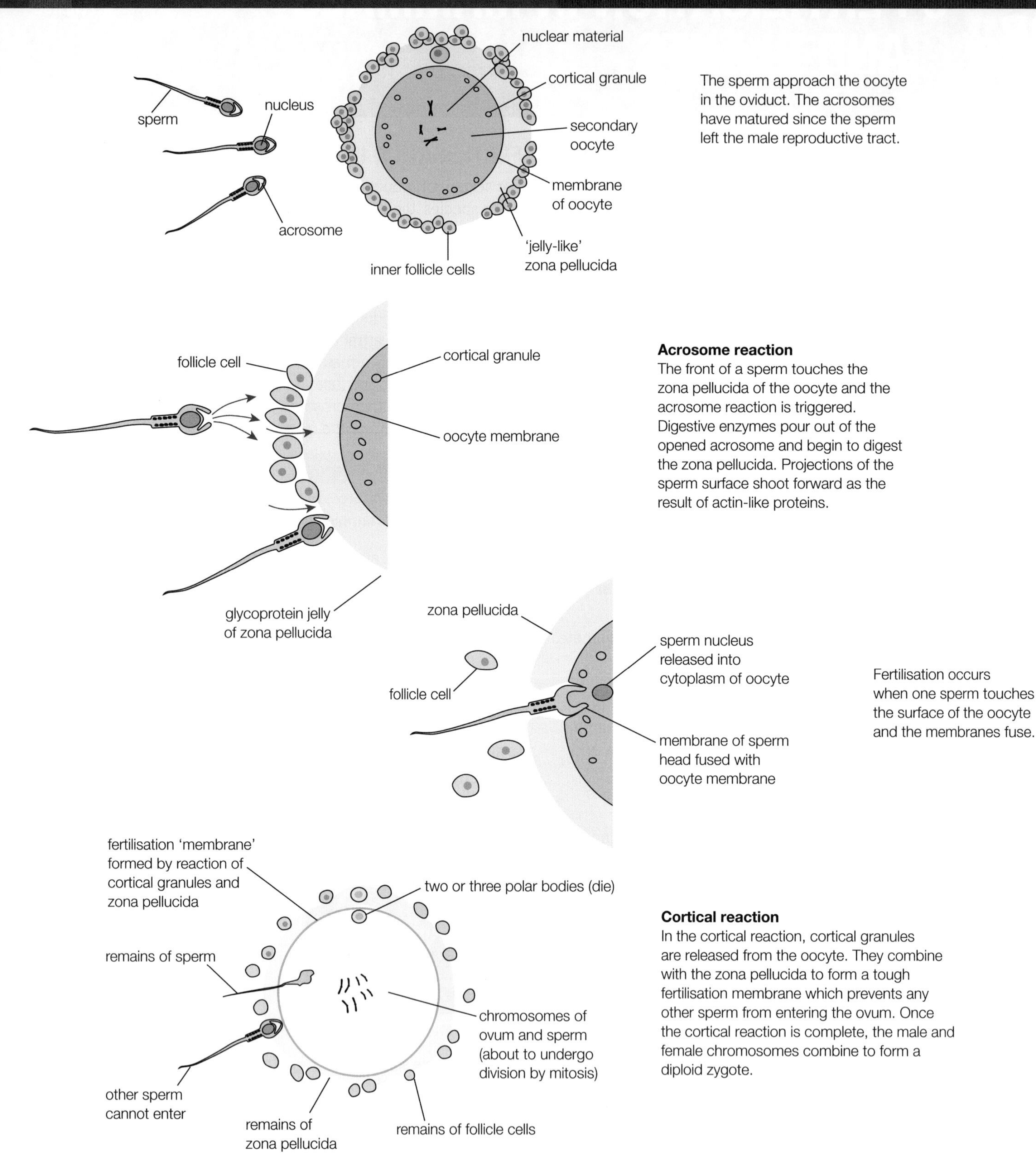

▲ **fig B** The acrosome reaction plays a vital role in the successful fertilisation of the egg and the cortical reaction ensures only one sperm is involved.

until the **cortical reaction** takes place. In the cortical reaction, cortical granules in the cytoplasm of the ovum release enzymes into the zona pellucida. These enzymes destroy the sperm-binding sites and also thicken and harden the jelly of the zona pellucida. This then forms a tough **fertilisation membrane** around the fertilised ovum. The fertilisation membrane now repels other sperm as the electrical charge returns to normal.

The head of the sperm enters the oocyte, but the tail region is left outside. Once the head is inside the ovum it absorbs water and swells, releasing its chromosomes to fuse with those of the ovum and forming a diploid zygote. At this point, fertilisation has occurred and a new individual has formed. Fertilisation is also referred to as **conception** in the case of humans.

LEARNING TIP

Remember that fertilisation is when the male and female nuclei fuse.

FERTILISATION IN PLANTS

The male gamete is contained within the pollen grain. The female gamete is embedded deep in the tissue of the ovary. The pollen grain lands on the surface of the stigma of the flower during pollination. The molecules on the surface of the pollen grain and the stigma interact. If they 'recognise' each other as being from the same species, the pollen grain begins to grow or **germinate**. Often the pollen grain will only germinate if it is from the same species, but a different plant. This helps to prevent self-fertilisation, which would reduce variety. Alternatively, pollen grains from the same plant may start to germinate, but be unable to penetrate the carpel.

A pollen tube begins to grow out from the tube cell of the pollen grain through the stigma into the style. The tip of the pollen tube produces hydrolytic enzymes to digest the tissue of the style, so the pollen tube can make its way down towards the ovule. The digested tissue acts as a nutrient source for the pollen tube as it grows. As the pollen tube grows down towards the ovary, the pollen tube nucleus and the generative nucleus travel down it. The generative nucleus divides by mitosis as it moves down the tube to form two male nuclei. These are the male gametes. The growth of the pollen tube is very fast due to the rapid elongation of the cell. Eventually the tip of the pollen tube passes through the micropyle of the ovule. Once the tube has entered the micropyle, the two male nuclei are passed into the ovule so that fertilisation can occur. Flowering plants undergo what is known as **double fertilisation**. One male nucleus fuses with the two polar nuclei to form the endosperm nucleus, which is triploid. The endosperm is involved in supplying the embryo plant with food when it begins to germinate. The other male nucleus fuses with the egg cell to form the diploid zygote (see **fig C**). At this point, fertilisation is complete and the development of the seed and the embryo within can begin.

CHECKPOINT

SKILLS INTERPRETATION

1. What is the importance of the reaction between pollen grains and the surface of the stigma in plants?
2. What part is played by enzymes in:
   (a) mammalian fertilisation?
   (b) fertilisation in flowering plants?
3. How is polyspermy prevented, and why is it important to stop it happening?

SUBJECT VOCABULARY

pollination the transfer of pollen from the anther to the stigma, often from one flower to another
external fertilisation the process of fertilisation in which the female and male gametes are released outside of the parental bodies to meet and fuse in the environment
internal fertilisation the fertilisation of the female gamete by the male gamete, which takes place inside the body of the female
mating the process by which a male animal transfers sperm from his body directly into the body of the female
acrosome reaction the reaction seen when the sperm reach the oocyte and enzymes are released from the acrosome and digest the follicle cells and the zona pellucida
polyspermy the fertilisation of an egg by more than one sperm
cortical reaction the reaction seen when cortical granules in the cytoplasm of the ovum release enzymes into the zona pellucida; these enzymes destroy the sperm-binding sites and also thicken and harden the jelly of the zona pellucida
fertilisation membrane the tough layer that forms around the fertilised ovum to prevent the entry of other sperm
conception the term used for fertilisation of the ovum in humans
germinate (of pollen) the process by which a pollen tube starts to grow out of the pollen grain to transfer the male nuclei to the ovule
double fertilisation the process that occurs in plants in which one male nucleus fuses with the two polar nuclei to form the triploid endosperm nucleus and the other fuses with the egg cell to form the diploid zygote

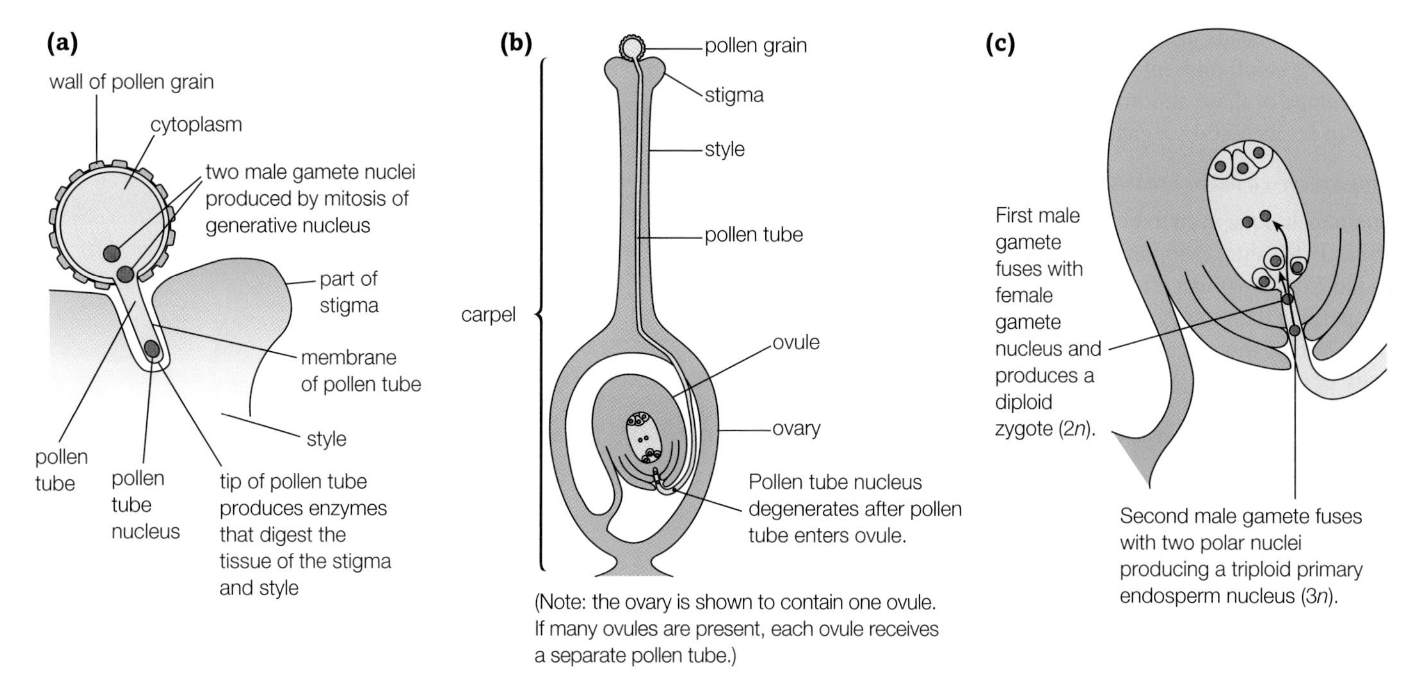

▲ **fig C** A summary of the events that follow pollination in a flowering plant and lead to fertilisation of the ovule.

TREATING MALE INFERTILITY

SKILLS CRITICAL THINKING, DECISION MAKING, INNOVATION, PERSONAL AND SOCIAL RESPONSIBILITY, CONTINUOUS LEARNING, INTELLECTUAL INTEREST AND CURIOSITY, ETHICS

Read the following extracts taken from a book about *in vitro* fertilisation. The book has been used with patients in infertility clinics and parts have been reproduced on an infertility support website. It looks at the problems of treating male infertility.

BOOK EXTRACT

… an infertile man

… the lowest number of sperm counted as normal is 20 million sperm per cm^3 of semen! Once the sperm count falls below this level, it begins to affect fertility. If the sperm count is just a bit below normal there are things which a man can do to increase the numbers – all of them very low tech! If the testes get too warm, the level of sperm production falls, so cool showers or baths, baggy underwear and loose clothing can help to increase the sperm count. Cutting out smoking can also help increase sperm numbers. If the count is really low, things are more difficult.

Numbers aren't everything

… the ability of sperm to fertilise eggs successfully depends on more than numbers. For the man to be fertile, his sperm need to have actively lashing tails and around 50% of them must swim forward in straight lines rather than round and round in circles.

It is also very important that the semen does not contain too many abnormal sperm. Every man produces a certain number of sperm with two heads instead of one, or with two tails, or with a break in the small midsection between the head and tail. But if the percentage of these abnormal sperm gets too high, then the chances of a successful pregnancy fall.

Injecting sperm – a major breakthrough

The main technique used to help men with abnormal sperm is known as ICSI (intra cytoplasmic sperm injection). Eggs are harvested from a woman after treatment with fertility drugs. A single sperm is then injected right into the cytoplasm of each egg cell. Two or three healthy embryos will then be returned to the body of the mother just as in normal IVF treatment.

There are two groups of patients for whom ICSI offers hope. ICSI can help men who have severe sperm problems, even if they cannot produce semen at all or have very few healthy sperm because it only takes one sperm to fertilise each egg. Also, couples who produce healthy eggs and sperm for IVF, but cannot achieve fertilisation can be helped, because whatever the problem with fertilisation it can be overcome by the direct insertion of a sperm into an egg. ICSI is widely used for at least 30% of all the couples who need IVF technology to conceive.

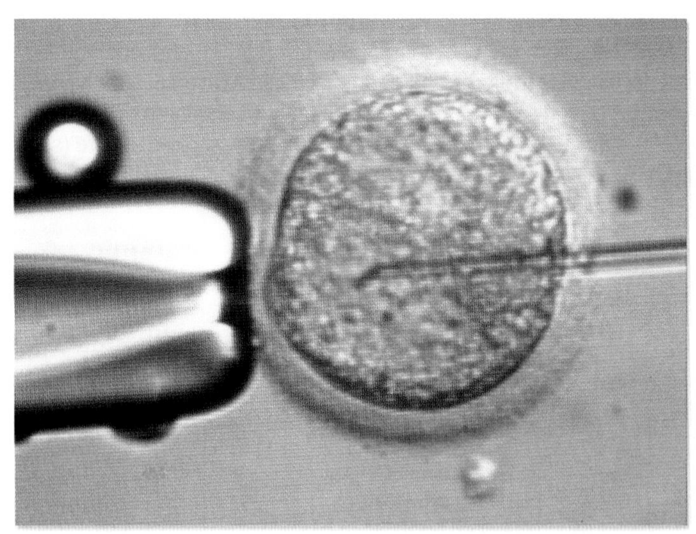

ICSI removes all the normal barriers to conception as the sperm cannot fail to reach the egg.

From: *In vitro fertilisation* by Ann Fullick. 2nd ed. 2009. Heinemann series 'Science at the Edge'

SCIENCE COMMUNICATION

1 What are the main problems with writing a book on complex biological issues such as infertility and infertility treatments that aims to inform interested readers and support people who really need the information?

2 The book contains many quotes from doctors, embryologists and research scientists involved in infertility treatments. Do you think this is a useful feature for a book of this type? Justify your response.

3 The book contains a case study of a couple with unexplained infertility who had treatment for many years before conceiving twins by IVF. They went on to have three more children naturally. Discuss the value of a case study like this in a book on infertility treatment.

SKILLS CRITICAL THINKING, PERSONAL AND SOCIAL RESPONSIBILITY, COMMUNICATION

WRITING SCIENTIFICALLY

To justify your response, you need to give clear evidence to support your answer.

BIOLOGY IN DETAIL

Now let us look at the biological basis of the problems of infertility. You have studied the structure of gamete cells and fertilisation. This will help you answer these questions.

4 Look at the ways in which a man can try to increase his sperm production and suggest biological reasons for each of the methods suggested.

5 Explain, referring to the normal process of fertilisation of an ovum by a sperm, why the problems described in this extract about low sperm numbers, abnormal sperm or inactive sperm would lower or remove the chance of a successful pregnancy.

ACTIVITY

Across the world, infertility affects about 15% of couples. Males are found to be solely responsible for 20–30% of infertility cases and contribute to 50% of cases overall. In the Middle East, the male factor is involved in 60–70% of infertile couples. Infertility can also be due to female factors or a combination of low fertility in both partners.

Research the causes and treatment of infertility, bearing in mind what you know about gamete cells, fertilisation and reproduction. Produce a visual aid to show the data in the most eye-catching and informative way possible.

SKILLS CREATIVITY, INNOVATION, CONTINUOUS LEARNING, INTELLECTUAL INTEREST AND CURIOSITY

SCIENTIFIC SOURCES

As you carry out your research, consider where the sources you find have come from, who wrote them and for whom they were written. Established scientific publications are good sources of reliable information, whereas other resources might be less dependable for a number of reasons. Think about what makes a source reliable and why.

3B EXAM PRACTICE

1 (a) The cell cycle includes interphase and mitosis. Mitosis has four phases: prophase, metaphase, anaphase and telophase. The photograph below shows plant root cells at various stages of the cell cycle.

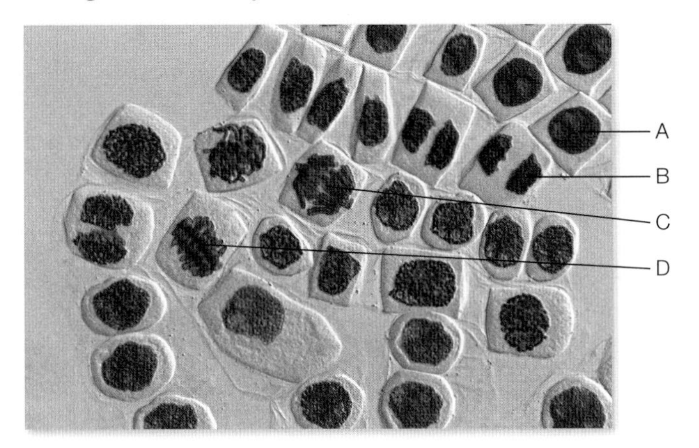

(i) Which cell is undergoing anaphase?

A cell A
B cell B
C cell C
D cell D [1]

(ii) Which cell is shown just before cytokinesis?

A cell A
B cell B
C cell C
D cell D [1]

(iii) State how many of the cells in the photograph are in metaphase. [1]

(iv) Calculate the mitotic index for this plant tissue. Show your working. [2]

(b) Describe the events that take place during prophase and metaphase of mitosis. [5]

(c) Name two structures that are produced during interphase. [2]

(Total for Question 1 = 12 marks)

2 The graphs below show changes in the DNA content of cells during the cell cycle in two different plants, A and B.

Plant A

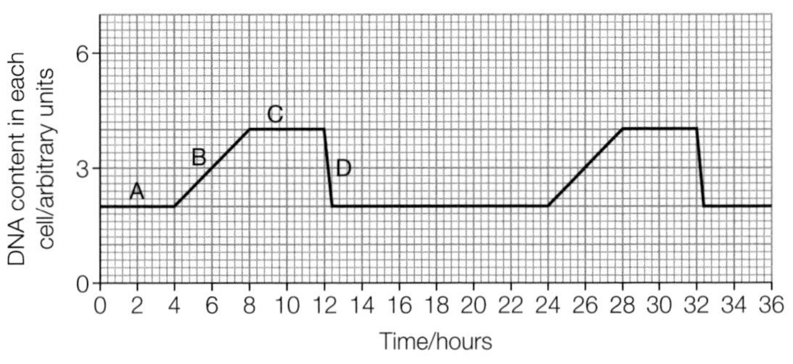

Plant B

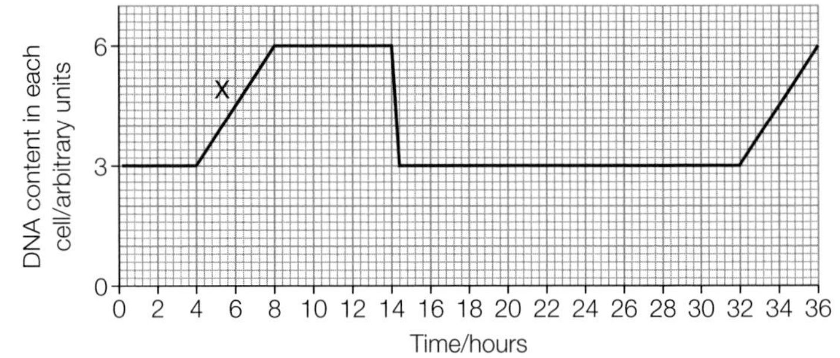

(a) At what point are the cells of plant A undergoing mitosis?

A A
B B
C C
D D [1]

(b) What is happening at point X in plant B?

A replication of the cell
B replication of DNA
C protein synthesis
D meiosis [1]

(c) Describe the events that are occurring inside the cells of plant A between 13 and 24 hours. [2]

(d) State **two** differences between the cell cycle of plant A and the cell cycle of plant B and suggest what might cause these differences. [3]

(e) Calculate the DNA content of each cell in plant B after meiosis. [2]

(Total for Question 2 = 9 marks)

3 Fertilisation involves the fusion of haploid nuclei.

(a) The diagram below shows a human sperm cell.

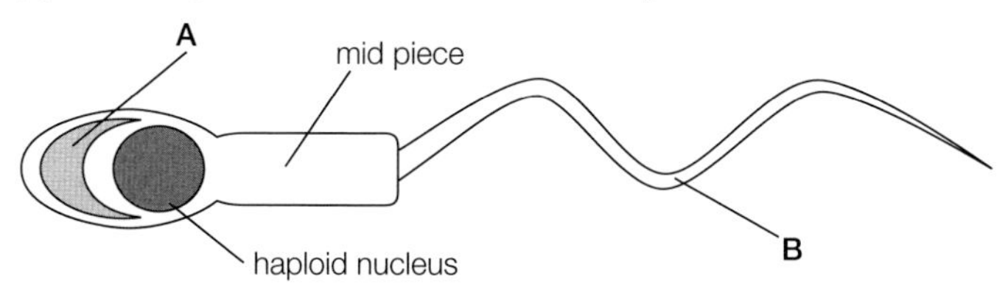

(i) Name the structures labelled A and B. [2]

(ii) Explain why the sperm has a haploid nucleus. [2]

(iii) Complete the table comparing human spermatozoa and a human ovum.

Feature	Spermatozoa	Ovum
size	50 μm	
number of chromosomes	23	
motility	very motile	
number needed for fertilisation		one

[4]

(b) Describe the changes in the female gamete from the point when a sperm releases its digestive enzymes to the point when the two nuclei fuse. [3]

(Total for Question 3 = 11 marks)

4 (a) The diagram below shows a section through a *Primula* (primrose) flower.

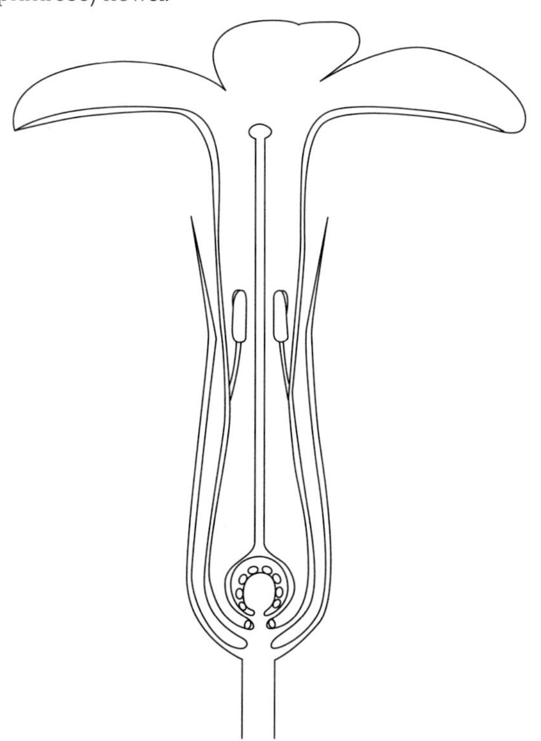

Pollination has occurred once pollen lands on the stigma. Describe the events that take place after pollination to ensure that fertilisation takes place. [4]

(b) An experiment was carried out to measure the rate of growth of a pollen tube in germinating pollen grains. Fresh pollen grains were placed in a 0.5 mol dm^{-3} sucrose solution, and kept at a temperature of 20 °C. The pollen tube growth rates were recorded at time intervals of 30 minutes for a period of three hours.

(i) Suggest why sucrose was added to the solution. [1]

(ii) The results of this experiment are shown in the table below.

Time (min)	Growth rate of pollen tube (mm per 30 min)
30	0.156
60	0.169
90	0.182
120	0.169
150	0.052
180	0.032

Draw a graph of these results. [4]

(iii) Describe how the growth rate of the pollen tube changed during the experiment. [2]

(iv) Calculate the length of the pollen tube at the end of the experiment. [2]

(c) Boron is known to affect the growth of pollen tubes from pollen grains. In one experiment, pollen grains were placed in two different media, one containing boron and one without boron.

The lengths of the pollen tubes were measured every 6 hours, for a total of 36 hours.

The results are shown in the table below.

Time (hours)	Mean length of pollen tubes (μm)	
	Medium without boron	Medium with boron
6	20	45
12	45	90
18	70	170
24	90	250
30	100	280
36	100	300

The mean growth rate of the pollen tubes without boron from 6 to 12 hours is 4.17 μm h^{-1}.

(i) Calculate the mean growth rate of the pollen tubes with boron from 6 to 12 hours.

Show your working. [2]

(ii) Compare the growth of pollen tubes in these two media. [2]

(Total for Question 4 = 17 marks)

TOPIC 3 CELL STRUCTURE, REPRODUCTION AND DEVELOPMENT

CHAPTER 3C DEVELOPMENT OF ORGANISMS

If a racehorse damages a tendon, a promising racing career could be over. However, several racehorses have recently won prestigious races after having stem cell treatment to help repair damaged tendons. The stem cells were taken from the horses themselves, grown on in the laboratory and then injected into the damaged area. In 2009, the winner of the Welsh Grand National had previously had stem cell treatment. In 2014, a trial of the same treatment began in human beings. Stem cells hold the exciting prospect of being able to help paralysed people to walk, blind people to see, and people threatened with degeneration of the brain to function normally.

In this chapter, you will learn how cells differentiate to become specialised. You will learn how some genes have multiple alleles and others are linked on the same chromosome so that they tend to be inherited together. You will also see how genes and the environment interact to affect the phenotype. You will see how genes can be switched on and off, and gene expression can be controlled at several stages during the transcription and translation. You will look at transcription factors, spliceosomes, and the exciting new science of epigenetics in which scientists are uncovering how the environment and genes interact. You will also learn about the different types of stem cell - totipotent, pluripotent and multipotent - and their roles in the body. Finally, you will discover how epigenetic modifications can result in totipotent stem cells becoming fully differentiated body cells. You will find out how pluripotent stem cells can be produced and the opportunities they provide to develop new medical advances.

MATHS SKILLS FOR THIS CHAPTER

- **Translate information between graphical, numerical and algebraic forms** (*e.g. considering the rate of gene expression*)

What will I study in this chapter?

- How cells differentiate to become specialised
- How multiple alleles can affect the phenotype
- How some genes can be linked together - autosomal linkage
- How the environment can alter the expression of genes
- The control of gene expression
- Factors that affect gene expression, including transcription factors, post-transcription changes in the mRNA and epigenetic modification
- That epigenetic modification is important in ensuring cell differentiation
- The differences between totipotent and multipotent stem cells
- The early development of the embryo to the blastocyst stage
- How society uses scientific knowledge to make decisions about the use of stem cells in medical advances

What prior knowledge do I need?

Chapter 2C

- Inheritance of monohybrid characteristics

Chapter 3A

- The ultrastructure of eukaryotic cells

Chapter 3B

- The cell cycle as a regulated process in which cells divide into two identical daughter cells
- The stages of mitosis
- How mitosis contributes to growth, repair and asexual reproduction

What will I study later?

Chapter 7C (Book 2: IAL)

- How genes can be switched on and off by steroid hormones

Chapter 8C (Book 2: IAL)

- Gene technology
- How drugs can be produced by genetically engineered organisms
- The way recombinant DNA is produced
- How recombinant DNA can be inserted into other cells using a variety of vectors
- How antibiotic resistance and replica plating can be used to identify recombinant cells

3C 1 CELL DIFFERENTIATION

SPECIFICATION REFERENCE 3.9(i) 3.9(ii) 3.18 3.21

LEARNING OBJECTIVES

- Know that a locus is the location of genes on a chromosome.
- Understand the linkage of genes on a chromosome.
- Understand how cells become specialised through differential gene expression, producing active mRNA leading to the synthesis of proteins which in turn control cell processes or determine cell structure in animals and plants.
- Understand how some phenotypes are affected by multiple alleles for the same gene, or by polygenic inheritance (many different genes).

In **Chapter 3B** you learned how, in human reproduction, the ovum and sperm meet to form a single new cell. Also, you learned that one cell divides many times to form an entire human being, made up of many different cells, tissues and organs. How does this happen? The genes in the DNA of a cell code for proteins. When a gene is expressed, active mRNA is produced and this leads to the synthesis of the protein it codes for. Many of these proteins are enzymes which control everything that happens in the rest of the cell: the compounds that are synthesised, the compounds that are broken down, even which genes are subsequently expressed. So the factors that control gene expression, control everything about a cell.

GENE EXPRESSION IN ACTION

In a multicellular organism, every cell contains the same genetic information but different cells perform different functions. They differentiate and develop into different tissues and organs. As this **cell differentiation** occurs, different types of cell produce more and more proteins which are specific to their cell type. Their shape and arrangement of the organelles also differ.

At the same time, almost all cells have a number of 'housekeeping' proteins in common. These are the proteins found in the structures which are common to most cells. Examples of such proteins include the structural proteins of the membranes and the enzymes needed in cellular respiration.

The fact that each cell type produces specific proteins that relate to the function of the cell means that different genes must be expressed in different types of cell. For example, the enzymes needed to produce insulin are only found in islet of Langerhans cells, in the pancreas. This provides scientists with a useful tool. The degree of differentiation between cells can be measured by comparing the proteins contained in cells and this enables scientists to work out which genes have been expressed and which have been suppressed. Scientists have recently discovered that, in human beings, the testes have the greatest variety of extra proteins. They have 999 proteins on top of the normal housekeeping proteins. The cerebral cortex of the brain has 318 extra proteins, and smooth muscle cells only express housekeeping proteins.

HOW DO GENES CONTROL THE PHENOTYPE?

In **Chapter 2C**, you discovered that each chromosome has several genes each of which codes for a particular protein. Each gene is found at a particular place on the chromosome and this location is known as the **locus** of the gene. Each gene has at least two different forms, known as alleles.

MULTIPLE ALLELES

In the examples you have studied so far, you have only looked at traits that are inherited as genes with just two possible alleles – for example, for pea shape and colour and the inheritance of cystic fibrosis and colour blindness. However, some features are determined by **multiple alleles**. This means there are more than two possible variants. No matter how many possible alleles there are, any one diploid individual will only inherit two of them. These alleles are still inherited in the same way, although the patterns of dominance may be more complex. One clear example of multiple alleles is the human ABO blood group system (see **Section 2C.2**). Here there are three possible alleles – A, B and O. These have a different pattern of dominance as well. Both A and B are dominant to O, so the O blood group is recessive, but A and B are **codominant**. This means both alleles are expressed and produce a protein. So different combinations of alleles give you different blood groups (see **table A** and **fig A**).

GENOTYPE	PHENOTYPE
OO	blood group O
AO or AA	blood group A
BO or BB	blood group B
AB	blood group AB

table A ABO blood groups – an example of inheritance through multiple alleles

▲ **fig A** Blood donations save lives. It is vital to know the blood group of each donor so the blood can be matched to the patient who needs it.

GENE LINKAGE

In the simple models of inheritance you have looked at so far, you considered a single trait inherited by a pair of alleles on the autosomes. You assumed that genes are inherited randomly, and that each gene, coding for an aspect of the phenotype, is inherited independently of all the other genes. The one big exception to this is sex linkage, where certain characteristics are inherited on the sex chromosomes and so it makes a difference if you are male or female.

However, all this is a very long way from the real mechanisms by which inheritance takes place. Hundreds and thousands of genes go to make up the genotype of any one individual and they are all passed on at the same time. Scientists have discovered that there are many cases where characteristics inherited on single genes are always associated with other characteristics, also carried on single genes. These genes are said to be linked. So how does **gene linkage** work?

LEARNING TIP

Remember that each person has pairs of chromosomes and pairs of alleles, therefore it is not possible to have more than two of the three alleles for blood group.

POLYGENIC TRAITS

Monohybrid genetic crosses involve only one gene locus. It is important to remember, however, that most traits in living organisms are determined not by a single gene but by several or many interacting genes. They are **polygenic**. Characteristics such as eye colour, weight and intelligence are determined by several different genes at different loci and, in many cases, interactions with the environment add further variety. So when you think about how genes affect the phenotype of an organism, remember monohybrid crosses are a very simple model that helps us to understand a much more complex reality.

Looking at how two different genes are inherited completely independently in a process called **digenic (dihybrid) inheritance** will help you understand gene linkage (see **fig B**). Digenic crosses are breeding experiments involving the inheritance of two pairs of contrasting characteristics at the same time. Although still a very long way from the complexity of real events, this goes one step closer to the living cell.

There are some occasions when the ratios are not what you expect. There can be several explanations for this:

- small sample size
- experimental error – especially when working with organisms such as *Drosophila*, which can escape or die relatively easily
- the process is random and so sometimes the unexpected happens
- unexpected ratios can mean the genes being examined are both on the same chromosome (they are linked).

You are now going to look at the process of gene linkage.

RATIOS IN DIHYBRID INHERITANCE

In the fruit fly *Drosophila*, grey bodies are dominant to ebony bodies, and long wings are dominant to very short vestigial wings. For a cross between two heterozygotes, you can use a Punnett square to determine the results of digenic inheritance. The crossing of two non-linked heterozygotes in a dihybrid cross always results in a 9 : 3 : 3 : 1 ratio of phenotypes (see **fig B**).

L represents the allele for long wings.
l represents the allele for vestigial wings.
G represents the allele for grey body.
g represents the allele for ebony body.

parental phenotypes: long wings grey body × long wings grey body

parental genotypes: LlGg × LlGg

gametes: LG Lg lG lg LG Lg lG lg

offspring genotypes:

Gametes	LG	Lg	lG	lg
LG	LLGG	LLGg	LlGG	LlGg
Lg	LLGg	LLgg	LlGg	Llgg
lG	LlGG	LlGg	llGG	llGg
lg	LlGg	Llgg	llGg	llgg

offspring phenotypes: long wings grey bodies : long wings ebony bodies : vestigial wings grey bodies : vestigial wings ebony body

9 : 3 : 3 : 1

fig B Crossing two heterozygotes in a dihybrid cross always results in a typical 9 : 3 : 3 : 1 ratio of parental phenotypes (the same as the parents) and recombinant phenotypes (different combinations of the phenotypes which result from inheriting the different alleles independently).

LEARNING TIP

Remember this ratio, and remember it is a ratio. Results of a cross may not look exactly like this ratio but will be similar.

LINKAGE IN FRUIT FLIES

When homozygous *Drosophila* with dominant broad abdomens and long wings are crossed with flies displaying recessive narrow abdomens and vestigial wings, all the offspring show the dominant phenotype, but possess the heterozygote genotype.

When the heterozygous flies are crossed, it is expected that a normal 9:3:3:1 ratio would be observed. This is the ratio of parental phenotypes (they look the same as the parents) and recombinant phenotypes (different combinations of appearance which result from inheriting the different alleles independently). Instead, scientists see a 3:1 ratio of dominant : recessive phenotypes, which is what we would expect in a monogenic cross. The explanation of this apparent discrepancy is that the genes which determine the width of the abdomen and the length of the wings are linked. This means they are sited on the same chromosome and inherited as if they were one unit (see **fig C**).

> **LEARNING TIP**
>
> Remember that in a closely linked cross, genes B and L are acting as if they are one gene: they always stay together.

B represents the allele for broad abdomen. b represents the allele for narrow abdomen.
L represents the allele for long wings. l represents the allele for vestigial wings.

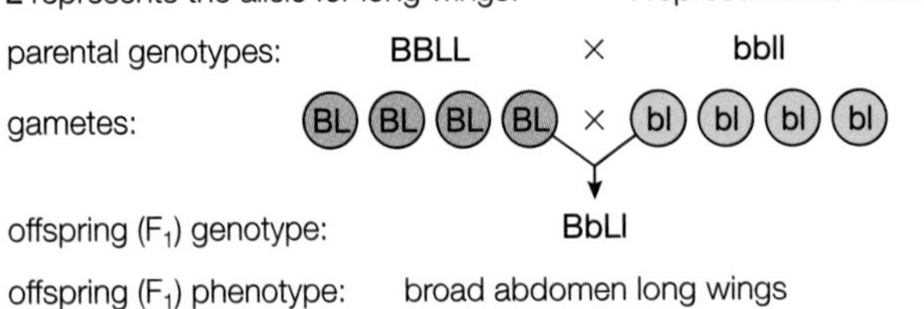

offspring (F1) phenotype: broad abdomen long wings

expected self cross of offspring (F1)

offspring (F1) genotypes: BbLl × BbLl

gametes: BL Bl bL bl × BL Bl bL bl

offspring (F2) genotypes:

	BL	Bl	bL	bl
BL	BBLL	BBLl	BbLL	BbLl
Bl	BBLl	BBll	BbLl	Bbll
bL	BbLL	BbLl	bbLL	bbLl
bl	BbLl	Bbll	bbLl	bbll

offspring (F2) phenotypes:

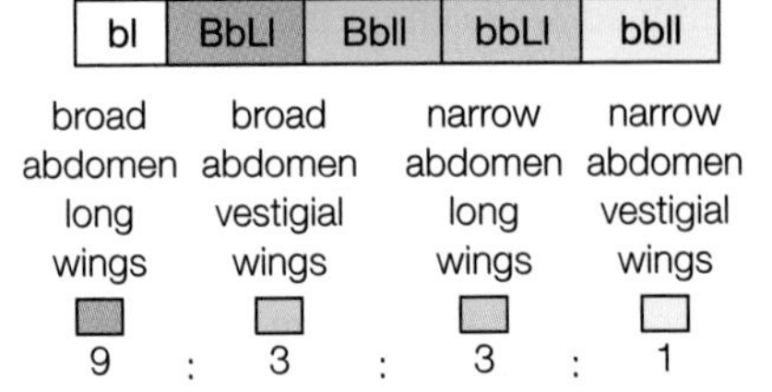

actual self cross of offspring (F1)

offspring (F1) genotypes: BLbl × BLbl

gametes: BL bl BL bl

offspring (F2) genotypes:

	BL	bl
BL	BBLL	BbLl
bl	BbLl	bbll

3 : 1

offspring (F2) phenotypes: broad abdomen long wings; narrow abdomen long wings

evidence of linked genes

▲ **fig C** These two genes are linked. They are inherited as if they are one unit because they are positioned close together on the same chromosome.

> **EXAM HINT**
>
> It is often helpful to draw a small sketch or diagram to help explain your answer to a question but the diagram must be labelled or annotated.

IDENTIFYING LINKED GENES

With completely unlinked genes, two genes are found on separate chromosomes and approximately equal numbers of gametes are formed which contain either the parental combinations of alleles (parental types) or a different combination of alleles (recombinant types). However, genes for different characteristics which are found on the same chromosome are linked. They are inherited, to a greater or lesser degree, as if they were a single gene. When genes are closely linked – for example, when they are located close to each other on the chromosome – then, when gametes are formed, recombination events which separate them rarely occur during meiosis. If the genes are more loosely linked – for example, when they are located further apart on the chromosome – then the number of recombination events in meiosis will be higher. Therefore, the tightness of the linkage of a pair of genes is related to how close together the linked genes are located on the chromosome. Genes that are very close together are less likely to be split during the crossing over stage of meiosis than genes that are further apart. If two or more genes are positioned very close together on a chromosome, they may be so tightly linked that they are never split up during meiosis and so the gametes formed will always be of the parental types. If the genes are further apart, crossing over between them is more likely to occur. Although in the majority of cases they will be passed on as a parental unit, sometimes they will be mixed and recombinant gametes produced, which will, in turn, be reflected in the offspring (see **fig D**).

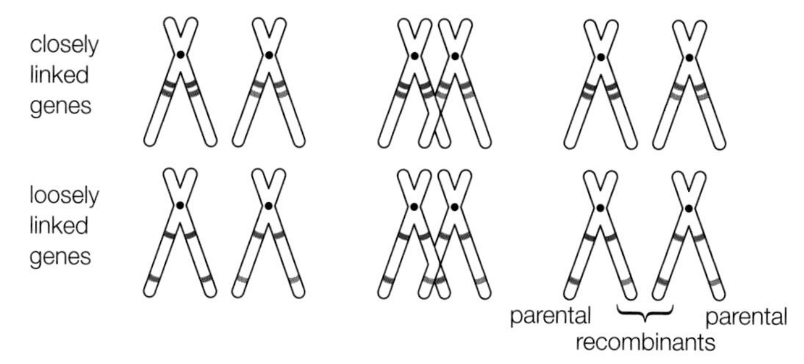

▲ **fig D** During meiosis, closely linked genes will be passed into the gametes as a single unit. Genes on the same chromosome but positioned further apart will form some recombinant gametes, with different mixtures of the alleles.

CHECKPOINT

SKILLS PROBLEM SOLVING

1. Use Punnett squares to help you answer this question.
   (a) A man has blood group A and his wife has blood group B. What are their possible genotypes?
   (b) If they are both heterozygotes, what is the likelihood that one of their children will be blood group O?
2. Explain the difference between a characteristic in the phenotype which has multiple alleles and a characteristic which is polygenic.
3. A scientist carries out a cross using fruit flies. It involves two characteristics which he expects to be inherited independently. The results are not what he expected. Suggest two possible reasons for these unexpected results.

SUBJECT VOCABULARY

cell differentiation the process by which a less specialised cell becomes more specialised for a particular function
locus place on a chromosome where any particular gene is found
multiple alleles more than two possible variants at a particular locus
codominant both alleles are expressed in the phenotype
gene linkage when genes for two different characteristics are found on the same chromosome and are close together so they are linked and inherited as a single unit
polygenic phenotypic characters determined by several interacting genes
digenic (dihybrid) inheritance the inheritance of two pairs of contrasting characteristics at the same time

2 INTERACTIONS BETWEEN GENES AND THE ENVIRONMENT

LEARNING OBJECTIVES

- Understand how phenotype is the result of an interaction between genotype and the environment.
- Understand how some phenotypes are affected by multiple alleles for the same gene, or by polygenic inheritance, as well as the environment, and how polygenic inheritance can give rise to phenotypes that show continuous variation.

EXAM HINT

Detailed questions can be asked about the effect of the environment.

Remember this is complicated and involves both the genetics of the organism and the conditions it is growing in.

Your genetic make-up is obviously very important in determining your phenotype or appearance. However, there is more than just your genotype involved. Genetically identical plants (clones) grow very differently when exposed to varying amounts of light, water and soil nutrients. They can be used to demonstrate very clearly that an organism's environment also has a big impact on its appearance. The ability of animals to grow as much as their genes allow also depends heavily on environmental factors such as the amount of food available. But it is more difficult to investigate the impact of environment on the phenotype with animals because of the difficulty of producing large numbers of cloned, genetically identical organisms, and the ethical aspects of experimentation.

The impact of environment is seen in Siamese cats, and certain rabbit breeds, that have dark 'points' on the ears, the muzzle and the paws (see **fig A**). The genotype of these animals suggests that they should have dark fur all over the body as a result of melanin which is produced in a process involving the enzyme tyrosinase. However, a mutation in Siamese cats results in a version of tyrosinase that is inactive at normal body temperatures and only works at lower temperatures. So in Siamese cats, the fur over the majority of the body is pale, but at the extremities – the ears, paws and nose where the temperature is lower – the enzyme is not denatured and as a result the fur is dark.

▲ **fig A** The environment of the animal's body interacts with the genotype to give the unique markings of a Siamese cat.

DID YOU KNOW?

E. COLI AND THE *LAC* OPERON

Environmental factors that affect the phenotype of an organism are often described in a general way, such as nutritional levels or temperature. But by looking at the phenotypes of bacteria, scientists have shown that it is possible in some cases to see the environmental effect take place at a biochemical level. In the 1950s and 1960s, two French geneticists, François Jacob and Jacques Monod, were working with the common gut bacterium *Escherichia coli*.

Lactose is a disaccharide found in milk that is broken down into galactose and glucose by the enzyme β-galactosidase. If there is no lactose present in the environment of the *E. coli* bacteria, then very few β-galactosidase enzyme molecules are found in each bacterial cell. But if lactose becomes available (which happens in your gut when you eat ice cream or drink a milkshake, for example) each bacterial cell will soon contain thousands of molecules of β-galactosidase and two other related enzymes. The presence of lactose induces the bacterial cells to make the β-galactosidase enzymes.

Jacob and Monod showed that this effect takes place at the level of the genes. They found a cluster of genes which function together to make the enzymes for breaking down lactose, and they called the cluster the *lac operon*. They developed a model of a regulatory gene which produces a repressor substance when there is no lactose present. This repressor binds with a particular region of DNA and blocks the binding of RNA polymerase; this prevents the code for the lactose-digesting enzymes being transcribed. So mRNA is not made, and neither is the lactose-digesting enzyme.

When lactose is present, it binds to the repressor substance, making it change shape. This new complex can't bind to the DNA so RNA polymerase has free access to the gene and transcription takes place, closely followed by translation at the ribosomes to produce the enzymes. The bacterial cell can now use lactose as a food source. So changes in the environment directly cause the change in the phenotype from not producing β-galactosidase to producing active enzyme (see **fig B**).

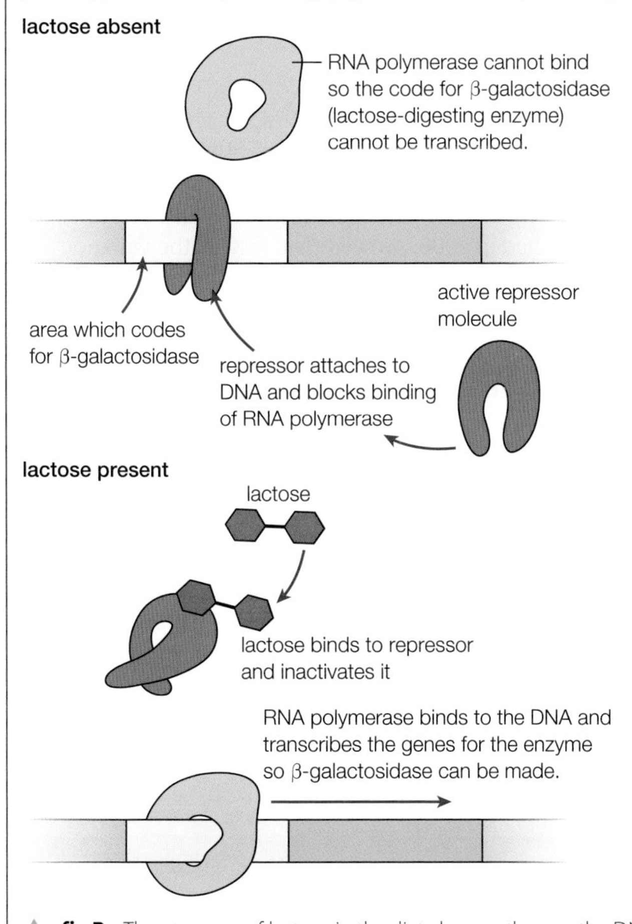

▲ **fig B** The presence of lactose in the diet changes the way the DNA is transcribed and translated. The presence of lactose switches on the production of the enzyme needed to digest it.

Operon systems like this give very sensitive control of the metabolism of the cell, allowing it to change in response to changes in the environment. They also provide an excellent research tool because scientists can change the nutrients which are given to microorganisms growing in cultures. This causes a change in their phenotypes and scientists can then investigate the mechanisms which cause these changes.

STUDYING VARIATION IN HUMANS

In mammals such as humans, the environment in the uterus affects the development of the fetus before birth. If the mother is malnourished, or the placenta is not very effective, the fetus may be deprived of vital nutrients and not achieve its full potential for growth. If a mother smokes, her fetus will be deprived of oxygen and this can affect the growth of both the body and the brain. Some drugs, and certain illnesses in a mother, can all have serious consequences for the phenotype of the fetus.

The genotype is also very important in determining the phenotype in human beings. But many characteristics, including our intelligence and our appearance, are the result of polygenic inheritance. This makes the situation even more complex. There are several difficulties in studying the interaction of genotype and environment in human beings. It is very important during any experiment on this interaction that all the organisms are subjected to the same conditions as much as possible. Then differences that are found between the organisms can be considered as genetic differences. However, in human beings, imposing conditions like these is impossible. Scientists need other ways of answering questions about the interaction of 'nature' and 'nurture', in other words the genetic make-up and the environment.

TWIN STUDIES

One strategy is to consider genetically identical individuals and try to separate the effect of genes and environment. In humans, this involves twin studies (as you saw in **Section 1C.2**). Identical twins are human clones – they have the same genetic material. Non-identical twins are just like normal brothers and sisters (siblings), with closely related but not identical DNA but, because they are the same age, they are more likely to have a similar environment than ordinary siblings. Ordinary siblings are useful as a control group. If they show a greater difference, it suggests that the environment has a stronger influence on that characteristic.

From the study summarised in **table A**, height appears to have a strong genetic component and is influenced relatively little by environmental factors. On the other hand, body mass also seems to be affected by external factors such as the family eating habits. IQ – one measure of intelligence – seems to be a combination of both, with environment playing a distinct and important role.

A team at University College, London studied 5000 pairs of twins aged between 8 and 11 years that were brought up together. Their results, published in 2008, showed that 77% of the variation in BMI and waist circumference of the children was caused by their genes and 23% by their home environment.

EXAM HINT

When using data from a table, always process the data. If you quote data without adding information, you will not gain marks.

TRAIT	IDENTICAL TWINS REARED APART	IDENTICAL TWINS REARED TOGETHER	NON-IDENTICAL TWINS	NON-TWIN SIBLINGS
Height difference	1.8 cm	1.7 cm	4.4 cm	4.5 cm
Mass difference	4.5 kg	1.9 kg	4.6 kg	4.7 kg
IQ score difference	8.2	5.9	9.9	9.8

table A These data show the results from a US study based on 19 pairs of identical twins reared apart, along with 50 pairs each of identical twins reared together, non-identical twins and non-twin siblings, by Newman, Freeman and Holzinger at the University of Chicago in 1937. Although these data are old, they are still relevant today.

We do not know everything about how genes influence human traits, but work like these twin studies helps improve our understanding. **Fig C** shows some conclusions about relative genetic and environmental influences taken from several similar studies.

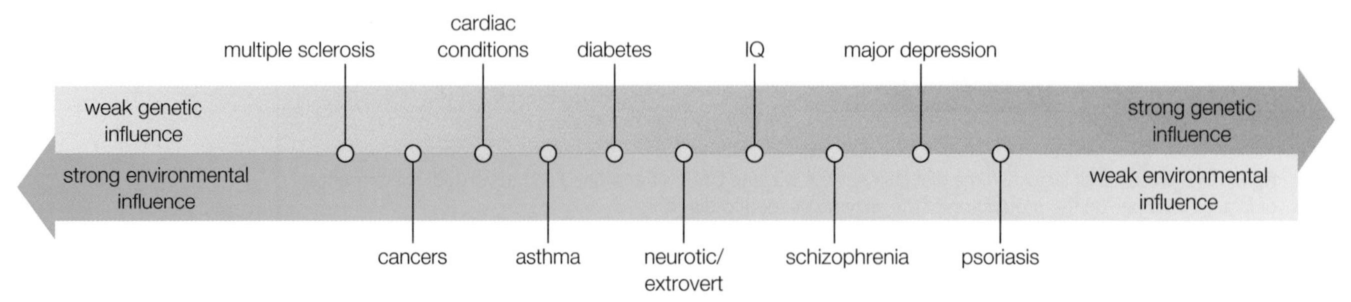

▲ **fig C** Nature versus nurture – the evidence suggests that both influence many of our characteristics.

DID YOU KNOW?

EPIDEMIOLOGICAL STUDIES

One way of separating environmental and genetic influences in humans is to look at very large numbers of individuals and consider a genetic feature together with information about lifestyle. For example, the Health Statistics Center of the West Virginia Bureau for Public Health in the US tracked the effect of a mother smoking during pregnancy by looking at data on all the live babies that were born in the region for the 10-year period 1989–98. These data – an enormous sample – were collected from West Virginia certificates of live birth, which include a question regarding the mother's smoking habits during pregnancy. The difference in birth weights and premature births between the babies born to mothers who are smokers compared to mothers who are non-smokers is quite striking and gives a clear picture of the impact of smoking on the phenotype of the baby (see **fig D**). The large sample size and 10-year time period of the study means that these data give a reliable assessment of the impact that smoking has on the phenotype of a baby.

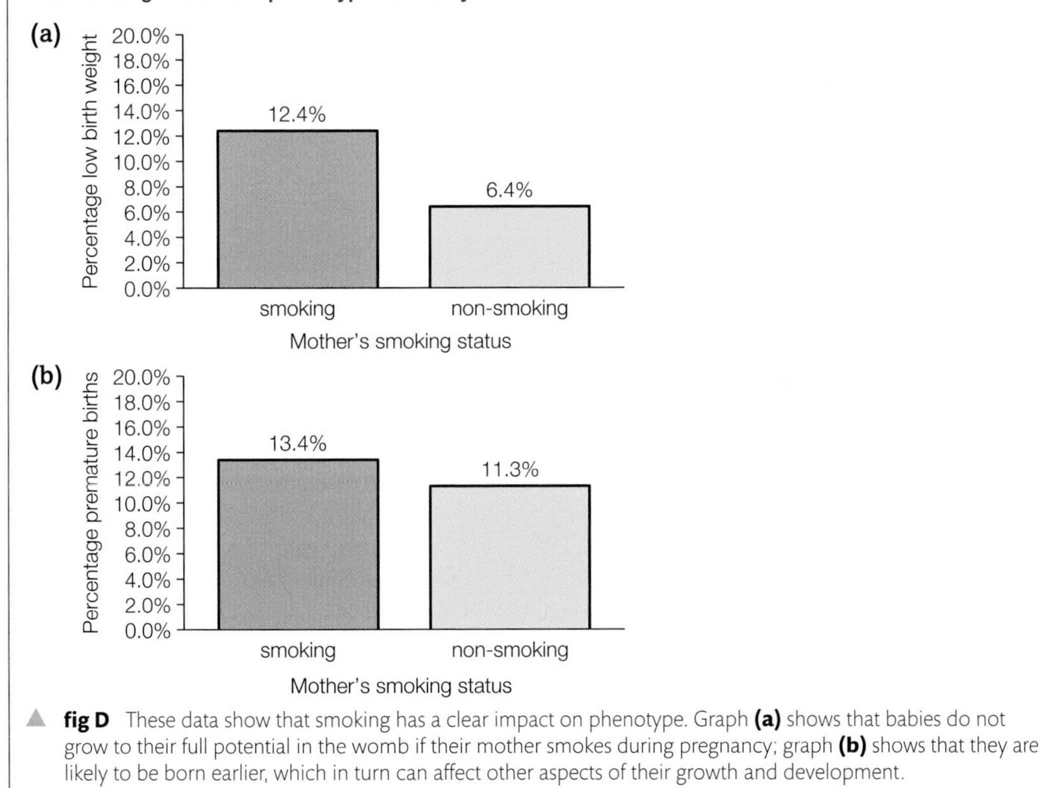

▲ **fig D** These data show that smoking has a clear impact on phenotype. Graph **(a)** shows that babies do not grow to their full potential in the womb if their mother smokes during pregnancy; graph **(b)** shows that they are likely to be born earlier, which in turn can affect other aspects of their growth and development.

Different species are often obviously very different from each other, but individuals of the same species can also show remarkable variation. As you have seen, some of the variation you see around you comes from the genotype of the organism, but some of it comes from the environment in which the organism grows. It is often useful to measure the variation in a population – it can help you decide whether a characteristic is inherited on a single gene or is polygenic, and how much the environment affects a characteristic.

DISCONTINUOUS AND CONTINUOUS VARIATION

In any population of organisms there are two types of variation – **discontinuous variation** and **continuous variation**. Discontinuous variation is shown by features that are either present or not, such as blood groups or sex (male or female). These features are generally determined by one or at most a very few genes, and the environment does not usually have an effect – you are either male or female, and your blood group is either A, B, AB or O. (There are exceptions of course – exposure to high levels of sex hormones in the uterus or rare chromosome mutations can make the sex of an individual difficult to determine, but usually discontinuous variation is very obvious.)

Characteristics that show continuous variation include weight and height in an animal species, or the number of leaves on a plant. In continuous variation, factors such as these are often determined by multiple genes (they are polygenic) but they are also very much affected by the environment.

LEARNING TIP

Remember that discontinuous variation has simple genetics, only one gene makes a difference: one difference, one gene. Continuous variation is often caused by a wide range of factors and genes.

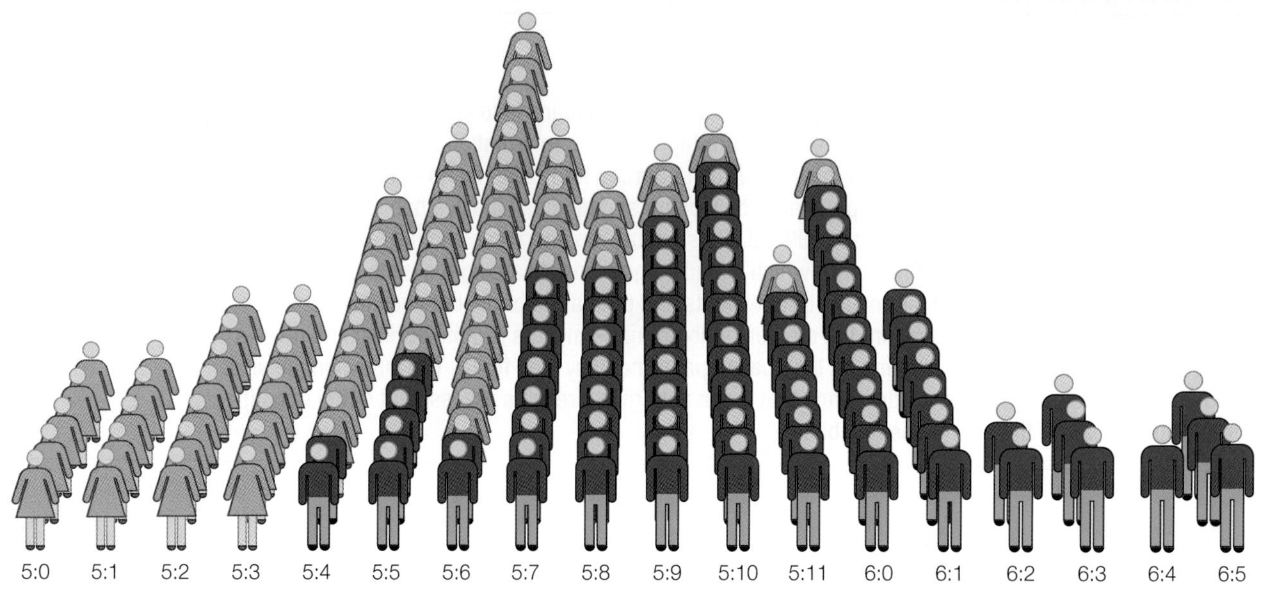

▲ **fig E** Height is a feature showing continuous variation, with clear sex differences but with overlap between taller women and shorter men.

EXAM HINT

When comparing two concepts such as continuous and discontinuous variation, a suitable table could be drawn with comparable factors in the same row.

STUDYING CONTINUOUS VARIATION

Height in humans has a very strong genetic component. Tall parents tend to have taller children than short parents. It is a polygenic feature: different genes affect different factors related to size, such as whether the person is male or female, the length of the bones in the legs or the size of the vertebrae. Genes also control the production of growth hormone and rate of bone growth. But height also has an environmental element. If an individual has a balanced diet throughout their growing years, then they are more likely to achieve their genetic height potential than if they are malnourished during one or more of their major growth periods. But if they are deprived of factors such as protein, or the calcium ions needed for bone growth in their diet, the body does not have the resources to fulfil the genetic instructions and so the person will not reach their full potential height. However, if someone has genes for shortness, they will not become tall no matter how much they eat. Height shows continuous variation because of the variety of factors that influence it.

When studying continuous variation in a population, you need to take large samples because chance can affect the results. If you take only a small sample from one specific area, you might, by random chance, pick a group of the shortest individuals in the population or those with the most vivid coloured plumage. You need to collect your sample randomly from as much of the organism's habitat as possible. If you collect from only one area, you might not get a result that is accurate for the whole species because a factor such as climate or diet may be different in that area. Data like these can be displayed using a graph or histogram, to show the frequency distribution of the characteristic clearly (see **fig F**).

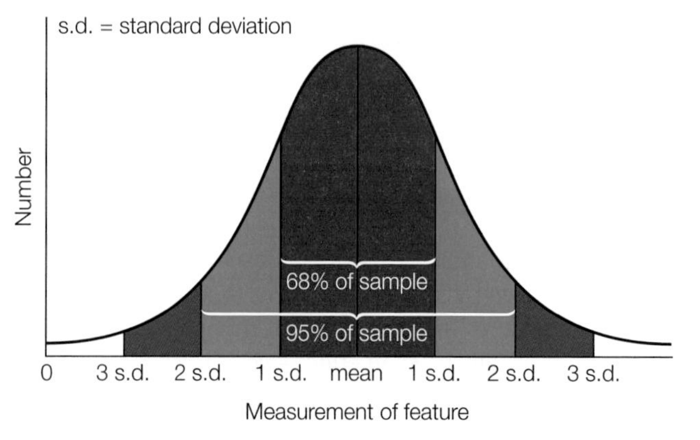

▲ **fig F** Most of the measurements in a normal distribution – for example, human height – are within three standard deviations from the mean of the sample. Graphs like these are very useful in biology.

CHECKPOINT

1. Explain the following observations in terms of the interaction of the genotype and the environment.
   (a) A patch of white fur is removed from the back of a rabbit which has a chocolate point pattern to its fur. A cool pack is kept on the area of skin as the fur grows back. The new fur which grows is dark.
   (b) If a Siamese cat needs surgery, an area of fur will be shaved off the body to leave a patch of bare skin. The fur that grows back over this skin is dark. It is not until after the next moult that the fur returns to its normal cream colour.
2. Why are data from identical twins reared apart even more useful than data from twins brought up together?
3. How reliable are the results from the study shown in **table A**? Explain your answer.
4. Explain why characteristics affected by both genotype and environment, especially polygenic characteristics, show continuous variation.

SKILLS DECISION MAKING

SUBJECT VOCABULARY

operon a unit consisting of linked genes which is thought to regulate other genes responsible for protein synthesis
discontinuous variation phenotypic features which are either present or not, usually inherited on one or at most a small number of genes.
continuous variation phenotypic features which show a huge range of values; they are usually polygenic and are also affected by environmental factors

3C 3 CONTROLLING GENE EXPRESSION

SPECIFICATION REFERENCE 3.18 3.19 3.20(i) 3.20(ii) 3.20(iii)

LEARNING OBJECTIVES

- Understand how cells become specialised through differential gene expression, producing active messenger RNA (mRNA) leading to the synthesis of proteins which in turn control cell processes or determine cell structure in animals and plants.
- Understand how one gene can give rise to more than one protein through post-transcriptional changes to mRNA.
- Understand how phenotype is the result of an interaction between genotype and the environment.
- Know how epigenetic modification, including DNA methylation and histone modification, can alter the activation of certain genes.
- Understand how epigenetic modifications can be passed on following cell division.

You have been looking at some of the enormous variation seen in living organisms, both between the different cells in the body of a single organism, and between different organisms. In this chapter, you are going to look at some of the mechanisms which bring about the variation you see.

CONTROLLING GENE EXPRESSION

You have around 20 000–25 000 individual genes on the chromosomes of your cells. In a differentiated cell, between 10 000 and 20 000 of those genes are actively expressed. Different combinations are expressed in different cells. This creates the variety of structure and function seen in cells of different tissues. The expression of a gene involves two key stages – transcription from DNA to messenger RNA (mRNA) and translation from mRNA to proteins (see **Section 2B.6**). Exerting controls at any of the stages of the process gives control over the expression of the genes (see **fig A**). The different proteins present in a cell, and the quantities of those different proteins, determines the type of cell and its function in the body. In addition, the proteins can be changed once they have been synthesised, giving another level of control over the expression of a gene.

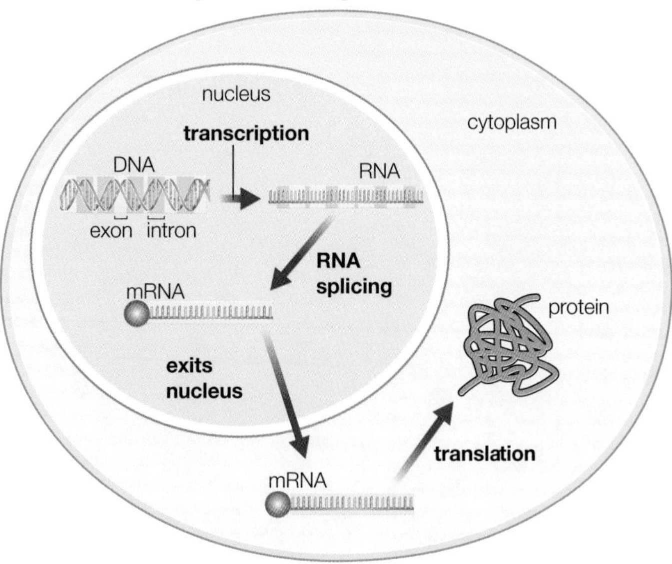

▲ **fig A** Gene expression can be controlled at any stage in the process of protein synthesis.

TRANSCRIPTION FACTORS AND THE CONTROL OF GENE EXPRESSION

The most common way of controlling gene expression is by switching on and off the transcription of certain genes. Transcription describes the process by which the genetic code of the DNA is copied to a complementary strand of mRNA before protein synthesis can occur. Messenger RNA transcription is a very effective point at which to control gene expression, because a single mRNA molecule results in the production of many protein molecules at a ribosome or polysome.

Transcription factors are proteins that bind to the DNA in the nucleus and affect the process of transcribing the genetic material. All transcription factors have regions that enable them to bind to specific regions on the DNA known as **promoter sequences**. Promoter sequences are usually found just above the starting point for transcription upstream of the gene. (See **Section 2B.6** to remind yourself of how DNA is transcribed in the nucleus of the cell.) Some transcription factors stimulate the transcription of a region of DNA simply by binding to a DNA promoter sequence. This stimulates the start of transcription of that area of the DNA.

Other transcription factors bind to regions known as **enhancer sequences** and regulate the activity of the DNA by changing the structure of the chromatin, making it more or less open to RNA polymerase. An open chromatin structure is associated with active gene expression; closed chromatin structures are associated with gene inactivity. In this way, transcription factors can either stimulate or prevent the transcription of the gene. These regulatory sites can be at the site of the gene or they can be a long distance (thousands of base pairs) from the gene they are controlling. This makes it more difficult for scientists to work out exactly what is happening.

Often several different transcription factors will be involved in the expression of a single gene, which gives many levels of control. Equally, a single transcription factor may control the activity of several different genes. It may stimulate the expression of one gene and suppress the expression of another. This is one way in which control over multiple genes is achieved. The control of transcription by transcription factors is the most frequent method of gene regulation. It means each gene can be expressed (switched on) or repressed (switched off) at different stages of the development of the organism, in different cell types and under different circumstances in the body (see **fig B**).

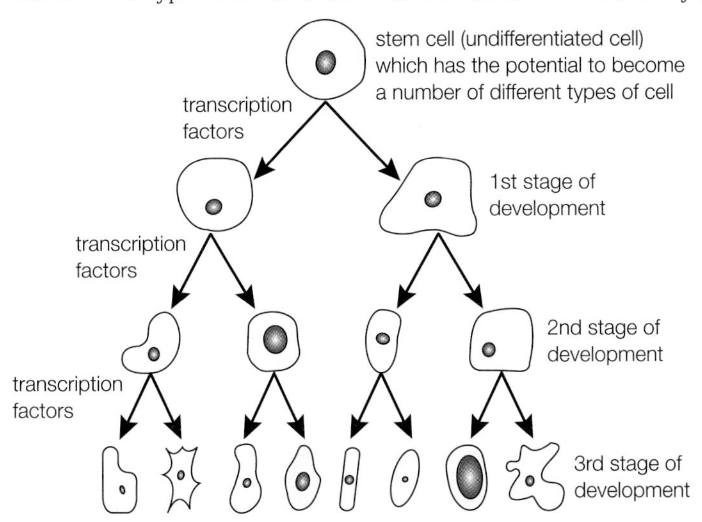

▲ **fig B** The wide variety of cells in an organism is the result of different transcription factors allowing different genes to be expressed or repressed in each cell type. This diagram gives you an idea of how many different types of cell can result from a single original as many different transcription factors are switched on and off.

This level of control of gene expression occurs in the nucleus where transcription factors control the transcription of regions of DNA into mRNA. However, there are many other ways in which the expression of a particular gene can be controlled in the cell.

RNA SPLICING

The mRNA produced in the nucleus results from the transcription of all the DNA making up a gene, including the **exons** and **introns** (the non-coding DNA). We now know that this mRNA is not quite finished when it is first transcribed. Several processes occur which modify it before it lines up on the ribosomes, so it is referred to as **pre-mRNA**. The modifications to the pre-mRNA always involve the removal of the introns and, in some cases, some of the exons are removed as well. Enzyme complexes called **spliceosomes** join together the exons that are to be transcribed and produce the mature, functional mRNA.

The spliceosomes may join the same exons in a variety of ways in a process known as RNA splicing. As a result a single gene may produce several versions of functional mRNA which is transcribed from the same section of DNA. These different versions of mRNA code for different arrangements of amino acids, which in turn produce different polypeptide chains and, therefore, different proteins. Ultimately, this can result in a single gene producing several different phenotypes. These post-transcriptional changes to mRNA lead to more variety in the phenotype than is coded for directly in the genotype. This is one of the ways in which the genotype can produce more proteins than there are genes.

One example of RNA splicing happens in the cells which form the inner ear of a chick (see **figs C** and **D**). As a result of spliceosome activity, a single gene gives rise to 576 different proteins that affect the sensitivity of the hairs in the inner ear. This enables chicks to hear a range of sounds from 50 to 5000 Hz.

LEARNING TIP

Be clear about the difference between introns and exons.

Remember, exons are expressed and introns are intrusions.

▲ **fig C** The hearing mechanism of chicks like these is based on 576 proteins that are all produced by RNA splicing from a single gene.

pre-mRNA
exon 1
intron
exon 2
spliceosome
removed intron
spliced exons
mature mRNA

▲ **fig D** RNA splicing to remove introns, and sometimes rearrange exons, enhances the expression of some genes.

POST-TRANSLATION CONTROL

Further modification of proteins may also occur after they have been synthesised. A protein that is coded for by a gene may remain intact or it may be shortened or lengthened by enzymes to give a variety of other proteins.

EPIGENETICS

Epigenetics is a relatively new area of research in biology. It studies genetic control by factors other than the base sequences on the DNA. So in a way, RNA splicing is a form of epigenetic control because it changes the mRNA and proteins produced from the original genetic code. Scientists are becoming increasingly aware of the role of epigenetic control in the normal development of specialised cells.

Scientists have known for a long time that the phenotype of an organism is the result of a combination of the genetic information in the cells and the environment in which they grow and develop (see **Section 3C.2**). They are also beginning to understand how environmental factors such as UV light levels and diet can influence the cell biochemistry. This can result in epigenetic changes future generations can inherit. Three intracellular systems that can interact to control genes in response to environmental factors include DNA methylation, histone modification and non-coding RNA.

DNA METHYLATION

The methylation of DNA (addition of a methyl $-CH_3$ group) is a widely used mechanism in epigenetics. The addition of the methyl group always occurs at the site where cytosine occurs next to a guanine in the DNA chain with a phosphate bond between them (a CpG site). The methyl group is added by a DNA methyltransferase enzyme. **DNA methylation** can also modify the structure of the histones, so it can have an epigenetic effect in more than one way.

DNA methylation always silences a gene or a sequence of genes. The methyl group changes the arrangement of the DNA molecule and prevents transcription from taking place. DNA methylation has been found to be extremely important in controlling gene expression and has a major role in many processes including embryonic development and X chromosome inactivation. In many specialised cells in adults, many genes are silenced by DNA methylation most or all of the time.

DNA demethylation is equally important. The removal of the methyl group enables genes to become active so they can be transcribed. Researchers are increasingly finding that problems with DNA methylation or demethylation are associated with diseases, including a number of human cancers (see **fig E**).

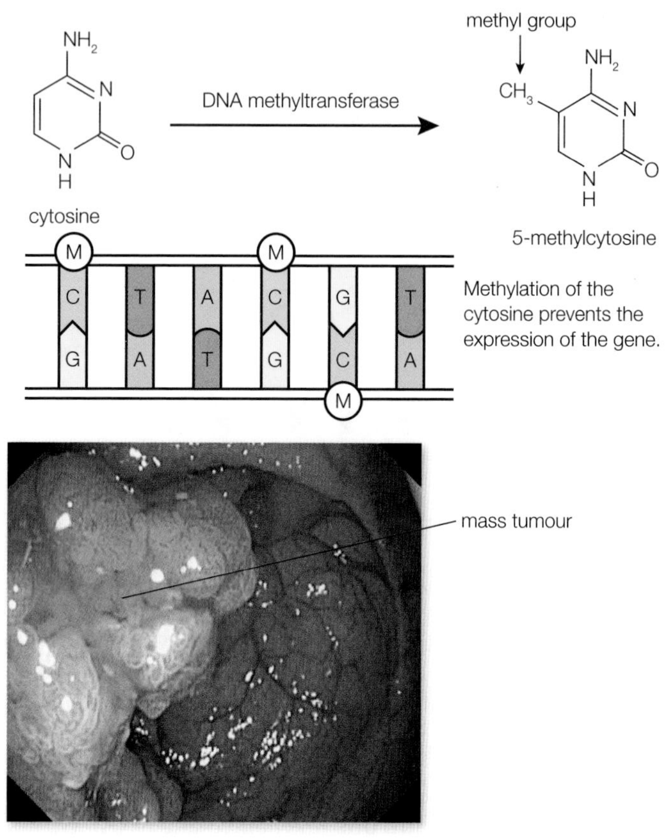

▲ **fig E** DNA methylation silences many genes. Scientists are finding demethylation of DNA may well be a factor in the growth of the tumour in some cancers, including ovarian and bowel cancers.

HISTONE MODIFICATION

Histones can be modified in a number of ways to affect the transcription of DNA and therefore gene expression. As you saw in **Section 3B.1**, histones are positively charged proteins. DNA helices wind around the histones to form chromatin. This DNA–protein complex makes up the chromosomes. The histones determine the structure of the chromatin. When the chromatin is densely supercoiled and condensed, the genes are not available to be copied to make proteins and this is known as **heterochromatin**. Active chromatin is more loosely held together, with uncoiled regions of DNA making more genes available for transcription so that new proteins can be made. This is one way in which cells of different types are produced.

Many different factors affect the modification of the histones, including steroid hormones. Modification processes include the following.

- **Histone acetylation** – an acetyl group ($-COCH_3$) is added to one of the lysines in the histone structure. Adding an acetyl group usually opens up the structure and activates the chromatin, allowing genes in that area to be transcribed. Removing an acetyl group produces heterochromatin again.
- **Histone methylation** – a methyl group ($-CH_3$) is added to a lysine in the histone. Depending on the position of the lysine, methylation may cause inactivation of the DNA or activation of a region. Methylation is often linked to the silencing of a gene and even whole chromosomes. For example, histone

methylation plays a role in the silencing of one of the X chromosomes in every cell in female mammals.

DID YOU KNOW?

HISTONE MODIFICATION IN ACTION

Moulting, the shedding of the exoskeleton of insects, is controlled by two hormones. Ecdysone, the 'moulting and metamorphosis' hormone, controls the events of the moult. It is a steroid hormone that was first extracted from the pupae of silkworms. Juvenile hormone controls the kind of moult that occurs. As juvenile hormone levels decrease, more adult characteristics occur. When there is no juvenile hormone, the insect becomes an adult.

The way in which ecdysone has its effect has been studied using the larvae of *Drosophila* (fruit fly) and *Chironomus* (midge). These insects have giant chromosomes in their salivary glands. They are 100 times thicker and 10 times longer than normal chromosomes and easily visible with the light microscope.

Banding is visible on these chromosomes due to supercoiled areas of DNA. When an insect is undergoing a moult, or when ecdysone is injected artificially into an insect, 'puffs' appear on the chromosomes (see **fig F**). These puffs appear to be pieces of genetic material from supercoiled areas that have been opened up and made available for transcription. It is thought that they carry information about the new proteins needed in a more adult stage of the life cycle. This supports the view that steroid hormones have a direct effect on the DNA of a cell. Scientists think that many steroid hormones act in a similar way. Unfortunately, not many organisms possess giant chromosomes, so the effect is not always easy to observe!

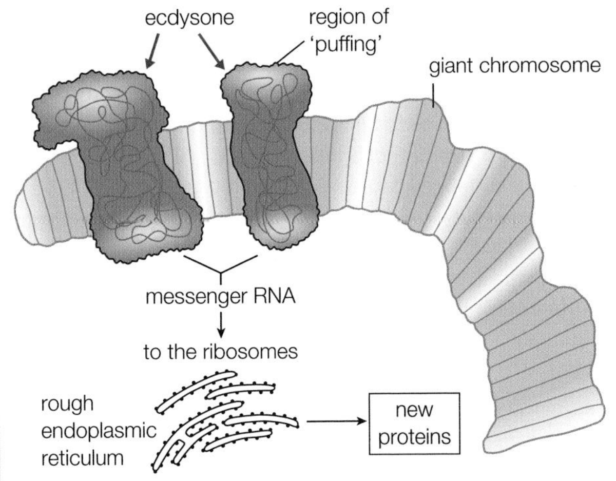

fig F Chromosome puffs appear on insect chromosomes as a result of the hormone ecdysone. They are followed by the synthesis of much RNA, which in turn results in the formation of new proteins.

NON-CODING RNA

About 90% of the human genome is transcribed into mRNA, but only about 2% of those RNA molecules code for proteins. Much of the rest of the **non-coding RNA (ncRNA)** seems to affect the transcription of the DNA code or modifies the products of transcription. Both genes and whole chromosomes can be silenced by ncRNAs. In female mammals, one of the X chromosomes in every cell is inactivated at random. This is largely due to the presence of an ncRNA called X-inactive specific transcript (Xist), which is produced by the active Xist gene on the inactive chromosome. The ncRNA coats one of the X chromosomes in female cells and deactivates it. The chromosome supercoils and condenses to form the stable, inactive Barr body. One X chromosome is inactivated in order to maintain the balance of gene products in males, XY, and females, XX. Another role for ncRNA in epigenetics seems to be in chromatin modification, where it acts on the histones to make areas of the DNA available or unavailable for transcription.

CELL DIFFERENTIATION

Cell differentiation occurs when unspecialised cells switch different genes on and off as needed to become specialised cells. Scientists are now aware that most of these changes are the result of epigenetic modification of the genetic material. Epigenetic modifications ensure that a wide range of very specific proteins are made within the cell as it differentiates for a specific function. The modification is often the result of DNA methylation or demethylation, or of histone modification.

These epigenetic changes may be in response to internal stimuli from the cell itself or in response to changes outside the cell (an external stimulus) which affect the inside of the cell. For example, the sex hormones produced at puberty trigger changes in many cells in the body through DNA methylation in the nuclei of the cells. The production of these hormones is a response to many factors including body mass which is affected by the environment – for example, by how much food is available. Once epigenetic changes have occurred within a cell, the modifications – the genes which are switched on or switched off – are passed on when the cell divides by mitosis. The process of cell differentiation can be summarised as follows:

- chemical stimulus (e.g. demethylation)/transcription factor
- certain genes activated/switched on
- mRNA produced from these genes
- translation of mRNA to form polypeptide/protein
- permanent modification of the cell.

LEARNING TIP

The mechanism of epigenetics relies on making the DNA coil less tightly for transcription or coil more tightly so it cannot be transcribed.

Make sure that you remember methylation of the DNA stops transcription while acetylation of the histones enables transcription.

CHECKPOINT

SKILLS ADAPTABILITY, ADAPTIVE LEARNING

1. Describe how RNA splicing enables a single gene to code for a number of proteins.
2. Explain what is meant by the term *epigenetics*.

SKILLS INTERPRETATION

3. Explain how the phenotype of an organism is the result of an interaction between the genotype and the environment.

SUBJECT VOCABULARY

transcription factor protein that binds to the DNA in the nucleus and affects the process of transcribing DNA into RNA
promoter sequence specific region on the DNA to which transcription factors bind to stimulate transcription
enhancer sequence specific region of DNA to which transcription factors bind and regulate the activity of the DNA by changing the structure of the chromatin
exons segments of a DNA or RNA molecule containing information coding for a protein or peptide sequence
introns segments of a DNA or RNA molecule containing information which does not code for a protein or peptide sequence
pre-mRNA mRNA that is transcribed directly from the DNA before it has been modified
spliceosomes enzyme complexes that act on pre-mRNA, joining exons together after the removal of the introns
DNA methylation methylation of DNA (addition of a methyl $-CH_3$ group) to a cytosine in the DNA molecule next to a guanine in the DNA chain and prevents the transcription of a gene
DNA demethylation removal of the methyl group from methylated DNA enabling genes to become active so they can be transcribed
heterochromatin densely supercoiled and condensed chromatin where the genes are not available to be copied to make proteins
histone acetylation addition of an acetyl group ($-COCH_3$) to one of the lysines in the histone structure, which opens up the structure and activates the chromatin, allowing genes in that area to be transcribed
histone methylation addition of a methyl group ($-CH_3$) to a lysine in the histone; methylation may cause inactivation or activation of the region of DNA, depending on the position of the lysine
non-coding RNA (ncRNA) 98% of the RNA, which does not code for proteins but affects the transcription of the DNA code, modifies the chromatin structure or modifies the products of transcription

3C 4 STEM CELLS

SPECIFICATION REFERENCE 3.17(i) 3.18 3.20(ii)

LEARNING OBJECTIVES

- Understand what is meant by the terms *stem cell*, *pluripotent* and *totipotent*, *morula* and *blastocyst*.
- Understand how cells become specialised through differential gene expression producing active mRNA, leading to the synthesis of proteins which, in turn, control cell processes or determine cell structure in animals and plants.
- Know how epigenetic modification, including DNA methylation and histone modification, can alter the activation of certain genes.

Fertilisation starts a complex series of events that will eventually lead to the birth of a fully formed new individual. In humans, the zygote (fertilised egg cell) has the potential to form all of the 216 different cell types needed for an entire new person. It is said to be **totipotent**. The future roles of individual cells are decided quite early in the life of an embryo.

THE EARLY STAGES OF DEVELOPMENT

The first stage of embryonic development is known as cleavage. Cleavage involves a special kind of mitosis where cells divide repeatedly without the normal interphase for growth between the divisions. The result of cleavage is a mass of small, identical and undifferentiated cells forming a hollow sphere known as a **blastocyst** (see **fig A**). In humans, this process takes about 5–6 days, and it takes place as the zygote is moved along the oviduct towards the uterus. One large zygote cell forms many small cells in the early embryo. The tiny cells of the early human embryo are known as **embryonic stem cells**. Stem cells are undifferentiated cells that have the potential to develop into many different types of specialised cell from the instructions in their DNA.

The very earliest cells in an embryo are totipotent like the zygote. By around the fourth day after fertilisation, they become a solid ball of 10–30 cells known as a **morula**. Each of the cells in the morula is still totipotent and has the potential to form every type of adult human cell. Within another day, the cells have divided more and formed a blastocyst. By the time the blastocyst is formed, the cells in the inner layer have already lost some of their ability to differentiate and the outer layer of cells goes on to form the placenta. The inner layer of cells can form almost all of the cell types needed in future, but not tissue such as the placenta. These cells are known as **pluripotent** embryonic stem cells. They have become pluripotent as a result of some genes already being permanently switched off.

LEARNING TIP

Remember that the zygote is the only truly totipotent stem cell.

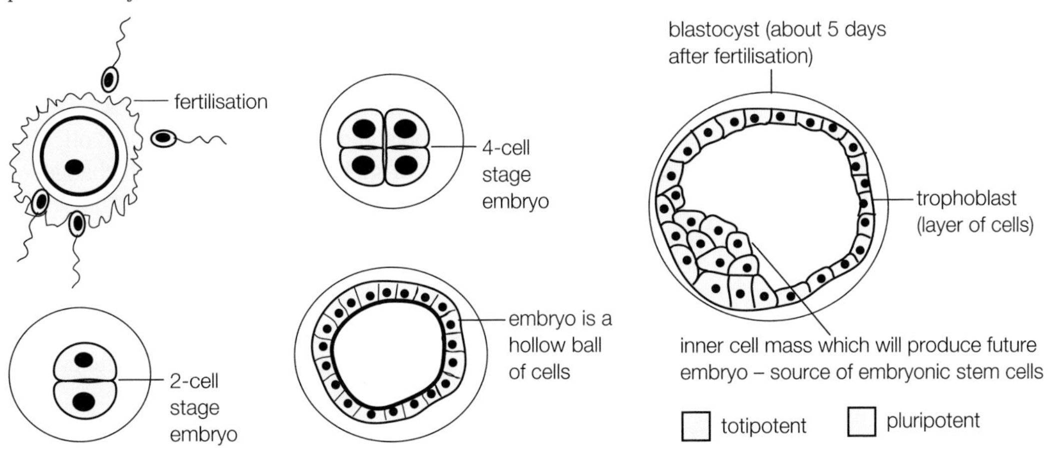

fig A In the very earliest stages of pregnancy any one of the cells has the potential to form an entire human being, but this ability is lost by the blastocyst stage.

TYPES OF STEM CELL

Stem cells come in a number of different types, with different levels of potency.

EMBRYONIC STEM CELLS

The earliest embryonic cells, such as those shown in **fig B**, are totipotent. By the blastocyst stage, when the embryo is implanted in its mother's uterus, the inner cells of this ball are pluripotent. Pluripotent stem cells change to become more specialised as the embryo develops. For example, they might form blood stem cells which give rise to blood cells, or skin stem cells which give rise to skin cells. By around three months of pregnancy, the cells have become sufficiently specialised that when they divide, they form only more of the same type of cell.

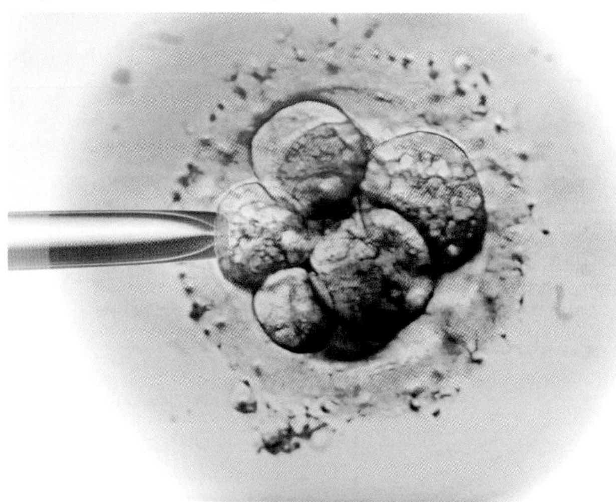

▲ **fig B** A 3-day-old human embryo: at this stage the cells are still totipotent.

UMBILICAL CORD STEM CELLS

The blood that drains from the placenta and umbilical cord after birth is a rich source of pluripotent stem cells. If this blood is frozen and stored, in theory those stem cells will be available throughout the life of the child should they or their family need them later for stem cell therapy. It may become possible to store stem cells from every newborn baby ready for when they might need them. However, it would take a lot of storage space to do this for everyone and would be expensive. Also, there is some evidence that the precursor cells of conditions like leukaemia are already present in the blood at birth. There is very little evidence of cord blood being used successfully to treat anyone and many scientists are unconvinced of the value of the process. The benefits are unproven at the moment, and the only way parents can currently store umbilical cord blood for their children is to pay to do it privately.

ADULT STEM CELLS

An adult human consists of many different types of highly specialised cell. However, some **adult stem cells (somatic stem cells)** remain as undifferentiated cells and are found among the normal differentiated cells in a tissue or organ. They can differentiate when needed to produce any one of the major cell types found in that particular tissue or organ. For example, white bone marrow contains stem cells that can form white blood cells. There are only a very small number of adult stem cells in each different tissue. They are difficult to extract and most of them form a very limited range of differentiated cells. They are said to be **multipotent**. They are difficult to grow in the laboratory.

EXAM HINT

The differences between totipotent, pluripotent and multipotent stem cells are important and have a bearing on how useful these cells can be in potential treatments.

Be sure you understand the differences.

THE DEVELOPMENT OF AN ORGANISM

Totipotent stem cells in the embryo become pluripotent stem cells in the blastocyst and finally become fully differentiated somatic cells in the body of the mature organism. How are these changes brought about?

The most common way of controlling gene expression is by switching on and off the transcription of certain genes. As you saw in **Section 3B.3**, this can be achieved in several ways including the action of transcription factors and epigenetic mechanisms such as DNA methylation, histone modification and ncRNAs. So, for example, during the process of differentiation, some parts of the chromosomes undergo supercoiling to prevent certain genes being transcribed, and other areas uncoil, opening them up for transcription so that new proteins are made. Some genes are activated and others are silenced. As development progresses, more genes are silenced in each cell. It is the combination of the particular genes that are activated or silenced that results in the different characteristics of fully differentiated mature cells.

AN EXAMPLE OF EPIGENETIC CONTROL IN HUMAN DEVELOPMENT

Haemoglobin is an essential molecule that carries oxygen around the body. Adult human haemoglobin contains two alpha and two beta globin chains. Fetal haemoglobin, which has a stronger affinity for oxygen than adult haemoglobin, contains two alpha and two gamma globin chains (see **Section 1B.2**).

During human development from embryo to fetus to baby, different versions of the globin genes are switched on and off. The levels of the different types of globin chain in the blood change through the 40 weeks of pregnancy and after the baby is born. Genes for the different proteins get switched on and off, and are activated in different tissues as development progresses (see **fig C**). Globin production moves from the yolk sac in the embryo to the liver in the fetus and then to the spleen. Finally, at the time of birth, the genes in the bone marrow have taken over this function almost completely.

The genes controlling the production of alpha globin are needed in both the fetus and mature baby. The genes controlling fetal gamma globin are very important during fetal development but need to be silenced around the time of birth. The adult beta globin genes need to be activated just before birth in the bone marrow. This is a clear example of epigenetics working in several tissues to change which proteins are made.

The mechanism which controls the genes that code for the different chains is still the subject of much active research. If scientists can work out how to reactivate the fetal haemoglobin gene in children and adults, they may be able to overcome the problems presented in genetic conditions such as sickle cell anaemia and thalassaemia, which affect the structure of adult haemoglobin. So far scientists have found evidence of some epigenetic control mechanisms.

- Histone acetylation appears to activate the gamma globin gene in the fetus.
- DNA methylation appears to be important in silencing the fetal gamma globin genes just before and after birth.
- Histone methylation appears to complement DNA methylation in silencing the fetal gamma globin.
- Non-coding RNAs have been associated with the process, but scientists are not yet sure what they do.
- A number of transcription factors are very important in the switch to the production of beta globin in the spleen and bone marrow as the fetus approaches full term and birth.

As we understand more about the control of cell differentiation, our ability to control stem cells will increase too.

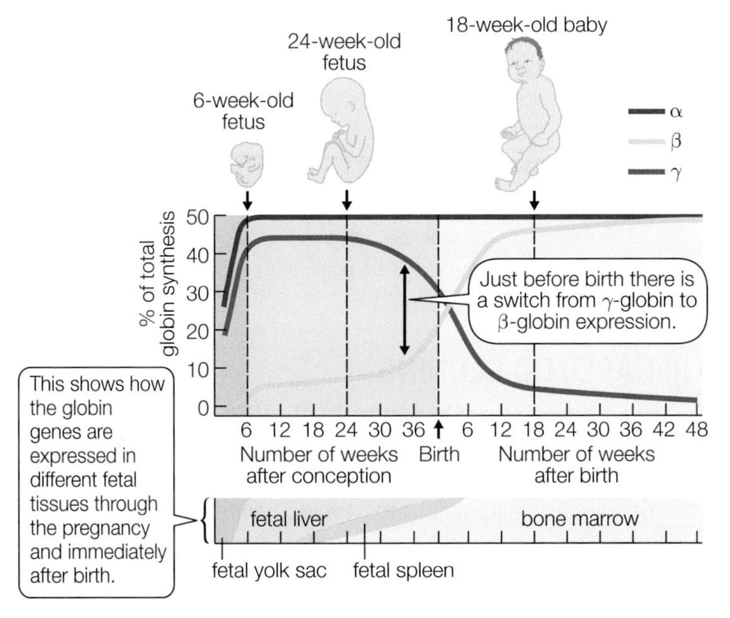

fig C Different genes for the production of globin molecules are switched on and off in different tissues during the development of a human being from an embryo to a fetus to a baby.

EXAM HINT

Remember that a table is an effective way to compare different types of cell.

CHECKPOINT

SKILLS INTERPRETATION

1. Compare embryonic and adult stem cells.
2. State the main differences between totipotent, pluripotent and multipotent cells

SKILLS CREATIVITY

3. Suggest reasons why scientists want to understand the control mechanisms behind the switching off and on of genes during the development of an organism.

SUBJECT VOCABULARY

totipotent an undifferentiated cell that can form any one of the different cell types needed for an entire new organism
blastocyst an early embryo consisting of a hollow ball of cells with an inner cell mass of pluripotent cells that will eventually form a new organism
embryonic stem cells the undifferentiated cells of the early human embryo with the potential to develop into many different types of specialised cell
morula an early embryo made up of a solid ball of 10–30 totipotent cells
pluripotent an undifferentiated cell that can form most of the cell types needed for an entire new organism
adult stem cells (somatic stem cells) undifferentiated cells found among the normal differentiated cells in a tissue or organ that can differentiate when needed to produce any one of the major cell types found in that particular tissue or organ
multipotent a cell that can form a very limited range of differentiated cells within a mature organism

3C 5 USING STEM CELLS

SPECIFICATION REFERENCE 3.17(ii)

LEARNING OBJECTIVES

- Be able to discuss the ways in which society uses scientific knowledge to make decisions about the use of stem cells in medical therapies.

Scientists are working hard to find ways of using stem cells to treat medical conditions where the patient's own cells are damaged or faulty.

About 30 years ago, we discovered in bone marrow stem cells which can form all the different types of blood cell. Bone marrow transplants are now used regularly in the treatment of certain cancers and immune system diseases. These transplants need to be taken from close relatives or from strangers who have matching immune systems, otherwise the body will reject and destroy the transplant.

Scientists also discovered in the bone marrow stem cells that can generate bone, fat, cartilage and fibrous tissue. In the 1990s, they discovered in the brain stem cells that can form the three main types of brain cell. Adult stem cells have been found in many different organs and tissues. In theory, these cells could be extracted from a patient and treated so that they develop into the new cells that the patient needs.

STEM CELL THERAPY

When stem cells were first cultured, scientists hoped that they could be used to produce new tissues but this has not yet been successful. It is very difficult to control the differentiation of the cells. Some of the early treatments did result in patients being cured of one condition, but they then developed cancer.

Much of the early work carried out by scientists on stem cells used cells harvested from early embryos. This process is allowed in many countries, but is unacceptable in others. The use of stem cells derived from embryos to develop new medical treatments has raised both practical and ethical difficulties.

The problems with embryonic stem cells have stimulated research into adult stem cells and these have been used successfully to produce new body parts, particularly the trachea. Stem cells from a patient are seeded onto a collagen-based framework, which may be from a donor or completely synthetic. The stem cells grow to form the required cells and the new trachea can be returned to the patient with no risk of rejection. It is hoped that other organs will be available soon (see **fig A**).

Another area of active research using adult stem cells is in the repair of hearts damaged by heart attacks. Adult stem cells are injected into the heart and, in some cases, the improvement in function has been significant. Some studies used adult stem cells from the bone marrow, others used them from the heart itself. Using the patient's own adult stem cells avoids the risk of rejection of new tissue. All of the studies so far have been on small numbers of patients and there is still much more research that needs to be done.

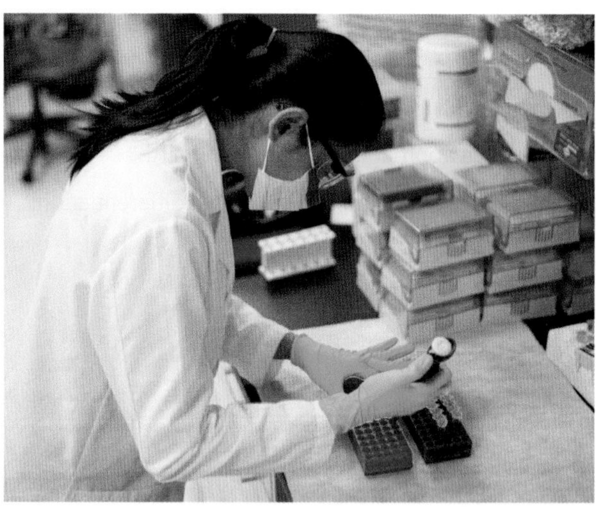

▲ **fig A** Clinics around the world are beginning to use stem cells to treat a wide range of diseases.

THERAPEUTIC CLONING

Somatic cell cloning or **therapeutic cloning** is an experimental technique that scientists hope to use in the future to produce large quantities of healthy tissue. This is one of the main ways in which scientists are using stem cells to develop new medical therapies. It is hoped that this could be used to treat people with diseases caused by faulty cells, such as type 1 diabetes or Alzheimer's disease.

The first step is to produce healthy cloned cells from the patient. This is done by removing the nucleus from one of their normal body cells and transferring it to a human ovum which has had its original nucleus removed. A mild electric shock is used to fuse the nucleus with the new cell and trigger development. The newly formed cell starts to develop and divide and produces a collection of identical cells with the same genetic information as the patient. This clone is simply a source of stem cells with genetic markers that match the patient perfectly. Stem cells are harvested and can then be cultured in a suitable environment so that they differentiate into the required tissue. These tissue cells can then be transferred to the patient, where they can do their job without the risk of the immune system rejecting them. **Fig B** summarises this process. This type of treatment is still very much at the experimental stage of development, because scientists are still trying to determine the exact triggers that control cell differentiation. Also, there is a shortage of donor eggs to use, and some people have ethical objections to the use of donor eggs in this way.

There is an additional factor that must be considered in the treatment of genetic diseases. The adult stem cell nucleus would need to be genetically modified *before* being added to the empty ovum, otherwise the cultured stem cells would carry the genetic mutation that caused the problem in the first place.

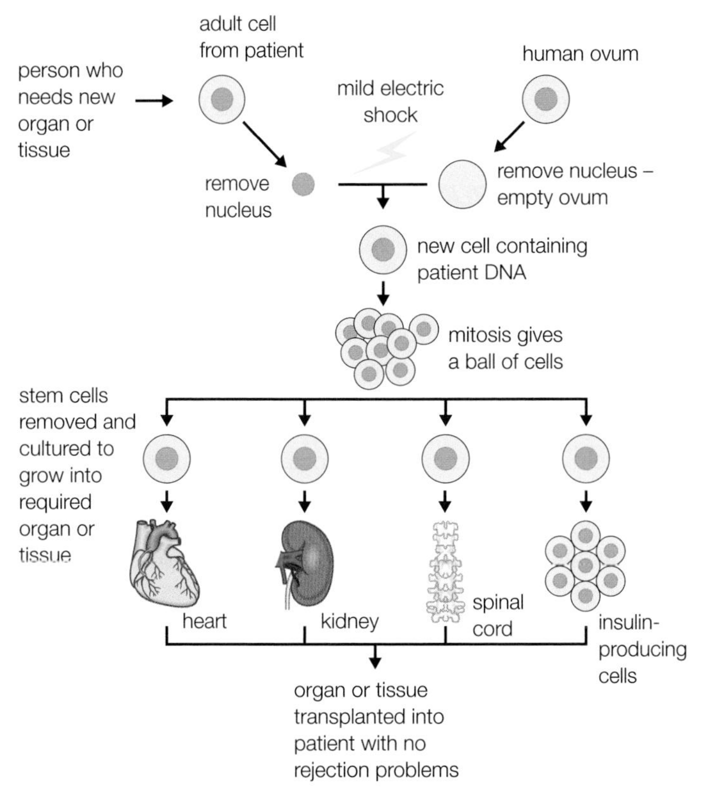

fig B Therapeutic cloning is still very experimental but this diagram shows how scientists and doctors hope it will be used in the future.

PITFALLS AND POTENTIAL BENEFITS OF STEM CELL THERAPY

There are problems that relate to the uses of all kinds of stem cell to develop new medical therapies. Society has to rely on the scientific knowledge presented to it to make decisions about the use of stem cells in this way. For example, at the moment no one is quite sure how the genes in cells are switched on or off to form specific types of tissue. However, our understanding of transcription factors and epigenetic mechanisms is increasing all the time and so the point when scientists can fully control the differentiation of embryonic and adult stem cells is getting closer.

There are also risks associated with stem cell therapy. There are concerns that stem cells could cause the development of cancers in the body. There is some evidence that people who have been given bone marrow transplants (stem cell treatment) to help them overcome leukaemia are at higher risk of developing other cancers later. However, there are also exciting potential advantages. At the moment, there are no cures for many of the conditions that stem cell therapy might solve. The ability to produce tailor-made cells to take over the function of damaged ones would be a revolution in medicine.

Organ transplants create their own problems. Glycoproteins on the surface of your cell membranes act as part of your cell recognition system. Your immune system recognises your own cells (self) and different cells (non-self) and destroys the non-self cells. This works well if you pick up an infection, but is potentially lethal if your immune system attacks a transplanted organ. After a transplant, people have to take immunosuppressant drugs for the rest of their lives, and this puts them at higher risk of infectious diseases. One advantage of using stem cell therapy is that it could avoid the risk of rejection. A mother's immune system does not attack and destroy her developing embryo, even though the markers on its cells are different from hers. Maybe new cells or organs that are created from embryonic stem cells would enjoy this same protection. If the stem cells can be cultured from the individual who needs them, rejection would be a thing of the past.

INDUCED PLURIPOTENT STEM CELLS

In 2006, a team of researchers in Japan made an astonishing breakthrough. They took adult mouse cells and, using genetic engineering techniques, reprogrammed them to become pluripotent again. Effectively, they produced stem cells without using an embryo. What is more, the **induced pluripotent stem cells (iPS cells)** renew themselves. The next stage was to repeat the process in humans. The team used harmless, genetically modified viruses to carry a group of four genes for specific transcription factors into skin cells taken from a 36-year-old and synovial fluid taken from a 69-year-old.

The induced pluripotent human stem cells appear to be very similar, although not identical, to human embryonic stem cells in their behaviour. The Japanese team even managed to make their iPS cells develop into brain cells and heart muscle cells.

One of the greatest potential benefits of this new technology is that it overcomes the ethical objections to using embryonic tissue as a source of stem cells. There is also no risk of rejection if cells from an individual are used to provide their own stem cells. It is not easy to persuade the cells to become pluripotent. Making them differentiate into the tissue you want is even more difficult. How well, and for how long, they will behave as pluripotent stem cells remains to be seen. Perhaps most worryingly, the new stem cells show a tendency to become cancerous very quickly because the genes used to make the cells pluripotent again are also strongly associated with cancer development.

Many scientists think that iPS cells may be the best way forward in stem cell medicine. As there are still major problems to overcome, there is a strong feeling in the scientific community that research into both types of stem cell should continue until it is clear which will produce the best treatments for patients in the future. So far, most of the successful work on medical therapies has been done using embryonic stem cells.

WHO COULD BENEFIT FROM STEM CELL THERAPY?

Stem cell therapy has the potential to treat a wide range of diseases caused by faulty cells. Here are some examples of areas where scientists feel pluripotent stem cells could give real therapeutic value.

PARKINSON'S DISEASE

Parkinson's disease is the second most common age-related brain disorder, and it affects millions of people around the world. Nerve cells in the brain that produce dopamine (dopamine neurones) stop working and are lost. As dopamine levels fall, people develop uncontrollable tremors in their hands and body. Their body becomes rigid and eventually they cannot move normally at all. Drug treatments have improved greatly in recent years, but there is still no long-term cure. Scientists hope that stem cell transplants will allow them to replace the lost brain cells and restore dopamine production, letting people return to a normal life.

Scientists have managed to get mouse stem cells to form dopamine neurones. They transplanted these cells into the brains of rats that had the symptoms of Parkinson's disease. The cells grew and released dopamine and the ability of the rats to control their movement improved. Scientists hope that eventually they will be able to develop pluripotent stem cells that could be implanted into the brain and take over dopamine production from the damaged and dead cells. Most scientists agree that pluripotent stem cells offer the best current hope of an effective long-term treatment for Parkinson's disease.

TYPE 1 DIABETES

Type 1 diabetes usually develops when people are young. The glucose-sensitive, insulin-secreting cells from the islets of Langerhans in the pancreas are destroyed or stop making insulin. This means the blood glucose concentration is uncontrolled. This can be very serious or even fatal. Although insulin injections work well enough, people affected by type 1 diabetes must monitor their food intake and blood glucose concentration and inject insulin regularly. Stem cell therapy could give them working pancreas cells again, restoring insulin production and therefore blood glucose control.

Scientists have succeeded in getting some mouse stem cells to form a group of cells that look and work just like insulin-producing tissue. Some of these cells were transplanted into mice with diabetes. They produced a rise in the blood concentration of insulin and improved control of blood glucose.

In 2014, scientists at Harvard University reported a major breakthrough discovery. They started with human embryonic stem cells and they developed mature human glucose-sensitive insulin-producing beta cells in the large quantities that they needed to use them in patients. They are now trialling the cells in animals with diabetes. If those trials are successful, the team hopes to use the new cells in human clinical trials within a few years. If islet cells can be developed from stem cells and be successfully transplanted into people with type 1 diabetes, the treatment could eventually free millions of people from the lifelong need to inject insulin.

DAMAGED NERVES

So far, there is no medical cure for damaged and destroyed nervous tissue in the brain and spine. These nerves do not usually regrow and so someone who suffers a major injury to their spine may be permanently paralysed below the location of the damage. Stem cells have been transplanted into mice and rats with damaged spines and the animals regained a certain amount of control and movement of limbs that had been paralysed. Examining the spinal cords of these rats and mice showed that the stem cells had grown into working adult nerve cells and the damaged spinal cords had at least partly rejoined, offering hope for future human treatments. Recently, cells from the brain area linked to the sense of smell were transplanted into a paralysed man and resulted in the return of some control. These were not stem cells, but it showed that progress in people is possible.

ORGANS FOR TRANSPLANTS

Many people die because their organs no longer function properly. Hearts, kidneys, livers and many other organs can be replaced by transplant, but only if there is a suitable donor organ available. There is a desperate need for new organs, preferably ones that will not cause rejection problems when they have been transplanted. In 2013, a team of researchers in Australia produced stem cells from human skin cells. They then manipulated the differentiation of these stem cells into minute functioning kidney units. Although they were only millimetres across, the mini-kidneys functioned and produced urine. Also in 2013, a team in Japan produced functioning three-dimensional liver buds from iPS cells. These connected with the circulatory systems of mice when they were transplanted into them. The hope is that, eventually, pluripotent stem cells will provide the huge supply of organs needed for transplantation globally each year.

STEM CELL SUCCESS

So far, the number of successful therapies using pluripotent stem cells has been very small. Many ideas are being developed and trialled in animals. Not many of these have had any impact on human health although recent results with age-related muscular degeneration are very positive. Scientists and doctors expect that numbers of successful treatments will increase dramatically in the next 10 years.

ETHICAL QUESTIONS

There have been some very powerful reactions to stem cell technology. As well as the many practical problems to be overcome before stem cell therapy becomes a standard treatment, society has many ethical issues to deal with.

The four ethical principles are as follows.

- Respect for autonomy – this means respect for individuals, by not performing procedures without consent.
- Beneficence – this means the aim of doing good, by giving medicine to relieve suffering, etc.
- Non-maleficence – this means doing no harm.
- Justice – this means treating everyone equally and sharing resources fairly, to avoid discrimination.

Many people think the new work on medical therapies using stem cells is a major breakthrough with the same potential to change healthcare as the discovery of antibiotics had more than 70 years ago. It may offer the hope of a cure to millions of people who currently have none.

People who are in favour of stem cell research suggest that after tissue lines are established, which come from a relatively small number of willingly donated embryos, the need to use further embryos will be reduced.

Also, many scientists feel that adult stem cells do not offer an effective alternative, because they are much more limited in their ability to form new and different tissues. They want research funding to be directed mainly at embryonic stem cell work.

However some people, including many religious groups, think it is ethically wrong to use cells from an embryo in this way. They think that no medical advances are worth the moral evil of using embryonic tissue as a source of stem cells. They do not consider any scientific arguments.

Many of these people feel that the use of adult stem cells and iPS offer an exciting and acceptable possible alternative and they campaign for research funding to be directed to projects using these ethically less-sensitive cells. In addition, the use of pluripotent stem cells from the umbilical cord of newborn babies may help to overcome many reservations or doubts. However, storing each baby's cord blood would be logistically and practically extremely difficult and very expensive, and there is little evidence yet that any therapies developed from the blood would work.

EXAM HINT

The ethical questions surrounding the use of stem cells and, in particular, how stem cells are sourced are major topics of discussion.

Be ready to present both positive and negative arguments for the use of stem cells.

Avoid the temptation to write a response that is unbalanced even if you have strong feelings about the use of embryos or embryonic cells.

THERAPEUTIC CLONING

In therapeutic cloning, the cloned cells created from adult cells are produced to provide pluripotent stem cells. Therefore, many people are very optimistic about the future potential of this new technology and see no major ethical issues to overcome. However, other people fear that if the cloning is allowed for therapeutic purposes it could be taken further, to produce a cloned baby. Also, many people have ethical or religious objections to using these cloned cells for research. So therapeutic stem cell cloning can raise even more ethical problems than embryonic stem cell research.

iPS CELLS

iPS cells are pluripotent so they can be turned into most cell types by carefully manipulating transcription and epigenetic factors. They come from the individual patient so there are no issues of rejection. They do not come from embryos and they are not capable of forming a new embryo. The ethical issues are all answered by using this technology. The biggest problem is that these cells are not so easy to grow and manipulate as natural pluripotent stem cells. Societies around the world will consider the scientific evidence to make decisions about the use of stem cells in medical therapies. However, in this field of science, the ethical views of a society will also affect their attitude to, and acceptance of, new stem cell treatments.

CHECKPOINT

SKILLS CRITICAL THINKING, CREATIVITY

1. What is the biggest scientific obstacle scientists need to overcome to enable them to develop useful therapies with pluripotent stem cells?
2. What are the advantages and disadvantages of umbilical blood and adult tissue as alternative sources of stem cells?
3. Choose one of the methods described above for producing stem cells. Prepare scientific arguments to explain why this technique should receive major funding in the next 10 years.

SUBJECT VOCABULARY

therapeutic cloning an experimental technique used to produce embryonic stem cells from an adult cell donor
induced pluripotent stem cells (iPS cells) adult cells that have been reprogrammed by the introduction of new genes to become pluripotent again

3C THINKING BIGGER

MICROBIOTA IN THE GUT

CRITICAL THINKING, ANALYSIS, DECISION MAKING, CREATIVITY, PERSONAL AND SOCIAL RESPONSIBILITY, CONTINUOUS LEARNING, INTELLECTUAL INTEREST AND CURIOSITY

Scientists have known for many years that bacteria living in our gut help our digestion. Recent research suggests that these bacteria may have many more important roles. For example, a scientific report in November 2017 suggested that a healthy gut microbiome could improve response to cancer treatments. It seems that our microscopic allies can influence the expression of genes and our phenotype. Maintaining a healthy balance of microbiota is essential for good health.

NEWS ARTICLE

We have many species of bacteria and Archaea inhabiting our gastric tract. They are regarded as mutualistic symbionts and they help us digest food and produce some nutrients such as vitamin K. These gut-dwelling bacteria are termed the gut microbiome.

Scientists have recently found that we have about 100 species of gut bacteria, and different people may harbour different proportions of these species. The actual number of bacteria is about 10 times the total number of cells in our bodies, but in total they weigh only between 200 g and 1500 g. Bacteria also live on our skin and in our noses and vaginas. We are a veritable ecosystem. Different types of bacteria can inhabit only specific areas of our bodies.

The bacteria we house have genes that are expressed. Human gut bacteria produce many chemicals and some of them act within us to help regulate our appetite. It is possible that people who have been exposed to too many antibiotics or who do not eat enough dietary fibre have an imbalance of gut bacteria, resulting in a deficiency of appetite-regulating chemicals that contributes to obesity.

A study published in *Nature* in August 2013 showed that people with more diverse gut microbiomes showed fewer signs of metabolic syndrome including obesity, high blood pressure and insulin resistance. Lack of certain types of gut bacteria may contribute to some autoimmune diseases such as rheumatoid arthritis, Crohn's disease and multiple sclerosis. Some gut bacteria can modify the production of neurotransmitters, hence some imbalances of gut bacteria diversity can contribute to depression.

Researchers at Imperial College, London, and at Johns Hopkins University in the USA, have shown that the gut biome in mice helps to regulate their blood pressure. Mice treated with antibiotics to knock out their gut bacteria showed a significant rise in blood pressure. This indicates that the gut microbiome is also involved in regulating aspects of the host's physiology by making chemicals that stimulate the production of specific enzymes involved in a metabolic pathway that leads to a reduction in blood pressure.

Some people suffer from unrestrained growth of a particular gut-dwelling bacterium, *Clostridium difficile*. This anaerobic bacterium is found in everyone and normally kept in check by other intestinal bacteria. When certain antibiotics are taken and some of the other gut bacteria are killed, *C. difficile* can multiply unchecked and produce toxins that cause severe diarrhoea and inflammation of the bowel. This may be fatal. These spore-forming bacteria can be spread on the hands of hospital staff, patients and their visitors, as well as on surfaces such as toilets, bedpans, door handles, curtains, clothes and floors.

Scientists have found that transplanting a small amount of faeces from a healthy person to a person suffering from a *C. difficile* infection can abolish the infection where antibiotics have failed to do so. Pharmaceutical companies are now producing capsules to make this transplant of bacteria easier to undertake, as the capsules can be swallowed. However, if someone receives a transplant of gut bacteria, they also need to adopt a fibre-rich diet that promotes the growth of those bacteria. This involves increasing their dietary fibre, mostly in the form of more fruit and vegetables.

Humans have about 20 000 genes in their genome. The collective genome of the gut microbiome may contain as many as 100 times more genes than our own genome. Many of these microbial genes code for proteins that we need, so it appears that our microbiomes influence our phenotypes. Our resident bacteria also help our immune system to develop and function properly. Because we rely on more than just our own genes, we may perhaps regard ourselves as superorganisms.

Adapted from: 'Me, myself, us. The human microbiome – looking at humans as ecosystems that contain many collaborating species could change the practice of medicine', *The Economist*, 18 August 2012

SCIENCE COMMUNICATION

This text is adapted from an article in *The Economist*.

1 The author uses terms such as *gastric tract*, *mutualistic symbionts* and *metabolic syndrome*.

(a) What does this suggest about the intended audience for this article?

(b) Look up the meaning of these terms and note them down.

(c) In the seventh paragraph, the author refers to 'transplanting a small amount of faeces'. What does the author mean? Do you think that *transplant* is the most suitable term to use?

SKILLS ANALYSIS, INTERPRETATION, COMMUNICATION, PERSPECTIVE TAKING

SKILLS CONTINUOUS LEARNING, INTELLECTUAL INTEREST AND CURIOSITY

BIOLOGY IN DETAIL

SKILLS CRITICAL THINKING, ANALYSIS, ADAPTIVE LEARNING, PERSONAL AND SOCIAL RESPONSIBILITY, CONTINUOUS LEARNING, INTELLECTUAL INTEREST AND CURIOSITY

Now you are going to think about the science in the article. You will be surprised how much you know already, but if you choose to do so, you can return to these questions later in your course.

2 You have already looked up the meaning of the term *mutualistic symbiont*. Explain why bacteria living in our intestines are described as mutualistic symbionts rather than as pathogens.

3 You studied prokaryotic cells earlier in this topic. Explain how the number of bacterial cells could be 10 times the number of our own cells but still only weigh about 1.5 kg.

4 Lifestyles in the Middle East and North Africa have changed significantly over the last 50 years. In particular, the diet has changed. People eat far more processed foods than before. How might this change in diet affect the gut microbiome and what potential effects could this have on the body?

ACTIVITY

SKILLS PERSONAL AND SOCIAL RESPONSIBILITY, EMPATHY/PERSPECTIVE TAKING, ASSERTIVE COMMUNICATION

Carry out some research about how gut bacteria can improve our health.

Start a multi-author blog to which other members of your class can contribute. The aim of the blog is to show that bacteria in our digestive systems are essential to health.

Remember that many people think of bacteria as organisms that cause disease. You need to convince them that bacteria are actually helpful and essential to good health.

You should remember who your audience is – your language and your descriptions must be written at a suitable level of detail.

Always ensure you read previous contributions to the blog so that you do not repeat what your classmates have already written.

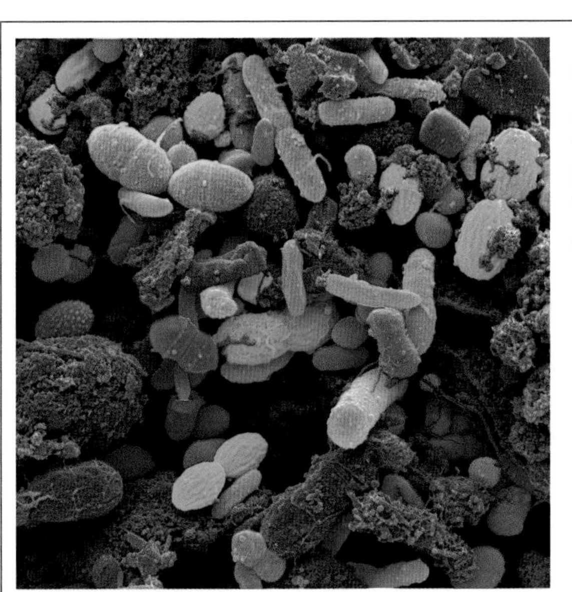

DID YOU KNOW?

This is a false-colour scanning electron micrograph of the bacteria found in human faeces. At least 50% of faecal material is bacteria that have been living in the gut. These bacteria have wide-ranging effects on our health.

3C EXAM PRACTICE

1 (a) The diagram below shows a pair of homologous chromosomes (a bivalent) during meiosis.

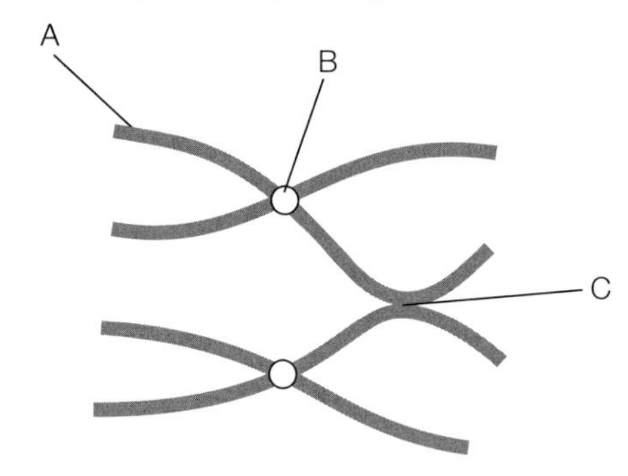

(i) What stage of meiosis is shown?

A prophase 1
B prophase 2
C anaphase 1
D metaphase 2 [1]

(ii) What are the correct names for structures A and B on the diagram? [1]

	A	B
A	chromatid	centriole
B	chromosome	centromere
C	chromatid	centromere
D	chromosome	centriole

(iii) Name the structure labelled C on the diagram. [1]

(b) In maize, the seeds can be yellow or white in colour. In addition, the seeds may have a smooth surface or a wrinkled surface. Each of these characteristics of maize seeds is an example of single-gene inheritance.

If a pure-breeding (homozygous) variety of maize with yellow, smooth seeds is crossed with a pure-breeding variety with white, wrinkled seeds, all the F_1 generation have yellow, smooth seeds.

The F_1 seeds were sown and the resulting plants were allowed to self-pollinate. The numbers of seeds resulting from this cross are shown in the table below.

Phenotype	Observed frequency	Expected frequency
yellow, smooth	2046	3150
white, smooth	851	1050
yellow, wrinkled	750	1050
white, wrinkled	1953	350
total number of seeds	5600	5600

(i) With reference to the events that occur during meiosis, explain why the results for the observed frequencies differ considerably from those for the expected frequencies. [5]

(ii) Calculate the proportion of non-parental types in the F_2 generation. [2]

(Total for Question 1 = 10 marks)

2 The ABO blood group system in humans is an example of multiple allele inheritance. Using this system, human blood can be classified into four possible blood groups: A, B, AB and O. The blood group of an individual is determined by a single gene pair.

(a) With reference to the inheritance of blood group in the ABO system, explain each of the following terms.

(i) codominance [2]

(ii) multiple allele inheritance [2]

(b) The family tree for a couple (P1 and P2) with three children is shown in the diagram below. The grandparents of the children and the blood group for each individual are also shown.

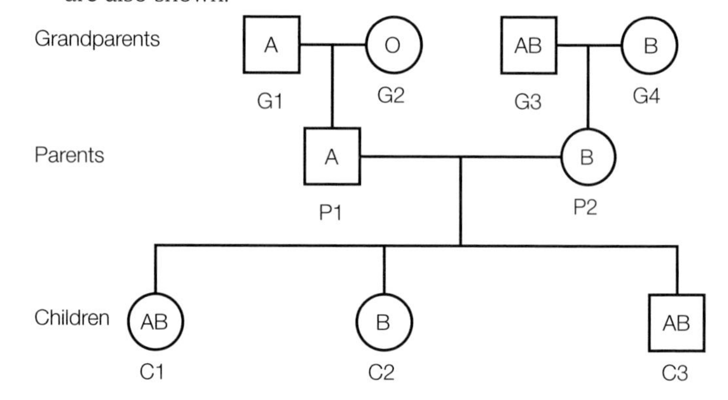

(i) State the genotype of each of the children. [2]

(ii) Calculate the probability that the next child born to this couple will have blood group O. Justify your answer using a genetic diagram. [4]

(Total for Question 2 = 10 marks)

3 A single stem cell can produce many genetically identical cells of different types.

(a) Name the process in which a stem cell becomes modified to perform a particular task.

A specialisation
B differentiation
C modification
D adaptation [1]

(b) The process that modifies stem cells can involve DNA methylation. Which of the following can DNA methylation **not** do?

A silence a gene
B induce a gene
C affect embryo development
D inactivate an X chromosome [1]

(c) There are adult stem cells in the human brain which can produce the different types of brain cell including nerve cells (neurones). Scientists are trying to find ways of growing adult brain stem cells in the laboratory.

(i) Name the type of cell division used to produce many genetically identical cells. [1]

(ii) Explain how genetically identical cells can be different. [1]

(iii) Suggest a reason why it might be useful to keep a supply of live stem cells from your brain in a laboratory. [1]

(d) Stem cells can come from embryos or from adult tissue.

(i) Suggest why research with embryonic stem cells is further advanced than research with adult stem cells. [2]

(ii) Some people think that research into and the medical use of embryonic stem cells is unethical and that we should await the results of research on adult stem cells.

Evaluate the benefits and the ethical issues associated with embryonic stem cell research. [4]

(Total for Question 3 = 11 marks)

4 During an infection, some white blood cells make glycoproteins which become part of their cell surface membranes. To make glycoproteins, the white blood cells must first synthesise proteins on the surface of their rough endoplasmic reticulum.

(a) Which of these processes enables one gene to produce many *different* proteins?

A pre-translational modification
B post-translational modification
C pre-transcriptional modification
D post-transcriptional modification [1]

(b) Describe how these newly made proteins are converted to glycoproteins on the cell surface membrane. [5]

(c) There are certain rare blood disorders which cause a shortage of white blood cells. One potential treatment would be to inject totipotent stem cells into individuals with these disorders.

(i) Explain what is meant by the term ***totipotent stem cell***. [2]

(ii) State why injecting totipotent stem cells may benefit a person with a shortage of white blood cells. [1]

(iii) Name **one** risk to the person receiving the stem cells. [1]

(Total for Question 4 = 10 marks)

5 (a) The diagram below shows two different stem cells and the differentiated cells that they can produce.

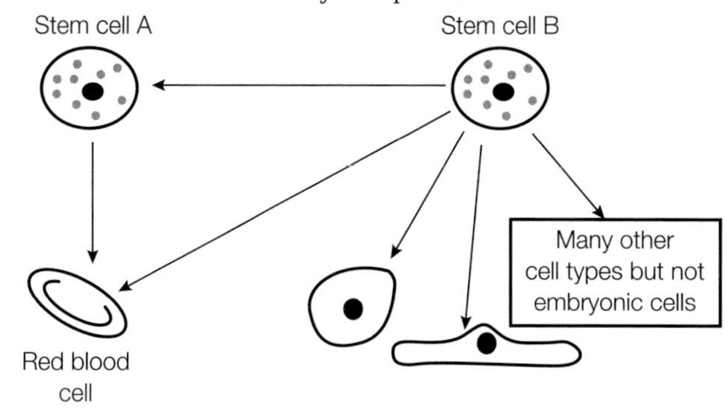

(i) Use the diagram to explain why stem cell B is described as *pluripotent*. [2]

(ii) Name **one** site where stem cell A may be found in an *adult* human. [1]

(iii) All the differentiated cells produced from stem cell B have the same genotype but have very different structures and functions. This is due to differential gene expression.

Explain how *differential gene expression* can enable cells which have the same genetic material to have very different structures and functions. [3]

(b) Three examples of how temperature affects organisms are given below. Which examples are due to differential gene expression? [1]

A	The rate of protein synthesis within a plant is temperature dependent.
B	The gender of turtles is determined by the temperature of the ground in which the eggs are laid.
C	Asexual reproduction is more rapid in bacteria if the temperature is higher.

(Total for Question 5 = 7 marks)

TOPIC 4 PLANT STRUCTURE AND FUNCTION, BIODIVERSITY AND CONSERVATION

CHAPTER 4A PLANT STRUCTURE AND FUNCTION

In this chapter, you will look at α- and β-glucose and discover how a small difference in bonding makes a huge difference to the structure of the molecules produced. You will study cellulose, the carbohydrate fundamental to plant cell walls. Plant cells have most of the features found in animal cells plus a few of their own. You will consider the structure and function of these features, highlighting some of the major differences between plant and animal cells. You will also be looking at how substances are transported around plants. You will consider the two main transport tissues, xylem and phloem, and compare their structure and functions. You will learn about transpiration and the transpiration stream, and you will discover which factors affect the rate of water uptake. You will also look at the translocation of sugars around the plant. The tissues that help to support a plant produce plant fibres used by people – for example, in clothing and paper. Plants also produce molecules that help them fight disease and we use some of them as medicines.

MATHS SKILLS FOR THIS CHAPTER

- **Carry out calculations using numbers in standard and ordinary form** (*e.g. use of magnification*)
- **Use expressions in standard form** (*e.g. when applied to areas such as the size of organelles*)
- **Use scales for measuring** (*e.g. graticule to measure size of cells*)
- **Make order of magnitude calculations** (*e.g. use and manipulate the magnification formula: magnification = size of image/size of real object*)
- **Use and manipulate equations, including changing the subject of an equation** (*e.g. magnification*)
- **Recognise and make use of appropriate units in calculations** (*e.g. rate of water uptake by a plant*)
- **Find arithmetic means** (*e.g. the mean number of stomata on different sides of a leaf, the mean rate of water uptake under different conditions*)
- **Translate information between graphical, numerical and algebraic forms** (*e.g. measurements of water uptake*)
- **Plot two variables from experimental or other data** (*e.g. graph of temperature against rate of transpiration using data from a practical investigation*)
- **Understanding that $y = mx + c$ represents a linear relationship** (*e.g. the relationship between rate of water uptake and the temperature of the surrounding air*)
- **Determining the intercept of a graph** (*e.g. graph showing data from a practical investigation into water loss from a plant*)
- **Calculating rate of change from a graph showing a linear relationship** (*e.g. rate of water uptake as volume taken up per unit time*)

What prior knowledge do I need?

Chapter 1A

- The importance of water to living things
- The structure of glucose and the storage polysaccharides amylose and glycogen

Chapter 2A

- The importance and structure of cell membranes
- How substances enter cells by diffusion, osmosis and active transport

Chapter 3A

- The way specimens are prepared for the light microscope (including staining) and the electron microscope
- The way specimens are magnified by the light microscope and the electron microscope
- How electron microscopy has increased our understanding of sub-cellular structures
- How the main sub-cellular structures of eukaryotic cells (including the nucleus, mitochondria and cell membranes) in plants and animals are related to their functions
- The main differences between animal and plant cells, including cell walls, chloroplasts and vacuoles

What will I study in this chapter?

- The structure of cellulose and the cell wall
- Specialised plant organelles including the details of chloroplasts and amyloplasts
- The arrangement of tissues in the stem and root
- Specialised cells making up xylem and phloem tissue in plants
- How the structure of the xylem and phloem are adapted to their functions in relation to their role in transport in a plant
- The effect of a variety of environmental factors on the uptake of water by plants
- How water is moved up through a plant from the roots to the leaves in the xylem tissue
- Translocation - the movement of sugars in the phloem
- The importance of water and inorganic ions (nitrate, calcium and magnesium ions) to plants
- How we use the fibres made by plants and the potential contribution to sustainability
- Plant-based medicines and how new drugs can be developed

What will I study later?

Chapter 5A (Book 2: IAL)

- The importance and biochemistry of photosynthesis
- The importance of the structural features of chloroplasts in photosynthesis

Chapter 7A (Book 2: IAL)

- The importance of the structural features of mitochondria in cellular respiration

Chapter 8B (Book 2: IAL)

- The importance of plant transport tissues in plant responses

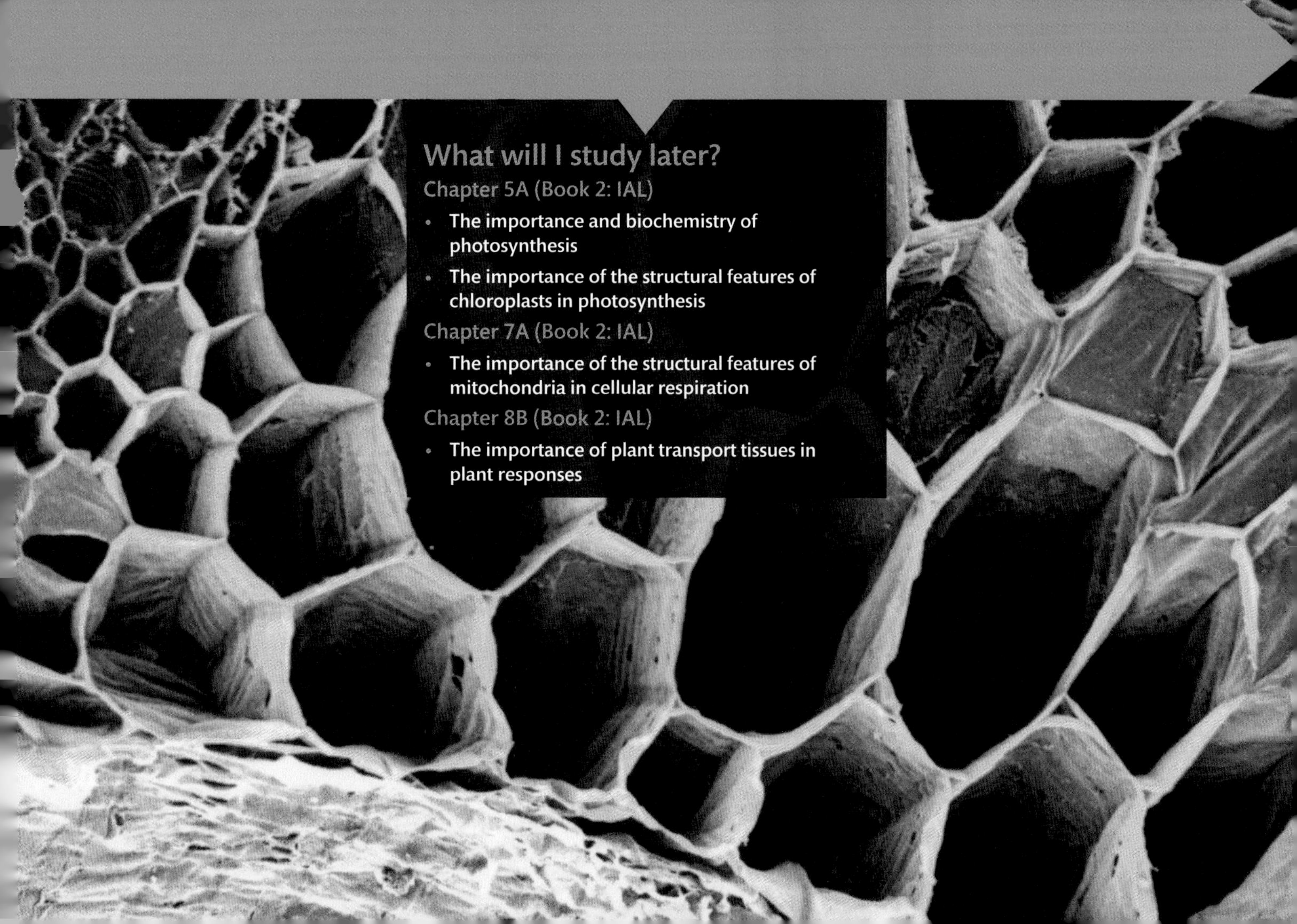

4A 1 THE CELL WALL

SPECIFICATION REFERENCE
4.1(i) 4.1(ii) 4.2 4.3 CP7

LEARNING OBJECTIVES

- Understand the structure and function of the polysaccharides starch and cellulose, including the role of hydrogen bonds between the β-glucose molecules in the formation of cellulose microfibrils.
- Know the structure and ultrastructure of plant cells including cell wall, chloroplast, amyloplast, vacuole, tonoplast, plasmodesmata, pits and middle lamella; understand the function of the structures.
- Know the appearance of plant organelles under the electron microscope.
- Understand how plant and animal cells compare.

Plants and animals are eukaryotes. A typical plant cell has many features which are the same as a typical animal cell (see **Sections 3A.2** and **3A.3**). They have many membranes and contain cytoplasm and a nucleus. Rough and smooth endoplasmic reticulum are spread throughout the cytoplasm, along with active Golgi apparatus. Mitochondria produce ATP, which is essential to the working of both the plant cell and the animal cell. However, there are several fundamental differences between plant and animal cells. Plant cells contain several kinds of organelle that are not found in animal cells and, most distinctively, plant cells have a cellulose cell wall (see **fig A**).

PRACTICAL SKILLS

CP7

Observing plant cells under the microscope

When you look at plant cells under the light microscope, or look at electron micrographs of plant cells, you will be using the skills you learned in Core practical 5, and applying them to plant cells instead of animal cells. You will find out about the features of plant cells and the organelles you will be observing in 4A.1, 4A.2 and 4A.3.

Remember when you make a drawing from a micrograph that you must always use pencil, you must draw what you see, you must label what you observe and you must give a scale or magnification. **Fig A** shows you the main features of plant cells as you will see them in both light and electron micrographs.

- The drawing beside the light micrograph is a good average plant cell, but it does not look like the cell and it has been coloured in. It is not a true observational drawing.
- The drawing of the electron micrograph is a good example of an observational drawing and this is what you should aim for.

(a) a light micrograph and drawing of a plant cell ×250

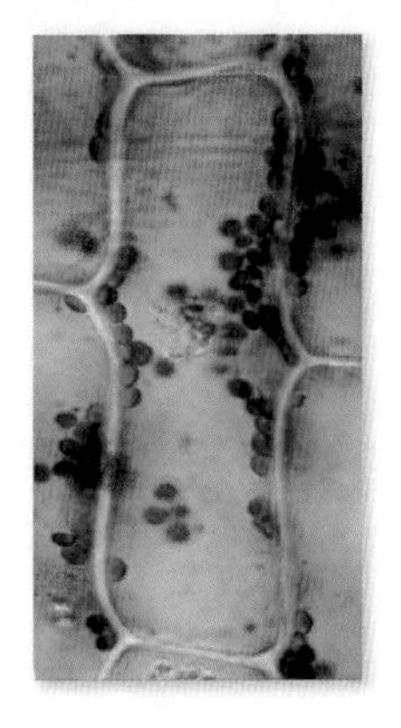

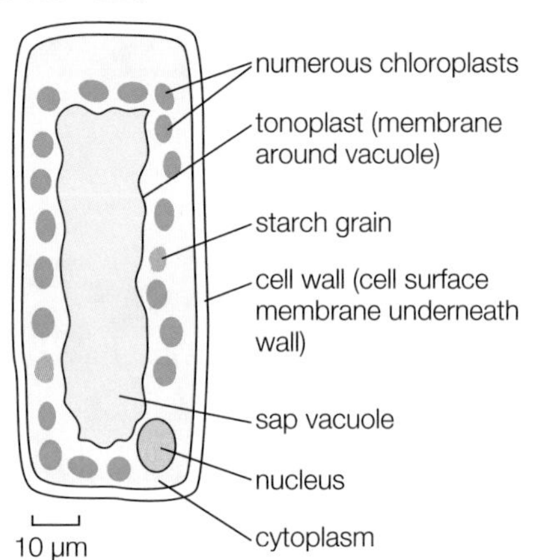

(b) an electron micrograph and drawing of a plant cell ×5000

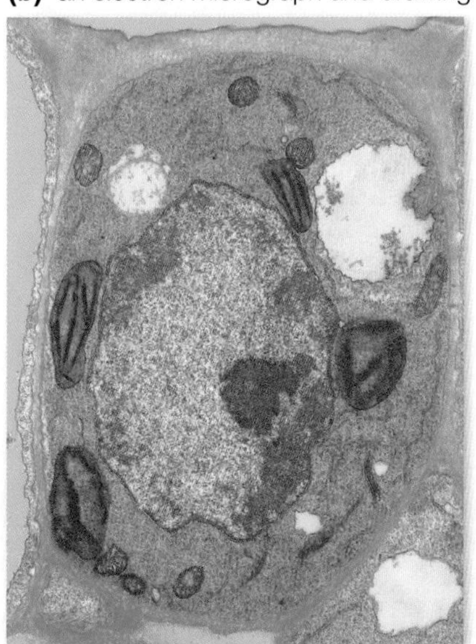

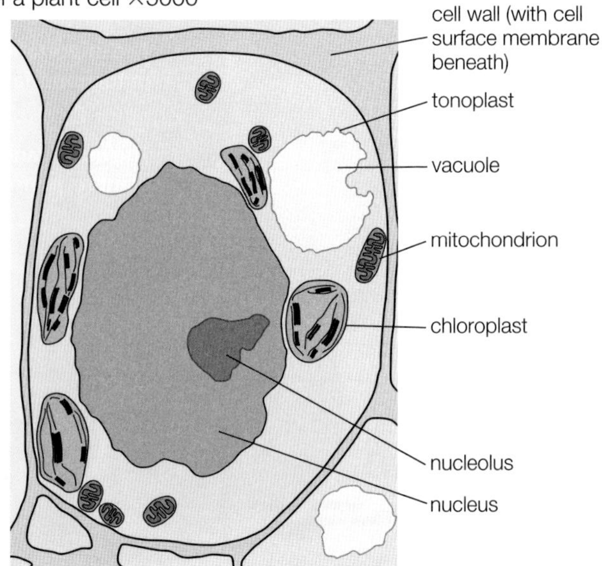

▲ **fig A** The light microscope shows us the major features of a plant cell; the electron microscope reveals the ultrastructure of the organelles.

THE PLANT CELL WALL

Animal cells can be almost any shape. Plant cells are usually more regular in their appearance. This is largely because each cell is bounded by a **cell wall**. You can visualise a plant cell as a jelly-filled balloon inside a small cardboard box. The cell wall (the cardboard box) is an important feature that gives plants their strength and support. It is mostly made of insoluble cellulose. The plant cell wall is usually freely permeable to everything that is dissolved in water – it does not act as a barrier to substances getting into the cell. However, **suberin** is added to the cell wall in cork tissues, and **lignin** is part of the cell wall structure in wood. These compounds reduce the permeability of the cell wall so that water and dissolved substances cannot pass through it.

The plant cell wall consists of several layers. The **middle lamella** is the first layer and is made when a plant cell divides into two new cells. It is mostly made of **pectin**, a polysaccharide that acts like glue and holds the cell walls of adjacent plant cells together. Pectin has lots of negatively charged carboxyl (–COOH) groups and these combine with positive calcium ions to make calcium pectate. The calcium pectate binds to the cellulose on either side. The cellulose microfibrils and the matrix build up on both sides of the middle lamella. To begin with, these walls are very flexible, with the cellulose microfibrils all being arranged in a similar direction. They are called **primary cell walls**.

Secondary thickening may occur as the plant ages. A **secondary cell wall** builds up, with the cellulose microfibrils laid densely at different angles to each other. This makes the composite material much more rigid. **Hemicelluloses** help to harden it further. In some plants, particularly woody perennials, lignin is then added to the cell walls to produce wood, which makes the structure even more rigid. Within the structure of a plant, there are many long cells with cellulose cell walls that have been heavily lignified. These are called **plant fibres** and people use them in many ways including for clothing, building material, ropes and paper (see **Section 4A.5**).

THE CHEMISTRY OF CELLULOSE

Cellulose is the main compound in plant cell walls. Cellulose is a complex carbohydrate and is similar to starch and glycogen which you studied in **Section 1A.3**. It consists of long chains of glucose joined by glycosidic bonds. However, glucose comes in two different forms (isomers), called α-glucose and β-glucose. The two isomers come from different arrangements of the atoms on the side chains of the molecule (see **fig B**).

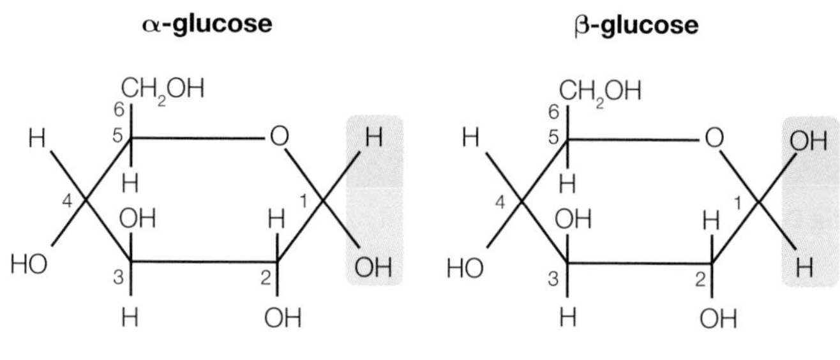

or, even more simply:

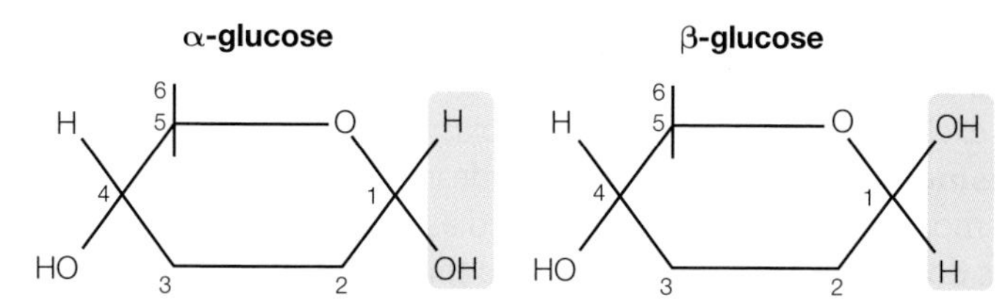

In these diagrams, the positions of carbon atoms are represented by their numbers only.
Note carefully the different arrangement of atoms around the carbon 1 atom in α-glucose and β-glucose.

▲ **fig B** The difference in structure between α-glucose and β-glucose may seem small, but it has a big impact on the function of the molecule.

LEARNING TIP

Remember that there is only a small difference between alpha and beta glucose, but the effect on the properties of the polymer is huge.

The different isomers make different bonds between adjacent glucose molecules, and this affects the polymers they create. In starch, the monomer units are α-glucose. In cellulose, they are β-glucose and they are held together by 1,4-glycosidic bonds (see **Section 1A.2**) where one of the monomer units has to be turned round (inverted) so the bonding can take place. This linking of β-glucose molecules means that the hydroxyl (–OH) groups stick out on both sides of the molecule (see **fig C**). Because of this, hydrogen bonds can be made between the partially positively charged hydrogen atoms of the hydroxyl groups and the partially negatively charged oxygen atoms in other areas of the glucose molecules. This is called cross-linking and it holds neighbouring chains firmly together.

β-glucose β-glucose

no inversion inversion

condensation hydrolysis

H_2O H_2O

portion of a cellulose molecule

hydrogen bonds

▲ **fig C** Cellulose molecules consist of β-glucose monomers joined by 1,4-glycosidic bonds.

Many of these hydrogen bonds are made, which makes cellulose a material with considerable strength. Cellulose molecules do not coil or spiral – they remain as very long, straight chains. In contrast, starch molecules, with 1,4- and 1,6-glycosidic bonds between α-glucose monomers, form compact globular molecules that are useful for storage. This difference in structure between starch and cellulose gives them very different properties. You can remind yourself of the chemistry of starch and why it is important as an energy source for animals by looking back to **Section 1A.3**.

Starch is an important source of energy in the diet for many animals. However, most animals do not possess the enzymes needed to break the 1,4-glycosidic bonds between the molecules of β-glucose, so they cannot digest cellulose. Ruminant animals use the cellulose-digesting enzymes from bacteria living in their gut to digest their food; termites use the enzymes in various protozoa (small eukaryotic organisms, see **Chapter 4B**). It is the cellulose in plant food that acts as roughage or fibre in the human diet – it is an important part of a healthy diet exactly because you can't digest it.

In the cell wall, groups of 10 000–100 000 cellulose molecules form microfibrils which can be seen under the electron microscope (see **fig D**). These cellulose fibrils are deposited in layers which are held together by a matrix of hemicelluloses and other short-chain carbohydrates. They act in the same way as glue, binding to each other and to the cellulose molecules. Mannose, xylose and arabinose are examples of the sugars involved. The combination of the cellulose microfibrils in the flexible matrix makes a **composite material**, combining the properties of both these materials in the plant cell wall. The cells are **turgid** (firm) most of the time, giving the strength to support the plant in a vertical position, yet the plant can wilt when water is in short supply and the cells become **flaccid** (floppy).

EXAM HINT

You may be asked to compare cellulose with other straight chain molecules such as collagen (a protein).

Despite its strength, don't forget that cellulose is a carbohydrate.

▲ **fig D** These cellulose microfibrils consist of thousands of cellulose chains held together by hydrogen bonds. The way they are arranged is different in primary and secondary cell walls, affecting both flexibility and strength.

PLASMODESMATA

Plant cells seem to communicate closely with each other even though they are encased in cellulose cell walls. In primary cell walls and in cell walls which do not have lignin in them, materials are exchanged through special cytoplasmic bridges called **plasmodesmata** (singular: plasmodesma) (see **fig E**). The plasmodesmata seem to be produced as the cells divide – the two cells do not separate completely, and threads of cytoplasm remain between them. These threads pass through gaps in the newly formed cell walls and signalling substances can pass from one cell to another through the cytoplasm. The interconnected cytoplasm of the cells is called the **symplast**. The cell walls are thinner in the region of the plasmodesmata.

When secondary thickening takes place, hemicelluloses and lignin are deposited in the cell wall making it thicker. In the areas around the plasmodesmata, this process doesn't happen, leaving thin areas of the cell wall called **pits**. There is no cytoplasm in the xylem cells but the pits allow water to move between the xylem vessels. They are important in maintaining a flow of water at even pressure through the plant.

Scientists are working hard to discover exactly how plant cells communicate through plasmodesmata. One clear piece of evidence comes from work with plant grafts: it shows that these intercellular junctions are vital in the life of plants. If a rose is grafted onto a root stock, the graft tissue will only start healthy cell division and grow after plasmodesmata bridges have established between the host tissue and the graft tissue.

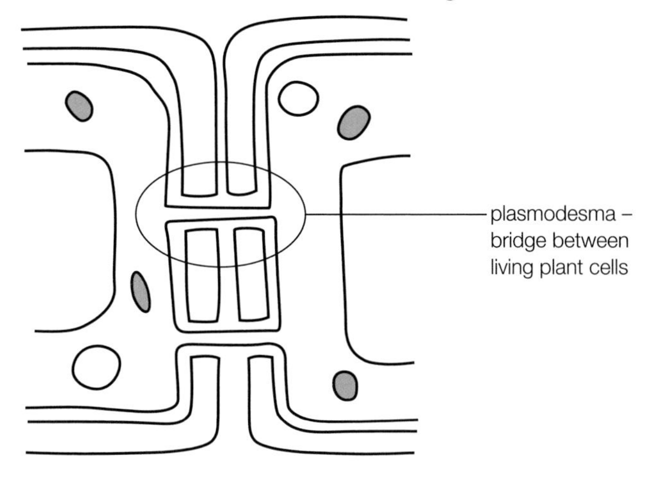

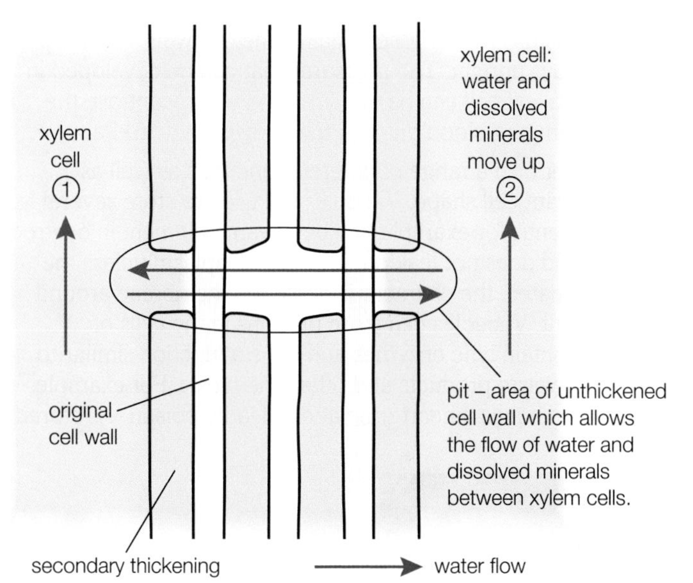

fig E Plasmodesmata provide a route for plant cells to communicate but scientists are still trying to find out exactly how this works.

CHECKPOINT

SKILLS INTERPRETATION

1. Describe the role cell walls play in the structure of a plant and explain how their structure is related to their function.
2. Describe how the plant cell wall changes as the cell grows and develops, and explain how this affects the cell.
3. Explain why plasmodesmata are an important feature of plant cell structure.

SUBJECT VOCABULARY

cell wall a freely permeable wall around plant cells, made mainly of cellulose
suberin a waterproof chemical that impregnates cellulose cell walls in cork tissues and makes them impermeable
lignin a chemical that impregnates cellulose cell walls in wood and makes them impermeable
middle lamella the first layer of the plant cell wall to be formed when a plant cell divides, made mainly of calcium pectate (pectin) that binds the layers of cellulose together
pectin a polysaccharide that holds cell walls of neighbouring plant cells together and is part of the structure of the primary cell wall
primary cell walls the first very flexible plant cell walls to form, with all the cellulose microfibrils orientated in a similar direction
secondary cell wall the older plant cell wall in which the cellulose microfibrils have built up at different angles to each other making the cell wall more rigid
hemicelluloses polysaccharides containing many different sugar monomers
plant fibres long cells with cellulose cell walls that have been heavily lignified so they are rigid and very strong
composite material a material made of two or more materials which combined together make a composite with different properties from either of the constituent materials
turgid swollen
flaccid floppy, soft
plasmodesmata cytoplasmic bridges between plant cells that allow communication between the cells
symplast all of the material (cytoplasm, vacuole, etc.) contained within the surface membrane of a plant cell
pits thin areas of cell wall in plant cells with secondary thickening, where plasmodesmata maintain contact with adjacent cells; in xylem vessels, where the cells are dead, they become simple holes through which water moves out into the surrounding cells

4A 2 PLANT ORGANELLES

SPECIFICATION REFERENCE 4.1(i) 4.1(ii) 4.2 CP7

LEARNING OBJECTIVES

- Know the structure and ultrastructure of plant cells including cell wall, chloroplast, amyloplast, vacuole, tonoplast, plasmodesmata, pits and middle lamella; understand how these relate to their functions.
- Know the appearance of plant organelles under the electron microscope.
- Understand how plant and animal cells compare.

EXAM HINT

When you compare similar structures such as plant and animal cells, don't forget to make a statement about each one. A suitable table is a good way to compare features in plant and animal cells.

In some ways, the ultrastructure of plant cells is very similar to that of animal cells, with their cytoplasm, nucleus, mitochondria, ribosomes, endoplasmic reticulum and more. But they can also be very different – plant cells may contain several kinds of organelle that are not found in animal cells. These include permanent vacuoles and chloroplasts.

PERMANENT VACUOLE

A vacuole is any fluid-filled space inside the cytoplasm which is surrounded by a membrane. Vacuoles occur quite frequently in animal cells, but they are only temporary, being created and destroyed when needed. In non-woody plant cells, the vacuole is a permanent structure with an important role. The vacuole can occupy up to 80% of the volume of a plant cell. It is surrounded by a specialised membrane called the **tonoplast** (see **fig A**). The tonoplast contains many different protein channels and carrier systems. It controls the movements of substances into and out of the vacuole and so it controls the water potential of the cell. The vacuole is filled with **cell sap**, a solution of various substances in water. This solution causes water to move into the cell by **osmosis** (see **Section 2A.3**), and this means the cytoplasm is kept pressed against the cell wall. This keeps the cells turgid (swollen) and the whole plant stays upright. The pressures that can be developed in this way are very large indeed. The pressure in a leaf cell can be up to 1500 kPa – in contrast, the pressure in a human artery when the heart is pumping blood out into the body is only 16 kPa.

The many different types of vacuole in plants perform a range of different functions as well as fulfilling the important role of maintaining the plant cell shape. Vacuoles are used to store several different substances. Many vacuoles store pigments; for example, the betacyanin pigment in beetroot is normally stored in the vacuoles of the cells and does not leak out into the cytoplasm unless the root is cut (see **Section 2A.1**). If the tissue is heated, the characteristics of the membrane around the vacuole will change and pigment will leak out. Vacuoles can store proteins in the cells of seeds and fruits and, in some plant cells, they contain lytic enzymes and have a function similar to lysosomes in animal cells. Vacuoles often store waste products and other chemicals. For example, digitalis – a chemical found in foxgloves that can act as a heart drug and a deadly poison – is stored in the vacuoles of the cells.

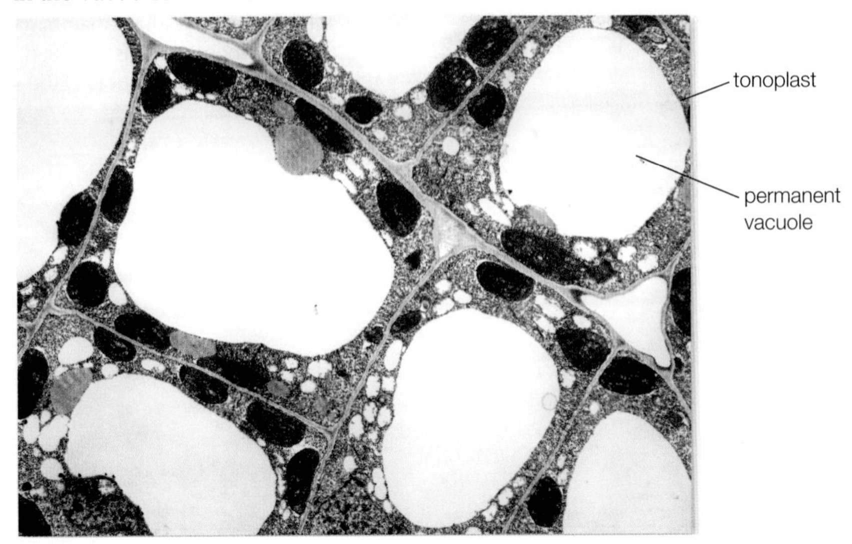

▲ **fig A** The tonoplast and the permanent vacuoles are important structures in the support systems of plants, but they also have many other functions.

CHLOROPLASTS

Of all the differences between plant and animal cells, the presence of **chloroplasts** in plant cells is probably the most important because they enable plants to make their own food. Not all plant cells contain chloroplasts – only cells from the green parts of the plant. However, almost all plant cells contain the genetic information to make chloroplasts. In some circumstances, different areas of a plant will become green and start to photosynthesise. The exceptions are parasitic plants such as desert hyacinth (*Cistanche tubulosa*). Cells in flowers, seeds and roots contain no chloroplasts and neither do the internal cells of stems or the transport tissues. In fact, the majority of plant cells do not have chloroplasts, but these organelles are very special and are unique to plants. Chloroplast structure is shown in **fig B**.

(a)

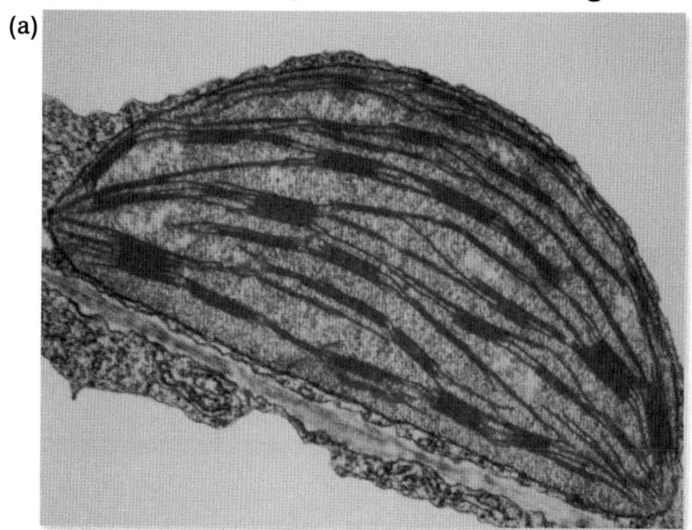

(b)

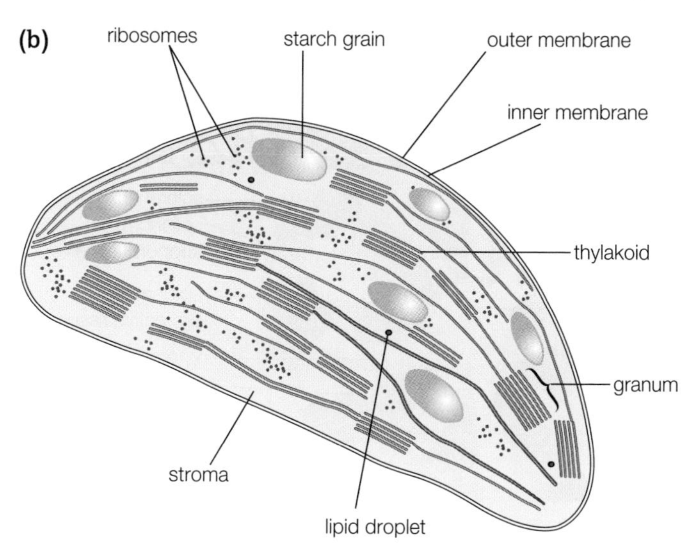

▲ **fig B** **(a)** Micrograph of a chloroplast; and **(b)** a labelled diagram to show structures in a chloroplast

Chloroplasts and mitochondria are similar in many ways. Like mitochondria, chloroplasts:

- are large organelles – they have a biconvex shape with a diameter of 4–10 μm and are 2–3 μm thick
- contain their own DNA
- are surrounded by an outer membrane
- have an enormously folded inner membrane that gives a greatly increased surface area where enzyme-controlled reactions take place
- are thought to have been free-living prokaryotic organisms that were engulfed by and became part of other cells more than 2000 million years ago.

However, there are also some clear differences. Chloroplasts:

- are the site of photosynthesis
- contain **chlorophyll**, the green pigment that is largely responsible for trapping the energy from light, making it available for the plant to use.

AMYLOPLASTS

Amyloplasts are another specialised plant organelle. They are colourless and store starch (see **Section 1A.3**). Remember starch is made of amylose and amylopectin joined together – this is where the amyloplasts get their name. The starch can be converted to glucose and used to provide energy when the cell needs it. Large numbers of amyloplasts are found in areas of a plant that store starch, for example potato tubers (see **fig C**).

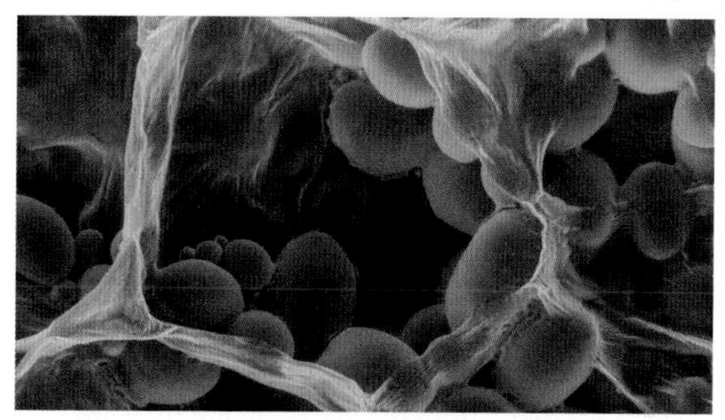

▲ **fig C** Amyloplasts in plant cells from a potato – they are full of stored starch.

CHECKPOINT

SKILLS INTERPRETATION

1. Describe how amyloplasts and chloroplasts differ.
2. Compare and contrast the structure of a typical plant cell with the structure of a typical animal cell.
3. Explain why chloroplasts are found only in particular parts of a plant. Suggest what happens to make part of a plant (e.g. a potato tuber) turn green when exposed to light.

SUBJECT VOCABULARY

tonoplast the specialised membrane that surrounds the permanent vacuole in plant cells and controls movements of substances into and out of the cell sap
cell sap the aqueous solution that fills the permanent vacuole
osmosis a specialised form of diffusion that involves the movement of solvent molecules down their water potential gradient
chloroplasts organelles adapted to carry out photosynthesis, containing the green pigment chlorophyll
chlorophyll the green pigment that is largely responsible for trapping the energy from light, making it available for the plant to use in photosynthesis
amyloplasts plant organelles that store starch

4A 3 PLANT STEMS

SPECIFICATION REFERENCE 4.4 4.5 4.6 CP7

LEARNING OBJECTIVES

- Know the similarities and differences between the structures, the position in the stem, and the function of: sclerenchyma fibres (support), xylem vessels (support and transport of water and mineral ions) and phloem (translocation of organic solutes).
- Understand how the arrangement of cellulose microfibrils and secondary thickening in plant cell walls contributes to the physical properties of xylem vessels and sclerenchyma fibres in plant fibres that can be exploited by humans.

Many plant cells are specialised and adapted for a specific role in the plant. This means they do not look like the 'typical' plant cell. They may be organised into tissues and organs that carry out a specific function in the plant. One example of a plant organ is the stem.

PROVIDING SUPPORT AND TRANSPORT

The primary function of a stem is support, to hold the leaves in the best position for obtaining sunlight for photosynthesis. Stems also support the flowers in a way that maximises the likelihood of pollination. The stem must provide flexible support, because plants are frequently blown by wind and battered by rain. Stems need to bend to endure the forces of the weather but still have the strength to stay upright.

The second major function of stems is the movement of materials around the plant. They provide the route along which the products of photosynthesis are carried – from the leaves where they are made to other parts of the plant where they are needed. Water moves steadily through the stems from the roots up to the leaves, and carries mineral ions which are needed for the synthesis of more complex chemicals.

Most stems are green because they contain chlorophyll. They carry out a small amount of photosynthesis but this is not a major function.

Not all plants have stems. The liverworts have a simple flat structure and the mosses have leaves which arise directly from a pad of rhizoids. Both of these groups have no specialised transport tissues and grow near the ground. However, the majority of the more complex plants do possess stems.

THE TISSUES THAT MAKE UP THE STEM

Stems contain many different tissues as you can see in **fig A**. They all have important roles, but you will be concentrating on the xylem, the phloem and the sclerenchyma.

The outer layer of the stem is the epidermis, which does not provide support but protects the cells beneath it. Much of the stem is packing tissue which consists of the most common type of plant cell, **parenchyma**. These are unspecialised cells, but they can be modified in several ways so they become suitable for storage and photosynthesis. For example, the outer layers of parenchyma cells in the stem may contain some chloroplasts. Some of the parenchyma in the stem is modified into collenchyma and sclerenchyma.

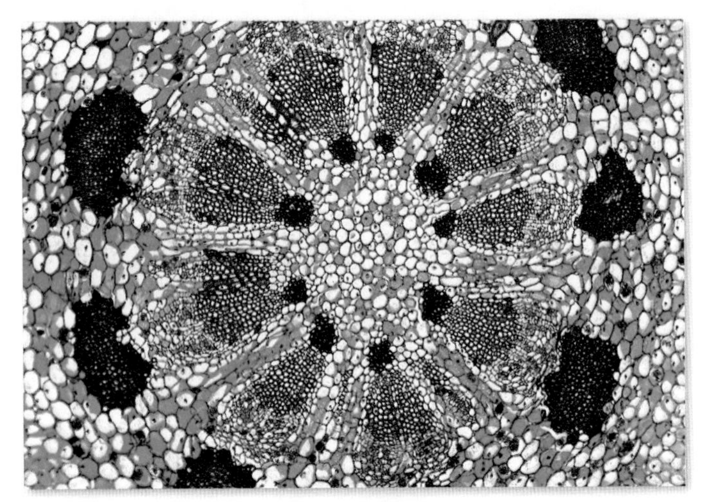

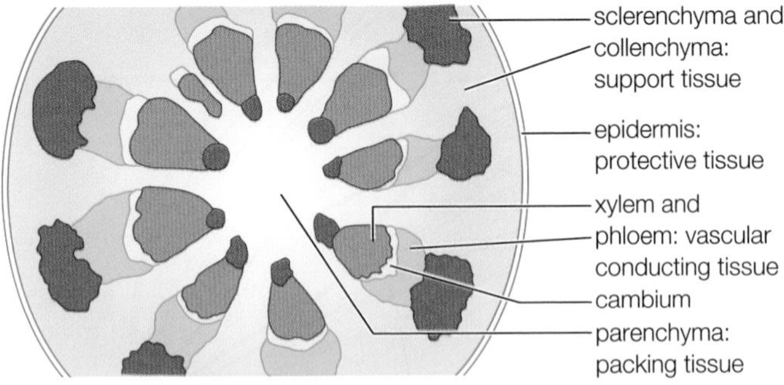

▲ **fig A** The distribution of the different tissues in the stem of a plant

Collenchyma cells have thick cellulose primary cell walls, which are even thicker at their corners (see **fig B**). This gives the tissue its strength. These cells are found around the outside of the stem, just inside the epidermis, and they give plenty of support but remain living, so they stretch as the plant grows and provide flexibility.

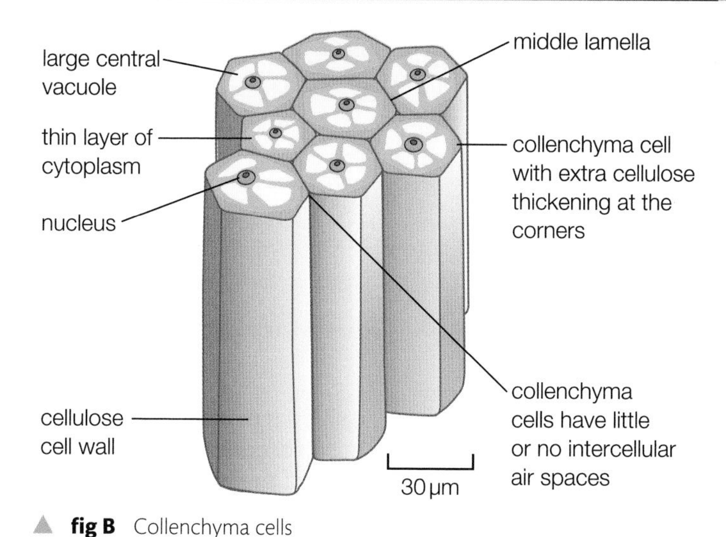

▲ **fig B** Collenchyma cells

SCLERENCHYMA

Sclerenchyma is a type of modified parenchyma (packing) tissue found in plant stems. It develops, as the plant grows, to support the increasing weight of the upper part of the plant. Sclerenchyma tissue is found around the vascular bundles in older stems and in leaves. All sclerenchyma cells have strong secondary walls made of cellulose microfibrils positioned at right angles to each other. Some sclerenchyma makes fibres, very long cells often found in bundles or cylinders around the outside of a stem or root. Lignin is deposited on the cell walls of these fibres in a spiral or a ring pattern, and this makes the fibres strong but also flexible. The strength of the fibres depends on their length and how much they are lignified (how much lignin they contain). When the fibre is lignified, the cell contents die because water cannot pass through lignin, and so the fibres become hollow tubes. Once this has happened, these cells can no longer grow, so plant growth has to be higher up the stem.

Sclerenchyma cells can also become completely impregnated with lignin and form **sclereids**. These very tough cells may be found in groups throughout the cortex of the stem or individually in plant tissue. For example, the gritty texture of pears is a result of individual sclereids in the flesh. Different aspects of sclerenchyma cells are shown in **fig C**.

TRANSPORT TISSUES IN PLANTS

Plants, like other large, multicellular organisms, need a system to transport substances to all the cells. The main transport tissues in plants are the **xylem** and **phloem** and they are found associated together in **vascular bundles** throughout the plant, including the stem, roots and leaves (see **fig D**).

- Xylem tissue carries water and dissolved mineral ions from the roots to the photosynthetic parts of the plant. The movement in the xylem is always upwards. Xylem consists of several different types of cell; most of the xylem cells are dead. Long tubular structures called xylem vessels are the main functional units of the xylem.
- Phloem is living tissue made of phloem cells which transport the dissolved product of photosynthesis (sucrose) from the leaves to where it is needed for growth or storage as starch. The flow through phloem can go both up and down the plant.
- **Cambium** is a layer of unspecialised cells which divide, giving rise to more specialised cells that form both the xylem and the phloem.

LEARNING TIP

Remember that xylem and phloem are tissues; they consist of several different types of cell.

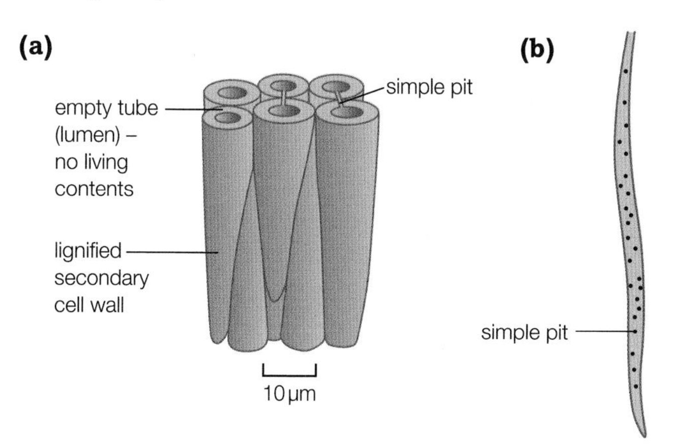

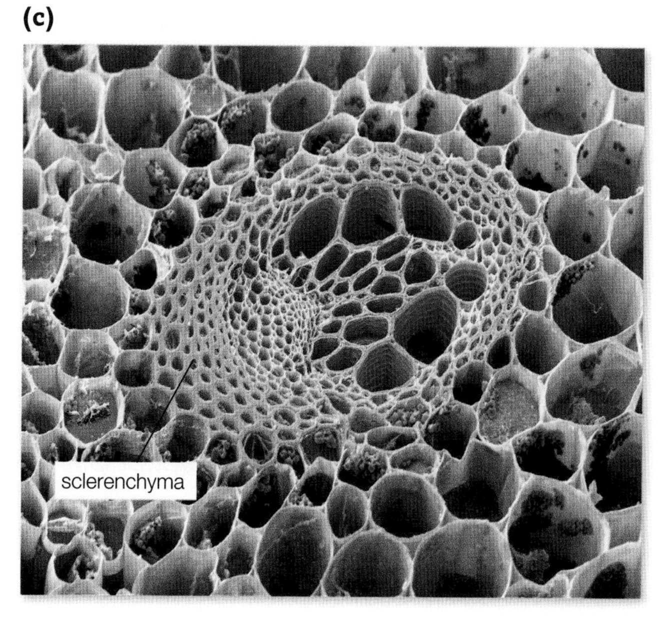

▲ **fig C** The structure of **(a)** sclerenchyma cell and **(b)** sclerenchyma fibres. The cross-section of a plant stem **(c)** shows the very thickened sclerenchyma cell walls.

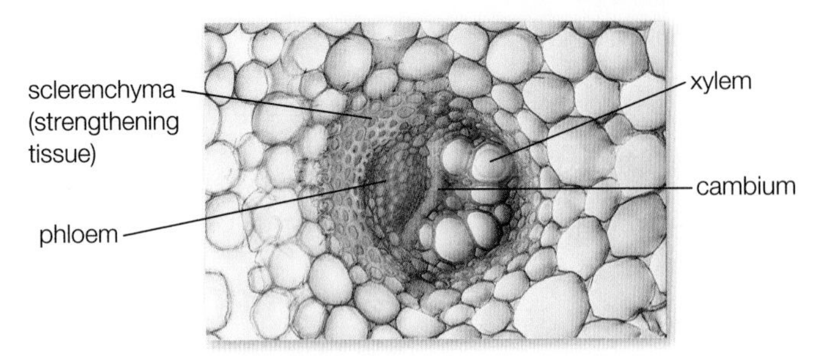

▲ **fig D** The arrangement of tissues in a vascular bundle combine strength and transport functions.

XYLEM

The xylem starts off as living tissue. The first xylem the plant makes is called the **protoxylem**. It can stretch and grow because the walls are not fully lignified. The cellulose microfibrils in the walls of the xylem vessels are arranged vertically in the stem. This increases the strength of the tube and allows it to resist the compression forces from the weight of the plant pressing down on it. Increasing amounts of lignin are incorporated into the cell walls as the stem ages and the cells stop growing. This means that the cells become impermeable to water and other substances. The tissue becomes stronger and more supportive, but the contents of the cells die. This lignified tissue is called the **metaxylem**. The end walls between the cells mostly break down so the xylem forms hollow tubes which go from the roots to the tip of the stems and leaves. **Fig E** summarises the formation of xylem vessels.

Water and mineral ions are transported from the roots to the leaves and shoots in the **transpiration stream**. Water moves out of the xylem into the surrounding cells through the specialised pits in the walls of the xylem vessels. The lignified xylem vessels are very strong and help to support the stems of plants, particularly in larger plants. In smaller, non-woody plants, the turgid parenchyma cells in the centre provide most of the support, as well as the sclerenchyma and collenchyma. This is why young plants wilt if too much water is lost. More xylem tissue is lignified to increase support as woody plants grow older. In trees, this lignified xylem makes up most of the trunk of the tree (the wood). The living cells around the cambium are on the outside of the trunk, just under the bark. A new ring of vascular tissue is made each year, so the growth rings of the tree provide a record of the xylem produced in each growing season.

LEARNING TIP

Do not confuse transpiration and the transpiration stream. The transpiration stream is a flow of water and solutes up the plant which is driven by the process of transpiration in the leaf.

▲ **fig E** The xylem vessels change from living cells to dead tubes of lignin as they develop.

PHLOEM

Mature phloem is a living tissue that transports food in the form of organic solutes around the plant. These molecules move from the leaves where they are made by photosynthesis to the tissues where they are needed. Materials in the phloem can be transported both up and down the stems in an active process called **translocation**.

The phloem consists of many cells joined to make very long tubes that run from the highest shoots to the end of the roots. However, the phloem cells do not become lignified and so the contents remain living. The walls between the cells become perforated creating specialised **sieve plates** and the phloem sap flows through the holes in these plates. The nucleus, the tonoplast and some of the other organelles break down as the gaps in the sieve plates are made. The phloem sieve tube becomes a tube filled with phloem sap and the mature phloem cells have no nucleus. They survive because they are closely associated with cells called **companion cells**. The companion cells are very active cells that have all the normal organelles, and they are linked to the sieve tube by many plasmodesmata. The cell membranes of companion cells have many infoldings that increase the surface area over which they can transport sucrose into the cell cytoplasm. They also have many mitochondria to supply the ATP needed for active transport. All the evidence suggests that companion cells support the sieve tube cells, which have lost most of their normal cell functions (see **fig F**).

EXAM HINT

Remember that phloem is living tissue; it transports food by an active process.
Xylem is dead; it acts as a simple pipe allowing the flow of water.

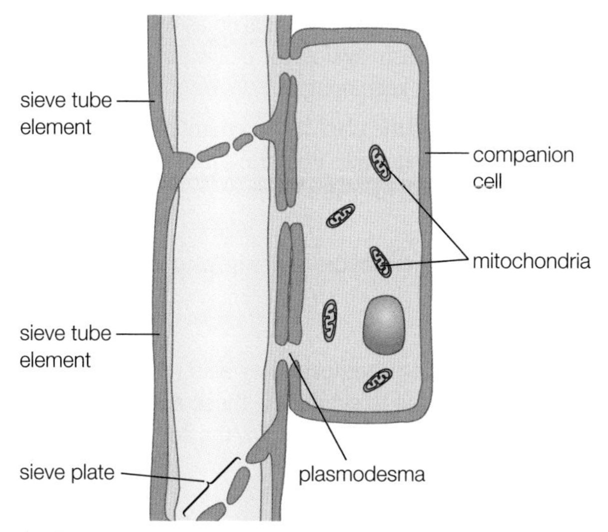

▲ **fig F** These are phloem vessels and companion cells - the tissue that moves sugars around plants.

SUPPORT AND TRANSPORT THROUGHOUT THE PLANT

Water, mineral ions and sugars need to be transported all over the plant. The support and strength provided by the sclerenchyma and the xylem are important in roots and leaves as well as in the plant stems. The roots need to be able to cope with the bending and straining forces as the plant moves in the wind and as the weight of leaves, flowers and fruit increase. The leaves need to be held flat so they can capture as much sunlight as possible. The main vein through the leaf has large xylem vessels and sclerenchyma to give as much support as possible.

So these tissues – xylem, phloem and sclerenchyma – are found throughout the plant. They are often found in vascular bundles. In the stem and root, the xylem is on the inside, the phloem on the outside and often a layer of strengthening sclerenchyma around that. The way these tissues are organised varies in the different areas of the plant, **fig G** shows them in the main vein of a leaf.

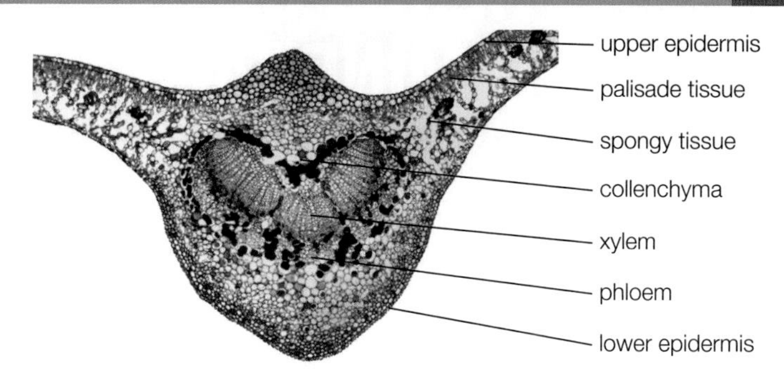

▲ **fig G** The arrangement of the xylem, phloem and sclerenchyma tissue varies in the different areas of the plant to give the support needed in different situations.

CHECKPOINT

1. State the main similarities and differences between xylem and phloem.

SKILLS CRITICAL THINKING

2. Explain how the different tissues in a stem enable it to carry out its function of support.
3. Describe how the support tissues change as a plant grows and explain what effect this has on the plant.
4. The bark of young trees contains a ring of vascular bundles. Gardeners and forestry workers protect young trees with plastic tubes around the lower part of the trunk. What do you think they are being protected from and why is this necessary?

SUBJECT VOCABULARY

parenchyma relatively unspecialised plant cells that act as packing in stems and roots to give support
collenchyma plant cells with areas of cellulose thickening that give mechanical strength and support to the tissues
sclerenchyma plant cells that have very thick lignified cell walls and an empty lumen with no living contents
sclereids sclerenchyma cells that are completely impregnated with lignin
xylem the main tissue transporting water and minerals around a plant
phloem the main tissue transporting dissolved food around the plant
vascular bundle part of the transport system of a plant, with phloem on the outside and xylem on the inside - often with strengthening sclerenchyma
cambium the layer of unspecialised plant cells that divide to form both the xylem and the phloem
protoxylem the first xylem the plant makes; it can stretch and grow because the walls are not fully lignified
metaxylem consists of mature xylem vessels made of lignified tissue
transpiration stream the movement of water up from the soil through the root hair cells, across the root to the xylem, then up the xylem, across the leaf until it is lost by evaporation from the leaf cells and diffuses out of the stomata down a concentration gradient
translocation the active movement of substances around a plant in the phloem
sieve plates the perforated walls between phloem cells that allow the phloem sap to flow
companion cells very active cells closely associated with the sieve tube elements that supply the phloem vessels with everything they need and actively load sucrose into the phloem

4A 4 THE IMPORTANCE OF WATER AND MINERALS IN PLANTS

SPECIFICATION REFERENCE 4.8

LEARNING OBJECTIVES

- Understand the importance of water and inorganic ions (nitrate, calcium ions and magnesium ions) to plants.

The phloem in plants transports the sugar which is made in the leaves by photosynthesis to all the cells in the plant. The sugar is needed for respiration and for building new materials, so the importance of the substances transported in the phloem is clear. But the xylem transports water and mineral ions from the soil to the rest of the plant. Why are these substances needed in plants?

WHY DO PLANTS NEED WATER?

Water is a vital element in all living organisms – most of the reactions of animal life take place in water, and plants are no different. In fact, plants are about 90% water (compared with humans who are about 65% water) so water is fundamental for their survival. As well as the need for water shared by all organisms, plants also have some further specific requirements.

- Plants need water for photosynthesis. They combine carbon dioxide and water to make glucose and oxygen, using energy from light. You will learn more about this in **Book 2 Chapter 5A**.
- Plants need water for support. Non-woody plants rely on the pressure that builds up as water moves into the vacuoles of the cells by osmosis. This forces the cytoplasm against the cell walls, making the cells rigid. If non-woody plants lack water, they wilt and cannot keep upright.
- Plants need water for transport. Mineral ions are carried around the plant in the xylem in a mass transport system which is powered by transpiration as water evaporates from the surfaces of cells in the leaves and diffuses out into the air. Sugars are carried around the plant in the phloem where they are dissolved in the water and moved by active transport.
- Plants need water to keep them cool – the evaporation of water from the leaves helps cool the plant.

WHY DO PLANTS NEED MINERALS?

Although plants can synthesise their own carbohydrates by photosynthesis, they also need other molecules such as proteins and fats. Certain minerals are needed to synthesise these and other substances essential for healthy growth. Plants must extract these minerals from the soil.

NITRATES

Nitrate ions are used to make amino acids and therefore proteins. These proteins include plant enzymes without which the cells could not function. Nitrates are also needed for the plant to make DNA and many hormones, as well as a range of other compounds in plant cells. When plants lack nitrates, the older leaves turn yellow and die, and growth is stunted. Eventually the plant dies.

CALCIUM

Calcium ions in the middle lamella of plant cell walls combine with pectin to make the calcium pectate which holds plant cells together. Calcium ions are also important in the permeability of membranes. When plants lack calcium, the growing points die back and the young leaves are yellow and crinkly.

MAGNESIUM

Magnesium ions are needed to produce the green pigment chlorophyll. Chlorophyll is vital to trap the light needed for photosynthesis, so if a plant cannot make enough chlorophyll it will eventually die. Magnesium is also needed for the activation of some plant enzymes and the synthesis of nucleic acids. Without magnesium, yellow areas develop on the older leaves and growth slows down.

EXAM HINT

In the examination, you may be asked to complete a table about the roles of minerals in plants. You may also need to write about the roles of minerals in plants as part of a longer question about plant growth or plant physiology.

DID YOU KNOW?

You can use nutrient solutions, each lacking a different mineral, to investigate the effect of mineral deficiencies on the growth of a plant (see figs A and B).

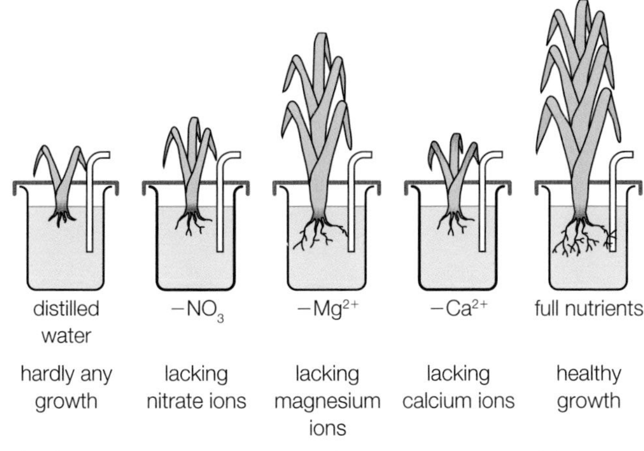

▲ **fig A** Experimental apparatus for investigating mineral deficiency in plants

▲ **fig B** These plants are each deficient in one mineral and show classic deficiency symptoms, particularly in the fastest-growing tissues such as the new leaves.

DID YOU KNOW?

MAKING THE MOST OF MINERAL UPTAKE

All plants take up and transport the minerals they need. But a few plants take up, transport and store a bigger range of minerals than usual. In the early 1970s, a tree was discovered in New Caledonia which produced a blue sap which contained 26% nickel in its dry mass. Other plants have been discovered that can accumulate large amounts of other metals such as cobalt, cadmium, zinc and even gold. The plants take up the minerals in an active process. In the 1980s, scientists suggested that plants might be used to extract certain metals from the earth. Now these special mineral-rich plants are used in several ways.

- In California, *Streptanthus polygaloides* plants are grown on nickel-rich soils where they take up so much nickel that it makes up as much as 1% of their dry mass. Farmers then burn the plants and the ash is smelted to produce the metal. The energy produced by burning the plants is used to generate electricity to power the extraction process and any excess electricity is sold to the local power company. Farming nickel pays better than growing wheat if you live on the right soil!
- Another exciting possibility is to use plants which take up and transport metal ions to clean up ground which has been contaminated by toxic heavy metals such as thallium and lead. Land that is polluted like this cannot be used for growing crops or grazing animals. However, the soil can be made safe to use again by using specialised plants, such as *Alyssum* species, which take up and transport the unwanted ions. They remove the toxic metals from the soil in a process called phytoremediation (see **fig C**).

▲ **fig C** Using plants to extract minerals from the earth is a green approach to soil clean-up.

CHECKPOINT

1. Describe **three** ways in which water is important to a plant.
2. Why are minerals important in a plant?
3. What would you need to consider when investigating mineral uptake by plants in the lab?

SKILLS PROBLEM SOLVING

4A 5 USING PLANT STARCH AND FIBRES

LEARNING OBJECTIVES

- Understand how the arrangement of cellulose microfibrils and secondary thickening in plant cell walls contributes to the physical properties of xylem vessels and sclerenchyma fibres in plant fibres that can be exploited by humans.
- Understand how the uses of plant fibres and starch may contribute to sustainability, including plant-based products to replace oil-based plastics.

People have always exploited plants to provide materials for building and clothing, for medicines, food and drinks, for dyes and for fuel. The structure of plants is adapted to their functions, and these same adaptations provide us with many of the raw materials of life. This section covers some of these uses.

Plants are central to the human diet: they provide the macronutrients of carbohydrates, lipids and proteins and also many micronutrients in the form of vitamins and minerals (see **fig A**). Plants also contain fibre in the cellulose walls of their cells; it cannot be digested but helps the working of the gut.

Some plants are grown as staple foods – the basic energy-supplying foods in the diet. Many of these have cells filled with amyloplasts. The plant may use these starch-storage organelles such as the storage tubers of the yam to survive difficult conditions. They may also be used to reproduce the species – for example, many seeds, such as wheat and rice, contain very rich stores of starch. This store of energy provides plenty of carbohydrate, some protein and oils as well as small amounts of valuable micronutrients and so it is a good food for people.

We use other plant products, including olives, sunflowers, linseed and many nuts, for the oils they contain. Pulses such as beans, peas, lentils, soya beans and chickpeas provide much of the protein requirement for people who eat little or no meat. Fleshy and succulent fruits including dates and bananas are also important foods, and are sources of sugars as well as vitamins.

In addition to eating many parts of plants ourselves, the animals that we raise as a source of food are fed on fodder plants such as grass.

▲ **fig A** Plant foods are a staple part of the human diet around the world.

FROM CONSTRUCTION TO CLOTHING

Food is only one way in which people use plants. You have already learned that the structure of cellulose fibres gives them great strength. These fibres may be further toughened and strengthened by lignin, turning them into what we recognise as wood. The properties of plant fibres and wood make them very useful to us.

PLANT FIBRES

We have used plant fibres such as hemp, jute, manila, flax and sisal for centuries to make ropes, paper and cloth. The fibres usually need to be extracted from the plant first. The fibres are very long sclerenchyma cells and xylem tissue and are usually very tough and strong. Cellulose and lignified cellulose are not easily broken down either by chemicals or by enzymes. Conversely, the matrix of pectates and other compounds around the fibres (including lignin) can usually be dissolved or removed.

Plant fibres have great **tensile strength** – they cannot easily be broken by pulling (under tension). This, along with their flexibility, makes them very useful. They usually exist in bundles of fibres which are much stronger than the individual cells.

PRACTICAL SKILLS CP8

Determining tensile strength

Tensile strength is the resistance of a material to breaking when it is under tension. In everyday language, this means how much can it be stretched before it breaks. Different types of fibre have different tensile strengths. Tensile strength is affected by the cross-sectional area of the fibre. You can investigate the tensile strengths of different fibres.

1 You will need to extract the fibres from the plants. There are several ways of doing this. The simplest is to soak plant stems and leave them until the tissue surrounding the fibres is soft and rotting. Then you can wash the rotten tissue away, leaving the fibres. This takes time and can be smelly.

2 To test the tensile strength of the fibres you will need to clamp them at both ends and apply a measured force to the middle of the fibre. For example, you can hang masses from the fibres. These may be very light masses, such as paperclips, or they may be relatively large, depending on the amount and type of fibre you are testing.

See if you can relate your results to the way the different plant fibres are used.

HOW FIBRES ARE PROCESSED TO MAKE PRODUCTS

Many traditional methods of producing fibres such as flax (often used for ropes) simply relied on the actions of natural decomposers to break down the material around the fibres. This is called retting. In many countries, natural retting has been replaced by manufacturing processes using chemicals and enzymes, which do the job much more quickly.

Probably the best known and most widely used of the natural fibres is cotton (see **fig B**). One of the great advantages of cotton is that it is produced in the form of almost pure fibres, packed around the seeds. So there is no need for retting or other treatment.

▲ **fig B** Cotton bushes produce 'ready-to-use' cotton fibres around their seeds.

Although single cotton fibre cells are very long, they are not long enough to be useful on their own. Spinning pulls out the short, single fibres and twists them together to form a long, apparently continuous thread. Spinning can be done on a small scale by individual people, but usually happens on a massive industrial scale. The resulting threads are then woven together to make a fabric. Similar processes are used with other plant fibres such as jute, sisal and hemp.

In the 20th century, synthetic fibres, for example nylon and polyester, were developed. They were new, exciting, quite cheap, very hardwearing and did not crease. But the limitations of these artificial fibres were known by the 21st century. Fabrics made from them do not 'breathe' and they do not absorb liquid, so they do not absorb body fluids such as sweat. They are made from chemicals from crude oil, a non-sustainable resource which gets increasingly expensive and is rapidly declining.

Sustainability – using materials which can be replaced – is an increasingly important idea. Plants are vital in developing sustainable resources. They take carbon dioxide from the atmosphere and lock it into their cell structures. People are realising that natural plant fibres are a much more sustainable alternative for a range of uses, from clothing fabrics to ropes and insulating materials. They can also be more comfortable to wear because they are more absorbent. The properties of many plant products mean that they are becoming increasingly important in providing what we need in the future.

WOOD

Wood is a composite material, made of lignified cellulose fibres embedded in hemicelluloses and lignin. The great benefit of a composite is that it has the properties of both materials. The cellulose fibres make the wood very resistant to compression (squeezing by weight) so it is excellent for weight-bearing in buildings and can be used in supporting columns as well as in horizontal beams. Wood also keeps some of the matrix flexibility and, because of the intermeshing cellulose fibres, it doesn't crack in the way that a stiff material does. So you can hammer a nail into it or cut out small pieces to make joints without damaging the strength of the wood.

Paper is usually made from fibres from wood. Wood fibres are not easy to extract because the matrix around the cellulose fibres contains much lignin. So wood is soaked in very strong alkalis such as caustic soda to produce a pulp which consists of cellulose and lignified cellulose fibres in water. Thin layers of pulp are then pressed onto frames where they dry to make paper.

Wood has many uses, from making baskets, fencing hurdles, boats, cricket bats or furniture to building homes (see **fig C**). Wood is also a good insulator so homes built mainly from wood need less heating in the winter and cooling in the summer than a brick house. Wood also locks up carbon dioxide and is a sustainable resource if it is managed carefully with replanting programmes. Even if it is burned, wood can be **carbon neutral** – taking in carbon as it grows and releasing it as it is burnt – but it has the great advantage of also being a renewable energy source.

▲ **fig C** Local wood and earth bricks are used as renewable resources to rebuild homes in Sri Lanka after the 2004 tsunami.

BIOPLASTICS: A SUSTAINABLE FUTURE?

During the last century, the use of natural materials in the developed world has declined with the development of new synthetic materials being produced from oil-based chemicals, particularly plastics. Plastics are synthetic polymers – long-chain molecules made of repeating units of small monomer molecules such as ethene and propene. Plastics vary from soft flexible solids with low melting points to hard brittle materials with very high melting points. They are used to make a wide range of products, from packaging to artificial joints and from cutlery to parts of cars. However, in the 21st century modern materials are being developed from natural products because the environmental problems caused by plastics are becoming increasingly clear.

Most plastics, such as polyethene and PVC, are polymers. These are made from petrochemicals originating from oil which is a non-renewable resource. These plastics cannot be broken down by decomposers – they are non-biodegradable – which has led to plastic pollution on a large scale (an example is shown in **fig D**). Some plastics can be melted down and recycled, but many cannot.

▲ **fig D** Plastics are found almost everywhere on the planet, from the Arctic circle to the Antarctic. They are causing enormous environmental damage – and most will not degrade and disappear.

BIOLOGICAL POLYMERS

Scientists are increasingly looking at the possibilities of producing and using **bioplastics** – plastics based on biological polymers such as starch and cellulose. These have two large potential benefits.

- They are a sustainable resource. The starch or cellulose comes from plants such as maize, wheat, potatoes and sugar beet. These plants can be grown easily to supply the needs of the bioplastics industry. When oil runs out, we will need another source of plastics.
- Bioplastics are biodegradable. Bacteria and fungi can usually break down bioplastics because they are based on biological molecules, though the process can be very slow.

Bioplastics are increasingly being used to replace traditional plastics in roles ranging from packaging to computers and mobile phones (see **fig E**). Bioplastics have existed for a long time, in fact one of the first widely used plastics was a bioplastic. In 1869, the American inventor John Wesley Hyatt Jr patented a compound made from cellulose which he used to coat non-ivory billiard balls. The problem was that it caught fire if put near a lighted cigar! But this was the beginning of celluloid, used extensively in photographic film and movie film.

In the 1920s, Henry Ford, the first person to mass-produce cars, experimented with plastics made from soya beans and even produced a plastic car in 1941. Then the Second World War (1939–45) and the growth of the petrochemical industry took attention away from bioplastics.

EXAM HINT

When asked to discuss or evaluate sustainable developments, remember that new developments usually have more than one outcome. When land is used to grow crops for sustainable developments such as bioplastics or biofuels, it is no longer available for growing food.

DIFFERENT TYPES OF BIOPLASTIC

Cellulose-based plastics are usually made from wood pulp (like that used in the paper industry). They are mainly used to make plastic wrapping for food. Cellophane has been a familiar bioplastic for many years.

Thermoplastic starch is the best known and most widely used bioplastic. It is made mainly from starch which is extracted from potatoes and maize. This is then mixed with other compounds such as gelatine, which change the properties of the starch. One of its main uses is in the pharmaceutical industry to make capsules to contain drugs. Thermoplastic starch is smooth, shiny and easy to swallow, yet it absorbs water and is readily digested – perfect for the job!

Other bioplastics include polylactic acid (PLA) which has very similar properties to polyethene but is biodegradable. It is mostly produced from maize or sugar cane. Uses of this bioplastic include computer casings, mobile phones and drinking cups. Poly-3-hydroxybutyrate (PHB) is a stiff biopolymer rather like polypropene. It is used in ropes, bank notes and car parts and is made mainly with products from the South American sugar industry.

We can burn bioplastics when their useful life is over. You might think that this is polluting and unnecessary when they will break down anyway. However, when bioplastics are broken down by decomposers they can produce methane, a greenhouse gas which is 25 times more potent than the carbon dioxide released when they are burned. So they can be very damaging to the environment in their own way. Also, the energy released during burning can be used to generate electricity and make more plastics.

Will bioplastics take over from oil-based plastics? The science and technology needed to produce them are becoming increasingly available. However, the plastics made from petrochemicals have extremely useful properties and it is not always easy to achieve these same properties in bioplastics. Economics and ethical considerations are also important. Bioplastics are still much more expensive than the oil-based alternatives. This is partly because the technology is still very new, and partly because of the economies of scale. About 150 times more conventional plastic is made each year worldwide than bioplastic. People need to be happy to pay more for a similar product to enable the bioplastics industry to develop. There are problems when we try to use crops such as maize, wheat, sugar cane and sugar beet for food, for biofuel and for bioplastics. Currently, there aren't enough crops to feed everyone. Who decides whether these limited crops are used for food to satisfy the immediate hunger of people around the world, or for biofuels or bioplastics? How can we work towards a sustainable future for everyone? These questions are more for society to answer than the science community.

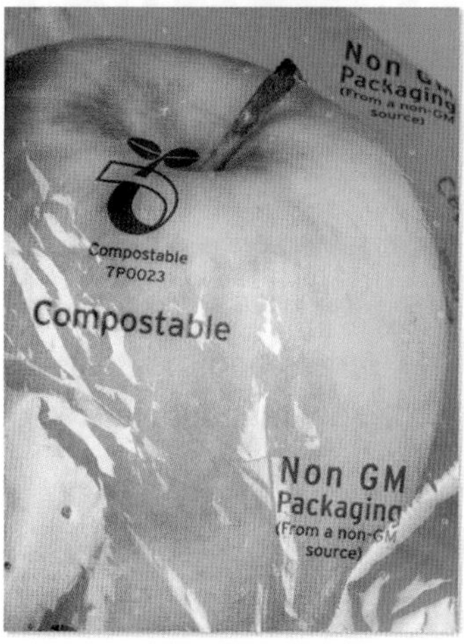

fig E Bioplastics can usually do the same job as synthetic polymers. They cause fewer environmental problems after use because they are biodegradable.

CHECKPOINT

1. State which features of plant cells make plants a useful source of food for people.
2. Explain why starch and cellulose are good starting points for the manufacture of bioplastics.
3. Describe **two** advantages and **two** disadvantages of using bioplastics rather than oil-based plastics.
4. Look for scientific comparisons between the performance of oil-based plastics and bioplastics. Which perform best? How important is this type of evidence when making decisions about using plastics, and what other factors might be considered?

SKILLS: INTELLECTUAL INTEREST AND CURIOSITY, CREATIVITY

SUBJECT VOCABULARY

tensile strength the resistance of a material to breaking under tension
carbon neutral a process where no net carbon is released into the atmosphere
bioplastics plastics based on biological polymers

4A 6 PLANT-BASED MEDICINES

SPECIFICATION REFERENCE 4.10 4.11 4.12 CP9

LEARNING OBJECTIVES

- Understand the conditions required for bacterial growth.
- Know that substances derived from plants can have antimicrobial and other therapeutic properties.

EXAM HINT

The potential for new medicines in undiscovered plant species is often given as one of the most important reasons for conservation of wildlife and habitats.

Plants produce a vast range of chemical substances. Some have the function of deterring animals that try to eat the plant, some have the function of destroying microorganisms that might cause disease. During centuries of general experimentation, people have found that some of these chemicals are also of great benefit in helping the human body fight discomfort and disease. Even today, World Health Organization studies show that 75–80% of the world population relies at least partly on plant-based medicines, particularly in rural and isolated areas such as the tropical rainforests. For much of this time, nobody has really known how plant cures work. Scientists are studying some of them to find out what they do in the body, so we can develop better treatments for illnesses.

PLANTS VERSUS MICROBES

There are more bacteria than any other type of organism. As you have learned, some types of bacteria cause human diseases such as tuberculosis and diphtheria. You may not know that bacteria can also cause plant diseases – in fact, millions of tonnes of crops are lost each year due to bacterial infections in plants. Fungal infections also have a terrible impact on the plants we grow for food or materials. In the same way that animals have defence systems against invading pathogens (see **Book 2 Topic 6**), plants have ways of protecting themselves against microbial attack.

LEARNING TIP

Remember that microbes need warmth, moisture and food.

BACTERIAL GROWTH

Bacteria reproduce by simply splitting in half in a process called **binary fission**. In ideal conditions, some bacteria can split into two cells every 20 minutes. For many types of bacteria, ideal conditions mean plenty of food, oxygen and water – and a warm temperature. Bacteria are so small they cannot be seen with the naked eye (see **Section 3A.4** to remind yourself of the size and structure of bacteria). To investigate bacteria, for example for scientific experiments or to investigate potential medicines, we need to **culture** them. This involves growing large numbers of the microorganisms so they can be measured in some way. They need to be given the correct level of nutrients and oxygen together with the ideal pH and temperature for them to grow. Bacteria are often grown on agar plates – the agar jelly contains all the nutrients the bacteria need to grow well. You will learn more about bacterial growth and disease in **Book 2 Topic 6**.

It is important to take great care when culturing microorganisms because:

- even if the microorganism you are planning to culture is completely harmless, there is always the risk of a mutant strain arising that may be pathogenic
- there is a risk of contamination of the culture by pathogenic microorganisms from the environment
- when you grow a pure strain of a microorganism, the entry of any other microorganisms from the air or your skin into the culture will contaminate it.

EXAM HINT

Remember the precautions taken during investigations that involve the growth of microbes. Describing and explaining such precautions are part of the evaluation of practical work that can be tested in the written examinations.

Health and safety precautions must always be followed very carefully when handling, culturing or disposing of microorganisms. It is important to use **aseptic techniques** to keep everything **sterile** and uncontaminated by other microorganisms. All the equipment must be sterile before the culture is started. It is particularly important that once a culture has grown, it does not leave the lab. The instrument used for inoculating the agar plate (adding the bacteria to the plate) must be sterilised in a Bunsen burner flame. You may also leave a Bunsen burner on a yellow flame on the lab bench to create convection currents to carry airborne bacteria away from the plates. All cultures should be disposed of safely by sealing them in plastic bags and sterilising them at 121 °C for 15 minutes under high pressure, before throwing them away. There are no ethical issues associated with the culturing of microorganisms from the perspective of the microorganisms themselves. However, the danger of accidentally infecting people, animals or plants in the local environment with pathogens should always be considered.

PRACTICAL SKILLS

CP9

Investigating the antimicrobial properties of plants

To investigate the antimicrobial properties of plants, you need a culture of bacteria grown using aseptic techniques. **Fig A** shows the main stages in this process. As an additional precaution against contamination, you can leave a lighted Bunsen burner on the bench to produce convection currents to carry airborne bacteria away.

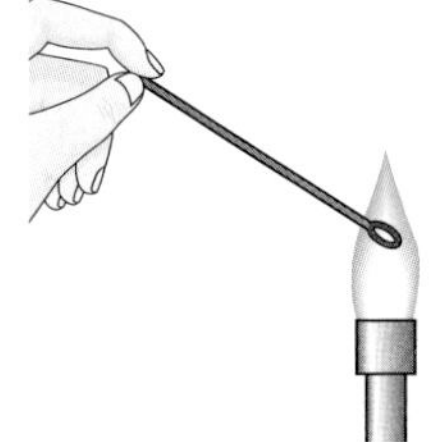

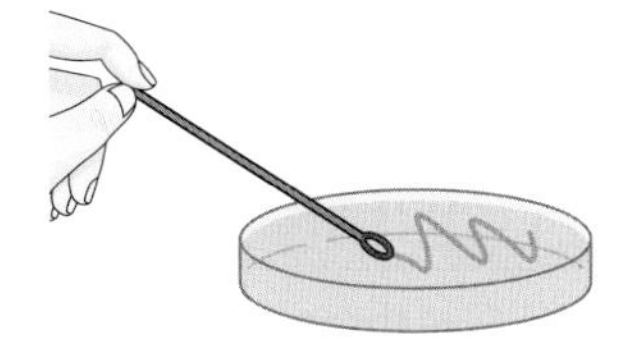

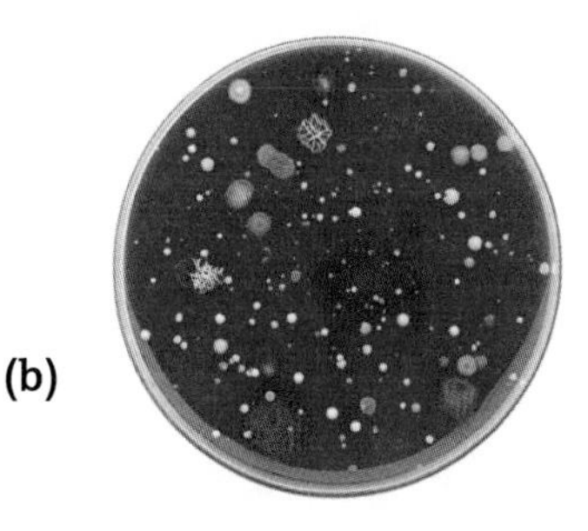

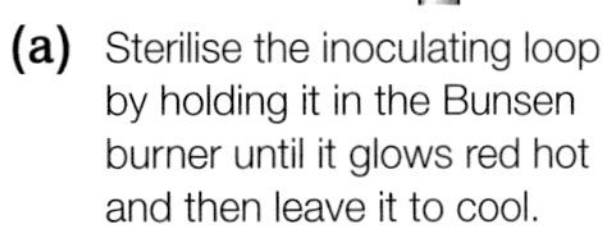

(a) Sterilise the inoculating loop by holding it in the Bunsen burner until it glows red hot and then leave it to cool.

Dip the sterilised loop in the suspension of the bacteria. Streak the loop across the surface of the agar, avoiding digging into the agar. Replace the petri dish lid, tape closed and label. Turn dish upside down.

(b)

▲ **fig A** **(a)** Inoculating an agar plate with a bacterial culture and **(b)** different types of bacteria cultured on an agar plate.

For your investigation, you will need to place small pieces of plant material, small circles of filter paper soaked in plant extracts and/or filter paper soaked in drugs based on plant material on the surface of the agar jelly after you have inoculated it with a bacterial culture. Any clear areas of jelly around these samples indicate that the bacteria are not growing and this suggests an antimicrobial chemical is present.

LEARNING TIP

Remember to prevent the bacteria in the air around you getting into your culture medium.

PLANT DEFENCES AGAINST MICROORGANISMS

Plants can provide an ideal place for microorganisms, such as bacteria and fungi, to grow. The only problem is that the microorganisms can damage and may even destroy the plants. Consequently, many plants have evolved chemical defences to kill any microbes which will invade and cause disease. These chemical defences can include both antiseptic compounds and antibiotics. For example, cotton plants (see **Section 4A.5 fig B**) produce a phenol called gossypol; it is an antiseptic which kills bacteria that might attack the seed. People are increasingly looking at antimicrobial chemicals from plants to provide drugs to treat bacterial diseases in humans.

ANTIMICROBIAL PLANT EXTRACTS

Some plants and fungi have been shown to have antimicrobial properties – they contain chemicals that kill bacteria and fungi. The antimicrobial properties of different plant extracts can be investigated in laboratories. Classically, agar culture plates are used to grow bacterial cultures with discs of filter paper soaked in plant extract placed on the agar. If the plant extract kills the bacteria, or stops them growing, you can see a clear area of agar around the disc.

EXTRACTING DRUGS FROM PLANTS

Plants produce other compounds besides antimicrobial chemicals. Some other compounds have effects from pain relief to destroying cancer cells. For example, salicylic acid is an everyday example of a drug which is derived from a species of willow. For centuries, willow bark was chewed or brewed up into a drink to relieve pain and fever. People even chewed on the anal glands of dead beavers to get pain relief. They knew beavers eat willow bark, and the pain-relieving compound becomes concentrated in the anal glands. Scientists discovered that the active ingredient in the bark was salicylic acid and developed a method to extract and purify it. Now we take a carefully measured dose of a closely related but safer compound, acetylsalicylic acid, in the form of a small white tablet called **aspirin** (see **fig B**).

▲ **fig B** Taking aspirin in tablet form is more convenient for pain relief than chewing willow bark or beaver anal glands.

One of the major advantages of extracting and purifying the beneficial drugs found in plants is that it is possible to give known, repeatable doses of the active ingredient. The levels of a chemical in any part of a plant will vary with the age of the plant, the season of the year or even the time of day. By extracting the chemicals and purifying them, an exact dose can be achieved every time. However, enormous amounts of plant material are needed for this. So scientists work to isolate healing chemicals from plants, analyse their chemical structure and then synthesise the drug on an industrial scale. In many cases, the original plant product is modified to make it even more effective.

The impact of these plant-sourced drugs is important for both individuals and wider society. People are ill less often, are less severely ill and are living longer. Moreover, parts of the world which could not develop because of the high prevalence of diseases are now developing due to the benefits of plant-based medicines. For example, malaria is a life-threatening disease spread by mosquitoes and is common in many tropical areas. Quinine, which comes from the cinchona tree, is used to prevent and treat malaria. The use of quinine made it possible for loggers and developers to work in the Amazon Basin. However, this means that the rich flora of the rainforests may well be destroyed before its true potential as a source of new medicines has been realised.

CHECKPOINT

1. (a) Give **three** factors which need to be in place for bacteria to grow well.

(b) What are aseptic techniques and why is it so important to use them when growing bacteria in the laboratory?

SKILLS INTERPRETATION

2. Some plants make antibacterial chemicals. Explain the advantages this gives to the plants.

SKILLS CREATIVITY

3. Give and explain **two** advantages of using manufactured drugs over plant extracts.

SUBJECT VOCABULARY

binary fission asexual reproduction in bacteria in which the bacteria split in half
culture growing microorganisms in the laboratory, providing them with the nutrients, oxygen, pH and temperature they need to produce large numbers so they can be observed and measured
aseptic technique method of carrying out a procedure to prevent contamination by unwanted microorganisms
sterile something free from living microorganisms and their spores
aspirin a widely used drug which relieves pain and reduces blood clotting and inflammation

7 DEVELOPING NEW DRUGS

SPECIFICATION REFERENCE 4.13

LEARNING OBJECTIVES

- Understand the development of drug testing from historic to contemporary protocols, including William Withering's digitalis soup, double-blind trials, placebo and three-phased testing.

In almost every culture around the world and throughout history, various plants have been used to treat diseases, sometimes more successfully than others. Now, in the 21st century, modern pharmaceutical companies recognise the value of many plant-based drugs. The way these medicines are developed and tested has changed dramatically over time.

WILLIAM WITHERING AND DIGITALIS SOUP

The story of William Withering (1741–99) and his development of an effective treatment for heart failure is a clear example from history of the way effective medicines can be extracted from plants. Foxgloves are wild woodland flowering plants found in the UK and Europe (see **fig A**). **Digitalis** is a chemical found in foxgloves that has been used as a poison for centuries. However, at the same time, there were many reports about the role of foxgloves in curing 'the dropsy'. This is an old name for the swelling (**oedema**) that results when the circulation is failing. Dropsy causes a long, slow death as organs like the kidneys fail, the legs swell and the lungs eventually fill with fluid. It wasn't until the work of William Withering in the 18th century that the medical potential of foxgloves was fully realised.

▲ **fig A** Foxgloves are the source of digoxin, a drug which can be a poison but also a cure.

William Withering was a British doctor and a keen botanist who published a book on plants. However, his real work was medicine and he was very successful. In 1775, a patient came to him with the symptoms of a serious heart condition. Withering had no effective treatment to offer so his patient went off to see a local 'wise woman', who used herbs to cure several conditions. Withering was very impressed when his patient recovered after drinking a soup made by the old woman, and he persuaded the woman to let him buy the recipe for her soup. It contained about 20 different herbs – Withering guessed that foxglove contained the active ingredient. Over the next 10 years, he tested a variety of potions made from foxgloves on 163 patients he was treating for dropsy at Birmingham (UK) General Hospital. Many of his patients got better, though some almost died of digitalis poisoning. He discovered that the side-effects of the drug included nausea, vomiting and worse. But when he got the dose right, the patient started to produce large quantities of urine as their kidneys recovered and removed the excess fluid and their heartbeat became stronger and more regular. By the end of his period of careful observations, Withering had discovered that the best treatment for dropsy and heart failure was to give the patient a soup made from the dried and powdered leaves of the foxglove. Boiling the leaves reduced the effect of the drug. Withering was extracting the digitalis chemical from the leaves of the foxgloves, and his digitalis soup successfully treated his patients. Drugs based on the chemicals in foxgloves, now called **digoxin**, are still in regular use by doctors today, about 230 years after William Withering carried out his ground-breaking work.

TESTING PROMISING NEW MEDICINES

In the past, herbal remedies were often used as a source of new drugs. However, this is not the way new drugs are usually found in the 21st century. Every medicine that comes onto the market today is the result of years of research and development (R & D). A new medicine has to be:

- effective – it cures, prevents or relieves the symptoms of the disease for which it is designed
- safe – non-toxic and without unacceptable side-effects
- stable – can be stored for some time and used under normal conditions
- easily taken into and removed from your body – able to get to its target in your body, and to be excreted (removed from the body) once it has done its job
- can be made on a large scale – can be manufactured in a very pure form, in large quantities and quite cheaply.

It takes about 10 years and about US$2.6 million to develop a new medicine that achieves all of these criteria.

One way scientists can look for new medicines is to investigate chemicals that bind to our protein receptors or to the active sites in our enzymes. Researchers often use computer models to fit new structures into the active site of enzymes or receptors that they think are important in disease processes. This may identify a useful starting point for further work.

When scientists think they have a compound that might make a useful medicine they will patent it. A patent gives the inventor the right to be the only one to make and sell their invention for the next 20 years. It is like a 'reward' for all the work that goes into discovering a possible new drug. The only problem is that quite a few of those 20 years will be taken up with more testing.

The ways in which a potential new drug is tested will vary from country to country, but to be used internationally they must be developed following strict criteria. The new compound is first tested on cell cultures, tissue cultures and whole organs in the lab. These tests are designed to see if the compound does what the scientists thought it would. Many chemicals fail at this stage because they don't work in living tissue or because they have harmful effects. But if the compound passes these tests it moves out of research and into development.

DRUG DEVELOPMENT AND ANIMAL TESTING

Before a drug can be tried on people you need a way of getting it into them – in other words, a good delivery system. This might be tablets, a liquid medicine, injections or a nasal spray. You also need to ensure the drug is stable so there is no risk of it breaking down to form something toxic or inactive before it works.
At this stage, the potential drug will be tested on animals to find out how it works in a whole organism. This will also show if the drug gets taken into the cells, if it is changed chemically in the body and if it is excreted safely.

Mammals are used which are as similar as possible to humans. The most widely used animals for initial tests are mice and rats (see **fig B**). Some tests must be carried out in two species, a rodent and a non-rodent. Animal testing is very expensive and time-consuming and is the centre of much ethical debate. Animals are replaced by tissue cultures and computer models wherever possible, the numbers of animals used are kept to a minimum and the tests used are refined to cause the minimum of distress. However, the information from computer modelling and from tests on cell or tissue cultures is not yet sufficient to move to testing drugs on people without animal testing. So the law states that animal testing must be carried out first.

EXAM HINT

Some people feel very strongly about the use of animals in such testing. Whatever your personal view, make sure you present a balanced argument in your answer.

▲ **fig B** Laboratory mice are widely used in drug testing and there are strict regulations to ensure they are kept in humane conditions.

Some people have ethical objections to the use of animals in this way. But the use of mice and rats is perceived as much less emotive than the use of dogs, cats or monkeys. The rodents provide valid models, the genetic make-up of the strains is well known and they are small and quite easy to keep in humane conditions.

CLINICAL TRIALS

If the animal testing has been successful, the very first human trials follow. A regulatory authority will usually take decisions about the testing and licensing of new medicines. They will only allow a drug to be trialled on people if they are happy with all the tests carried out so far. In drug trials, some of the people will be given a **placebo**. This is something that looks like the drug but has no active ingredient. The placebo is a control and it helps removes the possibility that people feel better just because they think they are getting a new drug.

In phase 1 trials, the new drug (or in some cases, placebo) is given to a small number of healthy volunteers. This is to check that the drug works as expected in the human body and doesn't cause any unexpected side-effects. At the same time, scientists continue looking at the effect of longer-term use of the drug in animal trials.

If a drug is successful in phase 1 human trials, it goes into phase 2 trials. This is when the new drug is used with patients affected by the target disease. Between 100 and 500 patient volunteers are given the new drug, and a similar number are given the best current treatment or, sometimes, a placebo. This is the first chance for scientists and doctors to see how the new medicine affects the disease in a real patient. The volunteer patients are closely monitored to find out more about the ideal dose, the effectiveness of the drug and any side-effects. Success at this stage means the new compound has a good chance of becoming a useful medicine.

Before a new drug is fully approved, it must be used on thousands of patients with the target disease. These are the phase 3 trials and over 5000 volunteer patients are used.

DOUBLE-BLIND TRIALS

Phase 2 and 3 trials are normally carried out as **double-blind trials**. This means neither the doctor/scientist nor the patient knows whether the patient is receiving the new medicine, a control medicine or a placebo. Patients often appear to respond to a treatment because they believe that it will do them good – there is much that is not understood about how the mind affects the body. This response is called the **placebo effect**. The use of a double-blind trial allows this to be measured. Instead of a placebo, a control medicine – the best-performing available treatment – is sometimes used to avoid any patient being denied treatment when they take part in the trial.

Phase 3 trials are used to confirm the effectiveness and safety of the new drug. The numbers of patients involved are large, so the trials also have a better chance of showing up any unexpected adverse side-effects. Patients are randomly allocated to receive the new medicine or the control/placebo. Data on effectiveness, side-effects and other information are collected and assessed to see if there are any statistically significant differences between the new medicine and the placebo or the currently available drug.

It is difficult to achieve a complete set of results in clinical trials, because many patients stop taking the medicine for various reasons. Some do not take it regularly.

In some trials, the new drug or drug combination is so successful that the trial is halted early. If the evidence shows that a new treatment is particularly effective it becomes unethical to deny the new treatment to the patients who are receiving the old treatment or placebo.

If a new drug is found to be safe and effective in phase 3 trials, the pharmaceutical company will try and get approval for the medicine to be sold. All the data submitted are assessed to evaluate the beneficial effects against the possible harmful effects of the medicine before a decision is made on whether or not to grant a licence for the product. Such a licence is needed before a new drug can be put on the market or an existing drug can be used to treat a different disease.

Even once a new medicine is being used to treat patients, trials continue. The medicine will be monitored for safety and effectiveness for as long as it is used. Any adverse reactions suffered by patients are reported and recorded to ensure that the benefits of using a medicine are always greater than the risks. **Fig C** summarises the main stages in modern drug development.

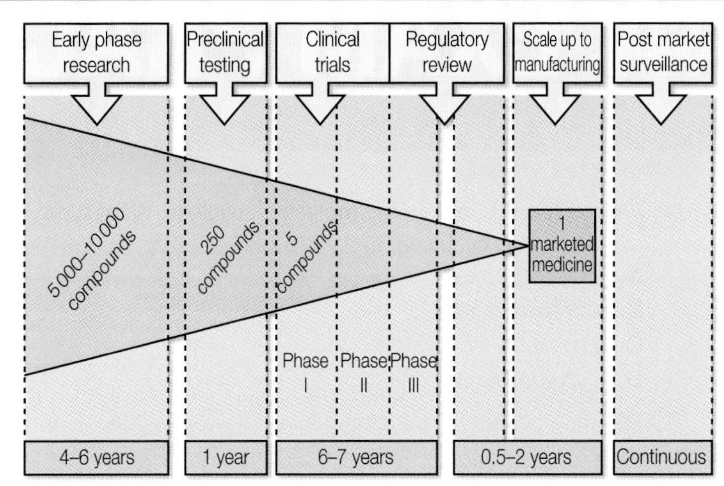

fig C This diagram summarises the main stages in a standard drug development protocol.

CHECKPOINT

1. Compare William Withering's methods of discovering digoxin with modern drug testing.
2. Draw a flow chart to summarise the main stages in the development of a new medicine from a plant thought to have medicinal properties.

SKILLS PERSONAL AND SOCIAL RESPONSIBILITY

3. Discuss the ethics of carrying out phase 1 human trials on healthy volunteers. Is there an alternative?
4. Sometimes a doctor may want to prescribe a drug before it has completed full human trials. Is it ever ethical to use a drug before it has undergone all stages of testing? Suggest **two** arguments for and **two** against this happening.

SUBJECT VOCABULARY

digitalis a chemical found in foxgloves that affects the beating of the heart
oedema swelling of the tissues due to fluid retention
digoxin a drug based on the chemical found in foxgloves that improves heart function
placebo an inactive substance resembling a drug being trialled which is used as an experimental control
double-blind trial a clinical drug trial where neither the doctor nor the patient knows whether the patient is receiving the new medicine, a control medicine or a placebo
placebo effect when patients appear to respond to a drug simply because they think it is doing them good

4A EXAM PRACTICE

1 (a) Plant cell walls contain the molecule cellulose. What type of molecule is cellulose?

A a disaccharide
B a polysaccharide
C a protein
D a glycoprotein [1]

(b) Identify the type of bond that holds two monomers together in cellulose.

A 1,4-glycosidic bond
B α 1,4-glycosidic bond
C β 1,4-glycosidic bond
D β 1,6-glycosidic bond [1]

(c) The diagram shows two glucose molecules. Complete the molecules.

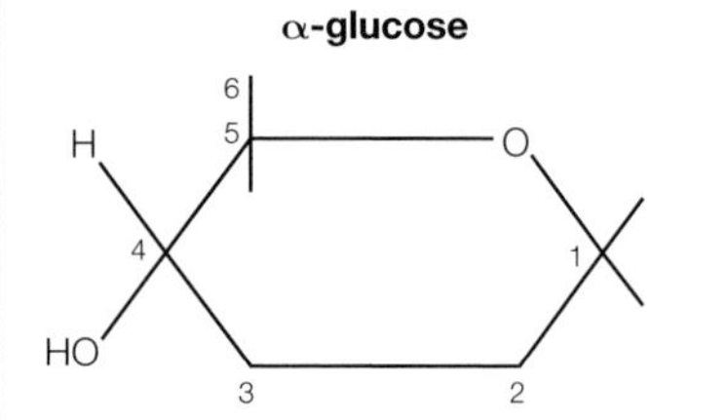

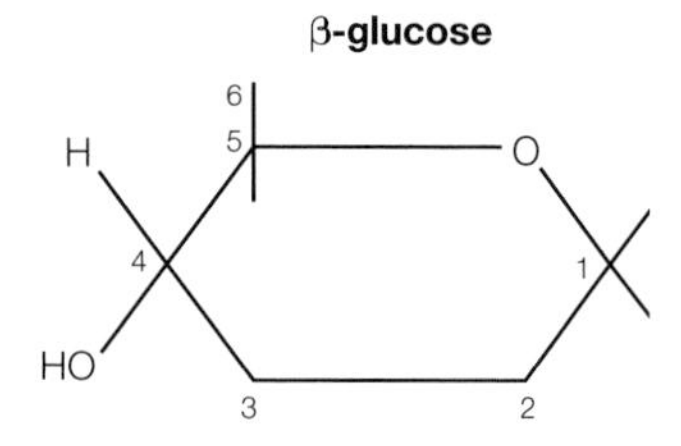

[2]

(d) Draw a diagram to show how the monomers of cellulose bind together to form cellulose. [3]

(Total for Question 1 = 7 marks)

2 The photograph below shows a chloroplast.

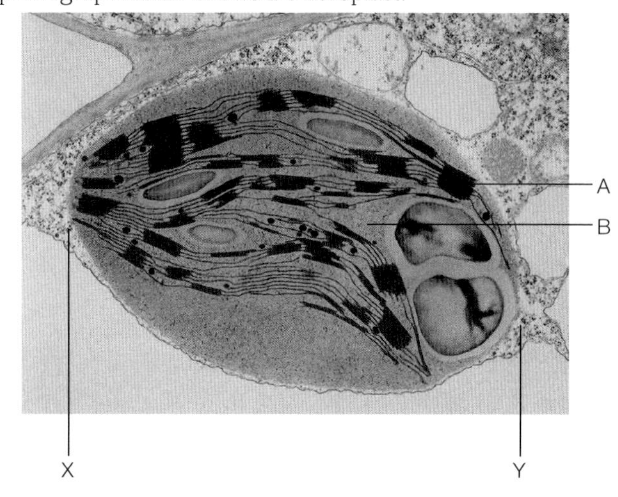

(a) What type of microscope was used to take this photograph?

A scanning electron microscope
B transmission electron microscope
C optical microscope
D light microscope [1]

(b) What feature do chloroplasts **not** have in common with mitochondria?

A 70S ribosomes
B DNA
C an envelope
D starch grains [1]

(c) What are the correct names for the structures labelled **A** and **B** on the diagram?

A thylakoid and stroma
B lamella and stroma
C stroma and granum
D granum and stroma [1]

(d) The actual length of the chloroplast between **X** and **Y** is 5 μm. Calculate the magnification of this diagram. Show your working. [3]

(e) Name **one** type of cell that contains chloroplasts. [1]

(f) Chloroplasts contain oil droplets and starch grains. Suggest a function for the oil droplets other than as a store of energy. [2]

(Total for Question 2 = 9 marks)

3 (a) Describe the arrangement of cellulose fibres in a plant cell wall. [2]

(b) The photograph below shows some onion cells as seen using the high power of a light microscope.

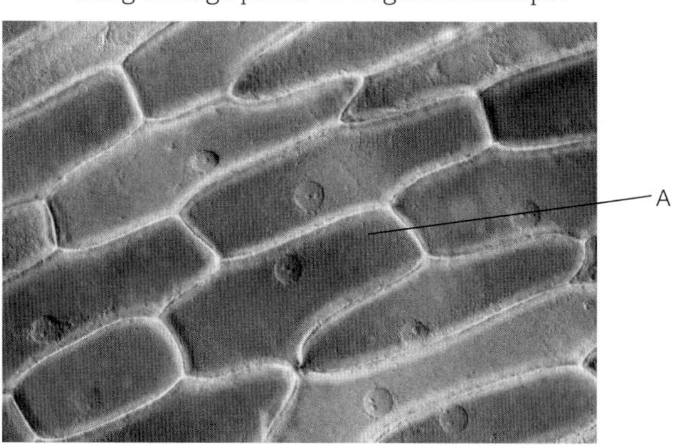

(i) Make an accurate labelled drawing of the cell labelled **A**. [6]

(ii) Each cell is approximately 40 μm in length. Calculate the magnification of your diagram. [2]

(Total for Question 3 = 10 marks)

4 (a) Draw a labelled diagram to show the tissues in a transverse section of a plant stem. [4]

(b) The vascular tissue is found in a central bundle in the roots. Suggest why the vascular tissue is found in a different arrangement in the stem and root. [3]

(c) Explain how xylem is adapted to its function. [4]

(Total for Question 4 = 11 marks)

5 (a) Humans make use of many plant products.

(i) Wood is often used to construct buildings. Explain how the properties of wood make it a good building material. [4]

(ii) Assess the advantages of clothing made from natural fibres such as cotton, compared to clothing made from synthetic fibres. [3]

(b) Evaluate the use of bioplastics as a substitute for plastics based on oil products. [4]

(Total for Question 5 = 11 marks)

6 (a) Complete the table to compare and contrast phloem tissue with xylem tissue.

Xylem	Phloem
transports water and minerals	
	living tissue
	transport up or down the stem
transport as mass flow	
transport depends on pull from transpiration	

[5]

(b) The movement of substances in the phloem is called translocation.

(i) Describe how the companion cells in phloem contribute to translocation. [2]

(ii) Explain how the companion cells are adapted to their function. [3]

(Total for Question 6 = 10 marks)

7 (a) Plants require a range of minerals. Complete the table to describe the function of minerals in plants.

Mineral	Function
magnesium	
	used to build the middle lamella of plant cell walls
nitrates	

[3]

(b) Describe how minerals are transported from the roots to the growing points in the plant. [3]

(c) Explain why the growing points of plants require:

(i) calcium [2]

(ii) nitrates. [2]

(Total for Question 7 = 10 marks)

8 A student wanted to investigate the rate of water uptake by a plant. Using a healthy potted plant he sealed the pot and soil in a plastic bag leaving the leaves exposed. The student assumed that the plant would take up water as it was lost from the leaves. He then placed the potted plant on a top pan balance and recorded the mass over 24 hours.

The experiment started at 8.00 in the morning. The results are shown in the table below.

Time from start/hours	Mass/g
0	895
4	889
8	884
12	878
16	877
20	876
24	873

(a) (i) Name the process that results in loss of water from the leaves.

A translocation
B transpiration stream
C transpiration
D translocation stream [1]

(ii) Why did the student seal the soil and pot in a plastic bag?

A to prevent evaporation of water from the soil
B to keep the soil in the pot
C to keep the top pan balance clean
D to prevent the roots from drying out [1]

(iii) Suggest a suitable control for this experiment. [2]

(b) Plot a graph showing the results. [4]

(c) Suggest why the rate of water uptake was lower towards the end of the experiment. [2]

(d) Describe how this procedure could be modified to investigate how temperature affects the rate of water uptake. [2]

(Total for Question 8 = 12 marks)

TOPIC 4 PLANT STRUCTURE AND FUNCTION, BIODIVERSITY AND CONSERVATION

CHAPTER 4B CLASSIFICATION

In 2012, scientists working in Papua New Guinea found the smallest known vertebrate to date – a tiny frog measuring 7.7 mm in length. *Paedophryne amauensis* feeds on tiny mites in the leaf litter of its rainforest home – and it can jump up to 30 times its own body length. DNA analysis shows that tiny frogs have evolved 11 times in different areas of the world, all filling a similar niche. In 2014, a new species of dead-leaf toad (*Rhinella yunga*) was discovered in the Peruvian Andes. In shape, colour and patterning, it resembles a dead leaf and, with the poison it exudes from glands on the back of its head, the toad looks similar to other toads of the same genus. It was only when scientists noticed that these toads lack eardrums that they realised they had discovered a new species. Finding new species is always exciting, but it becomes even more special when that new species is already endangered, such as the new species of orang-utan identified in November 2017.

Scientists used two different methods of identifying these new species – traditional observation of physical characteristics such as eardrums, and DNA analysis of the genome. In this chapter, you will find out more about how we classify the organisms in the world around us – and why it is important that we do so.

You will learn the main taxonomic groups of the living world including domains, kingdoms and species, and will begin to classify different organisms. You will consider the problems of defining a species in a way that is useful for all types of organism and evaluate the different ones in use. The use of DNA technology is having a major impact on our ability to identify organisms and work out how they are related to other species. There has been a long-running debate about the numbers of domains and kingdoms which should be used in classification – decide who you think is right!

MATHS SKILLS FOR THIS CHAPTER

- **Recognise and use expressions in decimal and standard form** (*e.g. when considering the number of base pairs in DNA and the proportion of those base pairs that may differ between species*)
- **Use scales for measuring** (*e.g. size and parts of different organisms for comparisons when classifying*)
- **Use ratios, fractions and percentages** (*e.g. regarding the proportion of base pairs shared in genes from different species*)

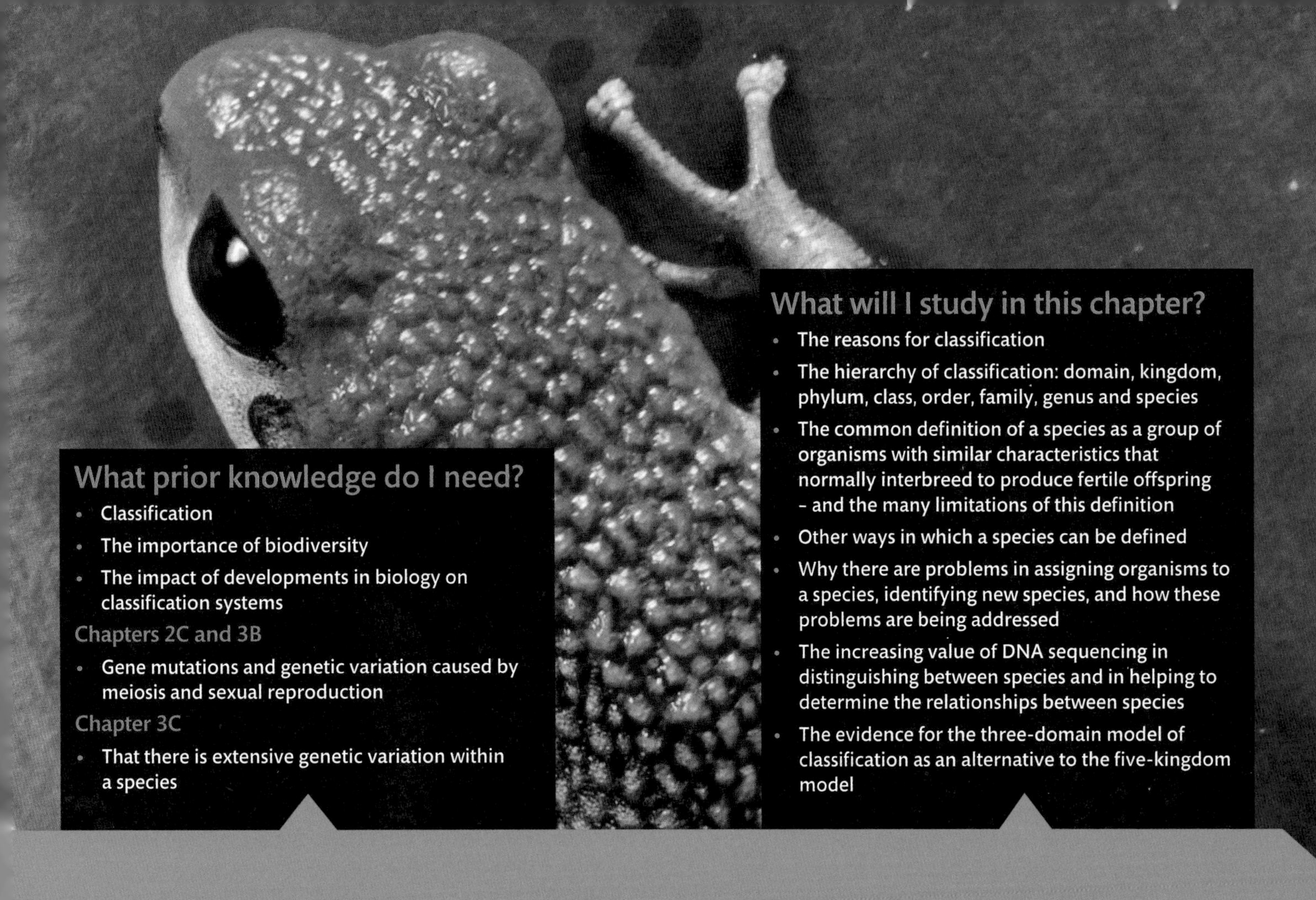

What prior knowledge do I need?

- Classification
- The importance of biodiversity
- The impact of developments in biology on classification systems

Chapters 2C and 3B

- Gene mutations and genetic variation caused by meiosis and sexual reproduction

Chapter 3C

- That there is extensive genetic variation within a species

What will I study in this chapter?

- The reasons for classification
- The hierarchy of classification: domain, kingdom, phylum, class, order, family, genus and species
- The common definition of a species as a group of organisms with similar characteristics that normally interbreed to produce fertile offspring – and the many limitations of this definition
- Other ways in which a species can be defined
- Why there are problems in assigning organisms to a species, identifying new species, and how these problems are being addressed
- The increasing value of DNA sequencing in distinguishing between species and in helping to determine the relationships between species
- The evidence for the three-domain model of classification as an alternative to the five-kingdom model

What will I study later?

Topic 4C

- Biodiversity
- How to measure biodiversity
- How species are well adapted to their habitat
- That variation within a species is important
- How new species arise as a result of natural selection
- How reproductive isolation can cause the formation of new species
- The concept of a gene pool and how the proportion of alleles can change within a population
- The need to conserve endangered species

Topic 5B (Book 2: IAL)

- The need to be able to classify organisms for practical investigations of populations in the field
- Understand the concepts of niche and succession

4B 1 PRINCIPLES OF CLASSIFICATION

SPECIFICATION REFERENCE 4.14(i) 4.15

LEARNING OBJECTIVES

- Understand that classification is a means of organising the variety of life based on relationships between organisms using differences and similarities in phenotypes and in genotypes, and is built around the species concept.

THE BACKGROUND TO BIODIVERSITY

Biodiversity is a measure of the variety of living organisms and their genetic differences. It is an important concept at the moment because the Earth's biodiversity is reducing rapidly. Many scientists think this may affect the future health of the planet. You will find out about biodiversity in more detail later in **Chapter 4C**. In this section you will be looking at some of the biology you need in order to understand biodiversity.

WHY CLASSIFY?

The result of millions of years of **evolution** is an enormous variety of living organisms. This great biodiversity (see **Chapter 4C**) means that there is a great variety of names. An organism may have different names not only in different countries, but even within different areas of the same country (see **fig A**). When biologists from different countries discuss an organism they need to be sure they are all referring to the same one. An internationally recognised way of referring to any living organism is essential. Biodiversity is a very important concept, and to quantify biodiversity we need a way of identifying the different groups of organisms. We classify the living world by putting organisms in groups based on their similarities and differences. Scientists can monitor changes in the populations of different types of organism if they know the numbers that there are in a particular habitat. It is also important for biologists to understand how different types of living organism are related to each other. A good classification system makes these ancestral relationships clear.

▲ **fig A** This plant is a rose in English, English, دْرو in Arabic, ρόδο in Greek, *rosa* in Spanish and *die Rosen* in German. The official classification Rosa is used and understood by biologists everywhere. The many different species of rose can be identified even more precisely, for example, *Rosa canina* (the Wild dog rose) and *Rosa acicularis* (the Arctic rose).

THE HISTORY OF TAXONOMY

Taxonomy is the science of describing, classifying and naming living organisms. This includes all of the plants, animals and microorganisms in the world and it is an enormous task. The aim of a classification system is to group organisms to accurately identify them and represent their ancestral relationships. From the time of the Greek philosopher Aristotle onwards, people put organisms into groups based mainly on their physical appearance or **morphology**. People often used **analogous features** to classify organisms. But such features may not have the same biological origin so this system can easily create misconceptions. For example, you might put wiggly, legless creatures including snakes, worms, slugs and eels in one classification group and flying animals such as bats, birds and flying insects in another group. A valid classification system must be based on careful observation and the use of **homologous structures** – that is, structures that genuinely show common ancestry.

In the 18th century, the Swedish botanist Carolus Linnaeus (1707–78) developed the first scientifically devised classification system. We still use many of his principles and his basic naming system today. However, we can now add many more modern techniques to the simple but detailed observation of organisms that he introduced.

THE MAIN TAXONOMIC GROUPS

The biggest taxonomic groupings are huge – the largest are the three **domains**, a grouping developed more recently which you will look at in more detail in **Section 4B.4**. The main taxonomic groups are, from the largest to the smallest: domain, **kingdom**, **phylum (division, for plants)**, **class**, **order**, **family**, **genus** and **species**.

The **Archaea** domain contains one kingdom:

- **Archaebacteria**: ancient bacteria thought to be early relatives of the eukaryotes. They were thought to be found only in extreme environments, but scientists are increasingly finding them everywhere – particularly in soil.

The Bacteria domain also contains one kingdom:

- **Eubacteria**: the true bacteria are what we normally think of when we are describing the bacteria that cause, for example, disease, and which are so useful in the digestive systems of many organisms and in recycling nutrients in the environment.

There are four kingdoms in the Eukaryota domain:

- **Protista**: a very diverse group of microscopic organisms. Some are heterotrophs – they need to eat other organisms – and some are autotrophs – they make their own food by photosynthesis. Some are animal-like, some are plant-like and some are more like fungi. Examples include *Amoeba*, *Chlamydomonas*, green and brown algae and slime moulds.

- **Fungi**: all heterotrophs – most are saprophytic and some are parasitic. They have chitin, not cellulose, in their cell walls.
- **Plantae**: almost all autotrophs, making their own food by photosynthesis using light captured by the green pigment chlorophyll. These include the mosses, liverworts, ferns, gymnosperms and angiosperms (flowering plants).
- **Animalia**: all heterotrophs that move their whole bodies around during at least one stage of their life cycle. These include the invertebrates (e.g. insects, molluscs, worms, echinoderms) and the vertebrates (e.g. fish, amphibians, reptiles, birds, mammals).

EXAM HINT

Make sure you know the features used to classify organisms into their kingdoms.

THE BINOMIAL SYSTEM

The binomial system of naming organisms was originally devised by Linnaeus. Biologists now use it universally. The way different organisms are classified is constantly under review as new data are discovered.

In the binomial system, every organism is given two Latin names – the word 'binomial' means 'two names'. The first name is the genus name and the second is the species or specific name which identifies the organism precisely. There are certain rules to writing binomial names:

- use italics
- the genus name has an upper-case letter and the species name a lower-case letter, e.g. *Homo sapiens* (human beings), *Bellis perennis* (common daisy)
- after the first use, binomial names are abbreviated to the initial of the genus and then the species name, e.g. *H. sapiens*, *B. perennis*.

A genus is a group of species that all share common characteristics so, for example, the genus *Vanessa* contains the Painted Lady *Vanessa cardui*, the Red Admiral *Vanessa atalanta* and the Indian Red Admiral *Vanessa indica*. These lovely butterflies have some very clear similarities, but enough differences for you to see why they are separate species (see **fig B**). It is not always so easy to tell the difference between species within a genus.

EXAM HINT

Remember that all members of the same genus have the same first name. Two species with the same second name do not belong to the same genus. They may be totally unrelated.

▲ **fig B** These two butterflies both belong to the genus *Vanessa*, but they are different species (*Vanessa atalanta* and *Vanessa cardui*).

LEARNING TIP

Remember the sequence of classification groups or taxa. It may help to make up a mnemonic such as: Desperate King Philip Came Over For Great Spaghetti.

Table A shows a number of different species with all of their levels of classification.

DOMAIN	Bacteria	Eukaryota	Eukaryota	Eukaryota
KINGDOM	Eubacteria	Animalia	Fungi	Plantae
PHYLUM/DIVISION	Proteobacteria	Chordata	Basidomycota	Magnoliophyta
CLASS	Gammaproteobacteria	Mammalia	Agaricomycetes	Liliopsida
ORDER	Enterobacteriales	Perissodactyla	Agaricales	Poales
FAMILY	Enterobacteriaceae	Equidae	Amanitaceae	Poaceae
GENUS	*Escherichia*	*Equus*	*Amanita*	*Oryza*
SPECIES	*Escherichia coli* *E. coli* common bacterium in the intestines	*Equus caballus* *E. caballus* domestic horse	*Amanita muscaria* *A. muscaria* fly agaric	*Oryza sativa* *O. sativa* rice

table A Full classification of four different organisms

CHECKPOINT

SKILLS PROBLEM SOLVING

SKILLS CONTINUOUS LEARNING

1. Explain why a classification system is needed in biology.
2. Draw a diagram to show the main groups of the most commonly used system of classification.
3. Discover the classification from domain to species of the following organisms: domestic cat, maize, honey bee and human being.

SUBJECT VOCABULARY

biodiversity a measure of the variety of living organisms and their genetic differences
evolution the process by which natural selection acts on variation to bring about adaptations and eventually speciation
taxonomy the science of describing, classifying and naming living organisms
morphology the study of the form and structure of organisms
analogous features features that look similar or have a similar function, but are not from the same biological origin
homologous structures structures that genuinely show common ancestry
domains the three largest classification categories: the Eukaryota, the Bacteria and the Archaea
kingdom the classification category smaller than domains; there are six kingdoms: Archaebacteria, Eubacteria, Protista, Fungi, Plantae and Animalia
phylum (division, for plants) a group of classes that all share common characteristics
class a group of orders that all share common characteristics
order a group of families that all share common characteristics
family a group of genera that all share common characteristics
genus a group of species that all share common characteristics
species a group of closely related organisms that are all potentially capable of interbreeding to produce fertile offspring
Archaea domain made up of bacteria-like prokaryotic organisms found in many places including extreme conditions and the soil; they are thought to be early relatives of the eukaryotes
Archaebacteria ancient type of bacteria found in many different environments
Eubacteria true bacteria (prokaryotic organisms)
Protista a kingdom in the five-kingdom classification system that contains all single-celled organisms, green and brown algae and slime moulds
Fungi a eukaryotic kingdom of heterotrophs with chitin in their cell walls
Plantae a mainly autotrophic eukaryotic kingdom containing mosses, liverworts, ferns, gymnosperms and angiosperms (the flowering plants)
Animalia a mainly heterotrophic eukaryotic kingdom including all the invertebrates and vertebrates

4B 2 WHAT IS A SPECIES?

LEARNING OBJECTIVES

- Understand that classification is a means of organising the variety of life based on relationships between organisms using differences and similarities in phenotypes and in genotypes, and is built around the species concept.

THE CONCEPT OF SPECIES

The concept of species is a very important one for biologists. We use species numbers to measure biodiversity (see **Chapter 4C**). We also look for changes in species to help us monitor the effect of both natural environmental changes and changes that result from human activity. Biologists look for both adaptations within a species and for changes in the numbers or types of species in an environment.

The concept of species is important in biology, so it makes sense that everyone works with the same model. However, this is not so easy. Species are defined in many different ways, and the best model changes with the circumstances and the type of organism being investigated.

THE MORPHOLOGICAL SPECIES CONCEPT

The definition of species that Linnaeus originally developed was a **morphological species model**, which was based solely on the appearance of the organisms he observed. For many years, scientists would look closely at the outer, and sometimes inner, morphology of the organisms and group them into species, genus etc. according to the extent of difference or similarity of the physical characteristics. Much of the classification we use now is based on morphology. This approach still works in many cases and you can see just by looking at an organism what it is – for example, you would never mistake a lion for a domestic cat. However, the appearance of an organism can be affected by many different things and there can be a huge amount of variation within a group of closely related organisms. In fact, in organisms that show **sexual dimorphism** – in which there is a great deal of difference between the male and female – the different sexes could be confused as different species in a morphological species model (see **fig A**).

▲ **fig A** Most people would not classify these two birds in the same species, unless they were seen mating, but the peacock and peahen are male and female peafowl.

DID YOU KNOW?

SPECIATION BASED ON JAWS

The shape and arrangement of jaws and teeth are often used to identify species. For example, in the alligatoridae, the upper jaw is wider than the lower jaw and so the lower teeth are almost completely hidden when the jaws are closed. In crocodylidae, the jaws are the same size and the teeth fit into each other so both upper and lower teeth are visible when the jaws are closed. However, most of us would not want to get close enough to decide!

▲ Is this a crocodile or an alligator?

Agaves and aloes are another example of the limitations of a morphological system of classification (see **fig B**). These similar looking plants are not at all similar in terms of their evolutionary relationship.

▲ **fig B** Agaves and aloes are both adapted to survive in similar desert conditions, but they come from different parts of the world and are not closely related.

THE REPRODUCTIVE OR BIOLOGICAL SPECIES CONCEPT

For many years, a morphological definition of a species was used almost without question. However, over time biologists moved to a basic model of a species based on the reproductive behaviour of the organisms. One widely used definition of a species is:

- a group of organisms with similar characteristics that interbreed to produce fertile offspring.

This definition of species overcomes issues such as sexual dimorphism and is regarded as a good working definition for many animal species, but it has limitations. One obvious limitation is that all the organisms in a species cannot attempt to interbreed to produce fertile offspring because they do not all live in the same area. So populations of organisms of the same species may not interbreed because they are in different places and not because they are different species.

In this species model, if two individuals from different populations mate, they are considered the same species if fertile offspring are produced and genes are combined or 'flow' from the parents to the offspring. So, for example, horses and donkeys look similar, but the offspring produced from a horse and a donkey is a mule, which is sterile (see **fig C**). The genes cannot flow to the next generation so they are not the same species. But the offspring produced between the largest horse and the smallest pony is fertile – they are extreme variants of the same species. However, this definition is not perfect. For example, lions and tigers are different species, but if a lion and tiger mate most of the offspring produced are fertile. To help overcome these limitations, two slightly more sophisticated definitions of species based on reproductive capability are:

- a group of organisms with similar characteristics that are all potentially capable of breeding to produce fertile offspring
- a group of organisms in which genes can flow between individuals.

A reproductive concept of species is a good working model for most animals, but it is much less helpful in classifying plants, which frequently interbreed with similar species to produce fertile offspring.

EXAM HINT

Make sure you can remember some examples of species that can produce hybrids.

(a)

(b)

(c)

▲ **fig C** When donkeys breed they produce young donkeys **(a)**, which grow up to produce more donkeys. When horses breed they produce foals **(b)**, which will produce more horses in the future. But if a horse and a donkey breed they produce an infertile mule **(c)** – so they are definitely separate species.

OTHER DEFINITIONS OF SPECIES

The definition of a species is constantly developing. Scientists now make decisions about which organisms belong in the same species and how they are related in a number of different ways. Some of these methods are much more sophisticated than simple observation. The fundamental chemicals of life such as DNA, RNA and proteins (see **Chapter 2B**) are almost universal. These chemicals are broadly similar across all species but differences are revealed when the molecules are broken down to their constituent parts. Scientists use these differences, in the science of **molecular phylogeny**, to build up new models of species and their relationships. But some of the different models of species are no better, and can be even worse than the original morphological model. They include:

- **Ecological species model** – based on the ecological niche occupied by an organism. This is not a very robust way of identifying species, as niche definitions vary and many species occupy more than one niche.
- **Mate-recognition species model** – a concept based on unique fertilisation systems, including mating behaviour. The difficulty is that many species will mate with or cross-pollinate other species and may even produce fertile offspring, but are nevertheless different species.
- **Genetic species model** – based on DNA evidence. This might seem the ultimate, reliable method of determining species, but people still have to decide how much genetic difference is needed for two organisms to be members of different species. Historically, collecting DNA was difficult and it took a long time and cost a lot of money to analyse. As DNA analysis continues to get faster and cheaper, this will ultimately become the main way of classifying organisms.

EXAM HINT

You need to understand these models. You may be asked to compare them or give advantages and limitations of each model. Don't forget that a table is a good way to make comparisons.

Ever-improving DNA analysis means species definitions and relationships will become increasingly important in classification. But for now, the biological definition of species combined with basic morphology is still widely used.

LIMITATIONS OF SPECIES MODELS

All the ways to define species have limitations, which include:

1. Finding the evidence – many living species have never been observed mating. This is particularly true if a new species is found that is similar to an existing species. Setting up a breeding programme is time-consuming, expensive and may not prove anything.
2. Plants of different but closely related species frequently interbreed and produce fertile hybrids. When should the hybrids themselves be regarded as a separate species?
3. Many organisms do not reproduce sexually. Any definition involving reproduction or reproductive behaviour is irrelevant for bacteria and the many protists, fungi and others that mainly reproduce asexually.
4. Fossil organisms cannot reproduce and do not usually have any accessible DNA, but they still need to be classified.

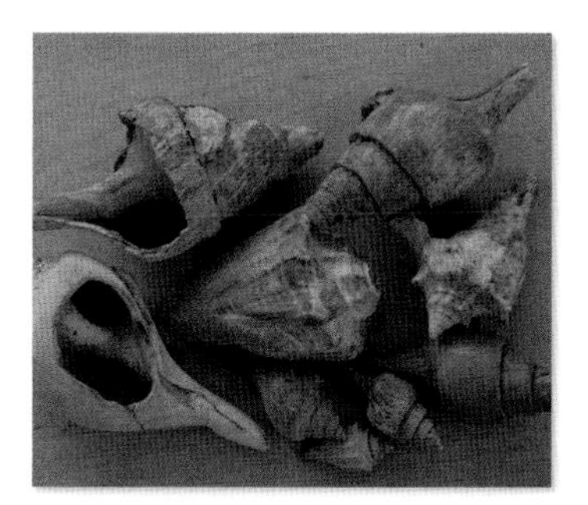

▲ **fig D** Breeding experiments or DNA analysis are not helpful in identifying shells and fossils.

IDENTIFYING A SPECIES

Despite all the problems, classifying organisms and identifying their species is still a widely used and extremely useful biological tool. Questions about identifying different organisms may be absolute, 'Is it species P or species Q?' Or they may be comparative, 'Is it a new species that has not been identified before, or just new to a particular scientist or area?' Information technology (IT) provides an ideal tool to help scientists answer these questions from simple identification apps to help you decide which bird, butterfly or orchid you have just seen, to the prospect of instruments that will be able to identify DNA in the field. IT is now very important in the process of classification.

As an example, the Natural History Museum in London, UK, is home to millions of specimens of different organisms from all over the world, which have been collected during several hundred years. Most of the species were identified by their external features many years ago, and details were recorded on handwritten and typewritten index cards, which are then stored in the museum's vast archives. New specimens are regularly sent to the museum for identification. To reduce the time spent searching the cards, scientists at the museum and the University of Essex are developing a system to scan and 'read' the card archives, and convert them into an internet-based database and a paper-based catalogue. This will make searching for a particular organism much easier and also give scientists around the world access to classification information while working in the field.

THE IMPORTANCE OF DNA IN CLASSIFICATION

In recent years, scientists have developed techniques that allow them to analyse the DNA and proteins of different organisms. In **DNA sequencing** the base sequences of all or part of the genome of an organism is revealed. DNA sequencing leads to **DNA profiling**, which looks at the non-coding areas of DNA to identify patterns. These patterns are unique to individuals, but the similarity of patterns can be used to identify relationships between individuals and even between species.

EXAM HINT

Questions testing your knowledge of the structure of DNA may include questions about the use of DNA.

DNA sequencing and profiling generates so much data that it would be impossible for individual scientists to go through it all searching for patterns. There is, however, a new science called bioinformatics. This involves the development of the software and computing tools needed to organise and analyse enormous quantities of raw biological data. Using bioinformatics, we can understand and use the information generated in DNA sequencing and profiling. You are going to discover some of the ways in which we can use this information to identify species and the relationships between them.

THE SAME...

Identifying species from their phenotype can be difficult. External conditions can result in major differences in the appearance of individuals of the same species. For example, red deer stags that live in woods and parkland have antlers that are much longer and broader than stags that roam highland mountainsides. They could easily be mistaken for different species, yet DNA evidence shows that they are the same.

...BUT DIFFERENT

In contrast, for many years the plant disease scab, which can destroy crops such as wheat and barley, was thought to be caused by a single fungus, *Fusarium graminearum*. Molecular geneticists in the United States have investigated the disease to try and help plant breeders and disease control specialists worldwide. DNA evidence shows that there are at least eight different species of *Fusarium* pathogens, which have a similar effect on crop plants. This evidence is based on the divergence of six different genes and the proteomic evidence of the proteins they produce (see **fig E**).

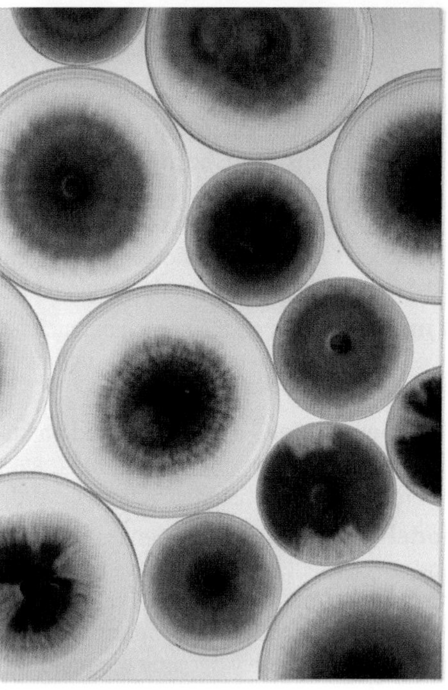

▲ **fig E** These cultures may all look the same, but DNA evidence shows that they are distinct species of fungi, all of which cause similar diseases in plants.

CHECKPOINT

SKILLS INTERPRETATION

1. Summarise the reasons why biologists need to classify organisms.
2. Why is it so important to be able to identify individual species?
3. Compare the advantages and practical difficulties of using classic morphology, reproductive capability and DNA analysis to decide if an organism belongs to a particular species.

SUBJECT VOCABULARY

morphological species model a species definition based solely on the appearance of the organisms observed
sexual dimorphism describes species where there is a great deal of difference between the appearance of the male and female
molecular phylogeny the analysis of the genetic material of organisms to establish their evolutionary relationships
ecological species model a species definition based on the ecological niche occupied by an organism
mate-recognition species model a species definition based on unique fertilisation systems, including mating behaviour
genetic species model a species model based on DNA evidence
DNA sequencing the process by which the base sequences of all or part of the genome of an organism is worked out
DNA profiling the process by which the non-coding areas of DNA are analysed to identify patterns

4B 3 DOMAINS, KINGDOMS OR BOTH?

SPECIFICATION REFERENCE 4.14(ii)

LEARNING OBJECTIVES

- Understand the process and importance of critical evaluation of new data by the scientific community leading to new taxonomic groupings, based on molecular evidence, including the three-domain system (Archaea, Bacteria and Eukaryota).

As you learned in **Sections 4B.1** and **4B.2**, for centuries classification has been based on detailed observations of morphology. For example, scientists might count the hairs on the foreleg of a fly or the petals of a flower to work out relationships between organisms. Now, biochemical relationships are increasingly being used to support or clarify relationships based on morphology. Scientists need to analyse the structures of many different chemicals as well as the DNA to identify the inter-relationships between groups of organisms. This analysis is called molecular phylogeny, and not all scientists interpret the results in the same way. Proteins are important molecules in these analyses.

The evidence from biochemical analysis may support or conflict with relationships based on morphology. For example, all green plants have similar complex pathways for making glucose from sunlight using chlorophyll, so it seems safe to assume they all developed from a common ancestor. In contrast, American porcupines and African porcupines occupy similar niches and look very similar, but biochemical analysis suggests that they are only very distantly related.

GEL ELECTROPHORESIS: A VITAL TECHNIQUE

The discovery that we can identify patterns in the DNA and RNA of different individuals and between different species has had enormous implications in many areas of science, including species identification and the development of models of the relationships between organisms over time. Patterns in DNA or RNA fragments are being used more and more in species identification. Comparisons between the amino acid sequences of similar proteins in different species or groups of organisms are also used to help us classify them or trace their developmental pathways. A useful technique in these processes is **gel electrophoresis**. This is a variation of chromatography which can be used to separate DNA and RNA fragments, proteins or amino acids according to their size and charge (see **fig A**).

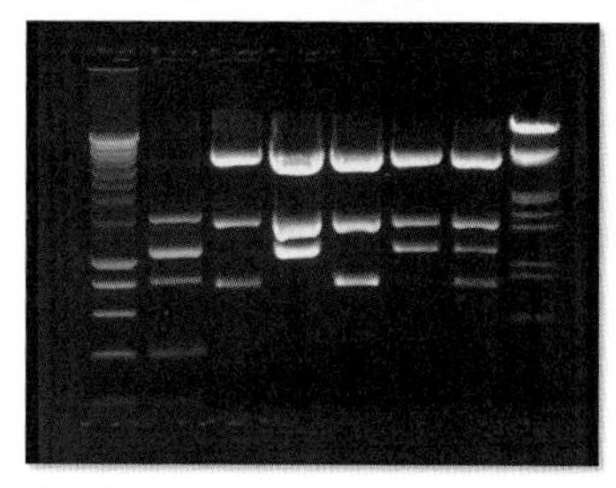

▲ **fig A** Gel electrophoresis can be used to identify patterns in the DNA, RNA and proteins of different organisms. This can help to determine both the species they belong to and their evolutionary links.

MORE BIOCHEMICAL RELATIONSHIPS

An understanding of the biochemical relationships between organisms is important in extending our knowledge of classification and relationships over time. Here are some other examples.

From evidence of comparative anatomy and embryology, the vertebrates and the echinoderms (an invertebrate group of animals that includes starfish and sea urchins) appear to come from one line of ancestors (see **fig B**). The annelid worms, molluscs and arthropods (including insects) appear to come from another line. Biochemical evidence seems to confirm this unlikely relationship. It shows that there are two different types of phosphagen. These are molecules that provide the phosphate group for the synthesis of ATP in muscles. Phosphocreatine occurs almost exclusively in the muscle tissue of vertebrates and echinoderms whilst phosphoarginine occurs in the other groups.

▲ **fig B** Starfish and sheep do not seem to have much in common, but biochemical and gel electrophoresis analysis shows that they are more closely related than you might think.

- Blood pigments are important in many animal groups. Analysis has shown that any group contains only one type of blood pigment – all vertebrates and many of the invertebrates have haemoglobin, all polychaete worms have chlorocruorin and all molluscs and crustaceans have haemocyanin. This allows scientists to build up a more detailed picture of the relationships between the different groups.
- Analysis of the sequence of amino acids in some specific proteins can help show the relationships within higher groups, such as a phylum. For example in mammals, analysis of fibrinogen (the protein involved in blood clotting) reveals how closely the different mammalian groups are related. Single amino acid changes are used to plot relationships.

A combination of DNA analysis, protein analysis and anatomical observations can bring some unlikely relationships to light – for example, the closest living relative to the hippopotamus appears to be the whale.

EXAM HINT

You could be asked to evaluate the use of molecular analysis in classification. Is it more reliable than morphological differences?

PHYLOGENETIC TREES

One way of showing the relationships between different groups of organisms is to use a **phylogenetic tree**. Phylogenetic trees are models of how different organisms are related. They may show the relationships between all living organisms, or just the relationships between a particular group of animals or plants. These trees were originally built up based on the morphology of organisms. Now scientists use DNA and amino acid evidence as a major tool to help them understand the relationships between groups of organisms. A phylogenetic tree shows a common ancestor which may no longer exist, and the tree branches as different species appear as a result of natural selection (see **fig C**).

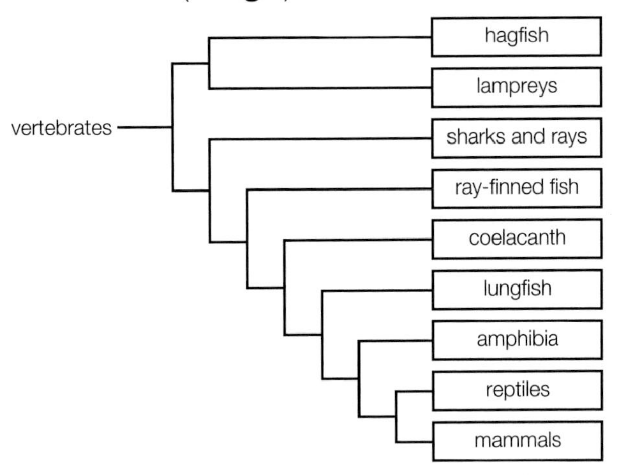

fig C This phylogenetic tree shows the relationships between a range of different animals.

TWO DOMAINS OR THREE?

For many years biologists divided living organisms into two large groups or domains. They named them the eukaryotes, cells with a complex cell structure (see **Sections 3A.2** and **3A.3**), and the prokaryotes, which included bacteria (see **Section 3A.4**). The theory was that eukaryotes developed from prokaryotes billions of years ago (see **fig D**). Some scientists think that chloroplasts became part of 'eukaryotic ancestor' cells first; others think that mitochondria were the first **endosymbionts**. It is possible that both processes happened at the same time because evidence to support one idea or another is almost impossible to obtain.

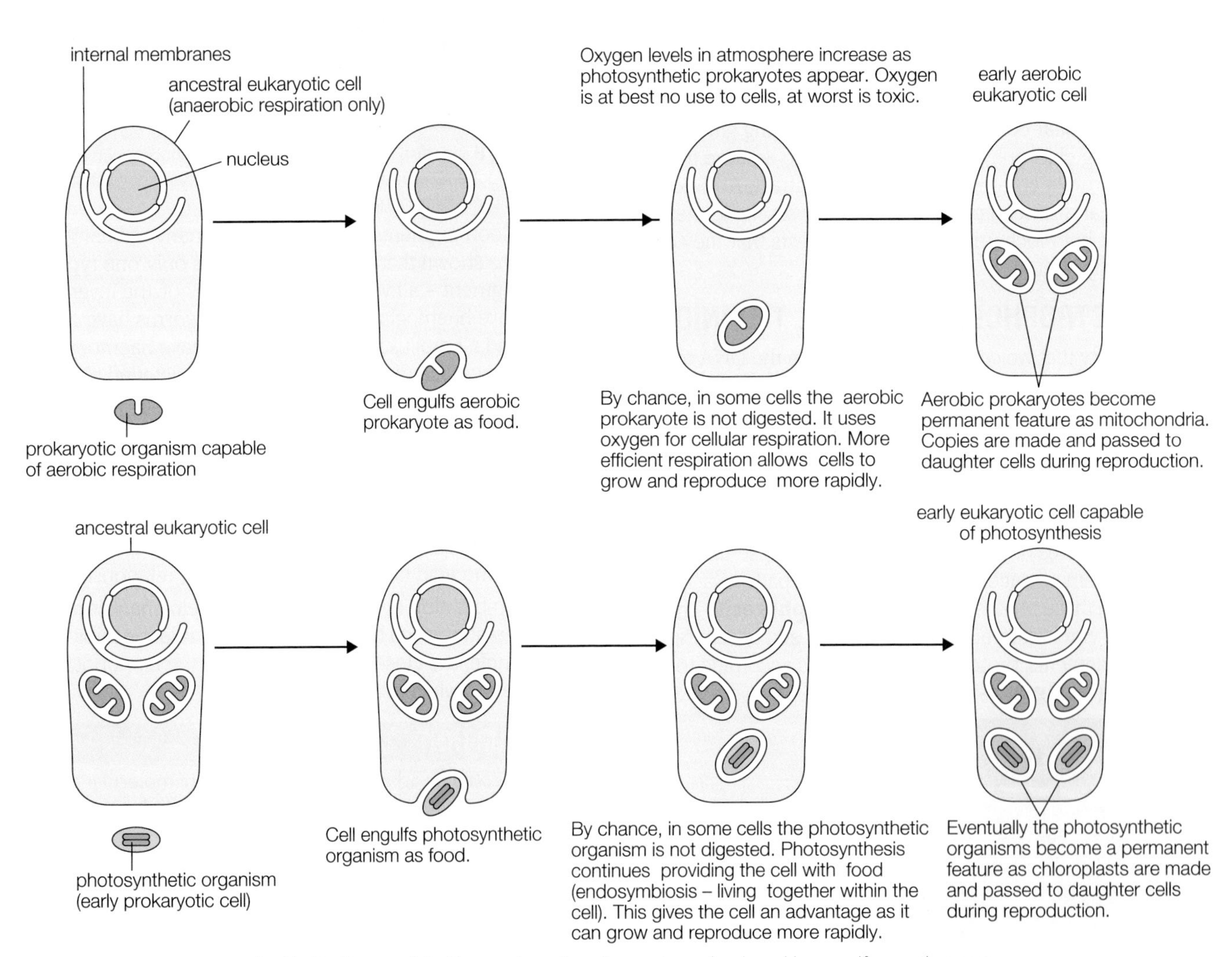

fig D One model of how eukaryotic cells may have developed by engulfing prokaryotes.

Scientists are using the techniques of molecular phylogeny to investigate this theory further. They have looked at the internal structures of the prokaryotes and eukaryotes, and compared them to the biochemistry of the proteins involved in their ribosomes and enzymes. Much of this analysis involves the use of bioinformatics to process the masses of data generated.

This meant a new theory developed that there are, in fact, three domains – two prokaryote domains, the Archaea and Bacteria, and the eukaryote domain, the Eukaryota. The Archaea and the Bacteria are as different from each other as they are from the Eukaryota (see **table A**). Genetic studies show that all three groups probably had a single common ancestor about three billion years ago. Some evidence suggests the Archaea are more closely related to Eukaryota, including us, than to the Bacteria – our last common ancestor was probably about two billion years ago.

CHARACTERISTIC	BACTERIA	ARCHAEA	EUKARYOTA
membrane-enclosed nucleus	absent	absent	present
membrane-enclosed organelles	absent	absent	present
peptidoglycan in cell wall	present	absent	absent
membrane lipids	ester-linked, unbranched	ester-linked, branched	ester-linked, unbranched
ribosomes	70S	70S	80S
initiator tRNA	formylmethionine	methionine	methionine
operons	yes	yes	no
plasmids	yes	yes	rare
RNA polymerases	1	1	3
ribosomes sensitive to chloramphenicol and streptomycin	yes	no	no
ribosomes sensitive to diphtheria toxin	no	yes	yes
some are methanogens	no	yes	no
some fix nitrogen	yes	yes	no
some conduct chlorophyll-based photosynthesis	yes	no	yes

table A Some of the cellular and molecular characteristics of the three domains of life on Earth

Apart from the ribosomes, scientists see two essential differences in the mass of data that has been generated about the three domains. Archaea replicate by binary fission which is controlled within a cell cycle. It is very similar to the cell cycle in eukaryotic cells but is different from replication in bacteria. Conversely, the membrane structure and the membrane proteins of the Archaea are unique – they are different from the bacteria and the eukaryotes, which have homologous structures. The Archaea have an ester link in their lipids, giving branched molecules that may provide extra strength in extreme environments. This supports a model which suggests a different origin for some of the cellular systems of the Archaea and Bacteria, but shows eukaryotes combining features of them both. Research which aims to establish these links has been published in peer-reviewed journals so that other scientists can repeat the procedures to verify the results. Most biologists now accept the three-domain theory but there is still debate about the domains and their origins (see **fig E**). However, evidence showing horizontal gene transfer between groups of organisms is increasing, so the idea of a complex interwoven network of ancestry also remains.

LEARNING TIP

Most scientists now believe that there are three domains rather than two. Decide on two or three pieces of evidence that can be used to justify the argument for three domains. Learn those pieces of evidence.

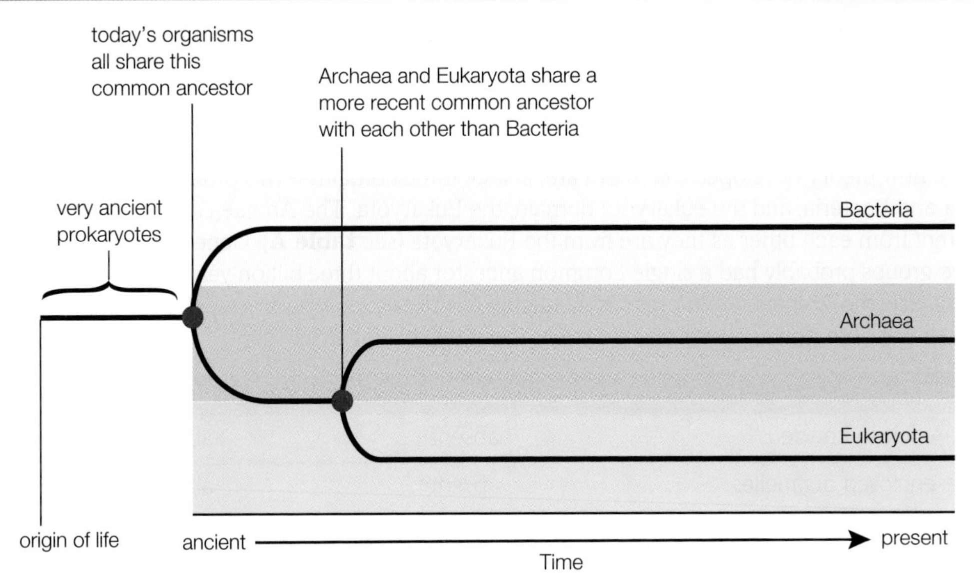

▲ **fig E** This phylogenetic tree shows possible relationships between the three domains of the living world – most but not all biologists think that the three domains share a common prokaryotic ancestor.

HOW MANY KINGDOMS?

When Linnaeus first worked out his classification system, he proposed all living things fitted into two large groups that he called the plant kingdom and the animal kingdom. Everything including fungi were fitted in – fungi counted as plants because they did not move about.

As technology and scientific knowledge increased, people began to see the world of microscopic organisms, including bacteria and the single-celled organisms we now call the Protista. The structure of fungi became clear and it is very different from the structure of plants. A new system emerged, based mainly on morphology, resulting in a five-kingdom classification. In this system, all of the prokaryotes are put together into one kingdom, the **Monera**. All the other single-celled organisms and the algae are grouped together in the Protista.

The five-kingdom system is still used, but increasingly biologists now use a six-kingdom classification system. This results from the work done using biochemical and DNA evidence in relation to the three domains. The prokaryotes are no longer regarded as a single group. The Archaea and the Eubacteria are two very different groups of organisms with very different biochemistry.

Here is a reminder of the six kingdoms of modern classification.

ARCHAEBACTERIA (ARCHAEA): PROKARYOTIC CELLS

These are ancient bacteria that have a wide variety of lifestyles and include the **extremophiles** – bacteria that can survive extreme conditions of heat, cold, pH, salinity and pressure. They normally reproduce asexually.

EUBACTERIA (BACTERIA): PROKARYOTIC CELLS

This kingdom includes the true bacteria and the cyanobacteria, which used to be called the blue-green algae. They normally reproduce asexually.

PROTISTA: EUKARYOTIC CELLS

This kingdom includes all the single-celled eukaryotic organisms, the green algae, the brown algae and the slime moulds. It is something of a catch-all for the organisms that do not fit in the other kingdoms. They mainly reproduce asexually.

FUNGI: EUKARYOTIC CELLS

This kingdom includes both unicellular organisms, for example yeasts, and multicellular organisms, for example toadstools and moulds. They are all heterotrophs. They reproduce both asexually and sexually.

PLANTAE: EUKARYOTIC CELLS

Almost all these organisms are multicellular autotrophs, making their own food by photosynthesis for which they need chlorophyll. They include the mosses, liverworts, ferns, gymnosperms and angiosperms (flowering plants).

ANIMALIA: EUKARYOTIC CELLS

The organisms in this kingdom are all multicellular and they are all heterotrophs. They include invertebrates and vertebrates. Sexual reproduction is common but some animals also reproduce asexually.

The models for the five-kingdom, six-kingdom and three-domain classification systems are summarised in **fig F**.

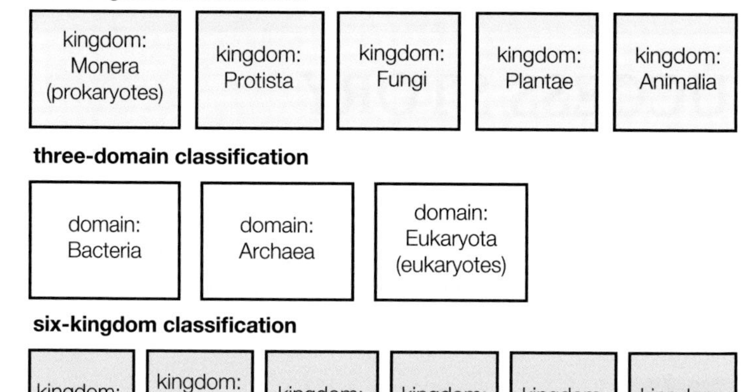

▲ **fig F** A six-kingdom classification makes more sense when the three domain system is also accepted.

CHECKPOINT

SKILLS INTERPRETATION

1. How is molecular phylogeny used to show genetic diversity and similarities between organisms?
2. Discuss the importance of the following in how we distinguish between species and construct relationships between species:
   (a) analysis of DNA, RNA and proteins
   (b) bioinformatics
3. Use the information on these pages along with other details you may find on the internet to help you explain the difficulties of drawing conclusions about how to classify the variety of organisms.
4. Discuss how the development of the three-domain model of classification affects other aspects of classification.

SUBJECT VOCABULARY

gel electrophoresis a method of separating fragments of proteins or nucleic acids based on their electrical charge and size
phylogenetic tree model used to show the relationships between different groups of organisms
endosymbionts organisms that live inside the cells or the body of another organism
Monera a kingdom in the five-kingdom classification system that contains the Archaea and Eubacteria
extremophiles bacteria that can survive extreme conditions of heat, cold, pH, salinity and pressure

4B THINKING BIGGER

REVIVING THE QUAGGA

SKILLS CRITICAL THINKING, PROBLEM SOLVING, ANALYSIS, INTERPRETATION

Until recently it was thought that the last quagga, a species similar to the plains zebra, had died in Amsterdam Zoo in 1883. In recent years DNA evidence suggested that the quagga was in fact a subspecies of the plains zebra, and a rebreeding programme in South Africa set out to restore the quagga to the African plains where it belongs.

ONLINE TRADE PUBLICATION

QUAGGA REBREEDING: A SUCCESS STORY

Until recently, it was believed that the last quagga died in Amsterdam Zoo in 1883. Today, however, this iconic animal is alive and back in the Western Cape. How was it possible to revive an animal from extinction? Keri Harvey speaks to the Quagga Project's Craig Lardner.

Contrary to popular belief, the quagga (*Equus quagga quagga*) is not a species in its own right. DNA analysis of quagga kept as museum specimens has proven that the extinct quagga was in fact a Burchell's or plains zebra with a colour variation, in which some of its leg and rump stripes disappeared. This also means that Burchell's or plains zebra still carry genes from the extinct quagga, though these may be more diluted now than before.

Vanishing stripes

Why exactly the Burchell's or plains zebra lost some of its stripes is unclear, but … differing colouration seems to provide optimal camouflage: the quagga in each area blend better into their specific surroundings. Another purported reason for the quagga's vanishing stripes, apart from camouflage and hence protection from predators, is tsetse flies. It has been suggested that the zebra's stripes repel tsetse flies and so too the diseases they carry. Because the quagga lived outside the tsetse fly areas, the distinct stripes became obsolete.

…When it was discovered that the Burchell's or plains zebra is a DNA match for the extinct quagga, the project set about attempting to 'rebreed' the quagga. This was done by selecting brownish zebra with reduced stripes and white tail bushes. In this way, the quagga genes could be concentrated to produce an animal that looks precisely like the 'extinct' quagga.

Only mitochondrial DNA was available from museum specimens and not nuclear and living DNA. For this reason, it was impossible to compare the rebred quagga to the original ones that became extinct. Nonetheless, the quagga in the Western Cape are believed to be the 'real thing', as it was in fact only coat pattern that distinguished a quagga from a Burchell's or plains zebra. Thus the Quagga Project seems to have succeeded in rectifying the tragedy that saw them being hunted to extinction.

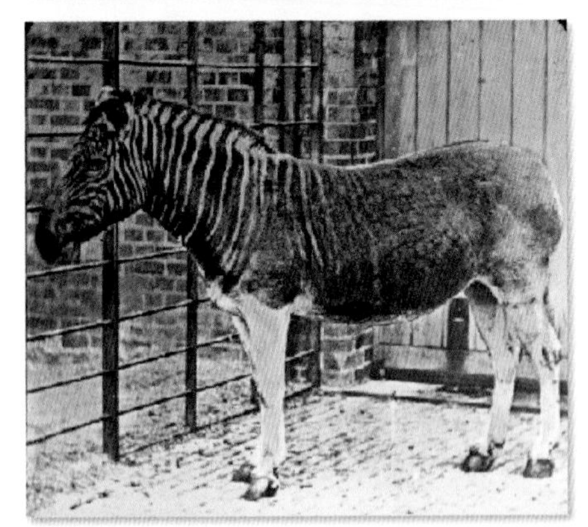

This stripe pattern on this restored quagga (top) is approaching the pattern seen in the only existing photo of a quagga (bottom), taken in London Zoo in 1870.

From *South African Farmer's Weekly* magazine.

* From Leonard, Jennifer A., Nadin Rohland, Scott Glaberman, Robert C. Fleischer, Adalgisa Caccone and Michael Hofreiter. 'A rapid loss of stripes: the evolutionary history of the extinct quagga.' *Biology Letters* 1, no. 3 (2005): 291–295.

SCIENCE COMMUNICATION

The article opposite is from *Farmer's Weekly*, which is published both in print and online in South Africa and aimed at farmers across Southern Africa. Consider the article and think about the type of writing being used. Try and answer the following questions:

1 Do you think this is a scientific piece of writing? Why or why not?

2 Using the information in this article, make a summary of what you now know about quaggas and how they have been rebred.

INTERPRETATION NOTE

Consider the format. This is a story told by a journalist after speaking to someone from the Quagga Project. Does this make you think the story is reliable? Why? Does anything make you wonder if the details are correct?

BIOLOGY IN DETAIL

Now let us look at the biology of this amazing story. You know about classification by anatomy and morphology, and how recent advances in DNA technology have enabled scientists to check the classification of species. Use these ideas to help you answer the following questions. You may find it useful to return to them later in your Biology course.

3 (a) Suggest why quaggas were classified as a separate species.

(b) Quaggas and plains zebras have now been classified as the same species. Explain how animals that look so different can be members of the same species.

4 Explain how DNA sequencing can be used to check the classification of species.

5 The quagga (*Equus quagga*) was described and named as a separate species in 1785. Burchell's zebra (*Equus burchelli*) was described and named in 1824. Some people believe that the Burchell's zebra should now be called *Equus quagga*. Discuss.

THINKING BIGGER TIPS

Remember the sequence of classification groups or taxa. It may help to make up an acronym such as: Desperate King Philip Came Over For Great Spaghetti.

ACTIVITY

Now read this extract*, which is an abstract from a peer-reviewed scientific journal.

Abstract

Twenty years ago, the field of ancient DNA was launched with the publication of two short mitochondrial (mt) DNA sequences from a single quagga (***Equus quagga***) museum skin....(Higuchi et al. 1984, *Nature* 312, 282–284). This was the first extinct species from which genetic information was retrieved. The DNA sequences of the quagga showed that it was more closely related to zebras than to horses. However, quagga evolutionary history is far from clear. We have isolated DNA from eight quaggas and a plains zebra (subspecies or phenotype ***Equus burchelli burchelli***). We show that the quagga displayed little genetic diversity and very recently diverged from the plains zebra...

....However, our results could be consistent with the quagga and the plains zebra being synonymized, as suggested earlier (e.g. Rau 1978; Groves & Bell 2004). Owing to priority, the correct name for plains zebras would thus be ***E. quagga***, with, according to Groves & Bell (2004) five living and one extinct subspecies, the quagga (***E. quagga quagga***)...

....We estimate that this divergence took place in the Pleistocene, about 120 000 to 290 000 years ago... (Dawson 1992). Therefore, the distinct coat colour of the quagga (Bennett 1980) must have evolved quite rapidly. Existing plains zebras show a geographical gradient in coloration with progressive reduction in striping from north to south, which has been explained as an adaptation to open country and for which the quagga represented the extreme limit of the trend (Rau 1974, 1978). Thus, the rapid evolution of coat colour in the quagga may be explained by either of two factors, or a combination of them: the disruption of gene flow owing to geographical isolation and/or an adaptive response to a drier habitat.

1 Compare and contrast the writing styles of the two pieces about the quagga.

2 Summarise the information about quaggas and the rebreeding programme you get from this paper and compare it to the information you got from the first article. How does the information differ? Which gave you the most information? Which was easiest to extract information from? Which did you find the most interesting?

3 From the information on quaggas in the above articles, put together a presentation for potential sponsors to support a fund-raising effort towards the reintroduction of the quagga onto the South African plains.

4B EXAM PRACTICE

1 (a) Why do scientists classify living things?
A To give scientists something to do.
B So that we can give names to living things.
C To understand which living things are our closest relatives.
D To understand the relationships between organisms. [1]

(b) In the five-kingdom classification of living things what is the correct sequence of taxonomic groups?
A Kingdom, Phylum, Class, Order, Family, Genus, Species
B Kingdom, Domain, Class, Order, Family, Genus, Species
C Kingdom, Phylum, Class, Family, Order, Genus, Species
D Kingdom, Phylum, Class, Order, Genus, Species [1]

(c) The diagram shows a single-celled organism.

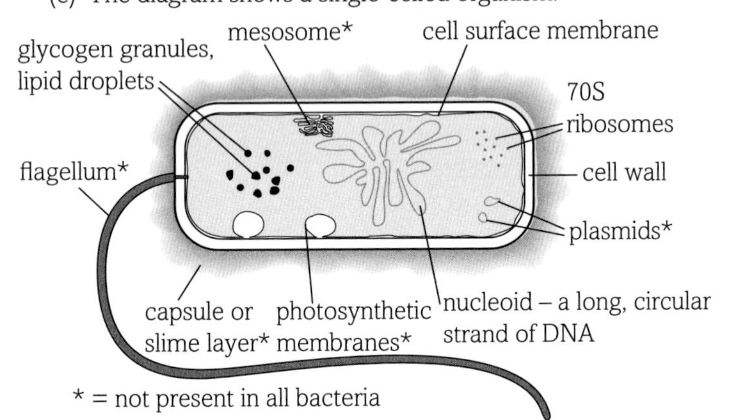

State three features shown in the diagram that tell you this organism is not a Eukaryote. [3]

(d) The earliest classification system used similarities in morphology and anatomy to place organisms into groups. Evaluate the use of these characteristics for classification. [5]

(Total for Question 1 = 10 marks)

2 (a) What is the correct way to write the scientific name for a human being?
A homo sapiens
B Homo sapiens
C *Homo sapiens*
D *homo Sapiens* [1]

(b) Discuss why scientists give each species a scientific name containing two words. [3]

(c) Explain what is meant by a species. [2]

(d) Explain how DNA sequencing can be used as a tool in taxonomy. [4]

(Total for Question 2 = 10 marks)

3 In the 1990s Carl Woese suggested a new way of grouping organisms into three domains.

(a) The table shows the three domains and gives some of the characteristics of each domain.

Domain	Some characteristics of each domain
P	True nucleus absent Small (70S) ribosomes present Smooth endoplasmic reticulum absent RNA polymerase made up of 14 subunits
Q	True nucleus present Large (80S) ribosomes present Smooth endoplasmic reticulum present RNA polymerase made up of 14 subunits
R	True nucleus absent Small (70S) ribosomes present Smooth endoplasmic reticulum absent RNA polymerase made up of 4 subunits

(i) Which letter, P, Q or R, represents the Eukaryotes? [1]

(ii) Which two letters represent the domains that are least closely related? [1]

(iii) Place a cross (✗) in the box that represents the Eukaryotes in the diagram below.

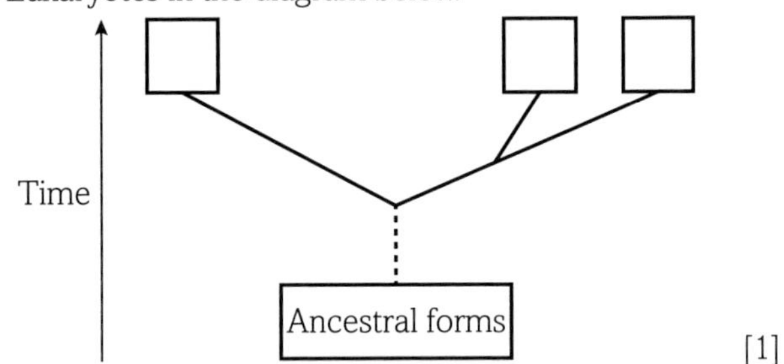

[1]

(b) One domain includes the plants and these have cells with a cell wall.

(i) Describe the structure of a plant cell wall. [4]

(ii) A student studied the cell wall arrangement between two adjacent plant cells. He noticed several features which he could not name. Two of these are described in the table below. Complete the table by writing in the name of each feature described.

Feature described	Name of feature
Site where there was no cell wall and the cytoplasm linked the two adjacent cells	
Dark line that is the boundary between one cell and the next cell	

[2]

(Total for Question 3 = 9 marks)

4 The scientist Carl Woese suggested that living organisms could be grouped into three domains. There are specific differences between the organisms in the three domains.

(a) What are the names of the three domains?
A Animalia, Archaea and Eukaryota
B Animalia, Bacteria and Prokaryota
C Archaea, Bacteria and Eukaryota
D Archaea, Eukaryota and Prokaryota [1]

(b) The table shows some characteristics of the three domains.

Characteristic	Domain		
	A	B	C
Mitochondria	Absent	Absent	Present
Cell wall with peptidoglycan	Yes	No	No
Amino acid carried on tRNA that starts protein synthesis	Formylmethionine	Methionine	Methionine
May contain chlorophyll	Yes	No	Yes
Sensitive to antibiotics	Yes	No	No

(i) Using the information in the table, suggest which of the domains A, B or C represents the Eukaryota. Give a reason for your answer. [2]

(ii) Many scientists believe that the Eukaryota domain is more closely related to the Archaea domain than to the Bacteria domain. Using the information in the table, suggest which of the domains A, B or C represents the Archaea domain. Give a reason for your answer. [2]

(c) Cells of the Eukaryota domain contain rough endoplasmic reticulum and Golgi apparatus. Both the rough endoplasmic reticulum and the Golgi apparatus are made up of membrane-bound sacs.

(i) Describe how you would recognise the Golgi apparatus as seen using an electron microscope. [3]

(ii) Describe the roles of rough endoplasmic reticulum and the Golgi apparatus in a cell. [3]

(d) 783 bases in a section of DNA coding for part of the RNA polymerase enzyme were sequenced in four different species. The numbers of differences within the sequences were recorded in the table:

	A	B	C	D
A	0	3	2	5
B	3	0	1	2
C	2	1	0	3
D	5	2	3	0

(i) How many amino acids are coded for by this length of DNA?

A 87

B 261

C 783

D 2349 [1]

(ii) State which species is most closely related to species A and justify your answer. [2]

(iii) Calculate the number of differences found between the two least closely related species as a percentage of the total number of bases. [2]

(Total for Question 4 = 16 marks)

5 (a) One concept of a species is the morphological species concept. Describe how scientists use this species concept in classification. [3]

(b) (i) An alternative to the morphological species concept is the biological species concept. What is the biological species concept? [3]

(ii) Explain two reasons why the biological species concept is sometimes difficult to apply. [4]

(c) State two characteristics that could be used to classify an organism into each of the following kingdoms:

(i) Fungi [2]

(ii) Protista [2]

(Total for Question 5 = 14 marks)

6 The fruit fly (*Drosophila melanogaster*) and the gorilla (*Gorilla gorilla*) are both members of the Animal kingdom.

(a) Complete the table giving their full classification.

Taxonomic rank	Gorilla	Fruit fly
Domain		
Kingdom	Animal	Animal
Phylum	Chordata	Arthropoda
Class	Mammalia	Insecta
	Primate	Diptera
Family	Hominidae	Drosophilidae
Genus		
Species		

[4]

(b) The similarity of proteins from different species can be established using antibodies that cause agglutination. These antibodies combine with the protein molecules sticking them together. An antibody that combines with the protein from one species will also combine with similar proteins from another related species. However, it will combine less well with the similar protein from a more distantly related species.

A scientist tested an agglutin manufactured to combine with proteins from species A on five other species. The results are shown in the table.

Species	Relative level of agglutination (%)
A	100
B	6
C	75
D	98
E	23
F	5

(i) The scientist concluded that species D is the most closely related species to A. Explain this decision. [1]

(ii) Explain why the agglutins were able to produce 98% agglutination in a different species. [3]

(iii) Explain why proteins found in more distantly related species produce less agglutination. [3]

[Total for Question 6 = 11 marks]

TOPIC 4 PLANT STRUCTURE AND FUNCTION, BIODIVERSITY AND CONSERVATION

CHAPTER 4C BIODIVERSITY AND CONSERVATION

Biodiversity is the variety of life - all the different types of organism, the variation between organisms of the same species and the ecosystems of which they are a part. In some places, there is great biodiversity - as seen in tropical rainforests and coral reefs. But in regions with more extreme climates, such as the lava fields in Iceland or in deserts around the world, there is less biodiversity. In extreme climates, organisms are more vulnerable when the climate changes.

In this chapter, you will define biodiversity and consider why it is so important. You will see how biodiversity can be measured in different ways, from species in a habitat to genes in a population. You will see how measuring the diversity of a habitat involves looking at the number of different species in an area and the relative numbers of the members of each species.

You will also learn about the genetics of populations. You will consider the factors affecting the gene pool - the sum of all the genes in a population at a given time - and use the Hardy-Weinberg equilibrium to measure changes in population genetics. You will consider the effects of different factors on the allele frequencies in a population, and how this can sometimes result in speciation. You will study the effects of genetic bottlenecks and the founder effect on subsequent populations.

Finally, you will look at the human impact on biodiversity and population sizes both on the land and in the oceans. You will discuss the idea that the sustainability of the world's resources depends on effective management of the conflict between human needs and the conservation of biodiversity.

MATHS SKILLS FOR THIS CHAPTER

- **Understand the principles of sampling as applied to scientific data** (*e.g. use an index of biodiversity to assess the biodiversity of a habitat*)
- **Use a given equation and substitute numerical values into algebraic equations** (*e.g. calculate an index of diversity*)
- **Use ratios, fractions and percentages** (*e.g. calculating genetic diversity and the heterozygosity index*)
- **Change the subject of an equation** (*e.g. Hardy–Weinberg equation*)
- **Substitute numerical values into algebraic equations** (*e.g. Hardy–Weinberg equation*)

What prior knowledge do I need?

Chapter 2C

- The construction of genetic crosses and pedigree diagrams

Chapters 2C and 3B

- Gene mutations and genetic variation caused by meiosis and sexual reproduction
- The nature of ecosystems and trophic levels
- That meiosis and mutations are sources of variation between individuals of the same species

Chapter 3B

- How random fertilisation during sexual reproduction brings about genetic variation

Chapter 4B

- Classification
- How human interactions with ecosystems can have both positive and negative effects on biodiversity
- Some of the benefits and challenges of maintaining local and global biodiversity

What will I study in this chapter?

- What biodiversity is, why it is important and how human activity can threaten it
- How biodiversity can be measured using an index of diversity
- How biodiversity can be assessed by looking at the variety of alleles in the gene pool of a population
- How genetic diversity can be assessed using a heterozygosity index
- How organisms can become very well adapted to their environment and the concept of niche
- How the Hardy–Weinberg equation can be used to monitor changes in allele frequencies
- How changes in allele frequencies may be the result of chance and not selection, including genetic drift, genetic bottlenecks and the founder effect
- How sustainability of resources depends on effective management of the conflict between human needs and conservation

What will I study later?

Topic 5 (Book 2: IAL)

- The effect of biotic and abiotic factors on the numbers and distribution of organisms in a habitat
- How the concept of niche accounts for distribution and abundance of organisms in a habitat
- The stages of succession from colonisation to the formation of a climax community
- The different types of evidence for climate change and its causes
- How knowledge of the carbon cycle can be applied to methods to reduce atmospheric levels of carbon dioxide
- The effects of climate change
- How a change in allele frequency can come about through gene mutation and natural selection
- How isolation reduces gene flow between populations, leading to allopatric or sympatric speciation
- How scientific conclusions about controversial issues can sometimes depend on who is reaching the conclusions
- How reforestation and the use of sustainable resources, including biofuels, are examples of the effective management of the conflict between human needs and conservation

4C 1 BIODIVERSITY AND ENDEMISM

SPECIFICATION REFERENCE 4.15 4.16

LEARNING OBJECTIVES

- Know that over time the variety of life has become extensive but is now being threatened by human activity.
- Understand what is meant by the terms *biodiversity* and *endemism*.

Biodiversity is an important word right now because the Earth's biological diversity, which has become extensive over millions of years of evolution, is reducing rapidly. At its simplest, biodiversity is a measure of the variety of living organisms and their genetic differences. Most scientists are agreed that loss of biodiversity is not a good thing, especially as there is growing evidence that some of the current loss is a direct result of human activity.

DEFINING BIODIVERSITY

Defining the term *biodiversity* is not easy. The number of different species is a useful basic measure of biodiversity, but the concept is more complex than this. The differences between individuals in a species, between populations of the same species, between communities and between ecosystems are all examples of biodiversity. Biodiversity can be assessed on different scales, from species level in a habitat to the genetic level within a population.

The Convention on Biological Diversity, the largest international organisation working on the subject, uses biodiversity as a term to describe the variety of life on Earth, from the smallest microbes to the largest animals and plants. They suggest the concept of biodiversity includes genetic diversity between individuals within a species and between different species, as well as the variety of different **ecosystems**.

WHY IS BIODIVERSITY IMPORTANT?

Does it really matter if there are fewer species of snails or beetles in the world or if an unknown plant species ceases to exist or if the genetic variation between the members of a rare population gets smaller and smaller? All the evidence suggests that it does.

All the organisms in an ecosystem are interdependent, and they can affect the physical conditions around them. Rich biodiversity allows large-scale ecosystems to function and self-regulate. These ecosystems are also interlinked on a larger scale across the Earth. If biodiversity is reduced in one area, the natural balance may be destroyed elsewhere. The air and water of the planet are purified by the actions of a wide range of organisms. Waste is decomposed and made non-toxic by many organisms, including bacteria and fungi. For example, microorganisms in soil and water convert ammonia into nitrate ions, which are then taken up and used by plants.

Photosynthesis by plants is important in stabilising the atmosphere and the world's climate. Land plants alone absorb more than a third of all the carbon dioxide that is produced by burning fossil fuels. Plants absorb vast amounts of water from the soil, which then evaporates into the atmosphere through transpiration, producing clouds that will produce rain. Therefore, plants help to determine where rain will fall. Plant roots, together with fungal mycelia, hold the soil together. This affects how water runs off or is absorbed by the soil and reduces the risk of flooding. Plant pollination, seed dispersal, soil fertility and nutrient recycling in systems such as the nitrogen and carbon cycles are vital for natural ecosystems and farming, and they all depend on a rich biodiversity.

Biodiversity provides the genetic variation that has allowed us to develop crops, livestock, fisheries and forests. It also enables further improvement by cross-breeding and genetic engineering. This variation will help us to cope with problems arising from climate change and disease. Plant biodiversity provides the potential of plants to produce chemicals that are important in many areas of human life, including new medicines.

The benefits of a biodiverse and healthy ecosystem are increasingly being assessed and valued as ecosystem services (see **fig A**).

▲ **fig A** Ecosystems such as this rainforest contain a huge range of biodiversity, with the potential to help humanity in many different ways.

ASSESSING BIODIVERSITY AT THE SPECIES LEVEL

Biodiversity can be measured in several ways. There are two main factors which need to be considered when measuring biodiversity at the species level. One is the number of different species in an area – the **species richness**. The other is the evenness of distributions of the different species – the **relative species abundance** of the different types of organism that make up the species richness. You will learn how to measure biodiversity yourself in **Book 2**.

SPECIES RICHNESS

Biodiversity varies enormously around the world in terms of numbers of species. The wet tropics are generally the areas of highest biodiversity. For example, it would be typical to find 150–280 tree species in 0.1 hectares of Amazon rainforest. Almost every tree is a different species. Imagine the numbers of other plants and animals associated with each type of tree and you can begin to appreciate the species richness of these areas.

As you move away from the wet tropics, the species diversity generally decreases. In temperate rainforests, tree species richness drops to 20–25 species in 0.1 hectares. Further north again, in the boreal forest in Scandinavia and Northern Canada, it falls to 1–3 species in 0.1 hectares. To highlight this, scientists have identified some **biodiversity hotspots** (see **fig B**) of unusual biodiversity. They occupy only 15.7% of the Earth's land surface, but are home to 77% of the Earth's terrestrial vertebrate species. Unfortunately, these areas are often areas with other resources that people want to use. For example, the Latin American rainforests have huge biodiversity, but are also a rich source of wood, gas, oil and minerals, and people are rapidly destroying the rainforests to access these resources.

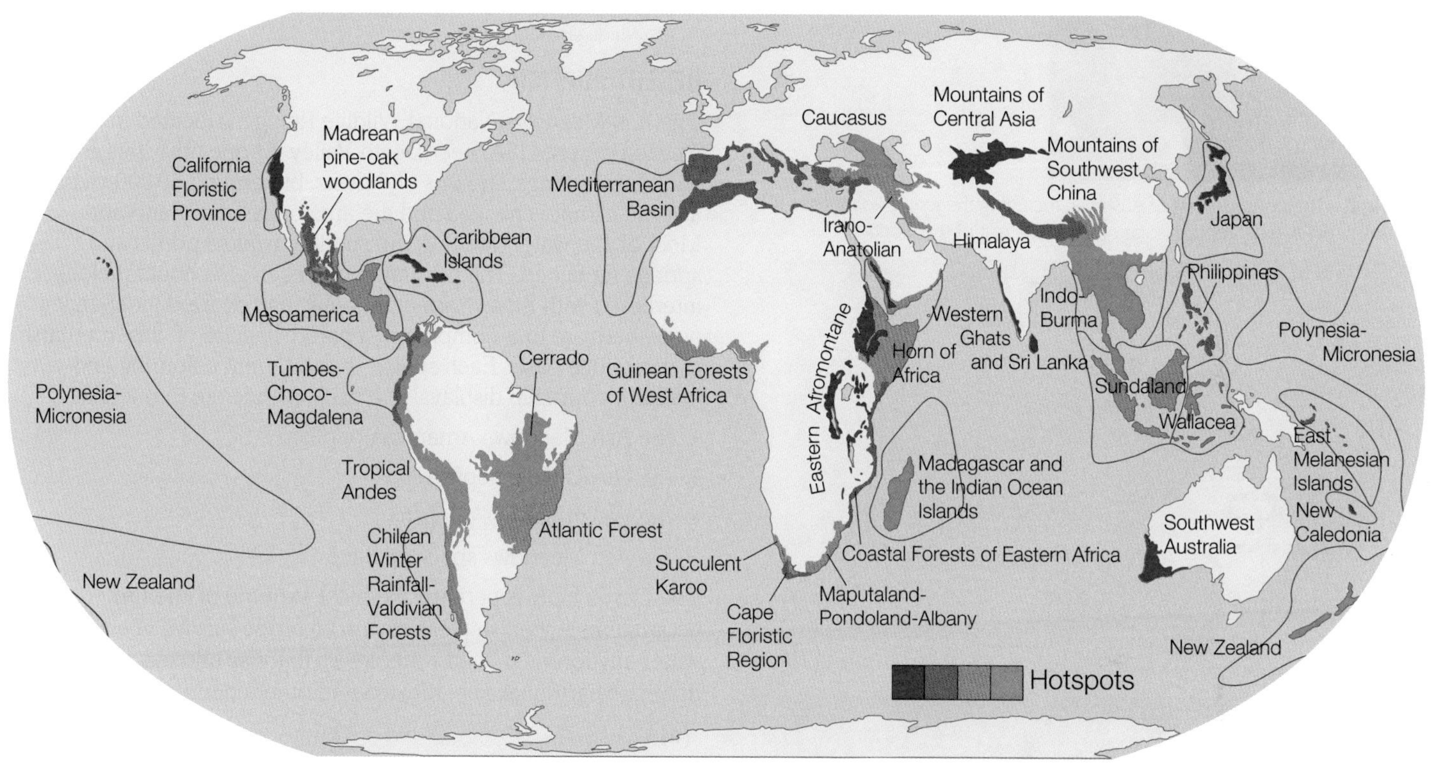

▲ **fig B** Known biodiversity hotspots around the world

Tropical regions also have areas of high marine and freshwater species richness including bacteria, plants, invertebrates such as corals and starfish, and vertebrates such as fish and mammals. Coral reefs are the marine equivalent of the tropical rainforests, and are the key areas of marine biodiversity. However, species richness is not the only important factor in a biodiversity hotspot.

EXAM HINT

It is very easy to write a lot of words and not say very much. If you use the correct terms, your response will be more directed and focused. Make sure you know how to use terms like *endemism*, *species richness* and *biodiversity*.

ENDEMISM

Another important way of measuring biodiversity is by looking at the number of **endemic** species in an area (see **fig D**). When a species evolves in geographical isolation – for example, on an island – and is found in only one place, it is said to be endemic to that place. So, for example, marsupial animals such as the kangaroos and the koala bear are well-known endemic animals of Australia (see **fig C**). Perhaps less well known, but also unique to particular countries, are the red slender loris (*Loris tardigradus*) of Sri Lanka, and the Cyprus Water Frog (*Pelophylax cypriensis*) and the Azraq Killifish (*Aphanius sirhani*) in Jordan. The areas of greatest biodiversity are not always the same as the areas with the greatest number of endemic species. This is why it is so difficult to prioritise areas for conservation.

There have been many ideas about why some areas have very rich biodiversity – in fact, around 125 different theories have been published. Most have been eliminated because they do not apply to all organisms or they are not supported by the evidence. The best current model suggests that:

- a very stable ecosystem allows many complex relationships to develop between species
- high levels of productivity (when photosynthesis rates are very high) can support more niches
- when organisms can grow and reproduce rapidly, it is more likely that mutations occur, leading to adaptations which allow organisms to exploit more niches.

▲ **fig C** The unique fauna of Australia, including the iconic koala bear, is the result of its geographical isolation from the rest of the world.

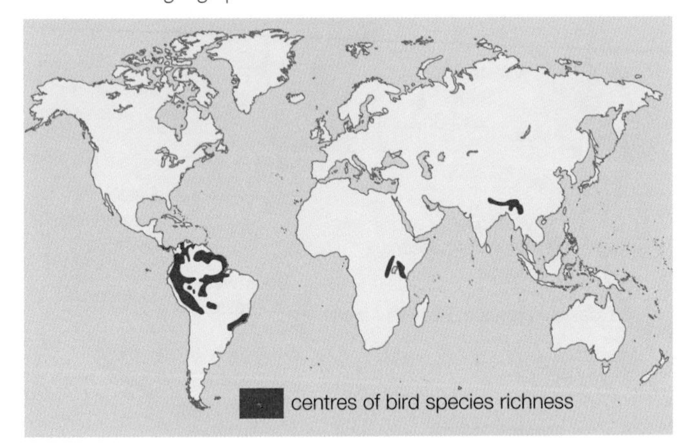

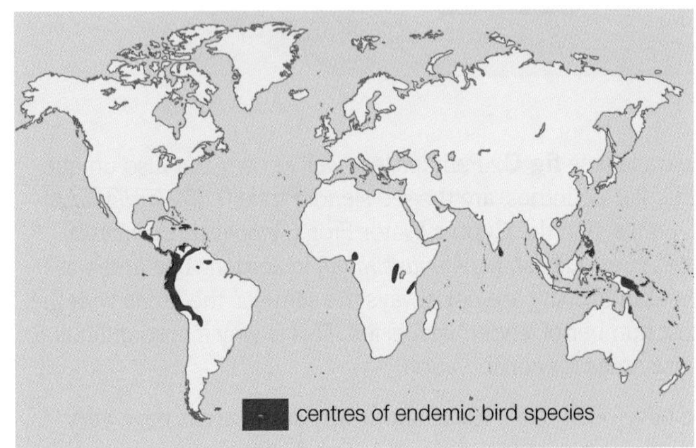

▲ **fig D** Hotspots of biodiversity can be measured by species richness and by endemic species.

ENDEMIC SPECIES OF MADAGASCAR

Madagascar is a large island off the coast of East Africa. Almost all the species found there are endemic to the island (see **fig E**). Organisms range from the amazing giant baobab trees to ring-tailed lemurs, from the bizarre elephant's foot plant to the tailless tenrec (a small mammal which can have over 30 babies at a time). The only species on the island that are not endemic have been taken to Madagascar by people in quite recent times. Imported species can cause many problems for the endemic species.

▲ **fig E** The ring-tailed lemur is just one of the unique species endemic to Madagascar.

DESERT PUPFISH

The Ash Meadows National Wildlife Refuge is located in the Nevada desert, USA, near Death Valley. At one time, this area had many springs, streams and rivers, but around 50 000 years ago the climate changed and the area became dry and arid. Most of the water dried up, but small individual ponds and springs remained. The fish trapped in each area could no longer interbreed with those from other areas and evolved independently. Now there are four completely separate species of desert pupfish in this refuge oasis. Each species has different colouring and different courtship displays, and each is endemic to one place:

- the Ash Meadows Amargosa pupfish
- the Devils Hole pupfish
- the Warm Springs pupfish
- the Ash Meadows speckled dace.

The Devils Hole is perhaps the most extreme of the tiny ecosystems – the fish there have adapted to survive in a warm water environment, with water levels that rise and fall when there are earthquakes in Mexico and other countries (see **fig F**).

(a)

(b)

▲ **fig F** **(a)** Devils Hole in the Nevada desert and **(b)** the endemic Devils Hole pupfish

SPECIES ABUNDANCE

The absolute number of species is not the only important factor in biodiversity. Relative species abundance is also significant. This is a measure of the relative numbers of the different types of organism.

Picture two areas of land in Europe, plot A and plot B (see **fig G**). There are five species of plant growing on each plot – grass, daisies, buttercups, dandelions and lady's bedstraw – in the proportions shown in **table A**.

part of plot A

plot B under a quadrat – a frame used to mark out a sampling area

▲ **fig G** Sections of plots A and B examined for **table A**

	GRASS	DAISIES	DANDELION	BUTTERCUPS	LADY'S BEDSTRAW
Plot A	95%	2%	1%	1%	1%
Plot B	30%	20%	15%	15%	20%

table A Showing species data for plot A and plot B

An area showing an even abundance of different species is considered to be more biodiverse than one containing the same number of species but dominated by one or two of those species. In this case, plot B is more diverse than plot A because the numbers of the different species are more evenly spaced. Plot A might be a well-tended lawn or city park, while plot B could be a more neglected version of the same sort of site.

The risks to biodiversity are not evenly spread around the world. Certain areas are much more vulnerable to damage and loss, particularly small isolated ecosystems such as islands, rainforests, coral reefs, bogs and wetlands. Many of these areas are also biodiversity hotspots so if they are damaged, many species will be lost. Every time a species becomes extinct, the biodiversity of the world decreases. On the other hand, every time a new species evolves, biodiversity increases.

Unless people understand the level of biodiversity in an area, it is impossible for them to recognise how important its loss might be, and so decide what to do about it. How can we put a number on biodiversity?

CHECKPOINT

SKILLS ANALYSIS

1. Using the maps in **fig D**, describe how the areas of high bird biodiversity and high bird endemism differ, and explain why this might be.

SKILLS CREATIVITY, INTERPRETATION

2. Suggest why endemic populations often have low genetic diversity.

SUBJECT VOCABULARY

ecosystems biological communities where organisms interact with one another and with their physical environment
species richness the number of different species in an area
relative species abundance the relative numbers of species in an area
biodiversity hotspot an area with a particularly high level of biodiversity
endemic a species that evolves in geographical isolation and is found in only one place

2 MEASURING BIODIVERSITY

SPECIFICATION REFERENCE 4.17 4.18 4.20(ii)

LEARNING OBJECTIVES

- Understand how biodiversity can be compared in different habitats using the formula to calculate an index of diversity (D):

$$D = \frac{N(N-1)}{\sum n(n-1)}$$

- Know how biodiversity can be measured within a habitat using species richness, and within a species using genetic diversity by calculating the heterozygosity index:

$$\text{heterozygosity index} = \frac{\text{number of heterozygotes}}{\text{number of individuals in the population}}$$

- Understand that changes in allele frequency can come about as a result of mutation and natural selection.

THE DIVERSITY INDEX

Scientists have developed many different ways of measuring the biodiversity of an ecosystem such as a pond (see **fig A**). Some are more useful than others, and all have limitations. In this example, both the species richness and the species abundance of an area are taken into account in a formula that gives a **diversity index** at the species level within a habitat:

$$D = \frac{N(N-1)}{\sum n(n-1)}$$

where

D = diversity index

N = the total number of organisms of all species

n = the total number of organisms of each individual species – the abundance of the different species

$\sum$ = the sum of all the values that follow. (You need to calculate $n(n-1)$ for each species and then add them together.)

▲ **fig A** Natural and artificial ponds both quickly build up a range of biodiversity as algae, bacteria, protists, plants and animals colonise the water.

WORKED EXAMPLE

A simple example based on some samples of animal life collected from a local pond.

SPECIES	NUMBER OF ORGANISMS COLLECTED (n)	($n-1$)	$n(n-1)$
dragonfly larvae	4	3	4 × 3 = 12
mosquito larvae	12	11	12 × 11 = 132
water boatman	3	2	3 × 2 = 6
tadpole shrimp	4	3	4 × 3 = 12
copepods	17	16	17 × 16 = 272
Total number of organisms (N)	40	39	40 × 39 = 1560

table A In a sample from a small pond, the species shown above were found.

You can calculate the diversity index for this pond as follows:

$$D = \frac{N(N-1)}{\sum n(n-1)}$$

$$D = \frac{1560}{12 + 132 + 6 + 12 + 272}$$

$$D = \frac{1560}{434}$$

Diversity index = 3.60

LEARNING TIP

Remember that big is better: it is always better to have a high value of biodiversity.

EXAM HINT

There are several different formulae which look very similar to this one. Make sure you know this one and how to use it.

HOW BIODIVERSITY VARIES

As you saw in **Section 4C.1**, there have been many ideas about why some areas have particularly rich biodiversity. The best current model suggests that high biodiversity is seen in:

- very stable ecosystems
- areas where there are high levels of productivity (when photosynthesis rates are very high)
- areas where organisms can grow and reproduce rapidly.

In general, when an environment has extreme environmental conditions (e.g. a desert) the biodiversity is low (see **fig B**). Any change in this extreme environment has a big impact on population numbers. This type of ecosystem tends to be unstable and very susceptible to change. A particularly severe frost, a flood or a new pathogen can devastate or even wipe out one or more populations. This type of environment also has some unfilled niches. This means that an incoming organism can become established very rapidly and overpower an existing species if they are competing for food or territories.

▲ **fig B** Relatively few species can survive in extreme environments like the desert, and small changes can threaten those organisms.

In less hostile environments, biodiversity can be very high. This results in a very stable ecosystem, because a new species moving in or out will have almost no effect.

As a result of these factors, some areas are more vulnerable to the loss of biodiversity than others. Biodiversity can be lost due to natural events such as a volcanic eruption or flooding, but also as a result of human activities. As you learned in **Section 4C.1**, small or isolated ecosystems are particularly vulnerable to damage and destruction. When these easily damaged areas are also biodiversity hotspots, there is a high risk of losing that biodiversity if a natural or human-produced disaster takes place. To help us decide which areas are most important for us to conserve, we need to measure biodiversity all around the world.

WHEN TO MEASURE BIODIVERSITY

Biodiversity is not constant. For example, the animal species in an area can vary with the time of day. Many bat species flying on a warm evening will not be visible the next morning. What is more, in the temperate and alpine areas of the world there are distinct seasons. This means that the picture of biodiversity in an area will change considerably through the year. The number of plant species in the same area of a UK woodland floor or meadow measured during the summer is different from that measured in winter, and has different biodiversity. Similarly, wetland feeding sites for migrating birds around the world are alive with aquatic and wading birds during some months of the year; but during others, they are quite empty.

BIODIVERSITY WITHIN A SPECIES

Biodiversity within an individual species is also a very important concept. The gene pool of a species is all of the genes in the genome, including all the different variants of each gene. Modern DNA analysis allows us to measure biodiversity on a different scale, at a genetic level – and we are discovering that genetic diversity within a species is very important.

GENE AND ALLELE FREQUENCY

You learned in **Chapter 2C** that mutations are changes in the DNA structure. Many mutations have no effect at all on the phenotype, while others may have useful, or damaging or lethal effects. Mutations can increase the gene pool of a population by increasing the number of different alleles available. The relative frequency of a particular allele in a population is called the **allele frequency**. If a mutation results in an advantageous feature, the allele will be selected for and so increase in frequency in the population. If the mutation is disadvantageous, natural selection will sometimes result in its removal from the gene pool. More frequently, it will be retained at a very low frequency unless it also transfers some benefits. A disadvantageous allele in one set of environmental conditions may become an advantageous allele if conditions change. The changes in allele frequency due to natural selection may lead to new species emerging.

MEASURING GENETIC BIODIVERSITY

The genetic variation within a population is an important measure of biological health and well-being – without variation, a population is vulnerable. Modern technology has made it possible to build up a clear model of **genetic diversity** within a population by analysing the DNA and comparing particular regions for similarities and differences.

Scientists look at the proportion of a population which is heterozygous for a given feature. Heterozygotes have more than one allele for a given feature – to remind yourself of the difference between homozygotes and heterozygotes, look back to **Section 2C.2**. When the DNA is analysed, only one band will show up if an organism is a homozygote. For heterozygotes, two bands appear, one for each allele. This can be used to calculate a **heterozygosity index** for the population for the particular DNA sequence. It is calculated using the equation:

$$\text{heterozygosity index} = \frac{\text{number of heterozygotes}}{\text{number of individuals in the population}}$$

The heterozygosity index is a useful measure of genetic diversity in a population. A high heterozygosity index reflects a high level of genetic variation and a potentially healthy population, whereas a low index suggests a population is in trouble.

1	2	3	4	5	6	7	8	9	10	11	12	13	14	15	16	17	18	19	20	21	22	23
—	—	—	—	—		—	—	—			—	—		—	—	—	—		—		—	—
	—				—			—	—	—			—			—	—	—		—		—

▲ **fig C** The DNA analysis of 23 European toads for a single gene shows some heterozygous individuals but most are homozygous.

LEARNING TIP

Remember big is better: a higher proportion of heterozygotes means better genetic diversity.

In the data in **fig C**, the DNA of 23 individuals of a species of European toad has been analysed. Five of the toads are heterozygous, and the rest of the population is homozygous for one allele or another. So:

$$\text{Heterozygosity index} = \frac{5}{23} = 0.217$$

This is not particularly high – for example the heterozygosity of several healthy populations of Siberian apricot trees in China was 0.639 and 0.774 for two different genetic regions. However, the heterozygosity of the toad population seems to be very healthy when compared to some of the animals with the biggest problems of genetic diversity.

EXAM HINT

Be ready to use this formula for assessing genetic diversity. And be ready to state whether it suggests high diversity and a healthy population or low diversity and a species at risk.

DID YOU KNOW?

CHEETAHS AND LOW GENETIC DIVERSITY

Scientists have discovered that cheetahs have very little genetic diversity. These beautiful members of the cat family are the fastest land animals on Earth (see **fig D**). Sadly, they are not only low in numbers, they have so little genetic biodiversity that they are in danger of being made extinct by a single disease or a small change in their environment. The heterozygosity index of different populations ranges from 0.0004–0.014. These extremely low ratios demonstrate that the animals are genetically all very similar. Scientists think that (for some reason) they must have been almost destroyed 10 000–20 000 years ago. As a result of this population bottleneck (see **Section 4C.5**), modern populations of cheetahs are descended from just a few original animals. Numbers are declining now because their habitat is disappearing and there are serious worries about the survival of the species.

▲ **fig D** The genetic diversity of these beautiful animals is dangerously low.

Models of the molecular phylogenetic relationships between related organisms based on DNA and other evidence are a useful tool for measuring biodiversity. For example, scientists at the Natural History Museum in London, UK, have built up contrasting maps of biodiversity based on both numbers of species and DNA similarities. The ones shown in **fig E** show bee populations – the most biodiverse areas for bees are Ecuador (highest species richness) and Kashmir (highest genetic diversity). This type of study can be hugely important for conservation work. If you are trying to conserve biodiversity with limited funding – and funding is always limited – you need to be confident that you have chosen the area with the highest biodiversity. Maps like these can be generated for overall biodiversity or for the diversity of particular groups of animals and plants. They can be produced for the whole world, for individual countries or for small local areas. The value of this type of data is that it can be used to highlight areas that need protection. If it is regularly updated, it also provides a way of monitoring changes in biodiversity anywhere.

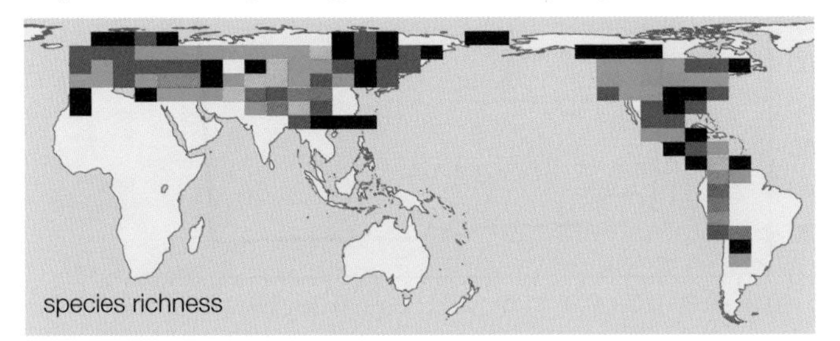

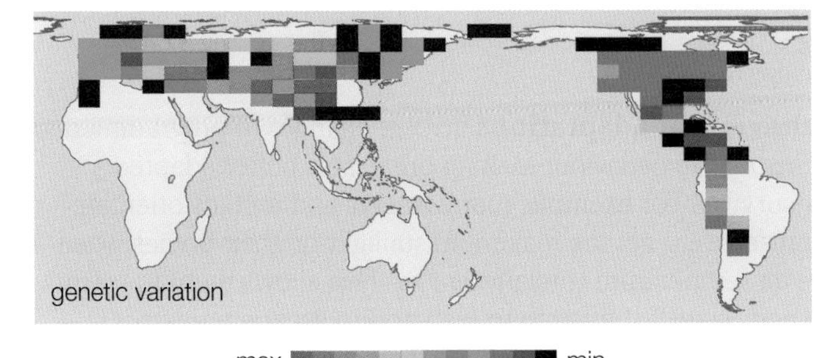

max min

▲ **fig E** These maps show the biodiversity of bees around the world measured by species richness and by genetic diversity.

THE ISOLATED ISLANDS OF HAWAII

The Hawaiian island populations show clearly how living organisms adapt to a particular niche or role in the community. They also demonstrate why it is important for scientists to look at biodiversity at the level of species numbers and abundance, and at the level of genetic diversity.

The Hawaiian Islands are very isolated – 4000 km from the nearest continental land mass and 1600 km from other islands. They have a great deal of biodiversity in terms of species numbers – 1000 species of native flowers, 10 000 species of insects, 1000 species of land snails and around 100 species of birds. But before people introduced them, there were no reptiles and only one species of mammal – a bat. Analysis of the DNA of the endemic populations shows that they are very closely related, even though some of them look very different. All those insect species seem to have evolved from only about 400 original species, while there appear to have been only seven founder species of land birds. So, in these isolated conditions, a small group of founder organisms have adapted and evolved to take advantage of the different ecological niches that were available to them. Places where endemism is common often have a rich biodiversity in terms of species numbers, but quite low genetic diversity. This is one reason why areas with many endemic populations are very vulnerable to the introduction of disease.

CHECKPOINT

1. The term *biodiversity* is often used in the media simply to indicate the number of species of living organisms. Explain why it gives a limited picture.

SKILLS ANALYSIS

2. In a small area of European woodland, a group of students counted the number of species present in a measured area, and the numbers of individuals of the different species.

SPECIES	NUMBER OF ORGANISMS COLLECTED (*n*)
holly	9
bramble	3
oak	3
butcher's broom	5
ivy	3
yew	1
Total number of organisms (*N*)	24

Calculate the index of diversity for this area of woodland and decide whether you think it has high or low biodiversity.

3. Using the information supplied here and other sources, discuss why it is important to measure both species richness and genetic diversity to give a full picture of biodiversity.

SUBJECT VOCABULARY

diversity index a way of measuring the biodiversity of a habitat using the formula:

$$D = \frac{N(N-1)}{\Sigma n(n-1)}$$

allele frequency the frequency with which a particular allele appears within a population

genetic diversity a measure of the level of difference in the genetic make-up of a population

heterozygosity index a useful measure of genetic diversity within a population expressed as:

$$\frac{\text{number of heterozygotes}}{\text{number of individuals in the population}}$$

4C 3 ADAPTATION TO A NICHE

SPECIFICATION REFERENCE 4.19

LEARNING OBJECTIVES

- Understand the concept of niche and be able to discuss examples of adaptations of organisms to their environment (behavioural, anatomical and physiological).

Every species is part of the complex system of interactions between the physical world and other living organisms. We call this system **ecology**. Each species exists in a specific **niche**.

The niche occupied by an organism is an important concept that is difficult to define. It describes the role of the organism in the community – rather like a job description or a way of life. You can consider different aspects of a niche, such as the food niche or the habitat niche. Some niches are very large and general, for example organisms that eat grass; some are very small and specific, for example organisms that feed by cleaning the teeth of other, larger organisms.

SUCCESSFUL ADAPTATION

A successful species is well adapted to its niche, meaning that individuals in that species have characteristics that increase their chances of survival and reproduction, and therefore of passing those characteristics on to the next generation. Adaptations may be of many different kinds, including anatomical, physiological and behavioural.

- **Anatomical adaptations** involve the form and structure of an organism. Examples include the thick layer of blubber in seals and whales, and the sticky hairs on the sundew plant that enable it to capture insects (see **fig A**).

fig A The sticky hairs of a sundew leaf are anatomical adaptations that enable the plant to capture insects. The plant can then digest the insects to supplement the nutrients in the poor soil of the bogs where it grows.

- **Physiological adaptations** involve the way the body of the organism works and include differences in biochemical pathways or enzymes. For example, diving mammals can stay under water for much longer than non-diving mammals without drowning. When a diving mammal is under water, its heart rate drops significantly, so that the blood is pumped around the body less often and the oxygen in the blood is not used as rapidly (see **fig B**). The main body muscles can work more effectively using anaerobic respiration than those of land-living mammals. This means that the oxygen-carrying blood is directed to the brain and the heart where it is still needed. This is called the mammalian diving response.

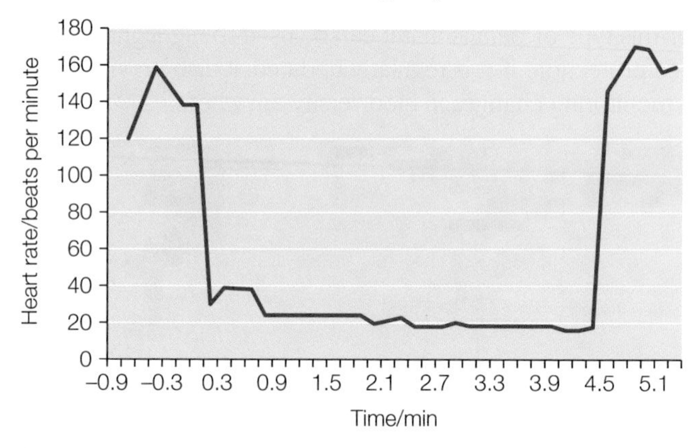

fig B The heart rate of a seal slows as it dives. This slowing is called bradycardia.

- **Behavioural adaptations** involve changes to programmed or instinctive behaviour making organisms better adapted for survival. For example, many insects and reptiles orientate themselves to get the maximum sunlight on their bodies when the air temperature is relatively low. This allows them to warm up and move fast enough to feed and to escape predators. When they get hot, they change their orientation to minimise their exposure to the sun, or shelter from it (see **fig C**). Social behaviour such as hunting as a team or huddling together for warmth can improve the survival chances of both individuals and groups. Migrating to avoid harsh conditions, courtship rituals and using tools are other examples of behavioural adaptations.

fig C Early in the day, lizards flatten themselves against the warm ground to heat up. When they start to get too hot, they will push themselves away from the ground so they can cool down more easily.

EXAM HINT

Make sure you know the difference between anatomical and physiological adaptations.

ADAPTATIONS FOR SURVIVAL

Successful adaptations enable species to exploit every possible habitat and the different niches within each habitat. Most organisms have a mixture of different types of adaptation that enable them to survive and succeed in their particular environment. Carnivorous fungi and camels are two examples.

FUNGAL CARNIVORES

Fungi are often considered saprophytes, breaking down dead material in the soil. However, some fungi are active carnivores. Nematode worms are found in huge numbers in the soil. Some groups of fungi have developed adaptations that enable them to capture and feed on these worms. Some produce sticky nets or adhesive pads to trap the worms, some live inside the living worms. *Arthrobotrys anchonia* is a fungus that actively lassoes nematodes. It traps the worms as they pass through loops of its hypha called constriction rings, involving both anatomical and physiological adaptations (see **fig D**). Three fungal cells form a ring and when a nematode moves into it, a combination of wall changes and the osmotic potential of the cells result in water moving in fast. Within 0.1 seconds the ring inflates and holds the nematode in its grip. Often the nematode puts its tail through another loop as it moves to try and escape. Within hours the fungus grows more hyphae that penetrate the body of the nematode and digest it, absorbing the nutrients and transporting them within the fungus. It is thought that these predatory fungi evolved from saprophytic fungi in a high-carbon, low-nitrogen environment. The nematodes provide the missing nitrates.

▲ **fig D** Some carnivorous fungi have adaptations that enable them to lasso their moving prey.

CAMELS: THE ULTIMATE SURVIVORS

Camels are large mammals. They live and breed in some of the hottest, driest and most inhospitable niches available. The incredible adaptations of the camel allow it to survive and thrive as a plant-eater in desert environments. The animal is vital in the survival of civilisations who make desert environments their home.

Camels have a huge range of anatomical, physiological and behavioural adaptations, and the combination of them all means they are the ultimate survivors: large mammals in niches where few other organisms can live, whatever their size (see **fig E**).

SOME ANATOMICAL ADAPTATIONS OF CAMELS

- Large eyes give good vision in many directions and are protected by long lashes which protect the eyes against the sandy environment.
- The nostrils are long and slit-like and the camel can close them to protect against sand and wind.
- The upper lip is split, hairy, extensible, slightly prehensile and sensitive. This allows it to identify and gather food, and avoid the thorns which are a protective adaptation of many desert plants.
- A hump on the back helps insulate the animal from the heat of sun. The fat concentrated in the hump allows easy evaporation of sweat over the rest of the skin surface as there is little fatty insulation under the skin. This is important in cooling down.
- The feet are large and flat with tough pads. These spread out the weight of the camel as it walks over sand and also prevent heat damage.
- Tough pads on the knees prevent damage from hot sand when the camel rests.

▲ **fig E** Camels are perfectly adapted to their desert niche. See how many of the anatomical adaptations of camels you can identify in this photo.

SOME PHYSIOLOGICAL ADAPTATIONS OF CAMELS

The physiological adaptations of camels have evolved to cope with several different challenges.

Thermoregulation

Camels face extremes of temperature in the desert: extreme heat in the day and cold at night. They have evolved some sophisticated physiological mechanisms to help them cope with these extremes without using all their resources.

- The supple skin is covered in fine hairs which can be erected to trap an insulating layer of air during cold nights.
- The camel can withstand a wide variation of internal (core) body temperature. Most mammals – and camels under ideal conditions – maintain their body temperature within a 2 °C range. Under heat stress, or when a camel is dehydrated and cannot afford to lose water in thermoregulation, a camel can allow its body temperature to vary over 6 °C. This saves energy and water as it does not need to produce so much sweat.
- Camels can lose 30% of their body weight through water loss and then make it up in 10 minutes drinking water, without changing the osmotic potential of the blood. Consequently, the blood does not become thick and unable to flow when the camel is extremely dehydrated.

- Camels have a hump on the back where most of their fatty tissue is found. This acts as a food store. As the fat is broken down to release energy during cellular respiration, water is produced as a by-product (see **Book 2 Topic 7**). This metabolic water means camels can survive longer than most mammals without actually drinking water.

Water balance

The desert niche of the camel means it often has to survive without water for considerable periods of time. It has some adaptations which mean it uses very little water and can restrict normal water losses when the supply is limited.

- Camel body tissues can withstand a loss of up to 30% of the body water without being damaged – it can go without water in the desert heat for up to 10 days.
- Camels can drink up to 180 litres of water in 24 hours without affecting the osmotic balance of their cells.
- The fat in the hump acts as an energy store. When this fat is broken down in aerobic respiration (see **Book 2 Topic 7**) water is produced as a waste product. For example, 20 kg of fat releases a total of just over 21 kg of water – and the hump of a camel can store around 35 kg of fat.
- Water is lost by sweating, and in urine and faeces. Camels can withstand big temperature variations which means that they minimise water loss by sweating. They have kidneys adapted to produce very concentrated urine – this reduces water loss and enables camels to drink quite salty water. The production of urine is also slowed down to a minimum when a camel is dehydrated – again reducing water loss.
- Camels can continue to produce dilute milk even when they are dehydrated. This enables them to breed successfully in the desert and also means they can provide a good source of food and drink for people in these dry environments.

DID YOU KNOW?

Dromedary camels have red blood cells which can withstand expanding to more than 240% of their original volume without bursting – most animals can only withstand 150% expansion before the cells burst. This allows the camel to drink so much that its body fluids are diluted but its blood cells are not damaged as a result.

SOME BEHAVIOURAL ADAPTATIONS OF CAMELS

The behavioural adaptations of camels are particularly clear in wild camels – the behaviour of domesticated camels is often directed by their human masters.

- When it is hot and camels are dehydrated, they sit down early in the morning before the ground warms up, with their legs tucked underneath so they absorb as little heat from the ground by conduction as possible.
- Camels, like many animals including insects and lizards, orientate themselves towards the sun so as little of their body surface area as possible is exposed to the heating rays.
- Groups of camels may lie down together, again minimising the amount of surface area of each camel exposed to the sun.
- Camels are browsers – they eat a wide range of vegetation and their height makes many shrubs and trees available to them which other animals cannot reach.

Perhaps the camel is the ultimate example of how organisms can be adapted to a very special and challenging niche.

EXAM HINT

If you describe an adaptation be sure you can explain how it helps survival.

CHECKPOINT

1. Explain the importance of the niche concept in understanding the adaptations of organisms.

SKILLS ANALYSIS

2. Looking at the data from **fig B,** answer the following questions.
   (a) What do the negative numbers on the *x*-axis represent?
   (b) How long did the recorded dive last?
   (c) What was the percentage depression of the heart rate? Based on this, how long would you predict that the dive might have lasted without the bradycardia?
   (d) State the type of adaptation involved in this response and explain why it is so important to the survival of the seal in its ecological niche.
3. How do the anatomical and physiological adaptations of camels help them to survive in their hot, dry, plant-browsing niche?

SUBJECT VOCABULARY

ecology the study of the relationships of organisms to one another and to their physical environment
niche the role of an organism within the habitat in which it lives
anatomical adaptations adaptations involving the form and structure of an organism
physiological adaptations adaptations involving the way the body of the organism works, including differences in biochemical pathways or enzymes
behavioural adaptations adaptations involving programmed or instinctive behaviour making organisms better adapted for survival

4 GENE POOL AND GENETIC DIVERSITY

LEARNING OBJECTIVES

- Understand how the Hardy–Weinberg equation can be used to see whether a change in allele frequency is occurring in a population over time.
- Understand that changes in allele frequency can come about as a result of mutation and natural selection.

In biological terms, a **population** is a group of individuals of the same species occupying a particular habitat and a particular niche within that habitat. The habitat of an organism is the place where it lives, and includes both the physical and biological elements of the surroundings. The niche of an organism is its place within the ecosystem, including its habitat and its effect on other organisms. You will learn more about these ecological concepts in **Book 2 Topic 5**.

POPULATION GENETICS

In population genetics, we take the gene as the unit of evolution and look at how the genetic composition of a population evolves over time. The sum total of all the alleles in a population at a given time is called the **gene pool**, and it will be millions or even billions of genes. Fortunately, it is usually the gene pool for a single trait that we look at. At any point in time, a population of organisms will have a particular gene pool, with different alleles occurring with varying frequencies, as you saw in **Section 4C.2**. Evolution can be considered as a permanent change in allele frequencies within a population. What does this mean and how is it measured?

ALLELE FREQUENCIES

The frequency of alleles in a population is not fixed. As the environment changes, so the frequency of different alleles changes through the process of natural selection and adaptation. For example, warfarin is a chemical that prevents the blood from clotting and it has been used as rat poison for about 70 years – the rats die of internal bleeding (see **fig A**). When warfarin was introduced, some rats already carried a mutation that, by chance, gave them resistance to the poison. The poison acted as a powerful **selection pressure** and resulted in a rapid increase in the frequency of the resistance allele.

Selection pressure is seen when a change occurs in an environment. Some individuals will have alleles which give them an advantage in the new conditions – as when rat poison containing warfarin is introduced to a habitat. Rats with the alleles that give resistance to warfarin will be more likely to survive and reproduce successfully, passing on those resistance alleles. The same factor has a negative selection pressure on rats with the susceptible alleles – they will be more likely to die and not reproduce so the frequency of the susceptible allele will be reduced in the population. Eventually, the majority of rats became resistant to warfarin and new, more powerful poisons had to be developed. Speciation involves a change in the allele frequencies within a population and is driven by selection pressures over time.

The number of individuals carrying a certain allele in a population determines the allele frequency. It describes what proportion of individuals carry a certain allele and is usually expressed as a decimal fraction of 1. The frequency of an allele in the population is not correlated with whether it codes for a dominant or recessive phenotype.

Take an imaginary gene with two possible alleles **A** and **a**. If all of the individuals in a breeding population of 100 diploid organisms are heterozygous, then in theory the frequency of each allele is 100/200 or 0.5. It is very rare for this to be the case, but it gives us a way of building a model.

fig A The presence or absence of the allele that gives rats resistance to warfarin is obvious only once they have been exposed to the poison.

We have a general formula that can be used to represent the allele frequencies for the dominant and recessive phenotypes in the gene pool of a population. The frequency of the dominant allele is represented by p and the frequency of the recessive allele is represented by q. The frequency of the dominant allele plus the frequency of the recessive allele will always equal 1:

$$p + q = 1$$

This simple equation is not very useful because it is usually almost impossible to distinguish between heterozygotes and dominant homozygotes based on their phenotype. This means that measuring the frequencies of heterozygotes and homozygotes in the population is not possible. However, we can readily observe the distribution of the recessive phenotypes in a population and from this we can calculate all the genotype frequencies using the Hardy–Weinberg equation. This is when the simple relationship $p + q = 1$ becomes very useful.

THE HARDY–WEINBERG EQUILIBRIUM

The amount of change that takes place in the frequency of alleles in a population indicates whether the population is stable (unchanging) or is evolving. In 1908, the British mathematician G.N. Hardy and the German physician W. Weinberg developed an equation independently of each other. This equation could be used to describe the mathematical relationship between the frequencies of alleles and genotypes within a stable theoretical population that is not evolving – these conditions are explained on the next page. The **Hardy–Weinberg equilibrium** theory states that in a population that is not evolving, the allele frequencies in the population will remain stable from one generation to the next if there are no other evolutionary influences. If the population is evolving, the allele frequencies will change from generation to generation and so the population is not in equilibrium. When a new species is formed, gene flow will be reduced because there will be no more breeding between the two original populations.

While the Hardy–Weinberg equation provides a simple model of a theoretical stable population, its main use is in calculating allele and gene frequencies in population genetics, providing a means of measuring and studying changes in species over time.

The algebraic equation developed by Hardy and Weinberg is expressed as:

p^2	+	$2pq$	+	q^2	= 1
frequency of homozygous dominant genotype in population		frequency of heterozygous genotype in population		frequency of homozygous recessive genotype in population	

fig B A striking phenotype such as this albino alligator makes it very easy to see that here is an animal with only recessive alleles.

USING THE HARDY–WEINBERG EQUATION

Since recessive phenotypes are easily observable (see **fig B**), we can measure their frequency and calculate allele frequencies that we can then use in the Hardy–Weinberg equation to estimate genotype frequencies.

The frequency of homozygous recessive individuals is represented as q^2. From this, q is easily obtained by finding the square root of q^2. The result gives the frequency of the recessive allele and by substituting this figure into our initial formula of $p + q = 1$, the frequency of the dominant allele p can be found.

WORKED EXAMPLE

In **Section 2C.3** you studied the inheritance of the albino trait. People who inherit the allele for the dominant pigmented trait may have the genotype **AA** or **Aa**. People who are albino are homozygous for the allele determining the recessive phenotype **aa**. Tests on a sample of North Americans showed that the frequency of albinos in the population was 1 in 20 000. This tells us that the frequency (q^2) of the homozygous recessive trait is 1/20 000 = 0.000 05, so we calculate the value of q:

$q^2 = 0.00005$

so $q = \sqrt{0.00005} = 0.007$

we know $p + q = 1$

so $p = 1 - 0.007 = 0.993$

By substituting these values into the expressions from the Hardy-Weinberg equation, the frequency of homozygous **AA** and heterozygous **Aa** genotypes can be calculated:

$p^2 + 2pq + q^2 = 1$

Frequency of homozygous **AA** = p^2

$p^2 = 0.993^2 = 0.986$

Frequency of heterozygous **Aa** = $2pq$

$2pq = 2(0.986 \times 0.007) = 0.014$

This gives us the frequencies for each of the three genotypes for albinism in the North American population. 98.6% of the population are homozygous for the dominant **A** allele, 1.4% are heterozygotes and 0.005% are homozygous for the recessive trait and are albinos. The allele frequencies must add up to 1, and the population percentages to 100%.

EXAM HINT

Always check that the numbers you have calculated make sense.

LEARNING TIP

The mathematics behind the Hardy-Weinberg equilibrium is very simple but few candidates seem able to use the equations correctly. Remember that you must look at the recessive phenotype because the frequency of the recessive phenotype in the population gives you q^2.

From there it is easy to work out q and then p.

CONDITIONS OF THE HARDY–WEINBERG EQUILIBRIUM

The Hardy–Weinberg equation describes the situation in a theoretical stable equilibrium, where the relative frequencies of the alleles and the genotypes stay the same over time. This implies that, when there are no factors that change the equilibrium, allele frequencies will remain constant within a population from generation to generation. In this theoretical population:

- there are no mutations
- there is random mating
- the population is large
- the population is isolated (no immigration or emigration – no organisms move in or move out)
- there is no selection pressure (all genotypes are equally fertile/successful).

In the real world, these conditions are almost never met. Deviations from the Hardy–Weinberg equilibrium show that species are continuously changing. The factors that result in deviations from the hypothetical equilibrium state are the selection pressures that bring about a long-term change in the gene pool, changing the allele frequencies in the population and driving speciation forward. Upsetting the gene-pool equilibrium results in the formation of new species.

MUTATIONS

For the allele frequency to remain stable in a population, no mutations must occur. Mutations involve changes in the genetic material, so the alleles are changed. As you know already, spontaneous mutations occur within a population all the time. Mutations in the somatic cells of animals will not be passed on to their offspring and may or may not affect the individual themselves. In animals, only mutations in the germ line cells – the cells that create the eggs and sperm – will affect the alleles of the next generation. In plants, the germ line cells are not fixed in the embryo. A mutation that takes place in a single plant stem as it grows can therefore become part of the gametes of a flower that forms on that stem (see **fig C**).

▲ **fig C** Ruby red grapefruit result from two separate mutations in the growing cells of grapefruit trees. One produced pink fleshed fruit. Several years later, another mutation produced a redder and much sweeter flesh. In both cases, the mutation produced a new dominant phenotype.

Although mutations occur continuously, they often do not affect a population very rapidly. In a single generation, each gene has between 1 in 10^4 and 1 in 10^9 chance of mutation. Recent work from the 1000 Genomes Project suggests that each of us has around 60 new mutations passed on from our parents. The clear majority of these will be recessive and will never be expressed. Occasionally, mutations arise that will bring benefits to an individual and will become established within the gene pool.

NON-RANDOM MATING

One of the most important requirements for a gene pool to remain in equilibrium is for random mating to occur. Random mating means that the likelihood of any two individuals in a population mating is independent of their genetic make-up. If mating occurs randomly, the frequency of the alleles in the population will stay the same.

Non-random mating occurs when some feature of the phenotype affects the probability of two organisms mating – and as a result natural selection takes place. For example, if a male animal displays in some way to attract the female, it is not random mating (see **fig D**). The male peacock with the most impressive tail, the buck with the largest antlers and most aggressive nature, the butterfly with the brightest wings and most spectacular display flight – these will appear to be more attractive than average to the females of the species and this applies a selection pressure. Consequently, they will be more likely to have the opportunity to mate and pass on their genes, ensuring that their offspring are likely to carry the alleles for these attractive characteristics. This is natural selection in action.

Within human populations, non-random mating is the normal situation. In every human culture, value judgements are used in the selection of a partner, by the individual or perhaps by the family or social groups to which the individual belongs. In different populations, different traits are regarded as desirable and so the gene pool shifts. For example, in the Kuna people of Panama, albinism is regarded as a very desirable trait and the incidence is approximately 1 in 200 people – in comparison, the average frequency globally is 1 in 17 000. This is a remarkable shift in the allele frequency within the population and shows the effect non-random mating can have on the gene pool equilibrium.

▲ **fig D** Male magnificent frigatebirds display their red throats to attract a mate. The female selects the most impressive one, so mating is not random.

POPULATIONS OF VARYING SIZES

The Hardy–Weinberg equation is only valid if it is applied to a large population – at least several thousand individuals. This is because maintaining genetic equilibrium depends on a random assortment of the alleles. Large populations containing many individuals usually have large gene pools – the chance of losing an allele by random events is reduced in a large population. For example, if allele **Z** occurs in 10% of the population and that population consists of 10 individuals, then only one individual will carry the allele. If that individual is lost through predation or disease, the allele is also lost, even if it codes for an advantageous phenotype. However, in a population of 5000 individuals, 500 will carry the advantageous allele and the likelihood of all of those organisms being destroyed is small. So there is a bigger chance of a potentially useful allele being maintained in the larger population. This is one reason why large, genetically diverse populations are needed to maintain biodiversity.

ISOLATION

If the Hardy–Weinberg genetic equilibrium is to be maintained, the population must exist in isolation. There should be no migration of organisms either into or out of the population. Of course, this is very rarely the case in the living world. Insects carry pollen from one population of flowers to another and the wind can carry it for miles. Male animals frequently leave their family groups and go in search of other populations to find a mate. Many simple organisms release their gametes directly into the water to be carried great distances before fertilisation occurs. In all of these cases, migration of genetic material into or out of the population is happening. Consequently **gene flow** occurs. This tends to make the different populations more similar, but constantly changes the allele frequencies within each individual population. When the gene flow between two populations is reduced, they become effectively isolated and this is when speciation is more likely to occur.

SELECTION PRESSURE

For Hardy–Weinberg equilibrium to apply, all alleles would have the same level of reproductive advantage or disadvantage, but we know this is not the case. Many alleles are neutral in their effect, but some alleles code for a phenotype which gives an advantage or disadvantage to the individual. If the environmental conditions change, a new selection pressure will be exerted, as you saw in the example of the rats and warfarin on **page 277**. As a result of the change in conditions, alleles which have been neutral in their impact may become advantageous or disadvantageous. Some of the individuals with different combinations of alleles will now have an advantage. They will be most likely to survive and pass on their successful alleles. The genetic make-up of the population will change over time. This is natural selection in progress. Selection pressure drives speciation as environments change.

CHECKPOINT

SKILLS ANALYSIS

1. In a population of plants, 95 out of 200 individuals express the homozygous recessive phenotype for hairy leaves. Calculate:
   (a) the percentage of the population that is homozygous for the recessive phenotype
   (b) the percentage of the population that is heterozygous
   (c) the frequency of homozygotes for the dominant phenotype in the population.
2. Grey fur in mice is recessive to brown fur. In a population of 150 mice, 126 of them have brown fur. Give the expected frequency of homozygous recessive, homozygous dominant and heterozygous individuals in the population of mice. Explain your workings.

SKILLS INTERPRETATION

3. Explain why the Hardy-Weinberg equilibrium is rarely observed in living organisms across the generations.

SUBJECT VOCABULARY

population a breeding group of individuals of the same species occupying a particular habitat and a particular niche
gene pool the sum total of all the genes in a population at a given time
selection pressure the effect of one or more environmental factors that determine whether an organism will be more or less successful at surviving and reproducing; selection pressure drives speciation
Hardy-Weinberg equilibrium the mathematical relationship between the frequencies of alleles and genotypes in a population; the equation used to describe this relationship can be used to work out the stable allele frequencies within a population
gene flow the migration of either whole organisms or genetic material into or out of a population and into another population, tending to make different populations more alike, but changing the allele frequencies within each individual population all the time

5 REPRODUCTIVE ISOLATION AND SPECIATION

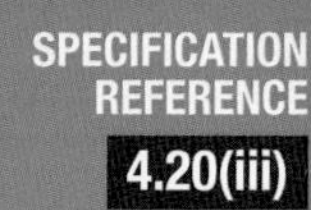

LEARNING OBJECTIVES

- Understand that reproductive isolation can lead to accumulation of different genetic information in populations, potentially leading to the formation of new species.

Changes in populations of organisms are very interesting and they can tell us a great deal about the process of natural selection. But a species is usually a much bigger entity than a single population, and often consists of a large number of populations spread across a country or even countries. So how is a new species formed?

ISOLATION AND SPECIATION

A species is a group of organisms sharing several features, which are capable of interbreeding to produce fertile offspring. **Speciation** is the formation of a new species. It happens as a result of the isolation of parts of a population. The important factor in the process is reproductive isolation, and the reduced gene flow between the different populations which is a consequence of that isolation. The two isolated populations experience different conditions, and this means that different selection pressures act on the different populations. As a result, natural selection acts in different directions on the two populations. Over a period of time, both the genotype and the phenotype of the isolated groups will change. This can continue to the extent where, even if members of the split population are reunited, they can no longer successfully interbreed.

Speciation can also occur as a result of **hybridisation** and this is particularly common in plants. Sometimes, two closely related species can breed and form fertile hybrids that are successful in their own right and may be better adapted to the niche. In some cases, these hybrids do not produce fertile offspring if they are crossed back to their parent plants, so a new species is formed. The new species may out-compete the parent plants. An example of this is seen in the UK with the hybrid plant formed between native English and imported Spanish bluebells.

ISOLATING MECHANISMS

For different species to evolve from an original species, different populations of the species usually have to become *reproductively isolated* from each other. This means that mating and, therefore, gene flow between them is restricted. There are several ways in which this can happen.

- Geographical isolation: a physical barrier such as a river or a mountain range separates individuals from an original population.
- Ecological isolation: two populations inhabit the same region, but develop preferences for different parts of the habitat.
- Seasonal isolation (also known as temporal isolation): the timing of flowering or sexual receptiveness in some parts of a population becomes different from the usual timing for the group. This can eventually lead to the two groups reproducing several months apart.
- Behavioural isolation: changes occur in the courtship ritual, display or mating pattern so that some animals do not recognise others as being potential mates. This might be due to a mutation that changes the colour or pattern of markings.
- Mechanical isolation: a mutation occurs that changes the genitalia of animals, so they can only mate successfully with some members of the group. Or it changes the relationship between the stigma and stamens in flowers, making pollination between some individuals unsuccessful.

Reproductive isolation is the most important factor in speciation.

ALLOPATRIC SPECIATION

Allopatric speciation occurs when populations become physically or geographically separated. Scientists recognise allopatric speciation as the main evolutionary process. Allopatric speciation is of enormous importance in the history of evolution, as great land masses moved and separated. The physical isolation of populations continues to occur as a result of natural changes, for example as islands form and disappear, as ice floes melt, rivers change course and lakes either dry up or appear – and so allopatric speciation continues to be very important. Some of the changes that result in allopatric speciation are the result of human interventions such as dams, roads and cities.

There are many examples of allopatric speciation. Some of the most striking are when organisms become completely isolated – for example when islands are formed. When a species evolves in geographical isolation and is found in only one place it is said to be endemic (see **Section 4C.1**).

ADAPTIVE RADIATION

Allopatric speciation is often followed by **adaptive radiation**. Adaptive radiation occurs when one species develops rapidly to form several different species, which all fill different ecological niches. There are some well-known examples of adaptive radiation.

AUSTRALIAN MARSUPIALS AND MONOTREMES

Australia is well known for its unusual fauna and flora. Perhaps most unusual are two groups of mammals, the **marsupials**, which protect their young in pouches, and the much rarer egg-laying **monotremes**. In the rest of the world, the **placental mammals** dominate. Until about 5.5 million years ago, Australia was joined to the rest of the world's land mass. At that time, the only mammals were marsupials and monotremes. After Australia separated from the other continents, the marsupials evolved to fill an enormous range of niches. These included the large herbivorous kangaroo with its wide-ranging niche and the koala with its eucalyptus tree niche. Others have carnivorous niches (e.g. the quoll and the Tasmanian devil). On other continents, placental mammals evolved and mostly replaced the marsupials and monotremes but these did not reach Australia until humans arrived, eventually bringing the other mammals with them.

DARWIN'S FINCHES

These birds provide a classic example of how different selection pressures result from the availability of a range of different niches and then lead to adaptive radiation. The finches were discovered by the great 19th century naturalist, Charles Darwin, on his voyage on HMS *Beagle*. On the Galapagos Islands near the equator, there are several feeding niches for birds (e.g. small seeds, large nuts, insects living in rotten bark). The original finches that arrived on the islands were of a single species. No one is quite sure how they got there because the islands are over 500 miles from land, but a small flock was probably carried there by a storm or a hurricane.

Within the birds that arrived at the islands, there would have been variation in alleles and characteristics. Different niches on the islands would have favoured individuals with different variations. For example, a bird with a slightly smaller, stronger bill would get more food by eating mainly seeds. This would enable it to thrive, reproduce and pass on its beak characteristics to its offspring. In the following generations, natural selection resulted in individuals with small strong beaks ideally adapted to eating seeds. Similarly, a finch with a longer, thinner beak might well be more successful probing dead wood for insects. It would therefore begin to feed almost exclusively in that way. By exploiting different niches, the finches avoided competing for the same scarce food resources. Over several million years, at least 14 different species of finch have emerged on the Galapagos Islands from one common ancestral species (see **fig A**).

Because food was such an important selection pressure, it was important to mate with a finch with a similarly shaped beak to pass on the advantageous characteristic. Mating with a finch that had a differently shaped beak would produce a variety of offspring less well adapted to feeding, so there was a selective pressure on choosing the right kind of mate. As a result, any phenotypic and behavioural changes that made choosing the right mate easier were also selected, so the different species look different. Although the finches specialise and feed on particular types of food, and vary considerably in size and appearance, DNA analysis has shown that genetically they are remarkably similar.

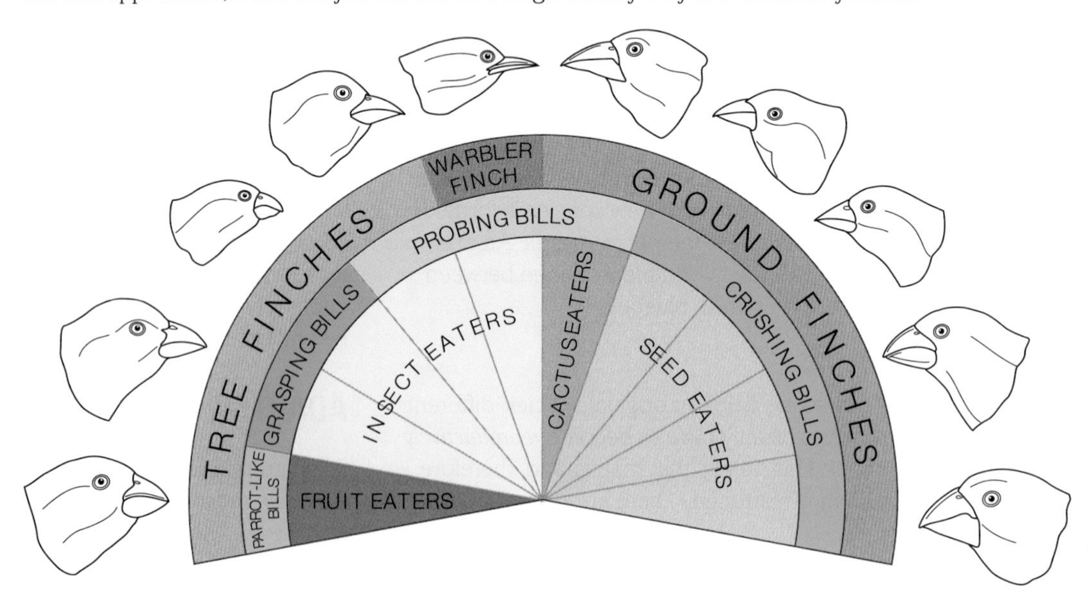

▲ **fig A** Over several million years, at least 14 species of finch developed from the original ancestor species. The anatomy of the beaks and adaptations for feeding of 10 of these finches is shown in this diagram.

LEARNING TIP

Sym means same (similar to simultaneous), so sympatric means same place.

SYMPATRIC SPECIATION

Sympatric speciation occurs between populations of a species living in the same place that become reproductively isolated by mechanical, behavioural or seasonal changes. Some gene flow continues as speciation happens – a very different model to allopatric speciation. Sympatric species are closely related and occupy overlapping ranges. Many scientists are reluctant to classify examples of sympatric speciation. They suggest that, when speciation occurs, there is always an adaptive pressure, including previously unknown microhabitats, which drive the formation of two species and produce a barrier between the populations. DNA evidence often shows that species which were originally thought to come from sympatric speciation actually show evidence of cross-breeding or geographical or niche separation at some point.

POPULATION BOTTLENECKS

A large population is needed to maintain a large and diverse gene pool. Even within a large population, allele frequencies are affected by factors such as mutations and non-random mating. If the size of the population is greatly reduced, so is the gene pool, and allele frequencies can change significantly.

The size of a population may be dramatically reduced by an environmental disaster, a new disease, hunting by humans or other very efficient predators, or habitat destruction. This is called a **population bottleneck** and it causes a severe decrease in the gene pool of the population. Many of the gene variants present in the original population are lost, so the gene pool shrinks and the allele frequency changes dramatically. In almost all cases, genetic diversity is greatly reduced.

After a catastrophic event, the remaining small population is vulnerable to the complete loss of some alleles, and a single mutation or new individual can have a bigger effect than usual as a result. As the population recovers, it may become so different from the original population genetically that it becomes a new species.

Cheetahs are the fastest land animals, capable of bursts of speed of up to 95 km hr^{-1}. Unfortunately, as you saw in **Section 4C.2**, they have very little genetic diversity. At the end of the last Ice Age, many of the largest mammalian species, including woolly mammoths, cave bears and giant deer, became extinct. DNA evidence shows that cheetahs may well have come very near to extinction at the same time. Although cheetah population numbers have recovered to some extent, their genetic diversity has not. The gene pool is very small because they are all descended from this ancestral population bottleneck. All cheetahs have about 99% of their alleles in common; there is little genetic diversity and the allele frequency is static. This means fertility is low and they are very vulnerable to any environmental change, such as climate changes or a new disease. The northern elephant seals shown in **fig B** show a similar effect but, in this case, the population bottleneck was the result of human actions.

THE FOUNDER EFFECT

The **founder effect** is the loss of genetic variation that occurs when a small number of individuals leave the main population and set up a separate new population, producing a voluntary population bottleneck. The alleles carried by the individuals who leave the main population are unlikely to include all the alleles, or at the same frequencies, as the original population. Any unusual genes in the founder members of the new population may become more frequent as the population grows.

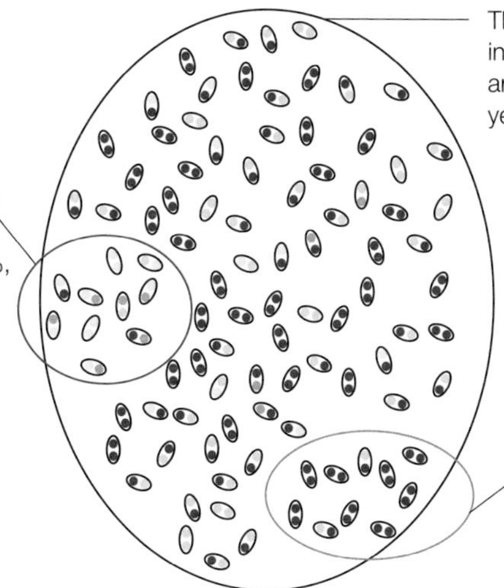

fig C A model of the founder effect

DID YOU KNOW?
HUMAN-IMPOSED BOTTLENECKS

Heavy hunting of northern elephant seals meant that by the end of the 19th century there were only about 20 individuals left. Their population has recovered to over 30 000 but their allele frequency remains low (see fig B). They have far less genetic diversity than southern elephant seals, which were much less intensively hunted and so have retained a much bigger gene pool.

fig B These northern elephant seals are all descended from around 20 individuals who survived human hunters. Consequently, the gene pool is very small even though the numbers are now quite large.

The founder effect is demonstrated clearly among the Amish, an American religious group who live in three isolated communities. One of the groups has a high frequency of a very rare genetic disorder called Ellis–van Creveld syndrome. This is because among the founder members were a married couple one of whom carried the gene. More cases of Ellis–van Creveld syndrome have been found in this one small population than in the whole of the rest of the world, providing a clear demonstration of the role of the founder effect in bringing about a dramatic change in allele frequencies in a population (see **Did you know?** below). The current population's gene pool is significantly different from that of the original population.

LEARNING TIP

Be clear about the difference between a population bottleneck and the founder effect. Both cause a limited gene pool, which means the population has allele frequencies that are different from the original or main population. A population bottleneck is imposed on a population by the loss of most members of the population through natural events or human intervention. The founder effect is the result of the isolation of a small group of organisms from the original population.

DID YOU KNOW?

THE FOUNDER EFFECT IN THE USA

Mr and Mrs Samuel King moved to Pennsylvania in 1744 as members of a small group of 200 people who founded an Amish community. They could have had no idea of the legacy they took with them.

One of them was heterozygous for Ellis–van Creveld syndrome. This rare genetic disease results in a type of dwarfism where the limbs are shortened, there are often extra fingers and toes, and people often die young as a result of associated heart problems. The Kings produced many children, who also had many children. This raised the frequency of the allele within the population to high above its normal level. Purely by chance, at least one other member of the founding group was heterozygous for this recessive gene. As a result of interbreeding in this isolated community, 1 in 14 individuals in the population now carry the gene. This leads to a distressingly high number of affected births. In 1964, there were 43 cases of Ellis–van Creveld syndrome in a population group of 8000. In the general population, the incidence of this genetic condition can be as low as 1 in 200 000 births.

CHECKPOINT

SKILLS ANALYSIS

1. If 43 individuals with Ellis-van Creveld syndrome are found in a population of 8000, use the Hardy-Weinberg formula to work out the frequency of the Ellis-van Creveld allele in the population and the numbers with the homozygous and heterozygous forms of the dominant phenotype.

SKILLS INTERPRETATION

2. Explain the difference between a population bottleneck and the founder effect.
3. The lions of the Ngorongoro crater in Tanzania are isolated from other populations of lions. In 1962, the population was almost wiped out by disease and only nine females and one male were left. The current population of crater lions is 75–125 animals. They have much lower genetic diversity than lions in the much bigger populations of the nearby Serengeti plains, and have problems of low fertility. What do these observations suggest about the gene pool and allele frequencies of these two populations of lions?

SUBJECT VOCABULARY

speciation the formation of a new species
hybridisation the production of offspring as a result of sexual reproduction between individuals from two different species
allopatric speciation speciation that occurs when populations are physically or geographically separated and there can be no interbreeding or gene flow between the populations
adaptive radiation a process by which one species develops rapidly resulting in several different species which fill different ecological niches
marsupials mammals that give birth to very immature young and then protect them in pouches
monotremes primitive mammals that lay eggs and feed their offspring with milk from mammary glands
placental mammals mammals that provide for the developing fetus during gestation through a placenta
sympatric speciation speciation that occurs between populations of a species in the same place; they become reproductively separate by mechanical, behavioural or seasonal mechanisms; gene flow continues between the populations to some extent as speciation occurs
population bottleneck the effect of an event or series of events that dramatically reduces the size of a population and causes a severe decrease in the gene pool of the population, resulting in large changes in allele frequencies and a reduction in genetic diversity
founder effect the loss of genetic variation that occurs when a small number of individuals become isolated, forming a new population with allele frequencies not representative of the original population

4C 6 CONSERVATION: WHY AND HOW?

SPECIFICATION REFERENCE 4.15 4.21

LEARNING OBJECTIVES

- Know that over time the variety of life has become extensive but is now being threatened by human activity.
- Be able to evaluate the methods used by zoos and seed banks in the conservation of endangered species and their genetic diversity, including scientific research, captive breeding programmes, reintroduction programmes and education.

The human population of the planet is over 7 billion and growing (see **fig A**). Every individual contributes some greenhouse gases to the atmosphere through activities such as breathing, using electricity or driving a car. Growing food such as rice, and keeping cattle for beef and milk also contribute. Everyone also produces bodily waste. It is estimated that over 200 million tonnes of human waste are left untreated every year and much of that goes into our seas, oceans, rivers and lakes. The pollution produced has a massive effect on ecosystems all around the world. And there are many other ways in which people affect their environment.

All over the world, people are constantly taking resources from the biosphere. Global ecosystems supply food, water, building materials, clothing, medicines and more for the global population. As the human population gets ever larger and we take more from the environment than we need, biological resources are being increasingly depleted. The extinction of species and loss of biodiversity in many areas of the world are clear examples of human influence on a wide range of ecosystems.

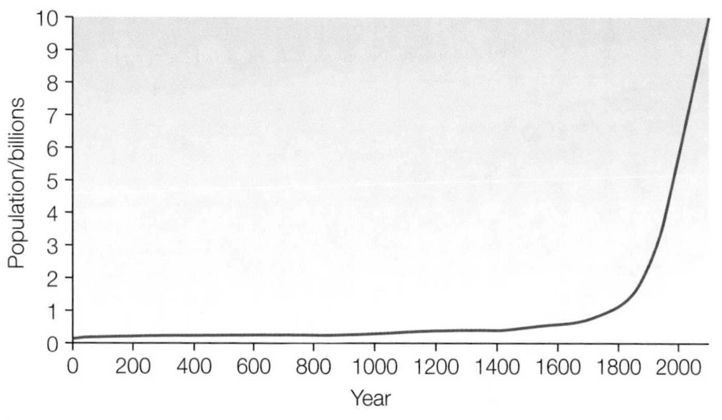

▲ **fig A** This graph shows the enormous increase in world population in recent centuries. This increase has been built in part on fossil fuels (see below) and so it is not sustainable.

THE HUMAN POPULATION EXPLOSION

As people learned to farm plants and animals to provide a reliable supply of food, so more children survived and populations grew, but on a fairly small and local scale. Later, tools and then machines enabled us to farm on a much bigger scale. Now we can change the environment with reservoirs, roads, canals, towns and cities (see **fig B**). We have developed medicines that keep us and our children alive and allow many of us to survive to great age. We have also developed engines that burn fossil fuels and release large quantities of exhaust gases into the atmosphere. We cut down huge areas of rainforest to grow crops or rear cattle. We have developed floating factories that can clear fish from whole areas of the oceans – very different from local fishing boats that provide for a family, a village or a local market.

All of these activities have been driven by basic desires to provide ourselves with food, shelter and goods. We appear to be having a major impact on many of the ecosystems of the Earth as an unintended consequence. Scientists have increasing evidence that we are affecting the global climate, biological resources and biodiversity around the world.

▲ **fig B** A city ecosystem is very different from a natural ecosystem. This shows Nairobi, the largest city in Kenya.

THE HUMAN THREAT TO BIODIVERSITY

Human effects on ecosystems are many and widespread. Biodiversity is essential at both the species and the genetic level. The extensive variety of life, which has evolved over millions of years, is now being threatened in many different ways because of human activities.

CLIMATE CHANGE

From individual observations to major scientific studies there is growing agreement that the Earth's climate is changing. Global temperatures are rising. This is seen in overall climate patterns and in the increasing number and frequency of 'extreme weather events' all over the world.

The world's climate has changed regularly over time. The fossil record shows us how often the world has gone through ice ages and periods of tropical heat or desertification. The difference is that this time, it is happening quickly and there is a growing body of evidence to suggest the changes are the direct result of human activities. These changes in climate mean that many plants and animals will no longer survive.

DEPLETION OF BIOLOGICAL RESOURCES

A growing world population means an increase in the demand for resources such as food, firewood and land. In the more economically developed parts of the world, people have gone way beyond simply fulfilling the basic needs of life. They eat far more food than they need and want a great variety of different foods, which may need to be imported from far away. As a result, biological resources are being depleted and ecosystems destroyed, both on land and in the oceans. Biodiversity is destroyed along with ecosystems.

CONSERVATION

Biodiversity is being reduced through activities such as overfishing, habitat reduction and as a result of climate change. However, many people around the world are working hard to reduce these trends and to conserve the biodiversity we still have. This can be tackled in many different ways.

Conservation means keeping and protecting a living and changing environment. It is an active process involving an enormous range of projects. Examples of conservation include:

- reclaiming land after industrial use
- helping to set up sustainable agriculture systems in the developed world
- protection of a single threatened species
- global legislation on pollution levels and greenhouse gas emissions.

Around the world, there are many animals and plants that are threatened with extinction. This often, but not always, results from human activities which cause habitat loss or climate change. There are two main ways of conserving animals and plants. We can conserve them outside their natural habitat, in zoos or seed banks. This is called ***ex-situ* conservation**. The alternative is ***in-situ* conservation**, which takes place in the natural habitat of the organism.

EX-SITU CONSERVATION

The United Nations Convention on Biological Diversity in 1992 defined *ex-situ* conservation as 'the conservation of components of biological diversity [living organisms] outside their natural habitats'. Sometimes, when an organism is threatened with extinction, there is not time to conserve their habitat or protect them *in situ* (on site). It is sometimes possible to conserve the species by removing some of the animals or plants from their natural habitat. At worst, this enables their genetic material to be conserved and at best a breeding population can eventually be returned to their natural habitat. *Ex-situ* conservation is always seen as a complementary approach to *in-situ* conservation, and ideally it occurs in the country where the threatened species originates.

EX-SITU CONSERVATION OF PLANTS

It has been estimated that 25% of the world's flowering plant species could disappear within the next 50 years. There are thought to be about 242 000 species of flowering plants now, so this would mean 60 500 species disappearing in less than one human lifetime.

Plants are of vital importance to all our lives. The genetic material of these extinct species would be lost forever. This would be a disaster for the plants and possibly also for human survival. Cross-breeding crop plants back to original wild plants, or using wild plants to supply genes for genetic engineering, are ways in which the long-term health of our crop plants can be maintained.

Botanic gardens (zoos for plants) maintain collections of many of the world's most interesting and unusual plants. In the 1960s, with a view to conservation, the Royal Botanic Gardens at Kew, UK, set up a seed bank that is now home to the seeds of around 80 000 plants (see **fig C**). It is called the Millennium Seed Bank and had two main aims. The first was to collect and conserve the seeds of the entire UK native flora – this was completed by 2009. Of the 1442 native UK plants, over 300 are already threatened with extinction. The second aim was to conserve the seeds of an additional 25% of the flora of the whole world by the year 2020, focussing particularly on the drylands (arid, subarid and subhumid regions) which are experiencing some of the most rapid loss of habitat. The drive to save and store seeds from as many plants as possible under ideal conditions has spread much farther than the UK. A seed bank was set up in Sri Lanka in 2011, and in 2017 the Abu Dhabi Plant Genetics Resources Centre was established to collect and document native plants from the United Arab Emirates.

▲ **fig C** Seeds like these, stored at low temperatures, will still be able to germinate and grow in hundreds of years – by which time it is hoped their habitats will have been restored and conserved.

A seed bank can preserve many plants in a state of effective suspended animation. Live seeds are collected from the wild, removed from the fruits and cleaned. They are screened using X-rays to be sure that they contain fully developed embryos. Then they are dried, put into jars and stored at between –20 and –40 °C. Many will survive and remain capable of germinating for up to 200 years. In general, the lifespan of a seed doubles for every 5 °C drop in temperature or 2% fall in relative humidity. Some of the seeds stored may even germinate in several thousand years' time.

Most plants make huge numbers of seeds, so they can be collected without damaging the natural population. Seeds are usually small, so large numbers of them can be stored quite cheaply in a small space. They contain all the genetic material of the plant, so they are a record of the genetic make-up of the species as well as a potential new plant for the future. There are now over 1000 seed banks around the world.

About 80% of the known species of plant could be stored in seed banks, but the seeds of some species do not store well. Unfortunately these include many crop plants such as mango, rubber, oak, avocado, cacao and coconut. These plants need to be conserved differently. They may be grown where they are found naturally, in field gene banks such as plantations, orchards and arboretums or as tissue cultures. In this way, the species is grown on, year after year. One problem is that field gene banks take up a lot of room and a lot of work. For example, the world potato collection at the International Potato Centre in Peru contains around 4100 different clones of potato, which all need to be planted annually. Using tissue cultures to conserve plants and growing plants on as needed takes up a lot less space and time, and allows more variety to be conserved.

EX-SITU CONSERVATION OF ANIMALS

It is not always possible to conserve animal species in the wild because the conditions that have put them under threat of extinction continue. Zoos and wildlife parks used to exist just for people to look at the animals, but today they are very important in animal conservation. In **captive breeding programmes** individuals of an endangered species are bred in zoos and parks in an attempt to save the species from extinction. Usually, the ultimate aim is to reintroduce the captive-bred animals into the wild to restore the original populations.

Reintroduction does not always work, but it can be successful in national parks or other protected areas. Species that have been saved by captive breeding and successfully reintroduced into protected areas in their own countries include Californian condors and Przewalski's horses. Captive breeding programmes for the white and black rhino, along with much conservation work in east Africa, gives hope that these amazing creatures will also be saved from extinction (see **fig D**).

▲ **fig D** The Mkomazi Game Reserve Rhino Sanctuary in Tanzania has been established to help build up the population of black rhinos and protect them from the poachers who have hunted them almost to extinction.

There are several problems with captive breeding and reintroduction programmes.

- There is not enough space or sufficient resources in zoos and parks for all the endangered species.
- It is often difficult to provide the right conditions for breeding, even if scientists know what those conditions are. For example, it is well known that the giant panda is difficult to breed even when conditions are ideal.
- Reintroduction to the wild will be unsuccessful unless the original reason for the species being pushed to the edge of extinction is removed.
- Animals that have been bred in captivity may have great problems in adjusting to unsupported life in the wild.
- When the population is small, the gene pool is reduced, and this can cause serious problems. Zoos try to overcome this by keeping detailed records of the genetic data of their breeding individuals. Sperm can be swapped with other zoos (for artificial insemination) to maximise genetic variation in the offspring.
- Reintroduction programmes can be very expensive and time-consuming, and they may fail.

EXAM HINT

You are likely to be asked about the difficulties faced by people who are trying to conserve species and habitats. Make sure you show that you understand that conservation is not an easy thing to do.

DID YOU KNOW?

The Frozen Ark project was set up in 1996 and plans to conserve the genetic resources of animals by conserving their DNA and viable cells. It is the animal equivalent of the Millennium Seed Bank. The Frozen Ark aims to save DNA samples from endangered species. DNA is very stable and can be stored frozen for hundreds of years. The 10 million samples take up very little room as each is very small. This gives scientists the possibility of increasing genetic variation in critically endangered species or even cloning extinct species once their habitat has been restored and conserved. The Frozen Ark does not replace *in-situ* and *ex-situ* conservation – it supports them.

SUSTAINABILITY

In an ideal world, we would not need seed banks, zoos and other methods of conserving biodiversity. People are increasingly trying to find ways to prevent the loss of biodiversity in the first place. Habitats and ecosystems can be conserved with less conflict by encouraging sustainable methods of land use. For example, illegal logging operations in rainforests use 'slash and burn' techniques (cutting down all the trees and burning the ground afterwards) to harvest wood to sell and clear the soil for farming. The soil is soon exhausted and biodiversity lost. However, if we harvest the trees selectively and replant for the future, biodiversity can be maintained while people continue to use the forest for income. This is sustainable forestry.

Sustainable agriculture includes farming methods that minimise damage to the environment and avoid monoculture. These are becoming increasingly important around the world. They involve using organic fertilisers where possible, minimising the use of artificial fertilisers and chemical pesticides, using biological pest control, maintaining hedgerows and planting in rotation to avoid the soil becoming exhausted. Large-scale farming is vital to provide the food we need but often sustainable methods such as using biological pest control can increase yields and improve profits while being cheaper in the long term and better for the environment than using expensive manufactured chemicals.

It is important to get our priorities right. People and politicians often have limited sympathy for ecology, so it is important to target spending accurately. Around the world, there is a growing understanding of the need for sustainable agriculture and sustainable tourism to conserve biodiversity while still providing the food and income that people need. Research continues into how food and other resources can be produced in a way that minimises loss of biodiversity, and even increases it again. Tourism can also be developed in a way that is sustainable, does minimal damage to the environment, provides jobs and money for local people and also conserves the environment. This maintains biodiversity at the same time.

Costa Rica is a good example of a country which has reversed its habitat loss and is working hard to conserve its rich biodiversity. More than 25% of the country is now protected in some way, and most of the electricity is generated using renewable resources. Costa Rica now earns a considerable proportion of its national income from sustainable ecotourism. From the President to the children in the schools, people in Costa Rica have become passionate about conserving the rich biodiversity of their country. As a result, ecotourism is growing, but in a managed way which does not destroy the biodiversity people want to see. The ecotourism industry now provides many people with jobs and contributes around 6% of the national GDP. This small country shows that it is possible to use natural resources to make money and conserve biodiversity. You can see some examples of Costa Rican biodiversity in **fig E**.

LEARNING TIP

Remember that sustainable conservation must work for the local people, and works best if local people are involved.

(a) **(b)**

(c)

▲ **fig E** Costa Rica covers only about 0.03% of the surface of the Earth, but it has more than 4% of the world's biodiversity. The country is determined to maintain this biodiversity. **(a)** Green basilisk lizard; **(b)** red-eyed tree frog; **(c)** coral palm.

THE IMPORTANCE OF EDUCATION

People have to work hard to reduce pollution and to support conservation programmes. To help them understand why it is important to conserve biodiversity they need to learn about:

- the impact of human activities on the natural world
- ways in which people can act to protect animals, plants and habitats.

Education is very important. When people learn about the damage they are doing to the environment, they often want to change. When they discover the great range of biodiversity in the world, and how it can be conserved and protected, they usually want to help. Zoos, national parks and seed banks are all important in educating people. They allow people to see animals and plants from other parts of the world. They explain different environments and how they are threatened by human behaviour. They show how threatened species can be protected.

Maintaining biodiversity is a major issue for the 21st century. Success or failure will affect the whole planet, and the potential consequences of failure could be devastating for everyone. In a global economy, we can all contribute to maintaining biodiversity by the choices that we make now.

DID YOU KNOW?

RESTORING THE FORESTS IN ETHIOPIA

Professor Legesse Negash is a pioneer in the propagation of Ethiopia's native trees (see fig F). Like so many African countries, Ethiopia has tremendous natural biodiversity that is threatened by both possible climate change and deforestation. By finding ways to propagate some of the indigenous trees, and storing the germlines in seed and tissue banks, Legesse Negash aims to protect the soil against erosion and help to save the biodiversity of the country. He has set up the Center for Indigenous Trees Propagation and Biodiversity Development in Ethiopia. The centre produces thousands of young trees, and also aims to educate farmers and train new young biologists to work in this important area. In this way, new trees can be planted when the older ones are harvested.

fig F Professor Legesse Negash planting a keystone indigenous tree species (*Ficus vasta Forssk*) at his Center for Indigenous Trees Propagation and Biodiversity Development in Ethiopia.

CHECKPOINT

SKILLS INTERPRETATION

1. Preservation means preserving something exactly as it is now. Explain how conservation differs from preservation.
2. Explain the difference between *ex-situ* and *in-situ* conservation.
3. Explain why it is much easier to conserve plant biodiversity than animal biodiversity.
4. Discuss the main advantages and disadvantages of captive breeding and reintroduction programmes.

SUBJECT VOCABULARY

conservation maintaining and protecting a living and changing environment
***ex-situ* conservation** the conservation of components of biological diversity (living organisms) outside their natural habitats
***in-situ* conservation** the conservation of ecosystems and natural habitats, and the maintenance and recovery of viable populations of species in their natural surroundings
captive breeding programmes programmes where individuals of an endangered species are bred in zoos and parks in an attempt to save the species from extinction, and if possible to reintroduce them to their natural wild environment

NEW BUT DISAPPEARING FAST

CRITICAL THINKING, PROBLEM SOLVING, ANALYSIS, DECISION MAKING, CREATIVITY, INNOVATION, PERSONAL AND SOCIAL RESPONSIBILITY, CONTINUOUS LEARNING, INTELLECTUAL INTEREST AND CURIOSITY, ETHICS

After years of speculation about some unusual orang-utans living in an area of northern Sumatra, scientists have identified a new species of great ape. Despite similarities to other orang-utans, it is a separate species. The new species lives in three distinct populations only 100 km from larger populations of the more common Sumatran orang-utan. However, there are only about 800 individuals of this new species and scientists are concerned about their continued survival. The following extract is part of the discussion from the scientific paper about the discovery of this new great ape.

SCIENTIFIC PAPER

Discussion

Due to the challenges involved in collecting suitable specimens for morphological and genomic analyses from critically endangered great apes, our description of *P. tapanuliensis* had to rely on a single skeleton and two individual genomes for our main lines of evidence. When further data become available, a more detailed picture of the morphological and genomic diversity within this species and of the differences to other *Pongo* species might emerge, which may require further taxonomic revision. However, it is not uncommon to describe species based on a single specimen and, importantly, there were consistent differences among orang-utan populations from multiple independent lines of evidence, warranting the designation of a new species with the limited data at hand.

With a census size of fewer than 800 individuals, *P. tapanuliensis* is the least numerous of all great ape species. Its range is located around 100 km from the closest population of *P. abelii* to the north (Figure 2A). A combination of small population size and geographic isolation is of particularly high conservation concern, as it may lead to inbreeding depression and threaten population persistence. Highlighting this, we discovered extensive runs of homozygosity in the genomes of both *P. tapanuliensis* individuals (Figure S3), pointing at the occurrence of recent inbreeding.

To ensure long-term survival of *P. tapanuliensis*, conservation measures need to be implemented swiftly. Due to the rugged terrain, external threats have been primarily limited to road construction, illegal clearing of forests, hunting, killings during crop conflict, and trade in orang-utans. A hydroelectric development has been proposed recently in the area of highest orang-utan density, which could impact up to 8% of *P. tapanuliensis's* habitat. This project might lead to further genetic impoverishment and inbreeding, as it would jeopardize chances of maintaining habitat corridors between the western and eastern range (Figure 1A), as well as smaller nature reserves, all of which maintain small populations of *P. tapanuliensis*.

From: Alexander Nater, et al. Morphometric, behavioral, and genomic evidence for a new orangutan species. *Current Biology* November 2017. http://www.cell.com/current-biology/fulltext/S0960-9822(17)31245-9

SCIENCE COMMUNICATION

SKILLS: ANALYSIS, INTERPRETATION, COMMUNICATION, CREATIVITY

The article comes from a scientific paper. The intended audience is other scientists.

1 (a) Select three phrases from the text that indicate the intended audience is not the general public.
(b) Look up the definition of these phrases and write them down.

2 Re-write the first paragraph of the article in a style that is more suitable for members of the public.

BIOLOGY IN DETAIL

You should be able to answer all the questions here - but you may wish to come back to them when you have completed more of your course.

3 List the threats to the survival of *P. tapanuliensis* described in the article.

4 Explain how it is possible that *P. tapanuliensis* has evolved as a separate species even though there are other orang-utans only 100 km away.

5 The article states that there is evidence of recent inbreeding. If conservation efforts are successful, what would be the effect of inbreeding as the population increases?

ACTIVITY 1

Consider all the risks to the survival of *P. tapanuliensis*. Don't forget that internal factors such as genetic diversity are important as well as external factors.

Devise a conservation plan to ensure its continued survival. Include both *in-situ* and *ex-situ* conservation techniques.

Remember to carefully evaluate the relative importance of each technique you suggest: how important it is and how likely it is to succeed.

WRITING SCIENTIFICALLY

Evaluate means review information and then form a conclusion, drawing on evidence including strengths, weaknesses, alternative actions, relevant data or information. You should come to a supported judgement of a subject's qualities. Remember to consider and refer to context.

ACTIVITY 2

Is it important to conserve *P. tapanuliensis* as a separate species?

Prepare for a debate with other members of your class. Divide the class into two groups. One group should argue for the continued survival of the species. The other group argue that its continued survival is not important.

Members of each group could adopt different roles such as: a geneticist, a conservationist, a local inhabitant, a representative from the hydroelectric company, a logger, and so on. Each student could prepare their argument and then carry out a debate during the next lesson.

4C EXAM PRACTICE

1 Biodiversity is an important concept in conservation.

(a) The diagrams below show four identically sized areas A, B, C and D. Different shapes represent different species.

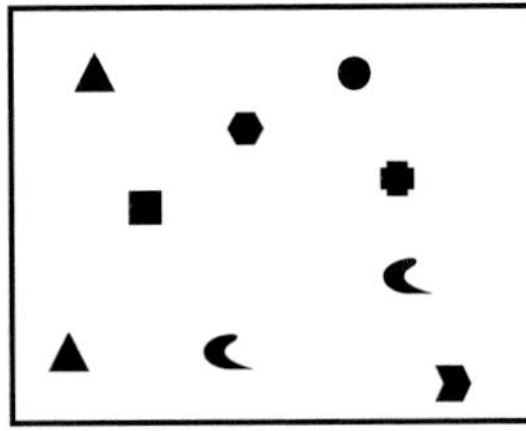

Area A

Area B

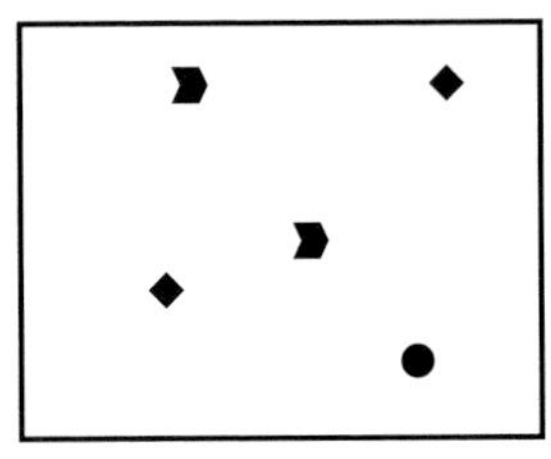

Area C

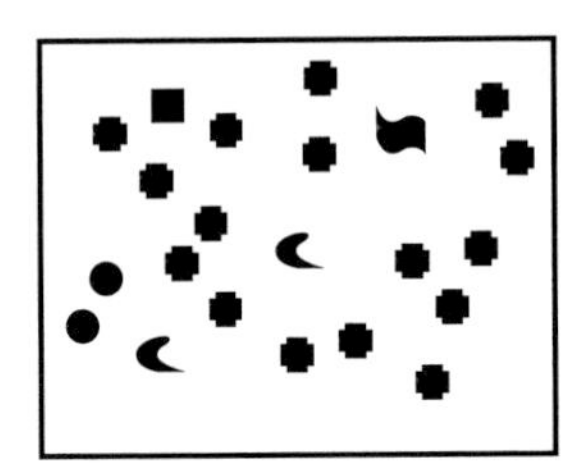

Area D

(i) Which area has the highest species richness?

A A
B B
C C
D D [1]

(ii) Which area contains an endemic species?

A A
B B
C C
D D [1]

(b) An area that is species rich may not always be considered to have high biodiversity. Explain the difference between species richness and biodiversity. [3]

(c) Zoos help to conserve species by having captive breeding programmes and reintroduction programmes. Describe how zoos use these programmes to help conserve rare species. [5]

(Total for Question 1 = 10 marks)

2 Plants like daisies, dandelions, clover and yarrow have broad leaves. They can be regarded as weeds on school playing fields as they compete with the narrower-leaved grasses like *Agrostis* and *Festuca*.

Selective herbicides can be used to kill the broad-leaved plants to allow the grass to grow better. These herbicides can take up to a week to have their effect.

A scientist surveyed the biodiversity of plants growing on a school playing field. One hundred quadrats (sampling areas) were used and the results aggregated.

These aggregated results were then put into a table and used to calculate an index of diversity.

Species	Number of plants growing in 100 $0.25\,m^2$ quadrats (n)	$n - 1$	$n(n - 1)$
daisy (*Bellis perennis*)	3000		
dandelion (*Taraxacum officinale*)	2000		
red clover (*Trifolium pratense*)	3500		
yarrow (*Achillea millefolium*)	2000		
grass (*Agrostis spp*)	40 000		
grass (*Festuca spp*)	4500		
bee orchid (*Ophrys apifera*)	1		
Total number of organisms (N)			

(a) (i) Complete the table by filling in the missing boxes. [3]

(ii) Use the information in the table to calculate the index of diversity given by the following formula:

$$D = \frac{N(N-1)}{\sum n(n-1)}$$ [3]

(b) A rare plant called a bee orchid (*Ophrys apifera*) was found in one part of the field.

(i) Assess whether 100 samples in this field was a large enough sample. [2]

(ii) Explain why the occurrence of this one bee orchid made no difference to the biodiversity as measured by the index of diversity. [1]

(c) Selective herbicides were sprayed on the playing field, and the scientist came back a week later to see how this treatment affected the biodiversity. Explain why the recalculated index of diversity was found to be 2.26. [3]

(Total for Question 2 = 12 marks)

3 (a) In the Hardy–Weinberg equation, $p^2 + 2pq + q^2 =$

A 0
B 1
C 0.5
D 2 [1]

(b) In the Hardy–Weinberg equation p^2 represents the:

A frequency of individuals in a population that have the dominant phenotype

B frequency of individuals in a population that have the homozygous dominant genotype

C frequency of individuals in a population that carry a dominant allele

D frequency of individuals in a population that have the homozygous recessive genotype. [1]

(c) Maple syrup urine disease (MSUD) is a recessive phenotype caused by a mutated allele. Individuals who have this disease cannot metabolise three amino acids (leucine, isoleucine and valine). They have neurological damage and urine with the characteristic maple syrup smell.

In a human population 1 out of 200 individuals are born with MSUD.

(i) Using the Hardy–Weinberg equation, where p represents 'normal' and q is the MSUD allele, calculate the values of p and q to three significant figures. [2]

(ii) Using your values for p and q, calculate the number of individuals expected in a population of 100 000 who are:

(I) homozygous dominant [2]

(II) heterozygous [2]

(III) homozygous recessive. [2]

(iii) Some Amish populations in North America have an MSUD allele frequency (q) of 0.0754. Suggest reasons for the differences in the occurrence of MSUD in both the general and the Amish populations. [4]

(Total for Question 3 = 14 marks)

4 Sickle cell disease is a recessive genetic condition which results in the deformation of red blood cells. An amino acid substitution, from glutamic acid to valine, occurs in the β-chain of haemoglobin, and at low oxygen levels the haemoglobin crystallises resulting in the red blood cell changing its shape.

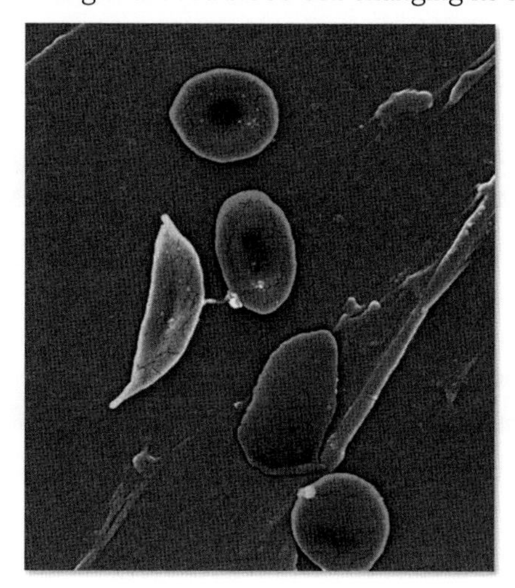

(a) Suggest what caused the amino acid substitution in the β-chain of haemoglobin. [1]

(b) Sickle cell anaemia is a serious condition causing oxygen deprivation and severe pain. Children who are homozygous for the sickle cell allele are likely to die before they are five years old unless treated. The allele is carried at various frequencies within different populations of humans as shown in the table.

Group of humans	Frequency of allele in population
African	1 in 3
African descent in America	1 in 13
Northern European	1 in 63

(i) Using the Hardy–Weinberg equation, calculate the percentage of people who might be expected to have sickle cell anaemia (homozygous for the sickle cell allele) in the African population. [2]

(ii) People who have at least one copy of the sickle cell allele have some protection against malaria. Untreated malaria can be fatal, as can sickle cell disease. Using this information and your own knowledge, explain why individuals with a heterozygous genotype are more common in regions where malaria occurs. [3]

(Total for Question 4 = 6 marks)

5 The mountain gorilla is an endangered species. There are about 800 individuals which live in two separate populations.

(a) Population bottlenecks are a threat to populations because:

A there are more individuals able to breed

B all allele frequencies are increased

C some alleles are likely to disappear from the population

D all allele frequencies are decreased. [1]

(b) An increase in which of these would not cause a population bottleneck?

A disease

B poachers

C predators

D available food [1]

(c) Suggest why such a small population size puts these animals at risk. [2]

(d) Describe three ways in which the local people could help to conserve the mountain gorilla. [3]

(e) Suggest how zoos could help in the conservation of mountain gorillas. [3]

(f) Explain why seed banks can be a more effective way to conserve plant species than botanic gardens. [4]

(Total for Question 5 = 14 marks)

MATHS SKILLS

In order to be able to develop your skills, knowledge and understanding in Biology, you will need to have developed your mathematical skills in a number of key areas. 10% of the marks in your exams will be for level 2 or higher maths. This is maths where you will need to make some kind of decision such as what equation to use or select what data to use from a graph or some other source. The maths will be tested in a biological context. Marks will be awarded for the correct answer, but you may also gain marks for showing your working – so always remember to set out your maths calculations clearly. This section gives more explanation and examples of some key mathematical concepts you need to understand. Further examples relevant to your International AS/A level Biology studies are given throughout the book.

ARITHMETIC AND NUMERICAL COMPUTATION

USING STANDARD FORM

Dealing with very large or small numbers can be difficult. To make them easier to handle, you can write them in the format $a \times 10^b$. This is called standard form.

To change a number from decimal form to standard form:

- Count the number of positions you need to move the decimal point by until it is directly to the right of the first number which is not zero.
- This number is the index number that tells you how many multiples of 10 you need. If the original number was a decimal, your index number must be negative.

Here are some examples:

DECIMAL NOTATION	STANDARD FORM NOTATION
0.000 000 012	1.2×10^{-8}
15	1.5×10^{1}
1000	1×10^{3}
3 700 000	3.7×10^{6}

USING RATIOS, FRACTIONS AND PERCENTAGES

Ratios, fractions and percentages help you to express one quantity in relation to another with precision. Ratios compare like quantities using the same units. Fractions and percentages are important mathematical tools for calculating proportions.

RATIOS

A ratio is used to compare quantities. You can simplify ratios by dividing each side by a common factor. For example 12 : 4 can be simplified to 3 : 1 by dividing each side by 4.

WORKED EXAMPLE

Divide 180 into the ratio 3 : 2

Our strategy is first to work out the total number of parts. Then divide 180 by the number of parts to find the value of one part.

Total number of parts = 3 + 2 = 5

Value of one part = 180 ÷ 5 = 36

Answer = 3 × 36 : 2 × 36 = 108 : 72

Check your answer by making sure the parts add up to 180:

72 + 108 = 180

EXAM HINT

You may be asked to work out the ratio of offspring from a genetic cross. This is level 2 maths because you need to calculate the ratio from a genetic diagram.

FRACTIONS

When using fractions, make sure you know the key strategies for the four operators.

Remember:

- the denominator is the number under the line in a fraction
- the numerator is the number over the line in a fraction
- the operator is what the fraction is doing (adding, subtracting, multiplying or dividing).

To add or subtract fractions, find the lowest common multiple (LCM) and then use the golden rule of fractions. The golden rule states that a fraction remains unchanged if the numerator and denominator are multiplied or divided by the same number.

WORKED EXAMPLE

$$\frac{1}{2} + \frac{1}{5} = \frac{5}{10} + \frac{2}{10} = \frac{7}{10}$$

To multiply fractions together, simply multiply the numerators together and multiply the denominators together.

WORKED EXAMPLE

$$\frac{2}{7} \times \frac{4}{9} = \frac{8}{63}$$

To divide fractions, simply invert or flip the second fraction and multiply.

WORKED EXAMPLE

$$\frac{2}{3} \div \frac{7}{9} = \frac{2}{3} \times \frac{9}{7} = \frac{18}{21} = \frac{6}{7}$$

PERCENTAGES

When using percentages, it is useful to recall the different types of percentage questions.

To increase a value by a given percentage, use a percentage multiplier.

WORKED EXAMPLE

Increase 30 mg by 23%

If we increase by 23%, our new value will be 123% of the original value. We therefore multiply by 1.23.

Answer = $30 \times 1.23 = 36.9$ mg

EXAM HINT

In the exam, this may be asked as: 'An enzyme-controlled reaction produced 30 mg of product at 20 °C. At 30 °C the product increased by 23%. Calculate the mass of product at 30 °C.' This is level 2 maths because you need to decide how to carry out the calculation.

To decrease a value by a given percentage, you need to focus on the part that is left over after the decrease.

WORKED EXAMPLE

Decrease 30 mg by 23%

If we decrease by 23%, our new value will be 100 − 23 = 77% of the original value. We therefore multiply by 0.77.

Answer = $30 \times 0.77 = 23.1$ mg

To calculate a percentage increase, use the following equation:

$$\text{Percentage change} = \frac{\text{difference between values}}{\text{original value}} \times 100$$

To calculate percentage decrease, use the same equation but remember that your answer should be negative. Also remember that a percentage increase or decrease requires two steps: calculate the change, then calculate the percentage.

WORKED EXAMPLE

The volume of a solution increased from 40 ml to 50 ml. Calculate the percentage increase.

Change in volume = 10 ml

$$\text{Percentage increase} = \frac{10}{40} \times 100 = 25\%$$

EXAM HINT

This could be asked in the context of conservation: 'A population of gorillas increased from 40 to 50. Calculate the percentage increase in population size.' This is level 2 maths because there are two steps in the calculation. You need to calculate the change, then you need to calculate the percentage.

ALGEBRA

USING ALGEBRAIC EQUATIONS

Using algebraic equations is a very important skill for finding the value of an unknown quantity. In the real world, letters are used to symbolise important variables such as the blood sugar level of a diabetic or the irregular heartbeat of a patient.

The key rule to remember when using equations is that any operation that you apply to one side of the equation must also be applied to the other side.

WORKED EXAMPLE

Find the value of x in the following equation: $7x - 6 = 36$

Adding 6 to each side gives $7x = 42$

Dividing each side by 7 gives $x = 6$

CHANGING THE SUBJECT OF AN EQUATION

It can be helpful to rearrange an equation to express or isolate the variable you are interested in. Always remember that any operation that you apply to one side of the equation must also be applied to the other side.

WORKED EXAMPLE

The diameter of a cell measured under the light microscope at magnification ×100 is 2 mm. Calculate the actual size.

You may remember the equation

image size = actual size × magnification

but note the question is asking us to find the actual size given the image size and magnification. We can rearrange the equation to suit our needs:

$$\frac{\text{image size}}{\text{magnification}} = \text{actual size}$$

$$\text{So actual size} = \frac{2}{100} = 0.02 \text{ mm}$$

HANDLING DATA

USING SIGNIFICANT FIGURES

Often when you do a calculation, your answer will have many more figures than you need. Using an appropriate number of significant figures will help you to interpret results in a meaningful way.

Remember the 'rules' for significant figures:

- the first significant figure is the first figure which is not zero
- digits 1–9 are always significant
- zeros which come after the first significant figure are significant unless the number has already been rounded.

Here are some examples.

EXACT NUMBER	TO ONE S.F.	TO TWO S.F.	TO THREE S.F.
45 678	50 000	46 000	45 700
45 000	50 000	45 000	45 000
0.002 755	0.003	0.002 8	0.002 76

UNDERSTANDING THE TERMS MEAN, MEDIAN AND MODE

There are three different measures of average that you should know how to calculate.

- The **mean** is calculated by adding up all of the values in the data set and dividing them by the number of values. It is sometimes called the arithmetical average. The mean takes into account each number of the data set equally and can be used for further statistical analysis such as calculating a standard deviation. However, a disadvantage of the mean is that it may be affected by extreme values.
- The **median** is the middle value when the values are arranged in order. The median of a data set is found by putting the values in order from lowest to highest and then finding the middle value. If there is an even number of values, the median is found by calculating the mean of the two middle values.
- The **mode** is the value that occurs most often. The mode of a data set is found by identifying the most frequent value. It may not be possible to calculate the mode if there are two or more values with the same highest frequency.

WORKED EXAMPLE

Find the mean, median and mode of the following data set:
7, 12, 18, 6, 2, 12

To find the mean, we add up all of the values in the data set and divide them by the number of values.

$$\text{mean} = \Sigma\frac{x}{n}$$

$$= \frac{(7 + 12 + 18 + 6 + 2 + 12)}{6}$$

$$= \frac{57}{6}$$

$$= 9.5$$

To find the median, we need to arrange the values in increasing order: 2, 6, 7, 12, 12, 18.

Since there is an even number of values, we need to look at the two middle values and find the mean. The third value is 7 and the fourth value is 12.

$$\text{median} = \frac{(7 + 12)}{2} = 9.5$$

To find the mode, we need to identify the value that occurs most frequently. The only number that occurs more than once is 12.

$$\text{mode} = 12$$

EXAM HINT

You will need to recall the equation to use and probably select the data to use from a table of results or some other source.

CALCULATING THE MEAN FROM FREQUENCY DATA

The mean can be calculated from frequency data by finding the sum of the individual values multiplied by their respective frequencies and then dividing by the total frequency.

WORKED EXAMPLE

The table below shows the results of a survey looking into the number of units of alcohol consumed in a week by a sample of patients. Find the mean number of units of alcohol consumed per week.

UNITS OF ALCOHOL CONSUMED IN A WEEK	NUMBER OF PATIENTS
0	4
2	7
4	12
6	9
8	15
10	23

$$\text{mean} = \frac{(0 \times 4) + (2 \times 7) + (4 \times 12) + (6 \times 9) + (8 \times 15) + (10 \times 23)}{70}$$
$$= \frac{0 + 14 + 48 + 54 + 120 + 230}{70}$$
$$= \frac{466}{70}$$
$$= 6.6571$$
$$= 6.7 \text{ to 1 d.p.}$$

EXAM HINT

You are given the data but this is level 2 maths because there are three steps in the calculation. You need to multiply units per week by the number of patients and then add up those values before dividing by the number of patients.

UNDERSTANDING MEASURES OF DISPERSION INCLUDING STANDARD DEVIATION AND RANGE

Two different sets of data may have similar averages but statisticians are interested in looking deeper into the data for meaningful differences in dispersion. For example, if one data set refers to patients who are given a new cancer drug and a second data set refers to patients who are given a placebo drug, it is very important to look for key differences in the dispersion of data, such as standard deviation and range, and not just at measures of average.

RANGE

The range of a set of data is the difference between the highest and lowest values in the set. To find the range, subtract the smallest value in the set from the largest value in the set.

STANDARD DEVIATION

Standard deviation is a measure of the dispersion or 'spread' of data around the mean.

- A low standard deviation indicates that the data have a narrow range and the points are closely grouped to the mean. This could indicate greater reliability.
- A high standard deviation indicates that the data points have a larger range and are less well grouped. This might indicate lower reliability.

To calculate the standard deviation, use the formula:

$$s = \sqrt{\frac{\Sigma(x - \bar{x})^2}{n}}$$

where s = standard deviation, x is an individual value, $\bar{x}$ = the mean value, n = the number of values.

TECHNIQUE

1 Calculate the mean of the data set by finding the sum of the values and then dividing by the number of values. This is $\bar{x}$.

2 For each data value, calculate the difference between the data value and the mean. Record these figures in a table.

3 Find the square of each of these differences. Record these figures in a new column in your table.

4 Find the sum of these squares. This is $\Sigma(x - \bar{x})^2$.

5 Divide this figure by the number of items in the data set. This is $\frac{\Sigma(x - \bar{x})^2}{n}$

6 Find the square root of your answer. This is the standard deviation.

WORKED EXAMPLE

A pupil investigates the effect that two newly developed fertilisers (A and B) have on the growth of potato crops. Fourteen 10 m² areas of a field were sectioned off and treated with either fertiliser A or B. The table below shows the yields of potatoes from the test areas following harvest.

(a) Calculate the mean and standard deviation for the test plot yields for fertilisers A and B.

(b) Interpret the results of your answers to (a).

	TEST PLOT YIELD/kg						
FERTILISER	PLOT 1	PLOT 2	PLOT 3	PLOT 4	PLOT 5	PLOT 6	PLOT 7
A	25	27	34	18	21	26	28
B	17	35	42	19	35	22	44

(a) To calculate the mean yield for A:

25 + 27 + 34 + 18 + 21 + 26 + 28 = 179

179/7 = 25.6 kg to 1 d.p.

To calculate the standard deviation for A:

$(25 - 25.6)^2 = 0.36$ $(27 - 25.6)^2 = 1.96$

$(34 - 25.6)^2 = 70.56$ $(18 - 25.6)^2 = 57.76$

$(21 - 25.6)^2 = 21.16$ $(26 - 25.6)^2 = 0.16$

$(28 - 25.6)^2 = 5.76$

Sum of squares = 157.72

EXAM HINT

You will not be expected to recall the equation for standard deviation. It is important that you set out any part of the calculation clearly. If your final answer is incorrect, you may still gain marks for using the correct technique.

Divide by the number of plots:

157.72/7 = 22.5314

Now calculate the square root of this value:

$\sqrt{22.5314} = 4.8$ to 1 d.p.

To calculate the mean yield for B:

17 + 35 + 42 + 19 + 35 + 22 + 44 = 214

214/7= 30.6 kg to 1 d.p.

To calculate the standard deviation for B:

$(17 - 30.6)^2 = 184.96$ $(35 - 30.6)^2 = 19.36$

$(42 - 30.6)^2 = 129.96$ $(19 - 30.6)^2 = 134.56$

$(35 - 30.6)^2 = 19.36$ $(22 - 30.6)^2 = 73.96$

$(44 - 30.6)^2 = 179.56$

Sum of squares = 741.72

$\sqrt{741.72/7} = 10.3$ kg to 1 d.p.

(b) Fertiliser B produces a greater yield of potato crop (19% increase from fertiliser A), however the variation in crop yield (as shown by the standard deviation) of plots treated with fertiliser B is much greater and so fertiliser A produces a more consistent crop yield.

UNDERSTANDING SIMPLE PROBABILITY

The term *probability* is used to talk about the likelihood of an event happening on a scale of 0 to 1. A probability of 0 means that it is impossible that an event will occur. A probability of 1 means that it is certain that an event will occur. You should be comfortable interpreting probabilities in a scientific context, such as the probability of developing a disease or inheriting a specific gene.

INTERPRETING A SCATTERGRAM

A scattergram is a useful way of representing the relationship between two variables. To draw a scattergram, first choose appropriate scales and label both axes. Then, use a pencil to draw a small point (a cross or sharp dot) for each pair of variables.

A scattergram can be used to interpret whether there is correlation between two variables. We say that there is correlation between two variables if when one variable changes, there is also a change in the other variable.

- If the points are distributed tightly around a line, the variables are strongly correlated.
- If the points are loosely distributed around a line, the variables are weakly correlated.
- If there is no pattern in the distribution of points, there is no correlation.

Correlation can be positive (as one variable increases, the other also increases) or negative (as one variable increases, the other decreases).

A scattergram may include one or more points that lie outside of the main spread of values. Such a point is called an outlier or anomaly and it can be ignored.

To draw a line of best fit, use a ruler to draw a straight line that passes as close as possible to all of the points. You can use a line of best fit to make estimations. This is called interpolation. The more closely correlated the variables, the more accurate your estimate is likely to be.

WORKED EXAMPLE

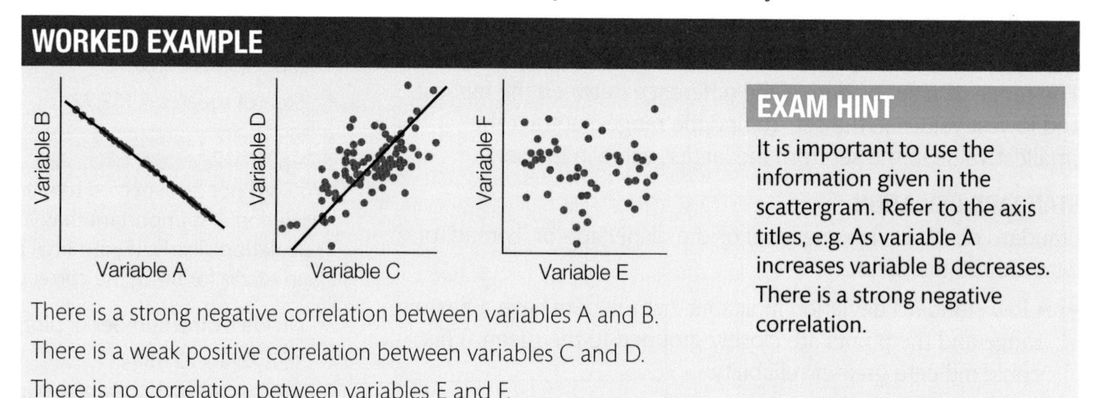

There is a strong negative correlation between variables A and B.

There is a weak positive correlation between variables C and D.

There is no correlation between variables E and F.

EXAM HINT

It is important to use the information given in the scattergram. Refer to the axis titles, e.g. As variable A increases variable B decreases. There is a strong negative correlation.

CONSTRUCTING HISTOGRAMS

Constructing frequency tables and histograms is often the first step to looking carefully at a set of raw continuous data and helps us to begin to look for patterns and behaviours in a data set. Histograms are very similar to bar charts but there are two differences.

- In a bar chart, each column represents a discrete category. The columns are of equal widths and always separated.
- In a histogram, the columns represent continuous data. The width of the columns is usually the same for each category. However, for more advanced work the widths may vary. The columns are always adjacent.

TECHNIQUE

1. Find the range of your values.
2. Choose the categories that you will use. Make sure that they are continuous (i.e. there are no gaps and there is no overlap between categories).
3. Create a frequency table.
4. Plot your data, ensuring that frequency is represented on the y-axis and that the categories are represented on the x-axis.

WORKED EXAMPLE

The weights of field mice found in a specified area of farmland to the nearest gram are shown below. Draw a histogram to represent the weights of field mice.

Weights of field mice (g): 42, 66, 75, 44, 52, 56, 60, 81, 64, 54, 37, 59, 47, 79, 66, 76, 53, 35, 40, 63, 56, 28, 43, 78, 83, 50, 38, 67, 68, 47, 52, 49, 32, 46, 72, 58, 58

To choose our categories, we first identify the range of the data. The highest value is 83 g and the lowest value is 28 g. The categories we choose need to at least cover this range. One sensible way of splitting this range is to use four categories each covering an interval of 20 g:

WEIGHT/g	FREQUENCY
20–39	5
40–59	18
60–79	12
80–99	2

We can now use this frequency data to draw a histogram.

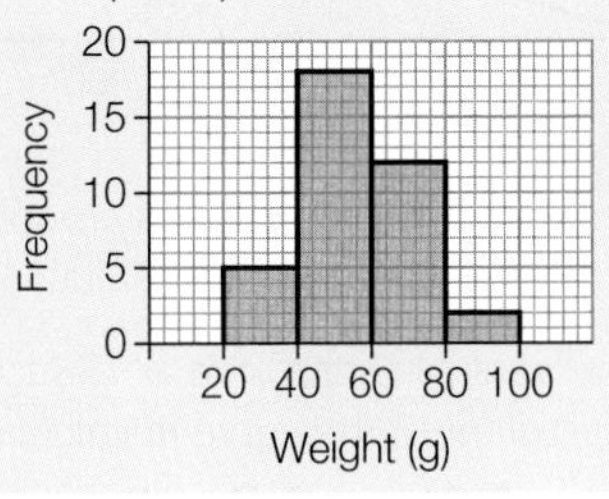

PRINCIPLES OF SAMPLING

When a scientist studies a population, it is not possible to study each organism in detail. Scientists therefore use sampling to estimate characteristics of the whole population by looking at a subset of individuals in the population. It is important that the sample chosen is representative of the habitat.

Once a suitable sample has been selected, it can be analysed. A measure of biodiversity that takes into account both the species richness and the species abundance of an area can be calculated using the following formula.

$$D = \frac{N(N-1)}{\sum n(n-1)}$$

where n is the number of individuals of a particular species (or the percentage cover for plants), and N is the total number of all individuals of all species (or the total percentage cover for plants).

GRAPHS

UNDERSTAND THAT $y = mx + c$ REPRESENTS A LINEAR RELATIONSHIP

Two variables are in a linear relationship if they increase at a constant rate in relation to one another. If you plotted a graph with one variable on the x-axis and the other variable on the y-axis, you would get a straight line. Any linear relationship can be represented by the equation $y = mx + c$ where the gradient of the line is m and the value at which the line crosses the y-axis is c. An example of a linear relationship is the relationship between degrees Celsius and degrees Fahrenheit, which can be represented by the equation $F = \frac{9}{5}C + 32$ where C is temperature in degrees Celsius and F is temperature in degrees Fahrenheit.

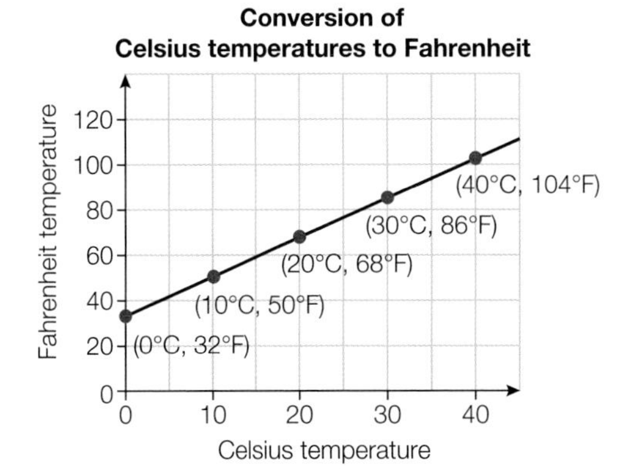

CALCULATE A RATE OF CHANGE FROM A GRAPH SHOWING A LINEAR RELATIONSHIP

The rate of change from a graph showing a linear relationship is the gradient, or steepness, of the line. It is a measure of the rate of change of one variable (represented on the x-axis) in relation to the other variable (represented on the y-axis).

TECHNIQUE

1 Draw a right-angled triangle anywhere on the line.

2 Use the following equation to calculate the rate of change:

$$\text{gradient} = \frac{\text{difference on } y\text{-axis}}{\text{difference on } x\text{-axis}}$$

3 State the unit for your answer.

DRAW AND USE THE SLOPE OF A TANGENT TO A CURVE AS A MEASURE OF A RATE OF CHANGE

A tangent is a straight line that just touches the curve at one point. The gradient of a curve at a given point is equal to the gradient of the tangent to the curve at that point.

TECHNIQUE

1 Use a ruler to draw a tangent to the curve.

2 Calculate the gradient of the tangent using the technique given for a linear relationship. This is equal to the gradient of the curve at the point of the tangent.

3 State the unit for your answer.

APPLYING YOUR SKILLS

You will often find that you need to use more than one maths technique to answer a question. In this section, we will look at two example questions and consider which maths skills are required and how to apply them.

WORKED EXAMPLE

Hydrogen peroxide is a toxic by-product of metabolism and is made in all living cells. Cells make the enzyme catalase in order to convert the toxin into water and oxygen. In order to study the effect of temperature on catalase activity, an experiment was set up using the equipment shown in the figure below. The volume of oxygen released in 30 seconds was measured at various temperatures using the gas syringe. The results of the experiment are shown in the graph below.

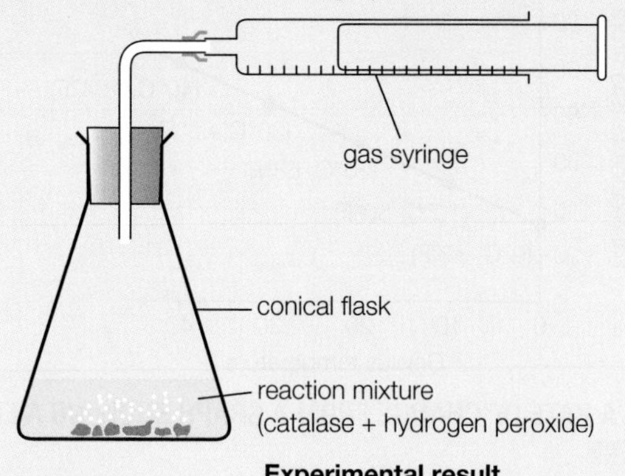

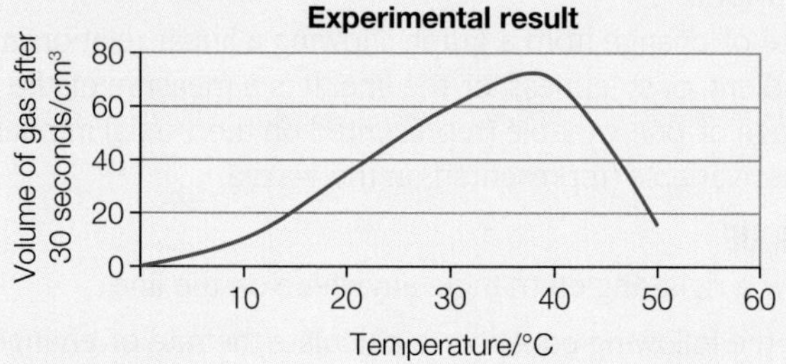

(a) *Calculate the percentage increase in volume of oxygen produced in 30 seconds at 40 °C compared to that produced at 10 °C.* (3 marks)

(b) *Calculate the rate of gas production at 20, 40 and 50 °C and interpret the results.* (5 marks)

(c) *A further experiment is carried out where the volume of oxygen is recorded over the entire time of the reaction at 10 °C and 40 °C. The results are shown below:*

	TOTAL VOLUME OF OXYGEN RELEASED/cm³									
TEMPERATURE /°C	**0 s**	**10 s**	**20 s**	**30 s**	**40 s**	**50 s**	**60 s**	**70 s**	**80 s**	**90 s**
10	0	2	5	9	16	22	28	33	35	35
40	0	7	15	27	33	35	35	35	35	35

Display both sets of results on a graph with an appropriate scale. (4 marks)

(d) *Calculate the difference in rate between the reactions at 30 seconds.* (4 marks)

(a) The volume of gas at 10 °C = 10 cm³ (1 mark)

The volume of gas at 40 °C = 70 cm³ (1 mark)

The percentage increase = $\frac{(70 - 10)}{10} \times 100 = 600\%$ increase (1 mark)

(b) 20 °C = $\frac{35}{30}$ = 1.17 cm³ s⁻¹ (1 mark)

40 °C = $\frac{70}{30}$ = 2.33 cm³ s⁻¹ (1 mark)

50 °C = $\frac{15}{30}$ = 0.5 cm³ s⁻¹ (1 mark)

The rate has doubled between 20 °C and 40 °C, but at 50 °C, the rate has decreased. (2 marks)

(c)

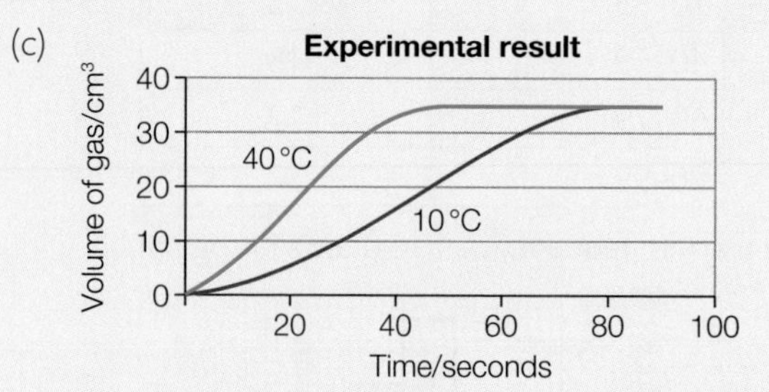

(marks awarded for: axes correct way round, axes labelled with units, points plotted correctly, suitable line drawn)

(d) We draw a tangent to each curve at 15 seconds so that we can use the gradient of the curve to calculate the rate.

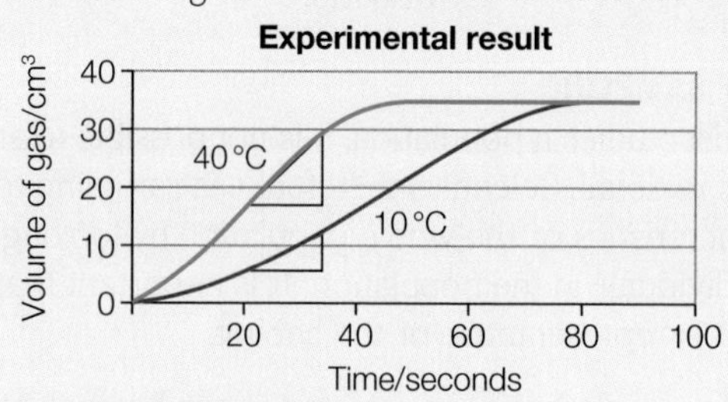

(1 mark for two correct tangents
1 mark for calculating the gradients
1 mark for calculating the rates of reactions
1 mark for calculating the difference)

We can then use the following equation to calculate gradient:

$$\text{gradient} = \frac{\text{difference on } y\text{-axis}}{\text{difference on } x\text{-axis}}$$

Rate of reaction at 10 °C at 30 seconds = $\frac{5}{13}$ = 0.38 cm³ s⁻¹

Rate of reaction at 40 °C at 30 seconds = $\frac{8}{10}$ = 0.8 cm³ s⁻¹

Difference in rate between reactions at 15 seconds = 0.8 − 0.38 = 0.42 cm³ s⁻¹

WORKED EXAMPLE

A photomicrograph of a T helper cell was taken using an electron microscope set at a magnification of ×50 000. In the image, several organelles were clearly identified and measured.

(a) *Calculate the actual object length of each organelle.* (4 marks)

ORGANELLE	IMAGE LENGTH/mm	OBJECT LENGTH/μm
nucleus	240	
endoplasmic reticulum	360	
lysosome	10	
mitochondrion	120	

(b) *A lysosome is a spherical organelle. Calculate the surface area and volume of a lysosome.* (3 marks)

(c) *Calculate the surface area to volume ratio of a lysosome.* (2 marks)

(a) The question tells us that the magnification is ×50 000.

We know that image size = actual size × magnification

To make it easier to use, we can rearrange this equation as actual size = $\frac{\text{image size}}{\text{magnification}}$

Actual length of nucleus = $\frac{240}{50\,000}$ = 0.0048 mm

Actual length of ER = $\frac{360}{50\,000}$ = 0.0072 mm

Actual length of lysosome = $\frac{10}{50\,000}$ = 0.0002 mm

Actual length of mitochondrion = $\frac{120}{50\,000}$ = 0.0024 mm

Before we can put these figures in the table, we need to convert to μm. 1 mm = 1000 μm so we need to multiply each figure by 1000.

ORGANELLE	IMAGE LENGTH/mm	OBJECT LENGTH/μm
nucleus	240	4.8
endoplasmic reticulum	360	7.2
lysosome	10	0.2
mitochondrion	120	2.4

(1 mark each)

(b) Recall the following formulae, where r is radius:

Surface area of sphere = $4\pi r^2$

Volume of sphere = $\frac{4}{3}\pi r^3$

From (a) you know that the diameter of the lysosome is 0.2 μm. This means that the radius must be 0.1 μm. (1 mark)

Surface area of lysosome = $4\pi(0.1)^2 = 4\pi \times 0.01 = 0.1257\ \mu m^2$ to 4 d.p. (1 mark)

Volume of lysosome = $\frac{4}{3}\pi(0.1)^3 = \frac{4}{3}\pi \times 0.001 = 0.0042\ \mu m^3$ to 4 d.p. (1 mark)

(c) It is simplest and most accurate to use the exact expressions from (b) involving π, rather than the final answers which have been rounded.

Surface area to volume ratio = $4\pi \times 0.01 : \frac{4}{3}\pi \times 0.001$

We can simplify by multiplying each side by 1000 and dividing each side by π:

Surface area to volume ratio = $40 : \frac{4}{3}$ (1 mark)

Now we can divide each side by 4 and multiply by 3 to get:

Surface area to volume ratio = 120 : 4 = 30 : 1 (1 mark)

PREPARING FOR YOUR EXAMS

IAS AND IAL OVERVIEW

The Pearson Edexcel International Advanced Subsidiary (IAS) in Biology and the Pearson Edexcel International Advanced Level (IAL) in Biology are modular qualifications. The IAS can be claimed on completion of the International Advanced Subsidiary (IAS) units. The International Advanced Level can be claimed on completion of all the units (IAS and IA2 units).

- International AS students will sit three exam papers. The IAS qualification can either be standalone or contribute 50% of the marks for the International Advanced Level.
- International A level students will sit six exam papers, the three IAS papers and three IAL papers.

The tables below give details of the exam papers for each qualification.

IAS Papers	**Paper 1: Unit 1** **Molecules, Diet, Transport and Health**	**Paper 2: Unit 2** **Cells, Development, Biodiversity and Conservation***	**Paper 3: Unit 3** **Practical Skills in Biology 1**
Topics covered	Topics 1–2	Topics 3–4	Topics 1–4
% of the IAS level qualification	40%	40%	20%
% of the IA level qualification	20%	20%	10%
Length of exam	1 hour 30 minutes	1 hour 30 minutes	1 hour 20 minutes
Marks available	80 marks	80 marks	50 marks
Question types	multiple choice short open open response calculation extended writing	multiple choice short open open response calculation extended writing	short open open response calculation
Mathematics	A minimum of 10% of the marks across all three both papers will be awarded for mathematics at Level 2 or above		

* This paper will contain some synoptic questions which require knowledge and understanding from Unit 1.

IAL Papers	**Paper 1: Unit 4** **Energy, the Environment, Microbiology and Immunity****	**Paper 2: Unit 5** **Respiration, the Internal Environment†**	**Paper 3: Unit 6** **Co-ordination and Gene Technology**
Topics covered	Topics 5–6	Topics 7–8	Topics 5–8
% of the IAL qualification	20%	20%	10%
Length of exam	1 hour 45 minutes	1 hour 45 minutes	1 hour 20 minutes
Marks available	90 marks	90 marks	50 marks
Question types	multiple choice short open open response calculation extended writing	multiple choice short open open response calculation extended writing	drawing short open open response calculation
Mathematics	A minimum of 10% of the marks across all three papers will be awarded for mathematics at Level 2 or above		

** This paper will contain some synoptic questions which require knowledge and understanding from Units 1 and 2.

† This paper will contain some synoptic questions which require knowledge and understanding from Units 1, 2 and 4.

EXAM STRATEGY

ARRIVE EQUIPPED

Make sure you have all of the correct equipment needed for your exam. As a minimum you should take:

- pen (a black ballpoint pen is best)
- pencil (HB)
- ruler (ideally 30 cm)
- rubber (make sure it's clean and doesn't smudge the pencil marks or rip the paper)
- calculator (scientific).

ENSURE YOUR ANSWERS CAN BE READ

Your handwriting does not have to be perfect but the examiner must be able to read it! When you're in a hurry it's easy to write key words that are difficult to decipher.

PLAN YOUR TIME

Note how many marks are available on the paper and how many minutes you have to complete it. This will give you an idea of how long to spend on each question. Be sure to leave some time at the end of the exam for checking answers. A rough guide of a minute a mark is a good start, but short answers and multiple choice questions may be quicker. Longer answers might require more time.

UNDERSTAND THE QUESTION

Always read the question carefully and spend a few moments working out what you are being asked to do. The command word used will give you an indication of what is required in your answer. It can be useful to highlight key words in the question.

Be scientific and accurate, even when writing longer answers. Use the technical terms you've been taught.

Always show your working for any calculations. Marks may be available for individual steps, not just for the final answer. Also, even if you make a calculation error, you may be awarded marks for applying the correct technique.

PLAN YOUR ANSWER

In questions marked with an *, marks will be awarded for your ability to structure your answer logically showing how the points that you make are related or follow on from each other where appropriate. Read the question fully and carefully (at least twice!) before beginning your answer.

MAKE THE MOST OF GRAPHS AND DIAGRAMS

Diagrams and sketch graphs can earn marks – often more easily and quickly than written explanations – but they will only earn marks if they are carefully drawn.

- If you are asked to read a graph, pay attention to the labels and numbers on the x- and y-axes. Remember that each axis is a number line.
- If asked to draw or sketch a graph, always ensure you use a sensible scale and label both axes with quantities and units. If plotting a graph, use a pencil and draw small crosses or dots for the points.
- Diagrams must always be neat, clear and fully labelled.

CHECK YOUR ANSWERS

For open-response and extended writing questions, check the number of marks that are available. If three marks are available, have you made three distinct points?

For calculations, read through each stage of your working. Substituting your final answer into the original question can be a simple way of checking that the final answer is correct. Another simple strategy is to consider whether the answer seems sensible. Pay particular attention to using the correct units.

SAMPLE EXAM ANSWERS

QUESTION TYPE: MULTIPLE CHOICE

The genetic pedigree diagram below shows the inheritance of Tay–Sachs disease in one family.

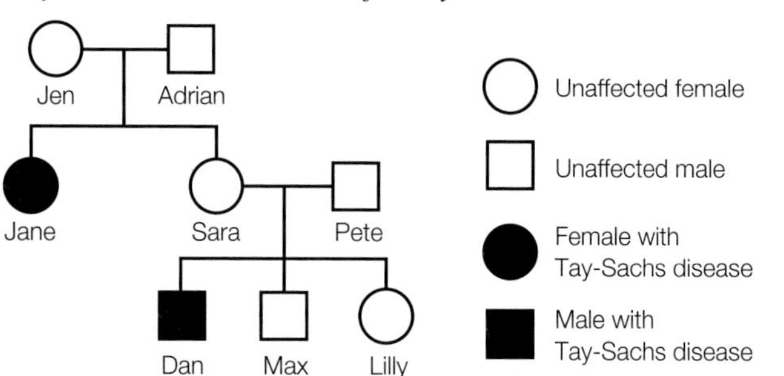

You are allowed to write on the exam paper, and it will help to do so. Start by writing the genotypes of the females with Tay–Sachs disease because they must be homozygous recessive. This means that their mums must have one recessive allele. Now put just one dominant allele on those females who are unaffected. Which female has only one allele? Make sure you do not confuse the two names which are similar (Jane and Jen)!

Put a cross (☒) in the box that correctly completes the statement.

The female whose genotype cannot be identified from the diagram is…

A Jane ☐

B Jen ☐

C Lilly ☐

D Sara ☐ (1)

Multiple choice questions always have one mark and the answer is given! For this reason students often make the mistake of thinking that they are the easiest questions on the paper. Unfortunately, this is not the case. These questions often require several answers to be worked out and an error in one of them will lead to the wrong answer being selected. The three incorrect answers supplied (distractors) will feature the answers that students arrive at if they make typical or common errors. The trick is to answer the question before you look at any of the answers.

Question analysis

- Multiple choice questions look easy until you try to answer them. Very often they require some working out and thinking.
- In multiple choice questions you are given the correct answer along with three incorrect answers (called distractors). You need to select the correct answer and put a cross in the box of the letter next to it.
- If you change your mind, put a line through the box (☒) and then mark your new answer with a cross (☒).

Average student answer

B Jen ☒

Jen cannot be the right answer since we know that she is unaffected (and therefore must have one dominant allele). She has given birth to Jane who has Tay–Sachs, and therefore Jen (the mother) must have a recessive allele as well. We know Jane must have two recessive alleles since she has Tay–Sachs disease. Sara is unaffected but has given birth to Dan who has Tay–Sachs disease, and therefore she, Sara, must be the same genotype as her mum Jen.

All this just leaves Lilly: We know she is unaffected, so must have at least one dominant 'normal' allele. However we have no idea whether she is homozygous or heterozygous. Thus Lilly is the correct answer.

COMMENTARY

This is an incorrect answer because:

- The student did not do the necessary working to find the correct answer. For a question like this, you should write the genotype of each person on the diagram.

If you have any time left at the end of the paper go back and check your answer to each part of a multiple choice question so that a slip like this does not cost you a mark.

QUESTION TYPE: SHORT OPEN

Cystic fibrosis and albinism are examples of recessive genetic disorders. Tay-Sachs disease is another example of a recessive genetic disorder.

Explain the meaning of the term recessive genetic disorder. (2)

The command word in this question is explain. This indicates that you will need to use 'therefore', 'so' or 'because' in your answer. Answer the question bit by bit: explain what makes an allele recessive, and then use a 'therefore' to explain how someone might suffer from this disorder.

Question analysis

- Generally one piece of information is required for each mark given in the question. There are two marks available for this question so make sure you make two distinct points.
- Clarity and brevity are the keys to success on short open questions. For one mark, it is not always necessary to write complete sentences.

Average student answer

The only way you are able to get the disease is if both your parents had the disease or both your parents are carriers. You have to be homozygous, two alleles the same. The recessive allele codes for the disease.

Misreading the question can lose you marks, as can answering in insufficient detail. One recessive allele does not code for the disease, but simply codes for a faulty protein.

COMMENTARY

This is an average answer because:

- The student will get one mark for remembering that people only suffer from a recessive genetic disorder if they inherit two copies of the recessive allele from their parents.
- The student has not explained what made the allele potentially cause a disorder: the version of the gene is faulty and does not code for a protein properly.

QUESTION TYPE: OPEN RESPONSE

Molecules are transported across the cell membrane in a number of different ways.

Describe the structure of a cell membrane. (3)

The command word in this question is describe. This means that you need to give an account of something. You do not need to include a justification or reason. Three marks are available so three distinct points need to be made. Remember that you can use bullet points or diagrams in your answer.

Question analysis

- With any question worth three or more marks, think about your answer and the points that you need to make before you write anything down. Keep your answer concise, and the information you write relevant to the question. You will not gain marks for writing down biology that is not relevant to the question (even if correct) but it will cost you time.

Average student answer

A cell membrane is made up of a phospholipid bilayer. Within this bilayer there are some proteins that span the membrane and others that are free to move within the membrane. Other features of the membrane include cholesterol, which sits within the bilayer, glycoproteins and glycolipids which are on the outer layers of the membrane and attached to either a protein or a lipid.

At this level, your answers need technical terms and clarity in expression otherwise you will find yourself losing marks.

COMMENTARY

This is an average answer because:

- The student has made five points, three of which meet the criteria needed to get full marks.
- The last sentence is poorly phrased: glycoproteins are molecules including a short carbohydrate group which is already attached to a protein.

QUESTION TYPE: EXTENDED WRITING

An investigation was carried out to study the effect of caffeine on the heart rate of a chicken embryo. The heart from a chicken embryo was removed and placed in a glucose solution. The heart rate was determined and recorded as a base heart rate. The experiment was repeated using glucose solutions containing five different concentrations of caffeine. The heart rate was determined and recorded as a percentage of the base heart rate for each solution.

Four marks are available so four points need to be made. If you have carried out the practical and written it up carefully (or corrected your write up using your teacher's feedback) then you should be well-prepared for this question.

Describe how this investigation could be carried out using Daphnia *instead of chicken embryos.* (4)

Question analysis

- There will be questions in your exams which assess your understanding of practical skills and draw on your experience of the core practicals. For these questions, think about:
  - how apparatus is set up
  - the method of how the apparatus is to be used
  - how readings are to be taken
  - how to make the readings reliable
  - how to control any variables.
- It helps with extended writing questions to think about the number of marks available and how they might be distributed. For example, if the question asked you to give the arguments for and against a particular case, then assume that there would be equal numbers of marks available for each side of the argument and balance the viewpoints you give accordingly. However, you should also remember that marks will also be available for giving an overall conclusion so you should be careful not to omit that.
- It is vital to plan out your answer before you write it down. There is always space given on an exam paper to do this so just jot down the points that you want to make before you answer the question in the space provided. This will help to ensure that your answer is coherent and logical and that you don't end up contradicting yourself. However, once you have written your answer go back and cross these notes out so that it is clear they do not form part of the answer.

Average student answer

By placing a Daphnia under a microscope, you will be able to determine the bpm by counting the heart beats in one minute. This will give you a control to compare against. Then by adding caffeine to the slide that the Daphnia is placed on, in regular increasing concentrations of caffeine, you should be able to calculate the heart rate of the Daphnia at different caffeine concentrations. By comparing against the control, this will allow you to note the differences in heart rate in relation to the concentration of caffeine.

Notice that the question says 'describe how this experiment could be carried out using *Daphnia*'. You need to adapt what you know already and apply it to this new situation.

COMMENTARY

This is an average answer because:

- Some important details have been missed. The student does not mention repeating the experiment to check for anomalies or controlling variables to ensure the results are valid.
- The student has not detailed how the heart rate is to be counted. The heart will beat between 100–200 times a minute so a sensible method would be by using a felt pen to place a dot on a piece of paper every time the heart beats.

QUESTION TYPE: CALCULATION

Age and gender are two factors that may influence the development of heart disease in an individual. The graph below shows the results of a survey in America on the incidence of heart disease in adults aged 18 and older.

The command word here is calculate. This means that you need to obtain a numerical answer to the question, showing relevant working. If the answer has a unit, this must be included.

Finding the numbers requires you to read the graph really carefully, paying close attention to the increments on the y-axis, as well as choosing the correct set of bars.

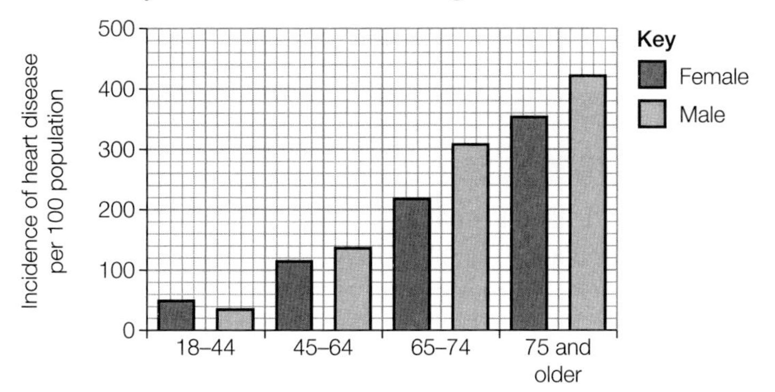

Calculate the increased risk that a man who is 75 or older has of developing heart disease, compared to a man aged between 18 and 44 years old. (2)

Question analysis

- The important thing with calculations is that you must show your working clearly and fully. The correct answer on the line will gain all the available marks. However, an incorrect answer can gain all but one of the available marks if the correct working is shown.
- Show the calculation that you are performing at each stage and not just the result. When you have finished, look at your result and see if it is sensible.
- At some point during your answer, you will need to do some kind of sum, and the skills are to decide
  - which numbers you need
  - which operation you need.

Average student answer

$410 - 15 = 395. \frac{395}{15} = 26.3$, so 26.3 times.

The student has not read the graph correctly. Five of the small sections on the y-axis equal 100, so each small section of the y-axis corresponds to 20. Therefore, the calculation of difference is actually 420 − 35, which is 385.

To work out the increased risk, you divide the difference by the risk for 18–44-year-old men.

It's a good idea when you finish the question to check whether you need to put in the units. In this question, there are none as it's simply a case of 'multiples of risk'.

COMMENTARY

This is an average answer because:

- The student has misread the graph so has used incorrect figures in the calculation.
- However, the correct technique has been used to work out the increased risk, so a mark would be awarded for using the correct method.

COMMAND WORDS

The following table lists the command words used across the IAS/IAL Science qualifications in the external assessments. You should make sure you understand what is required when these words are used in questions in the exam.

COMMAND WORD	THIS TYPE OF QUESTION WILL REQUIRE STUDENTS TO:
ADD/LABEL	Requires the addition or labelling to stimulus material given in the question, for example, labelling a diagram or adding units to a table.
ASSESS	Give careful consideration to all the factors or events that apply and identify which are the most important or relevant. Make a judgement on the importance of something, and come to a conclusion where needed.
CALCULATE	Obtain a numerical answer, showing relevant working. If the answer has a unit, this must be included.
COMMENT ON	Requires the synthesis of a number of factors from data/information to form a judgement. More than two factors need to be synthesised.
COMPARE AND CONTRAST	Looking for the similarities **and** differences of two (or more) things. Should not require the drawing of a conclusion. Answer must relate to both (or all) things mentioned in the question. The answer must include at least one similarity and one difference.
COMPLETE/RECORD	Requires the completion of a table/diagram/equation.
CRITICISE	Inspect a set of data, an experimental plan or a scientific statement and consider the elements. Look at the merits and/or faults of the information presented and back judgements made.
DEDUCE	Draw/reach conclusion(s) from the information provided.
DERIVE	Combine two or more equations or principles to develop a new equation.
DESCRIBE	To give an account of something. Statements in the response need to be developed as they are often linked but do not need to include a justification or reason.
DETERMINE	The answer must have an element which is quantitative from the stimulus provided, or must show how the answer can be reached quantitatively.
DEVISE	Plan or invent a procedure from existing principles/ideas.
DISCUSS	Identify the issue/situation/problem/argument that is being assessed within the question. Explore all aspects of an issue/situation/problem. Investigate the issue/situation/problem, etc. by reasoning or argument.

COMMAND WORD	THIS TYPE OF QUESTION WILL REQUIRE STUDENTS TO:
DRAW	Produce a diagram either using a ruler or freehand.
ESTIMATE	Give an approximate value for a physical quantity or measurement or uncertainty.
EVALUATE	Review information then bring it together to form a conclusion, drawing on evidence including strengths, weaknesses, alternative actions, relevant data or information. Come to a supported judgement of a subject's qualities and relation to its context.
EXPLAIN	An explanation requires a justification/exemplification of a point. The answer must contain some element of reasoning/justification; this can include mathematical explanations.
GIVE/STATE/NAME	All of these command words are really synonyms. They generally all require recall of one or more pieces of information.
GIVE A REASON/REASONS	When a statement has been made and the requirement is only to give the reasons why.
IDENTIFY	Usually requires some key information to be selected from a given stimulus/resource.
JUSTIFY	Give evidence to support (either the statement given in the question or an earlier answer).
PLOT	Produce a graph by marking points accurately on a grid from data that is provided and then drawing a line of best fit through these points. A suitable scale and appropriately labelled axes must be included if these are not provided in the question.
PREDICT	Give an expected result or outcome.
SHOW THAT	Prove that a numerical figure is as stated in the question. The answer must be to at least one more significant figure than the numerical figure in the question.
SKETCH	Produce a freehand drawing. For a graph, this would need a line and labelled axes with important features indicated; the axes are not scaled.
STATE WHAT IS MEANT BY	When the meaning of a term is expected but there are different ways of how these can be described.
SUGGEST	Use your knowledge and understanding in an unfamiliar context. May include material or ideas that have not been learnt directly from the specification.
WRITE	When the question asks for an equation.

GLOSSARY

70S ribosomes the ribosomes found in the mitochondria and chloroplasts of eukaryotic cells and in prokaryotic organisms

80S ribosomes the main type of ribosome found in eukaryotic cells, consisting of ribosomal RNA and protein, made up of a 60S and 40S subunit; they are the site of protein synthesis

ACE inhibitors drugs which block the production of angiotensin

acrosome the region at the head of the sperm that contains enzymes to break down the protective layers around the ovum

acrosome reaction the reaction seen when the sperm reach the oocyte and enzymes are released from the acrosome and digest the follicle cells and the zona pellucida

activation energy the energy needed for a chemical reaction to get started

active site the area of an enzyme that has a specific shape into which the substrate(s) of a reaction fit

active transport the movement of substances into or out of the cell using ATP produced during cellular respiration

adaptive radiation a process by which one species develops rapidly resulting in several different species which fill different ecological niches

adenine a purine base found in DNA and RNA

adenosine triphosphate (ATP) a molecule that acts as the universal energy supply molecule in cells; it is made up of the base adenine, the pentose sugar ribose and three phosphate groups

adult stem cells (somatic stem cells) undifferentiated cells found among the normal differentiated cells in a tissue or organ that can differentiate when needed to produce any one of the major cell types found in that particular tissue or organ

allele frequency the frequency with which a particular allele appears within a population

alleles versions of a gene, variants

allopatric speciation speciation that occurs when populations are physically or geographically separated and there can be no interbreeding or gene flow between the populations

amino acids the building blocks of proteins consisting of an amino group ($-NH_2$) and a carboxyl group ($-COOH$) attached to a carbon atom and an R group that varies between amino acids

amniocentesis a type of prenatal screening which involves removing a sample of amniotic fluid at around 16 weeks of pregnancy, culturing the fetal cells found and analysing them for genetic diseases

amylopectin a complex carbohydrate made up of α-glucose monomers joined by 1,4-glycosidic bonds with some 1,6-glycosidic bonds so the molecules branch repeatedly

amyloplasts plant organelles that store starch

amylose a complex carbohydrate containing only α-glucose monomers joined together by 1,4-glycosidic bonds so the molecules form long unbranched chains

anabolic reaction a reaction that builds up (synthesises) new molecules in a cell

anaerobic respiration cellular respiration that takes place in the absence of oxygen

analogous features features that look similar or have a similar function, but are not from the same biological origin

anaphase the third stage of active cell division where the centromeres split so chromatids become new chromosomes; they are moved to the opposite poles of the cell, centromere first, by contractions of the microtubules of the spindle

anatomical adaptations adaptations involving the form and structure of an organism

aneurysm a weakened, bulging area of artery wall that results from blood collecting behind a blockage caused by plaques

angina a condition in which plaques are deposited on the endothelium of the arteries and reduce the blood flow to the cardiac muscle through the coronary artery; it results in pain during exercise

Animalia a mainly heterotrophic eukaryotic kingdom including all the invertebrates and vertebrates

anion a negative ion

anthers male sex organs in plants that produce the male gametes contained in the pollen

anticoagulant a substance that interferes with the manufacture of prothrombin in the body

anticodon a sequence of three bases on tRNA that are complementary to the bases in the mRNA codon

antihypertensive drug which reduces high blood pressure

antioxidants molecules that inhibit the oxidation of other molecules which can lead to chain reactions that may damage cells

antisense strand (template strand) the DNA strand which acts as a template for an mRNA molecule

aorta the main artery of the body; it leaves the left ventricle of the heart carrying oxygenated blood under high pressure

apoptosis (programmed cell death) the breakdown of worn-out, damaged or diseased cells by the lysosomes

Archaea domain made up of bacteria-like prokaryotic organisms found in many places including extreme conditions and the soil; they are thought to be early relatives of the eukaryotes

Archaebacteria ancient type of bacteria found in many different environments

artefacts things observed in a scientific investigation that are not naturally present; they occur as a result of the preparation or investigation

arterial system the system of arteries in the body

arteries vessels that carry blood away from the heart

arterioles the very smallest branches of the arterial system, furthest from the heart

aseptic technique method of carrying out a procedure to prevent contamination by unwanted microorganisms

asexual reproduction the production of genetically identical offspring from a single parent or organism

aspirin a widely used drug which relieves pain and reduces blood clotting and inflammation

atheroma another term for a plaque formed on the arterial lining

atherosclerosis a condition in which yellow fatty deposits build up (increase in amount) on the lining of the arteries, causing them to be narrowed and resulting in many different health problems

ATP see adenosine triphosphate

ATPase an enzyme that catalyses the hydrolysis of ATP, releasing energy to move carrier systems and drive metabolic reactions

atrial systole when the atria of the heart contract

autosomes chromosomes which carry information about the body but do not determine the sex of an individual

bacilli rod-shaped bacteria

bacteriophage virus that attacks bacteria

behavioural adaptations adaptations involving programmed or instinctive behaviour making organisms better adapted for survival

beta blockers drugs which block the response of the heart to hormones such as adrenaline

biased when someone is unfairly for or against an idea (e.g. when a scientist is paid by someone with a vested interest in a specific result - they may receive benefit from the outcome)

bicuspid valve (atrioventricular valve) the valve between the left atrium and the left ventricle that prevents backflow of blood into the atrium when the ventricle contracts

bilayer a double layer of closely packed atoms or molecules

binary fission asexual reproduction in bacteria in which the bacteria split in half

biodiversity a measure of the variety of living organisms and their genetic differences

biodiversity hotspot an area with a particularly high level of biodiversity

bioplastics plastics based on biological polymers

blastocyst an early embryo consisting of a hollow ball of cells with an inner cell mass of pluripotent cells that will eventually form a new organism

body mass index (BMI) a calculation to determine if you are a healthy weight by comparing your weight to your height in a simple formula

Bohr effect the name given to changes in the oxygen dissociation curve of haemoglobin that occur due to a rise in carbon dioxide levels and a reduction of the affinity of haemoglobin for oxygen

breathing (ventilation) the process in which physical movements of the chest change the pressure so that air is moved in or out

buffer a solution which resists changes in pH

cambium the layer of unspecialised plant cells that divide to form both the xylem and the phloem

capillaries tiny vessels that spread throughout the tissues of the body

capsule a layer formed from starch, gelatin, protein or glycolipid, found around the outside of some bacteria

captive breeding programmes programmes where individuals of an endangered species are bred in zoos and parks in an attempt to save the species from extinction, and if possible to reintroduce them to their natural wild environment

carbaminohaemoglobin the molecule formed when carbon dioxide combines with haemoglobin

carbon neutral a process where no net carbon is released into the atmosphere

carbonic anhydrase the enzyme that controls the rate of the reaction between carbon dioxide and water to produce carbonic acid

cardiac cycle the cycle of contraction (systole) and relaxation (diastole) in the heart

cardiac muscle the special muscle tissue of the heart, which has an intrinsic rhythm and does not fatigue

cardiovascular diseases diseases of the heart and circulatory system, many of which are linked to atherosclerosis

cardiovascular system he mass transport system of the body made up of a series of vessels with a pump (the heart) to move blood through the vessels

carrier protein a protein that moves a substance through the membrane in active transport using energy from the breakdown of ATP or in passive transport such as facilitated diffusion down a concentration gradient

catabolic reaction a reaction which breaks down substances within a cell

catalyst a substance that speeds up a reaction without changing the substances produced or being changed itself

cation a positive ion

causation when a factor directly causes a specific effect

cell cycle a regulated process of three stages (interphase, mitosis and cytokinesis) in which cells divide into two genetically identical daughter cells

cell differentiation the process by which a less specialised cell becomes more specialised for a particular function

cell membrane the selectively permeable membrane which surrounds the cytoplasm of a cell, acting as a barrier between the cell contents and their surroundings

cell sap the aqueous solution that fills the permanent vacuole

cell wall a freely permeable wall around plant cells, made mainly of cellulose

centrioles bundles of tubules found near the nucleus and involved in cell division by the production of a spindle of microtubules that move the chromosomes to the ends of the cell

centromere the region where a pair of chromatids are joined and which attaches to a single strand of the spindle structure at metaphase

chiasmata the points where the chromatids break during recombination

chlorophyll the green pigment that is largely responsible for trapping the energy from light, making it available for the plant to use in photosynthesis

chloroplasts organelles adapted to carry out photosynthesis, containing the green pigment chlorophyll

chorionic villus sampling a type of prenatal screening where a small sample of embryonic tissue is taken from the developing placenta and the cells tested for genetic diseases

chromatid one strand of the replicated chromosome pair that is joined to the other chromatid at the centromere

chromatin the granular combination of DNA bonded to protein found in the nucleus when the cell is not actively dividing

chromosomal mutations changes in the position of entire genes within a chromosome

circulation the passage of blood through the blood vessels

class a group of orders that all share common characteristics

clones genetically identical individuals resulting from asexual reproduction in a single parent

cocci spherical bacteria

codominance in heterozygotes, where both alleles at a gene locus are fully expressed in the phenotype

codominant both alleles are expressed in the phenotype

codon a sequence of three bases in DNA or mRNA

collagen a strong fibrous protein with a triple helix structure

collenchyma plant cells with areas of cellulose thickening that give mechanical strength and support to the tissues

colloid a suspension of molecules that are not fully dissolved

companion cells very active cells closely associated with the sieve tube elements that supply the phloem vessels with everything they need and actively load sucrose into the phloem

complementary base pairs complementary purine and pyrimidine bases which align in a DNA helix, with hydrogen bonds holding them together (C–G, A–T)

complementary strand the strand of RNA formed that complements the DNA acting as the coding strand

composite material a material made of two or more materials which combined together make a composite with different properties from either of the constituent materials

concentration gradient the change in the concentration of solutes present in a solution between two regions; in biology, this typically means across a cell membrane

conception the term used for fertilisation of the ovum in humans

condensation reaction a reaction in which a molecule of water is removed from the reacting molecules as a bond is formed between them

conjugated proteins protein molecules joined with or conjugated to another molecule called a prosthetic group

conservation maintaining and protecting a living and changing environment

conservative replication a model of DNA replication which suggests that the original double helix remains intact and in some way instructs the formation of a new, identical double helix made up entirely of new material

continuous variation phenotypic features which show a huge range of values; they are usually polygenic and are also affected by environmental factors

correlation a strong tendency for two sets of data to change together

cortical reaction the reaction seen when cortical granules in the cytoplasm of the ovum release enzymes into the zona pellucida; these enzymes destroy the sperm-binding sites and also thicken and harden the jelly of the zona pellucida

covalent bonds bonds formed when atoms share electrons; covalent molecules may be polar if the electrons are not shared equally

cristae the infoldings of the inner membrane of the mitochondria which provide a large surface area for the reactions of aerobic respiration

crossing over (recombination) the process by which large multi-enzyme complexes cut and re-join parts of the maternal and paternal chromatids at the end of prophase I, introducing genetic variation

culture growing microorganisms in the laboratory, providing them with the nutrients, oxygen, pH and temperature they need to produce large numbers so they can be observed and measured

cyanide a metabolic poison that stops mitochondria working

cyclin-dependent kinases (CDKs) enzymes involved in the control of the cell cycle by phosphorylating other proteins, activated by attachment to cyclins

cyclins small proteins that build up during interphase and are involved in the control of the cell cycle by their attachment to cyclin-dependent kinases

cystic fibrosis (CF) a serious genetic disease caused by a recessive allele which affects the production of mucus by epithelial cells

cytokinesis the final stage of the cell cycle before the cell enters interphase again – division of the cytoplasm at the end of mitosis to form two independent, genetically identical cells

cytoplasm a jelly-like liquid that makes up the bulk of the cell and contains the organelles

cytosine a pyrimidine base found in DNA and RNA

degenerate code a code containing more information than is needed

deletion a type of point mutation in which a base is completely lost

denaturation the loss of the 3D shape of a protein (e.g. caused by changes in temperature or pH)

deoxygenated blood blood that has given up its oxygen to the cells in the body

deoxyribonucleic acid (DNA) a nucleic acid that is the genetic material in many organisms

deoxyribose a pentose sugar that is part of the structure of DNA

diastole when the heart relaxes and fills with blood

diffusion the movement of the particles in a liquid or gas down a concentration gradient from an area where they are at a relatively high concentration to an area where they are at a relatively low concentration

digenic (dihybrid) inheritance the inheritance of two pairs of contrasting characteristics at the same time

digitalis a chemical found in foxgloves that affects the beating of the heart

digoxin a drug based on the chemical found in foxgloves that improves heart function

dipeptide two amino acids joined by a peptide bond

diploid (2*n*) a cell with a nucleus containing two full sets of chromosomes

dipole the separation of charge in a molecule when the electrons in covalent bonds are not evenly shared

disaccharide a sugar made up of two monosaccharide units joined by a glycosidic bond, formed in a condensation reaction

discontinuous variation phenotypic features which are either present or not, usually inherited on one or at most a small number of genes.

dissociation splitting of a molecule into smaller molecules, atoms, or ions, especially by a reversible process

disulfide bond a strong covalent bond produced by an oxidation reaction between sulfur groups in cysteine or methionine molecules, which are close together in the structure of a polypeptide

diuretics drugs which increase the volume of urine produced

diversity index a way of measuring the biodiversity of a habitat using the formula:

$$D = \frac{N(N-1)}{\Sigma n(n-1)}$$

DNA demethylation the removal of the methyl group from methylated DNA enabling genes to become active so they can be transcribed

DNA helicase an enzyme involved in DNA replication that 'unzips' the two strands of the DNA molecules

DNA ligase an enzyme involved in DNA replication that catalyses the formation of phosphodiester bonds between the nucleotides

DNA methylation methylation of DNA (addition of a methyl $-CH_3$ group) to a cytosine in the DNA molecule next to a guanine in the DNA chain and prevents the transcription of a gene

DNA polymerase an enzyme involved in DNA replication that lines up the new nucleotides along the DNA template strands

DNA profiling the process by which the non-coding areas of DNA are analysed to identify patterns

DNA sequencing the process by which the base sequences of all or part of the genome of an organism is worked out

domains the three largest classification categories: the Eukaryota, the Bacteria and the Archaea

dominant a characteristic which is expressed in the phenotype whether the individual is homozygous or heterozygous for that allele

double circulation system a circulation that involves two separate circuits, one of deoxygenated blood flowing from the heart to the gas exchange organs to be oxygenated before returning to the heart, and one of oxygenated blood leaving the heart and flowing around the body, returning as deoxygenated blood to the heart

double fertilisation the process that occurs in plants in which one male nucleus fuses with the two polar nuclei to form the triploid endosperm nucleus and the other fuses with the egg cell to form the diploid zygote

double-blind trial a clinical drug trial where neither the doctor nor the patient knows whether the patient is receiving the new medicine, a control medicine or a placebo

duodenum the first part of the gut after the stomach

ecological species model a species definition based on the ecological niche occupied by an organism

ecology the study of the relationships of organisms to one another and to their physical environment

ecosystems biological communities where organisms interact with one other and with their physical environment

electron microscope a tool that uses a beam of electrons and magnetic lenses to magnify specimens up to 500 000 times life size

embryonic stem cells the undifferentiated cells of the early human embryo with the potential to develop into many different types of specialised cell

end products the final products of a chemical reaction

endemic a species that evolves in geographical isolation and is found in only one place

endocytosis the movement of large molecules into cells through vesicle formation

endoplasmic reticulum (ER) A 3D network of membrane-bound cavities in the cytoplasm that links to the nuclear membrane and makes up a large part of the cellular transport system as well as playing an important role in the synthesis of many different chemical substances

endosymbionts organisms that live inside the cells or the body of another organism

endosymbiotic theory a theory that suggests that mitochondria and chloroplasts originated as independent prokaryotic organisms that began living symbiotically inside other cells as endosymbionts

enhancer sequence specific region of DNA to which transcription factors bind and regulate the activity of the DNA by changing the structure of the chromatin

enzymes proteins that act as biological catalysts for a specific reaction or group of reactions

epidemiology the study of patterns of health and disease, to identify causes of different conditions and patterns of infection

epigenetics the study of changes in gene expression (active versus inactive genes) that does not involve changes to the underlying DNA sequence but affects how cells read genes

epithelial tissues tissues that form the lining of surfaces inside and outside the body

ester bonds bonds formed in a condensation reaction between the carboxyl group (-COOH) of a fatty acid and one of the hydroxyl groups (-OH) of glycerol

esterification the process by which ester bonds are made

eubacteria true bacteria (prokaryotic organisms)

evaluate to assess or judge the quality of a study and the significance of the results

evolution the process by which natural selection acts on variation to bring about adaptations and eventually speciation

***ex-situ* conservation** the conservation of components of biological diversity (living organisms) outside their natural habitats

exhalation breathing out

exocrine glands glands which produce substances and secrete them to where they are needed through a small tube called a duct

exocytosis the movement of large molecules out of cells by the fusing of a vesicle containing the molecules with the surface cell membrane; the process requires ATP

exons segments of a DNA or RNA molecule containing information coding for a protein or peptide sequence

external fertilisation the process of fertilisation in which the female and male gametes are released outside of the parental bodies to meet and fuse in the environment

extracellular enzymes enzymes that catalyse reactions outside of the cell in which they were made

extremophiles bacteria that can survive extreme conditions of heat, cold, pH, salinity and pressure

facilitated diffusion diffusion that takes place through carrier proteins or protein channels

facultative anaerobes organisms that use oxygen if it is available, but can respire and survive without it

family a group of genera that all share common characteristics

fatty acids organic acids with a long hydrocarbon chain

fertilisation membrane the tough layer that forms around the fertilised ovum to prevent the entry of other sperm

fertilisation the fusing of the haploid nuclei from two gametes to form a diploid zygote in sexual reproduction

fetal haemoglobin a form of haemoglobin found only in the developing fetus with a higher affinity for oxygen than adult haemoglobin

fibrin an insoluble protein formed from fibrinogen by the action of thrombin that forms a mesh of fibres that trap erythrocytes and platelets to form a blood clot

fibrinogen a soluble plasma protein which is the precursor of the insoluble protein fibrin

fibrous proteins proteins that have long, parallel polypeptide chains with occasional cross-linkages that produce fibres but with little tertiary structure

flaccid floppy, soft

flagella many-stranded helices of the contractile protein flagellin found on some bacteria; they move the bacteria by rapid rotations

fluid mosaic model the current model of the structure of the cell membrane including floating proteins forming pores, channels and carrier systems in a lipid bilayer

founder effect the loss of genetic variation that occurs when a small number of individuals become isolated, forming a new population with allele frequencies not representative of the original population

fungi a eukaryotic kingdom of heterotrophs with chitin in their cell walls

gametes haploid sex cells that fuse to form a new diploid cell (zygote) in sexual reproduction

gametogenesis the formation of the gametes by meiosis in the sex organs

gametophyte generation the haploid generation in plants that gives rise to the gametes by mitosis

gated channels protein channels through the lipid bilayer of a membrane that are opened or closed, depending on conditions in the cell

gel electrophoresis a method of separating fragments of proteins or nucleic acids based on their electrical charge and size

gene a sequence of bases on a DNA molecule; it contains coding for a sequence of amino acids in a polypeptide chain that affects a characteristic in the phenotype of the organism

gene flow the migration of either whole organisms or genetic material into or out of a population and into another population, tending to make different populations more alike, but changing the allele frequencies within each individual population all the time

gene linkage when genes for two different characteristics are found on the same chromosome and are close together so they are linked and inherited as a single unit

gene pool the sum total of all the genes in a population at a given time

generative nucleus the male nucleus that will fuse with the female nucleus

genetic diversity a measure of the level of difference in the genetic make-up of a population

genetic screening when whole populations are tested for a genetic disease

genetic species model a species model based on DNA evidence

genome the entire genetic material of an organism

genotype the genetic make-up of an organism with respect to a particular feature

genus a group of species that all share common characteristics

germinate (of pollen) the process by which a pollen tube starts to grow out of the pollen grain to transfer the male nuclei to the ovule

globular proteins large proteins with complex tertiary and sometimes quaternary structures, folded into spherical (globular) shapes

glucose a hexose sugar

glycerol propane-1,2,3-triol, an important component of triglycerides

glycogen a complex carbohydrate with many α-glucose units joined by 1,4-glycosidic bonds with many 1,6-glycosidic bonds, giving it many side branches

glycoproteins conjugated proteins with a carbohydrate prosthetic group

glycosidic bond a covalent bond formed between two monosaccharides in a condensation reaction, which can be broken down by a hydrolysis reaction to release the monosaccharide units

Golgi apparatus stacks of membranes that modify proteins made elsewhere in the cell and package them into vesicles for transport, and also produce materials for plant cell walls and insect cuticles

gonads the sex organs in animals

Gram staining a staining technique used to distinguish types of bacteria by their cell wall

Gram-negative bacteria bacteria that have no teichoic acid in their cell walls; they stain red with Gram staining

Gram-positive bacteria bacteria that contain teichoic acid in their cell walls and stain purple/blue with Gram staining

graticule a series of lines in the eyepiece of a microscope which help you measure specimens accurately

guanine a purine base found in DNA and RNA

haemoglobin a red pigment that carries oxygen and gives the erythrocytes their colour

haemophilia a sex-linked genetic disease in which one of the factors needed for blood to clot is not made in the body

haploid (*n*) a cell with a nucleus containing one complete set of chromosomes

Hardy–Weinberg equilibrium the mathematical relationship between the frequencies of alleles and genotypes in a population; the equation used to describe this relationship can be used to work out the stable allele frequencies within a population

hemicelluloses polysaccharides containing many different sugar monomers

heterochromatin densely supercoiled and condensed chromatin where the genes are not available to be copied to make proteins

heterogametic an individual who produces two types of gamete each containing different types of sex chromosome – in humans, this is the male

heterozygosity index a useful measure of genetic diversity within a population expressed as:

$$\frac{\text{number of heterozygotes}}{\text{number of individuals in the population}}$$

heterozygote an individual where the two alleles coding for a particular characteristic are different

hexose sugar sugar with six carbon atoms

high blood pressure blood pressure that is regularly more than 140/90 mmHg; this increases your risk of developing CVDs

high-density lipoproteins (HDLs) lipoproteins which transport cholesterol from body tissues to the liver and can help reduce risks of CVDs

histone acetylation the addition of an acetyl group ($-COCH_3$) to one of the lysines in the histone structure, which opens up the structure and activates the chromatin, allowing genes in that area to be transcribed

histone methylation the addition of a methyl group ($-CH_3$) to a lysine in the histone; methylation may cause inactivation or activation of the region of DNA, depending on the position of the lysine

histones positively charged proteins involved in the coiling of DNA to form dense chromosomes in cell division

homogametic an individual who produces gametes that contain only one type of sex chromosome – in humans, this is the female

homologous pairs matching pairs of chromosomes in an individual which both carry the same genes, although they may have different alleles

homologous structures structures that genuinely show common ancestry

homozygote an individual where both alleles coding for a particular characteristic are identical

hybridisation the production of offspring as a result of sexual reproduction between individuals from two different species

hydrogen bonds weak electrostatic intermolecular bonds formed between polar molecules containing at least one hydrogen atom

hydrolysis a reaction in which bonds are broken by the addition of a molecule of water

hydrophilic a substance with an affinity for water that will readily dissolve in or mix with water

hydrophobic a substance that tends to repel water and that will not mix with or dissolve in water

hydrostatic pressure the pressure exerted by a fluid in an equilibrium

hypertension high blood pressure, regularly measuring over 140/90 mmHg, which increases your risk of developing CVDs

hypertonic solution a solution in which the osmotic concentration of solutes is higher than that in the cell contents

hypotonic solution a solution in which the osmotic concentration of solutes is lower than that in the cell contents

***in-situ* conservation** the conservation of ecosystems and natural habitats, and the maintenance and recovery of viable populations of species in their natural surroundings

incipient plasmolysis the point at which so much water has moved out of the cell by osmosis that turgor is lost and the cell membrane begins to pull away from the cell wall as the protoplasm shrinks

independent assortment (random assortment) the process by which the chromosomes derived from the male and female parent are distributed into the gametes at random

induced pluripotent stem cells (iPS cells) adult cells that have been reprogrammed by the introduction of new genes to become pluripotent again

induced-fit hypothesis a modified version of the lock-and-key model of enzyme action where the active site is considered to have a more flexible shape; after the substrate enters the active site, the shape of that site changes around it to form the active complex; after the products have left the complex, the enzyme returns to its inactive, relaxed form

inferior vena cava the large vein that carries the returning blood from the lower parts of the body to the heart

inhalation breathing in

initial rate of reaction the measure taken to compare the rates of enzyme-controlled reactions under different conditions

insertion a type of point mutation in which an extra base is added into a gene, which may be a repeat or a different base

insulin a hormone made in the pancreas involved in the regulation of blood sugar levels

internal fertilisation the fertilisation of the female gamete by the male gamete, which takes place inside the body of the female

interphase the period between active cell divisions when cells increase their size and mass, replicate their DNA and carry out normal metabolic activities

intracellular inside the cell

intracellular enzymes enzymes that catalyse reactions within the cell

introns segments of a DNA or RNA molecule containing information which does not code for a protein or peptide sequence

ionic bonds bonds formed when atoms give or receive electrons; they result in charged particles called ions

isomers molecules that have the same chemical formula, but different molecular structures

isotonic solution a solution in which the osmotic concentration of the solutes is the same as that in the cells

isotopes different atoms of the same element, with the same number of protons but a different number of neutrons; isotopes have the same chemical properties

karyotype a way of displaying an image of the chromosomes of a cell to show the pairs of autosomes and sex chromosomes

kingdom the classification category smaller than domains; there are six kingdoms Archaebacteria, Eubacteria, Protista, Fungi, Plantae and Animalia

left atrium the upper left-hand chamber of the heart that receives oxygenated blood from the lungs

left ventricle the chamber that receives oxygenated blood from the left atrium and pumps it around the body

leucocytes white blood cells; there are several different types which play important roles in defending the body against the entry of pathogens and in the immune system

light microscope (optical microscope) a tool that uses a beam of light and optical lenses to magnify specimens up to 1500 times life size

lignin a chemical that impregnates cellulose cell walls in wood and makes them impermeable

lipids a large family of organic molecules that are important in cell membranes and as an energy store in many organisms; they include triglycerides, phospholipids and steroids

lipoproteins conjugated proteins with a lipid prosthetic group

lock-and-key hypothesis a model that explains enzyme action by an active site in the protein structure that has a very specific shape; the enzyme and substrate slot together to form a complex in the same way as a key fits in a lock

locus place on a chromosome where any particular gene is found

longitudinal studies scientific studies which follow the same group of individuals for many years

low-density lipoproteins (LDLs) lipoproteins which transport lipids around the body

lumen the central space inside the blood vessel

lung surfactant a special phospholipid that coats the alveoli and prevents them from collapsing

lysosomes organelles full of digestive enzymes used to break down worn-out cells or organelles or digest food in simple organisms

macromolecule a very large molecule often formed by polymerisation

magnification a measure of how much bigger the image you see is than the real object

marsupials mammals that give birth to very immature young and then protect them in pouches

mass transport system an arrangement of structures by which substances are transported in the flow of a fluid with a mechanism for moving it around the body

mate-recognition species model a species definition based on unique fertilisation systems, including mating behaviour

mating the process by which a male animal transfers sperm from his body directly into the body of the female

megagamete the female gamete, the egg cell, in plants

megakaryocytes large cells that are found in the bone marrow and produce platelets

megaspores the result of meiosis in plants that develop into the female gametes, ovules

meiosis a form of cell division in which the chromosome number of the original cell is halved, leading to the formation of the gametes

mesosomes infoldings of the cell membrane of bacteria

messenger RNA (mRNA) the RNA formed in the nucleus that carries the genetic code out into the cytoplasm

metabolic chain (metabolic pathway) a series of linked reactions in the metabolism of a cell

metabolism the sum of the anabolic and catabolic processes in a cell

metadata analysis (meta-analysis) when data from all the available studies in a particular area are analysed

metaphase plate (equator) the region of the spindle in the middle of the cell along which the chromatids line up

metaphase the second stage of active cell division where a spindle of overlapping protein microtubules forms and the chromatids line up on the metaphase plate

metaxylem consists of mature xylem vessels made of lignified tissue

micelles a spherical aggregate of molecules in water with hydrophobic areas in the middle and hydrophilic areas outside

microgametes the male gametes produced in plants, the pollen grains

microspores the result of meiosis in plants that develop into the spore (pollen) containing the male gametes

middle lamella the first layer of the plant cell wall to be formed when a plant cell divides, made mainly of calcium pectate (pectin) that binds the layers of cellulose together

mitochondria rod-like structures with inner and outer membranes that are the site of aerobic respiration

mitosis the process by which a cell divides to produce two genetically identical daughter cells

mitotic index the ratio between the number of cells in a tissue sample that are in mitosis and the total number of cells in the sample

molecular activity (turnover number) the number of substrate molecules transformed per minute by a single enzyme molecule

molecular phylogeny the analysis of the genetic material of organisms to establish their evolutionary relationships

Monera a kingdom in the five-kingdom classification system that contains the Archaea and Eubacteria

monohybrid cross a genetic cross where only one gene for one characteristic is considered

monolayer a single closely packed layer of atoms or molecules

monomer a small molecule that is a single unit of a larger molecule called a polymer

mononucleotides molecules with three parts - a 5-carbon pentose sugar, a nitrogen-containing base and a phosphate group - joined by condensation reactions

monosaccharide a single sugar monomer

monotremes primitive mammals that lay eggs and feed their offspring with milk from mammary glands

monounsaturated fatty acid a fatty acid with only one double covalent bond between carbon atoms in the hydrocarbon chain

morphological species model a species definition based solely on the appearance of the organisms observed

morphology the study of the form and structure of organisms

morula an early embryo made up of a solid ball of 10–30 totipotent cells

multifactorial disease a disease which results from the interactions of many different factors - not from one simple cause

multiple alleles more than two possible variants at a particular locus

multipotent a cell that can form a very limited range of differentiated cells within a mature organism

mutagen anything that increases the rate of mutation

mutation a permanent change in the DNA of an organism

myocardial infarction (heart attack) the events which take place when atherosclerosis leads to the formation of a clot that blocks the coronary artery entirely and deprives the heart muscle of oxygen, so it dies; it can stop the heart functioning

myoglobin a respiratory pigment with a stronger affinity for oxygen than haemoglobin.

niche the role of an organism within the habitat in which it lives

non-coding RNA (ncRNA) 98% of the RNA, which does not code for proteins but affects the transcription of the DNA code, modifies the chromatin structure or modifies the products of transcription

non-communcable conditions diseases which are not caused by pathogens and cannot be spread from one person to another

non-overlapping code a code where each codon codes for only one thing with no overlap between codons

non-reducing sugars sugars that do not react with Benedict's solution

nucleic acids/polynucleotides polymers made up of many nucleotide monomer units that carry all the information needed to form new cells

nucleoid the area in a bacterium containing the single circular loop of coiled DNA

nucleolus an extra-dense region of almost pure DNA and protein found in the nucleus; it is involved in the production of ribosomes and control of growth and division

nucleosomes dense clusters of DNA wound around histones

nucleus an organelle containing the nucleic acids DNA (the genetic material) and RNA, as well as protein, surrounded by a double nuclear membrane with pores

obligate aerobes organisms that need oxygen for respiration

obligate anaerobes organisms that can only respire in the absence of oxygen and are killed by oxygen

oedema swelling of the tissues due to fluid retention

oligosaccharides molecules with between 3 and 10 monosaccharide units

oocyte a cell in an ovary which may form an ovum if it undergoes meiotic division

operon a unit consisting of linked genes which is thought to regulate other genes responsible for protein synthesis

order a group of families that all share common characteristics

organ system a group of organs working together to carry out particular functions in the body

organelles sub-cellular bodies found in the cytoplasm of cells

organs structures made up of several different types of tissue to carry out a particular function in the body

osmosis a specialised form of diffusion that involves the movement of solvent molecules down their water potential gradient

osmotic concentration a measure of the concentration of the solutes in a solution that have an osmotic effect

ova the haploid female gametes in animals (singular = ovum)

ovaries the female sex organs in both animals and plants; they produce the female gametes called ovules in plants and ova in animals

ovules the haploid female gametes in plants

oxygenated blood blood that is carrying oxygen

oxyhaemoglobin the molecule formed when oxygen binds to haemoglobin

pancreatic duct the duct from the pancreas which carries digestive enzymes made in the pancreas into the duodenum

parenchyma relatively unspecialised plant cells that act as packing in stems and roots to give support

partially permeable membrane a membrane which only allows specific substances to pass through it

passive transport transport that takes place as a result of concentration, pressure or electrochemical gradients and involves no energy from a cell

pathogens microorganisms that cause disease

pectin a polysaccharide that holds cell walls of neighbouring plant cells together and is part of the structure of the primary cell wall

pentose sugar a sugar with five carbon atoms

peptide bond the bond formed by condensation reactions between amino acids

peptidoglycan a large, net-like molecule found in all bacterial cell walls made up of many parallel polysaccharide chains with short peptide cross-linkages

peripheral arteries arteries further away from the heart but before the arterioles

phagocytosis the active process when a cell engulfs something relatively large such as a bacterium and encloses it in a vesicle

phenotype the physical traits, including biochemical characteristics, expressed as a result of the interactions of the genotype with the environment

phenylketonuria (PKU) a recessive genetic disorder where those affected lack the enzyme needed to digest the amino acid phenylalanine; the amino acid builds up in the blood and causes severe brain damage

phloem the main tissue transporting dissolved food around the plant

phosphodiester bond bond formed between the phosphate group of one nucleotide and the sugar of the next nucleotide in a condensation reaction

phospholipids chemicals in which glycerol bonds with two fatty acids and an inorganic phosphate group

phylogenetic tree model used to show the relationships between different groups of organisms

phylum (division, for plants) a group of classes that all share common characteristics

physiological adaptations adaptations involving the way the body of the organism works, including differences in biochemical pathways or enzymes

pili thread-like protein projections found on the surface of some bacteria

pinocytosis the active process by which cells take in tiny amounts of extracellular fluid by tiny vesicles

pits thin areas of cell wall in plant cells with secondary thickening, where plasmodesmata maintain contact with adjacent cells; in xylem vessels, where the cells are dead, they become simple holes through which water moves out into the surrounding cells

placebo an inactive substance resembling a drug being trialled which is used as an experimental control

placebo effect when patients appear to respond to a drug simply because they think it is doing them good

placenta (plant) the pad of special tissue that attaches the plant ovule to the ovary wall

placental mammals mammals that provide for the developing fetus during gestation through a placenta

plant fibres long cells with cellulose cell walls that have been heavily lignified so they are rigid and very strong

plant stanols and sterols similar in structure to cholesterol, these compounds can help reduce blood cholesterol in those consuming them

Plantae a mainly autotrophic eukaryotic kingdom containing mosses, liverworts, ferns, gymnosperms and angiosperms (the flowering plants)

plaques yellowish fatty deposits that form on the inside of arteries in atherosclerosis

plasmids small, circular pieces of DNA that code for specific aspects of the bacterial phenotype

plasmodesmata cytoplasmic bridges between plant cells that allow communication between the cells

plasmolysis the situation when a plant cell is placed in hypertonic solution when so much water leaves the cell by osmosis that the vacuole is reduced and the protoplasm is concentrated and shrinks away from the cell walls

platelet inhibitory drugs drugs used to prevent blood clots forming by preventing platelets clumping together

platelets cell fragments involved in the clotting mechanism of the blood

pluripotent an undifferentiated cell that can form most of the cell types needed for an entire new organism

point mutation (gene mutation) a change in a single base of the DNA code

polar lipids lipids with one end attached to a polar group (e.g. a phosphate group) so that one end of the molecule is hydrophilic and one end is hydrophobic

polar molecule a molecule containing a dipole

pollen the spore which contains the haploid male gametes of plants

pollen tube a tube that grows out of a pollen grain down the style, into the ovary and through the micropyle of the ovule to carry the generative nucleus (which divides to form two male nuclei) to the ovule

pollination the transfer of pollen from the anther to the stigma, often from one flower to another

polygenic phenotypic characters determined by several interacting genes

polymer a long-chain molecule made up of many smaller, repeating monomer units joined together by chemical bonds

polypeptide a long chain of amino acids joined by peptide bonds

polyploidy a cell or an organism with more than two sets of chromosomes

polysaccharide a polymer consisting of long chains of monosaccharide units joined by glycosidic bonds

polysomes groups of ribosomes, joined by a thread of mRNA, that can produce large quantities of a particular protein

polyspermy the fertilisation of an egg by more than one sperm

polyunsaturated fatty acid a fatty acid with two or more double covalent bonds between carbon atoms in the hydrocarbon chain

population a breeding group of individuals of the same species occupying a particular habitat and a particular niche

population bottleneck the effect of an event or series of events that dramatically reduces the size of a population and causes a severe decrease in the gene pool of the population, resulting in large changes in allele frequencies and a reduction in genetic diversity

pre-mRNA the mRNA that is transcribed directly from the DNA before it has been modified

precision measurements with only slight variation between them

precursor a biologically inactive molecule which can be converted into a closely related biologically active molecule when needed

preimplantation genetic diagnosis testing the cells of an embryo produced by IVF to check for genetic diseases before it is implanted into the uterus of the mother

prenatal screening screening of an embryo or fetus before birth

primary cell walls the first very flexible plant cell walls to form, with all the cellulose microfibrils orientated in a similar direction

probability a measure of the chance or likelihood that an event will take place

promoter sequence specific region on the DNA to which transcription factors bind to stimulate transcription

prophase the first stage of active cell division where the chromosomes are coiled up and consist of two daughter chromatids joined by the centromere; the nucleolus breaks down

prosthetic group the molecule incorporated in a conjugated protein

proteases protein-digesting enzymes

prothrombin a large, soluble protein found in the plasma that is the precursor to an enzyme called thrombin

Protista a kingdom in the five-kingdom classification system that contains all single-celled organisms, green and brown algae and slime moulds

protoplasm the cytoplasm and nucleus combined

protoxylem the first xylem the plant makes; it can stretch and grow because the walls are not fully lignified

pulmonary arteries the blood vessels that carry deoxygenated blood from the heart to the lungs

pulmonary circulation carries deoxygenated blood to the lungs and oxygenated blood back to the heart

pulmonary veins the blood vessels that carry oxygenated blood back from the lungs to the heart

purine base a base found in nucleotides that has two nitrogen-containing rings

pyrimidine base a base found in nucleotides that has one nitrogen-containing ring

recessive a characteristic which is only expressed when both alleles code for it; in other words, the individual is homozygous for the recessive trait

red-green colour blindness a sex-linked genetic condition which affects the ability to distinguish tones of red and green

reducing sugars sugars that react with blue Benedict's solution and reduce the copper(II) ions to copper(I) ions giving an orangey-red precipitate

relative species abundance the relative numbers of species in an area

reliable evidence which can be repeated by several different scientists

resolution (resolving power) a measure of how close together two objects must be before they are seen as one

ribonucleic acid (RNA) a nucleic acid which is the genetic material in some organisms and is involved in protein synthesis

ribose a pentose sugar that is part of the structure of RNA

ribosomes the site of protein synthesis in the cell

right atrium the upper right-hand chamber of the heart that receives deoxygenated blood from the body

right ventricle the lower chamber that receives deoxygenated blood from the right atrium and pumps it to the lungs

risk the probability that an event will take place

risk factors factors which affect the risk of an event happening

RNA polymerase the enzyme that polymerises nucleotide units to form RNA in a sequence determined by the antisense strand of DNA

rough endoplasmic reticulum (RER) endoplasmic reticulum that is covered in 80S ribosomes and which is involved in the production and transport of proteins

saturated fatty acid a fatty acid in which each carbon atom is joined to the one next to it in the hydrocarbon chain by a single covalent bond

scanning electron micrographs (SEMs) micrographs produced by the electron microscope that have a lower magnification than TEMs, but produce a 3D image

sclereids sclerenchyma cells that are completely impregnated with lignin

sclerenchyma plant cells that have very thick lignified cell walls and an empty lumen with no living contents

secondary cell wall the older plant cell wall in which the cellulose microfibrils have built up at different angles to each other making the cell wall more rigid

selection pressure the effect of one or more environmental factors that determine whether an organism will be more or less successful at surviving and reproducing; selection pressure drives speciation

semiconservative replication the accepted model of DNA replication in which the DNA 'unzips' and new nucleotides align along each strand; each new double helix contains one strand of the original DNA and one strand made up of new material

semilunar valves half-moon shaped, one-way valves found at frequent intervals in veins to prevent the backflow of blood

sense strand the DNA strand that carries the code for the protein to be produced

septum the thick muscular dividing wall through the centre of the heart that prevents oxygenated and deoxygenated blood from mixing

serotonin a chemical that causes the smooth muscle of the blood vessels to contract, narrowing them and cutting off the blood flow to the damaged area

sex-linked diseases genetic diseases that result from a mutated gene carried on the sex chromosomes – in human beings, on the X chromosome

sex-linked traits characteristics which are inherited on the sex chromosomes

sexual dimorphism describes species where there is a great deal of difference between the appearance of the male and female

sexual reproduction the production of offspring that are genetically different from the parent organism or organisms by the fusing of two sex cells (gametes)

side-effect a secondary, usually undesirable effect of a drug or medical treatment

sieve plates the perforated walls between phloem cells that allow the phloem sap to flow

single circulation system a circulation in which the heart pumps the blood to the organs of gas exchange and the blood then travels on around the body before returning to the heart

smooth endoplasmic reticulum (SER) a smooth tubular structure similar to RER, but without the ribosomes, which is involved in the synthesis and transport of steroids and lipids in the cell

solute a substance in a solution, dissolved in the solvent

speciation the formation of a new species

species a group of closely related organisms that are all potentially capable of interbreeding to produce fertile offspring

species richness the number of different species in an area

specificity the characteristic of enzymes that means that each enzyme will catalyse only a specific reaction or group of reactions; this is due to the very specific shapes which come from the tertiary and quaternary structures

spermatozoa (sperm) the haploid male gametes in animals

spindle a set of overlapping protein microtubules running the length of the cell, formed as the centrioles pull apart in mitosis and meiosis

spirilla bacteria with a twisted or spiral shape

spliceosomes enzyme complexes that act on pre-mRNA, joining exons together after the removal of the introns

sporophyte the diploid main body of the plant

sporophyte generation the diploid generation in plants that produces spores by meiosis

starch a long-chain polymer formed of glucose monomers

start codon the sequence of bases which indicates the start of an amino acid chain – TAC; this is the code for the amino acid methionine

statins drugs that lower the level of cholesterol in the blood

stent a metal or plastic mesh tube that is inserted into an artery affected by atherosclerosis to hold it open and allow blood to pass through freely

sterile something free from living microorganisms and their spores

stop codon one of three sequences of bases which indicate the end of an amino acid chain

stroke an event caused by an interruption to the normal blood supply to an area of the brain which may be due to bleeding from damaged capillaries or a blockage cutting off the blood supply to the brain, usually caused by a blood clot

suberin a waterproof chemical that impregnates cellulose cell walls in cork tissues and makes them impermeable

substitution a type of point mutation in which one base in a gene is substituted for another

substrate the molecule or molecules on which an enzyme acts

sucrose a sweet-tasting disaccharide formed by the joining of glucose and fructose by a 1,4-glycosidic bond

superior vena cava the large vein that carries the returning blood from the upper parts of the body to the heart

surface area to volume ratio (sa : vol) the relationship between the surface area of an organism and its volume

sympathetic nerve inhibitors drugs which inhibit sympathetic nerves, keeping arteries dilated

sympatric speciation speciation that occurs between populations of a species in the same place; they become reproductively separate by mechanical, behavioural or seasonal mechanisms; gene flow continues between the populations to some extent as speciation occurs

symplast all of the material (cytoplasm, vacuole, etc.) contained within the surface membrane of a plant cell

systemic circulation carries oxygenated blood from the heart to the cells of the body where the oxygen is used, and carries the deoxygenated blood back to the heart

systole the contraction of the heart

taxonomy the science of describing, classifying and naming living organisms

teichoic acid a chemical substance found in the cell walls of Gram-positive bacteria

telophase the fourth stage of active cell division where a nuclear membrane forms around the two sets of chromosomes, the chromosomes unravel and the spindle breaks down

temperature coefficient (Q_{10}) the measure of the effect of temperature on the rate of a reaction

tendinous cords (valve tendons, heartstrings) cord-like tendons that make sure the valves are not turned inside out by the large pressure exerted when the ventricles contract

tensile strength the resistance of a material to breaking under tension

testes the male sex organs in animals that produce the male gametes (sperm)

therapeutic cloning an experimental technique used to produce embryonic stem cells from an adult cell donor

thrombin an enzyme that acts on fibrinogen, converting it to fibrin during clot formation

thromboplastin an enzyme that sets in progress a cascade of events that leads to the formation of a blood clot

thrombosis a clot that forms in a blood vessel

thymine a pyrimidine base found in DNA

tissues groups of specialised cells carrying out particular functions in the body

tonoplast the specialised membrane that surrounds the permanent vacuole in plant cells and controls movements of substances into and out of the cell sap

totipotent an undifferentiated cell that can form any one of the different cell types needed for an entire new organism

transcription the process by which the DNA sequence is used to make a strand of mRNA in the nucleus

transcription factors proteins that bind to the DNA in the nucleus and affect the process of converting, or transcribing, DNA into RNA

transfer RNA (tRNA) small units of RNA that pick up specific amino acids from the cytoplasm and transport them to the surface of the ribosome to align with the mRNA

translation the process by which the DNA code is converted into a protein from the mRNA strand made in the nucleus

translocation the active movement of substances around a plant in the phloem

transmission electron micrographs (TEMs) micrographs produced by the electron microscope that give 2D images like those from a light microscope, but magnified up to 500 000 times

transpiration stream the movement of water up from the soil through the root hair cells, across the root to the xylem, then up the xylem, across the leaf until it is lost by evaporation from the leaf cells and diffuses out of the stomata down a concentration gradient

tricuspid valve (atrioventricular valve) the valve between the right atrium and the right ventricle that prevents backflow of blood from the ventricle to the atrium when the ventricle contracts

triose sugar a sugar with three carbon atoms

triplet code the code of three bases that is the basis of the genetic information in the DNA

true breeding a homozygous organism which will always produce the same offspring when crossed with another true-breeding organism for the same characteristic

tube nucleus the male nucleus that will control the production of the pollen tube in fertilisation

turgid swollen

turgor the state of a plant cell when the solute potential causing water to be moved into the cell by osmosis is balanced by the force of the cell wall pressing on the protoplasm

ultrastructure the detailed organisation of the cell, only visible using the electron microscope

unit membrane a lipoprotein membrane which is composed of two protein layers enclosing a less dense lipid

unsaturated fatty acid a fatty acid in which the carbon atoms in the hydrocarbon chain have one or more double covalent bonds in them

uracil a pyrimidine base found in RNA

valid an investigation which is well designed to answer the question being asked

variation differences between organisms which may be the result of different genes or the environment they live in

vascular bundle part of the transport system of a plant, with phloem on the outside and xylem on the inside - often with strengthening sclerenchyma

veins vessels that carry blood towards the heart

venous system the system of veins in the body

ventricular systole when the ventricles of the heart contract

venules the very smallest branches of the venous system, furthest from the heart

vertebrates animals with a backbone or spinal column; they include mammals, birds, reptiles, amphibians and fish

vesicles membrane 'bags' that hold secretions made in cells

vibrios comma-shaped bacteria

villi finger-like projections of the lining of the duodenum and small intestine which increase the surface area for the absorption of digested food

water potential a measure of the potential for water to move out of a solution by osmosis

whole-chromosome mutations the loss or duplication of a whole chromosome

xylem the main tissue transporting water and minerals around a plant

zona pellucida a layer of protective jelly around the unfertilised ovum

zygote the cell formed when two haploid gametes fuse at fertilisation

INDEX

References in **bold** are to definitions.

Acknowledgements

The authors and publisher would like to thank the following individuals and organisations for their kind permission to reproduce copyright material:

Photographs

(Key: b-bottom; c-centre; l-left; r-right; t-top)

123RF.com: 235, Atic12 212, Daria Filimonova 198, Elena Sushytska 142, Hskoken 255r, Liliya Sayfeeva 234, Roman Kharlamov 132; **Alamy Stock Photo:** Andrew Walmsley 290, Ann and Steve Toon 260b, B.A.E. Inc. 69, BSIP SA Photos 153, Cuelmages 26-27, Custom Life Science Images 23, Frans Lanting/Mint Images Limited 283, Interfoto 150, James King-Holmes 286, Kunz Rolf E/Prisma by Dukas Presseagentur GmbH 270, Patrick J. Endres viit, 246-247, Phototake Inc 156t, Robertharding 231, Scott Camazine 31b, Stone Nature Photography 268b, The Natural History Museum 260c; **Anthony Short:** 4, 6, 7, 64, 66, 248, 249, 252t, 252c, 253, 264-265, 266, 268c, 269l, 269r, 271, 272t, 274l, 274r, 279l, 288l,r,c, vil, vir; **BBC:** Professor Legesse Negash 289; **Centers for Disease Control and Prevention:** Janice Haney Carr/ Sickle Cell Foundation of Georgia:Jackie George, Beverly Sinclair 293; **Getty Images:** Andrzej Wojcicki 168-169, Chris Jackson 287, Ed Reschke/Photolibrary ixt, 154c, 223c, Geoffrey Gilson Photography/Moment 2-3, Konrad Wothe/Minden Pictures 133, Paula Bronstein 236c, Ralph Slepecky/ Visuals Unlimited, Inc 161t, Roland Birke 28, Victor Habbick Visions/Science Photo Library 42; **Photo Take Inc.,:** ISM ixc, 223b; **Photoshot:** Oceans-Image 34; **Reuters:** Khaled Abdullah 194; **Science Photo Library:** 41, 46, AMI Images 183, Andrew Lambert Photography 10, 20, Asa Thoresen 163l, Athenais, ISM 45, 72, Biophoto Associates 39, 163r, 180c, 184l,r,bl,br, 220-221, 224, 244l, 275b, Chuck Brown 37, CNRI 166r, D. Phillips 38, 185, David M. Phillips 84, Dr Elena Kiseleva 74-75, Dr Gopal Murti 154t, Dr Keith Wheeler 228, Dr. Gopal Murti 157, 166t, 167, Dr. Jeremy Burgess 226, 227l, Paul Rapson 237, Dr. Kari Lounatmaa 79, Dr. Stanley Flegler/Visuals Unlimited, Inc. 85l, r, Dr.Rosalind King 165b, Gastrolab 206, Herve Conge, ISM viib, 190, 230, Hybrid Medical Animation 100-101, Innerspace Imaging 163c, J.C. Revy 84, 140l, r, Jackie Lewin, Royal Free Hospital 147, James King-Holmes 188, Juan Gaertner 109, Look at Sciences 170, Louise Hughes 102, Martin Shields 12, Medimage 13, Michael Abbey 84, NIBSC 166l, Pascal Goetgheluck 227r, 255b, pixologicstudio 50-51, Power and Syred 229, 244r, Professor P. Motta & D. Palermo 174, Professors P. Motta & T. Naguro 156b, Richard J. Green 165c, Science Source 87, Sovereign, ISM 47, Steve Gschmeissner 35, 35, 151, 152, 175t, 175r, 217, Thomas Deerinck, Ncmir 148-149, Wally Eberhart/Visuals Unlimited 130, Zephyr 71, 210; **Shutterstock:** Alexey Repka 251, Ana del Castillo 8, Anyaivanova 242, Artfully photographer 57, Betty Cadmus 252br, Brent Hofacker 12, By Geza Farkas 252bl, Covenant 268l, Dr Alan Lipkin 279r, Eastern light photography 252b, Eveleen 135, Evenfh 285, Frank Wasserfuehrer 255c, Kateryna Kon 134, Lebendkulturen de 88, Marilyn Barbone 240, Naypong 205, NeCoTi 80, nobeastsofierce 192-193, Olga_i 88, Oprea George 275c, Steve Wiechman 278, Tonanakan 277, Valentyn Volkov 14, Vera Zinkova 177, Yurchanka Siarhei 124-125, Rich Carey 236t, atiger 241; **The University of California:** Alex McPherson, Irvine/ National Institute of General 103; **U.S. Department of Agriculture:** Agricultural Research Service 254; **Veer(Corbis):** Carolina Biological/Visuals Unlimited 175l, enjoylife25 268r, gbrouwer 272b, Marilyna viic, 120,Prochasson Frederic 5, Roland Birke 28.